Hydrocarbon Contaminated Soils and Groundwater

Volume 2

Edward J. Calabrese &

Paul T. Kostecki

Editors

LEWIS PUBLISHERS
Boca Raton   Ann Arbor   London   Tokyo

Library of Congress Cataloging-in-Publication Data

(Revised for vol. 2)

Hydrocarbon contaminated soils and groundwater.

 "Proceedings from the First Annual West Coast Conference on
Hydrocarbon Contaminated Soils and Groundwater (February 1990, Newport
Beach, CA)"—Vol. 1, pref.
 Vol. 2 contains proceedings of the second annual West Coast Conference
on Hydrocarbon Contaminated Soils and Groundwater, March 1991.
 Includes bibliographical references and indexes.
 1. Oil pollution of soils—Congresses. 2. Oil pollution of water—
Congresses. 3. Water, Underground—Pollution—Congresses. I. Kostecki,
Paul T. II. Calabrese, Edward J., 1946–   . III. West Coast Conference
on Hydrocarbon Contaminated Soils and Groundwater.
TD879.P4H83   1991   628.5'5   91-1991
ISBN 0-87371-383-4 (v. 1)
ISBN 0-87371-603-5 (v. 2)

LEWIS PUBLISHERS, INC.
121 South Main Street, Chelsea, MI 48118

Printed in The United States of America 1 2 3 4 5 6 7 8 9 0

Preface

Hydrocarbon contaminants in soil and groundwater have emerged as immense environmental issues to both the public and private sectors in the past ten years. This has had an enormous impact on national resources: time, manpower, and money. One need only look at the problems associated with underground storage tanks to appreciate the impact. Of the seven to eight million underground storage tanks in the United States, only between 350,000 and 400,000 are corrosion protected; the balance must be removed or retrofitted by 1998.

Estimates for compliance with the Environmental Protection Agency regulations governing underground storage tanks, including spill and overfill protection, along with leak detection, will approach 3.5 billion dollars. However, when removal costs, new tank purchases, and remediation of soil and groundwater are factored in, the expenditures could approach 100 billion dollars over the next 20 years.

Not only are the costs high, the issues surrounding contaminated soil and groundwater are challenging and far-reaching. In most states, an intricate maze of federal, state, and local regulations must be negotiated to achieve a successful cleanup. In addition, site-specific characteristics make each site unique with nonuniformity and inconsistency the rule rather than the exception. The paramount concern is always with protecting the public's health and safety. No one person or discipline is capable of coordinating and adequately responding to all of the issues. It takes teamwork from multifaceted professionals who are well trained and educated in scientifically sound techniques and applications.

The group that assembled at the second annual West Coast Conference on Hydrocarbon Contaminated Soils and Groundwater in March, 1991, represented many of the skilled, professional disciplines that are essential to helping solve the many issues surrounding contaminated soil and groundwater. Regulators, hydrogeologists, and engineers responsible for sampling, assessing, and remediating contaminated sites, risk assessors and risk communicators were all present, each to educate their peers on their specialized piece in the whole process. What emerged is a body of knowledge that will be helpful for years to come. With each piece of shared knowledge comes another helpful hint to facilitate the environmental community's quest to provide low-cost, permanent solutions to another contaminated site. And with every successful idea which worked in one circumstance, that information can be transferred to another site at another time; another bridge can be built. Thus, a scientifically sound, economically rational set of processes can be developed which will allow environmental protection at a cost that will not put this country at an economic disadvantage.

This book, which contains the proceedings of the conference, should be recognized and used as a valuable tool to help find solutions to real world issues and obstacles at sites with soil and groundwater contamination. It is an up-to-date source of technical and regulatory information, written by practicing professionals in the environmental field who have found solutions to assist them in their daily activities and in their particular disciplines. I trust that it will be of use to you, the reader, too.

Tony Maggio
Groundwater Technology, Inc.

Acknowledgments

We wish to thank all the agencies, organizations, and companies that sponsored the conference. Without their generosity and assistance, the conference and this book would not have been possible.

Sponsors

California Department of Health Services
The Earth Technology Corporation
EPA/Office of Underground Storage Tanks
Groundwater Technology, Inc.
Hydro-Fluent, Inc.
ICF Technology Incorporated
Lawrence Livermore National Laboratory
McLaren/Hart
James M. Montgomery Consulting Engineers, Inc.
Radian Corporation
Shell Oil Company
TPS Technologies, Inc.
Western States Petroleum Association

Supporters

Applied Geosciences, Inc.
ARCO Products Company
IT Corporation
Port of Los Angeles
RETEC
Southern California Edison Company
Southern California Gas Company
Toxic Treatments (USA), Inc.

In addition, we express our deepest appreciation to the members of the Advisory Committee who volunteered their valuable time to provide guidance and encouragement.

Advisory Committee

Richard Beauregard
Radian Corporation

Mark Cousineau
Applied Geosciences, Inc.

Jay Dablow
Hydro-Fluent, Inc.

Ralph De La Parra
Southern California Edison Company

Frank Fossatti
Shell Oil Company

John Hills
Anaheim Public Utilities Department

Jon Holmgren
IT Corporation

Richard Jenkins
James M. Montgomery Consulting Engineers, Inc.

Raymond Kane
McLaren/Hart

Dorothy Keech
Chevron Oil Field Research Company

Phil La Mori
NOVATERRA

James Lousararian
TPS Technologies, Inc.

Tony Maggio
Groundwater Technology, Inc.

Paula Mills
The Earth Technology Corporation

Lynne Preslo
ICF Technology Incorporated

Donald Rice
Port of Los Angeles

Jim Richey
ARCO Product Company

Larry Sasadeusz
Southern California Gas Company

Kim Savage
US EPA/Region 9

Hans Stroo
RETEC

Wendell Suyama
Hughes Aircraft Company

Lucien Truhill
Orange County Chamber of Commerce

Michael Wang
Western States Petroleum Association

Kim Ward
State Water Resources Control Board

Jeff Wong
California Department of Health Services

Edward J. Calabrese is a board certified toxicologist who is professor of toxicology at the University of Massachusetts School of Public Health, Amherst. Dr. Calabrese has researched extensively in the area of host factors affecting susceptibility to pollutants, and is the author of more than 240 papers in scholarly journals, as well as 10 books, including *Principles of Animal Extrapolation, Nutrition and Environmental Health,* Vols. I and II, *Ecogenetics, Safe Drinking Water Act: Amendments, Regulations and Standards, Petroleum Contaminated Soils,* Vols. I and II, and *Ozone Risk Communication and Management.* He has been a member of the U.S. National Academy of Sciences and NATO Countries Safe Drinking Water committees, and most recently has been appointed to the Board of Scientific Counselors for the Agency for Toxic Substances and Disease Registry (ATSDR). Dr. Calabrese also serves as Chairman of the International Society of Regulatory Toxicology and Pharmacology's Council for Health and Environmental Safety of Soils (CHESS).

Paul T. Kostecki, Associate Director, Northeast Regional Environmental Public Health Center, School of Public Health, University of Massachusetts at Amherst, received his PhD from the School of Natural Resources at the University of Michigan in 1980. He has been involved with risk assessment and risk management research for contaminated soils for the last six years, and is coauthor of *Remedial Technologies for Leaking Underground Storage Tanks,* coeditor of *Soils Contaminated by Petroleum Products* and *Petroleum Contaminated Soils,* Vols. 1, 2, and 3, and of *Hydrocarbon Contaminated Soils,* Vol. 1, and *Hydrocarbon Contaminated Soils and Groundwater,* Vol. 1. Dr. Kostecki's yearly conferences on hydrocarbon contaminated soils draw hundreds of researchers and regulatory scientists to present and discuss state-of-the-art solutions to the multidisciplinary problems surrounding this issue. Dr. Kostecki also serves as Managing Director for the International Society of Regulatory Toxicology and Pharmacology's Council for Health and Environmental Safety of Soils (CHESS), as Executive Director of the newly formed Association for the Environmental Health of Soils (AEHS), and as Editorial Advisor to the *Journal of Soil Contamination* and *SOILS* magazines.

Contents

PART I
PERSPECTIVES ON HYDROCARBON CONTAMINATION

1. Solvable Problems of Hydrocarbon Contaminated Soils: An Industry Perspective, *Evan C. Henry* 3

2. A Cost-Effective Approach to Regulating Contaminated Soil: Set De Minimis Concentrations for Eight Different Exposure Scenarios, *James D. Jernigan, Rena Bass, and Dennis J. Paustenbach* 11

PART II
REGULATORY

3. Canadian Approach to Establishing Cleanup Levels for Contaminated Sites, *Connie Gaudet, Amanda Brady, Mark Bonnell, and Michael Wong* 49

4. Innovative Hydrocarbon Contaminated Soil Treatment Techniques in Nevada: A Regulatory Perspective, *Jim Najima and Allen Biaggi* 67

5. Regulatory Complexities Associated with the Cleanup of Petroleum Contaminated Sites, *Michael R. Anderson* 75

6. Evaluation of the Use of the Toxicity Characteristic in Cleanup Standard Selection, *J. David Thomas* 91

PART III
ENVIRONMENTAL FATE AND MODELING

7. The Influence of Methanol in Gasoline Fuels on the Formation of Dissolved Plumes, and on the Fate and Natural Remediation of Methanol and BTEX Dissolved in Groundwater, *J. F. Barker, C. E. Hubbard, L. A. Lemon, and K. A. Vooro* 103

8. Effects of Oxygenated Fuels on Groundwater
   Contamination: Equilibria and Transport Considerations,
   William G. Rixey and Ira J. Dortch 115

9. Estimating the Multimedia Partitioning of Hydrocarbons:
   The Effective Solubility Approach, *Donald Mackay and
   Wan Ying Shiu* 137

PART IV
SAMPLING AND SITE ASSESSMENT

10. Application of Aerial Infrared Survey Techniques in
    Detailing Soil Contamination, *J. M. Fernandez* 157

11. The Use of a Portable Infrared Analyzer to Perform Onsite
    Total Petroleum Hydrocarbon Analysis, *Roy A. Litzenberg,
    Richard H. Oliver, and James J. Severns* 169

12. Real-World Soil/Waste Standards: Solution to Differential
    Analytical Results Among Laboratories, *David J. Carty and
    Janick F. Artiola* 181

13. Site Assessment, Remediation, and Closure Under the
    Oregon UST Matrix, *Robert A. Dixon* 201

PART V
REMEDIATION ASSESSMENT AND DESIGN

14. Bioremediation of PAH Compounds in Contaminated Soil,
    Raymond C. Loehr 213

15. A State-of-the-Art Review of Remedial Technologies
    for Petroleum Contaminated Soils and Groundwater:
    Data Requirements and Efficacy Information, *Melitta Rorty,
    Lynne M. Preslo, Raymond A. Scheinfeld, and Mary E.
    McLearn* 223

16. The Need for a Laboratory Feasibility Study in
    Bioremediation of Petroleum Hydrocarbons, *W. T.
    Frankenberger, Jr.* 237

17. Technological Limits of Groundwater Remediation: A Statistical Evaluation Method, *William A. Tucker and Erin F. Parker* 295

18. Development of Post-Treatment Cleanup Criteria for In Situ Soils Remediation, *Phil La Mori and Jon La Mori* 311

PART VI
REMEDIATION CASE STUDIES

19. Microencapsulation of Hydrocarbons in Soil Using Reactive Silicate Technology, *Tom McDowell* 327

20. The CROW© Process and Bioremediation for In Situ Treatment of Hazardous Waste Sites, *Lyle A. Johnson, Jr. and Alfred P. Leuschner* 343

21. Onsite Thermal Treatments: A Case Study in Multisource Petroleum Contamination, *Robert Blanton and Jeffrey Powell* 359

22. Full-Scale, In Situ Bioremediation at a Superfund Site: A Progress Report, *Michael R. Piotrowski* 371

23. Steam Injection to Enhance Removal of Diesel Fuel from Soil and Groundwater, *John F. Dablow III* 401

24. Biological Treatment: Soil Impacted with Crude Oil, *Robert S. Skiba, Nancy Gilbertson, and James J. Severns* 409

25. Passive Recovery Trench Design Under the Influence of Tidal Fluctuations, *C. Y. Chiang, P. D. Petkovsky, R. A. Ettinger, I. J. Dortch, C. C. Stanley, R. W. Hastings, and M. W. Kemblowski* 417

26. NAPL Removal from Groundwater by Alcohol Flooding: Laboratory Studies and Applications, *Glen R. Boyd and Kevin J. Farley* 437

27. Cost-Effective Alternative Treatment Technologies for
    Reducing the Concentrations of Methyl Tertiary Butyl Ether
    and Methanol in Groundwater, *Kim N. Truong and
    Charles S. Parmele* 461

28. The Use of Modern Onsite Bioremediation Systems to
    Reduce Crude Oil Contamination on Oilfield Properties,
    W. W. Hildebrandt and S. B. Wilson 487

29. Use of Government Environmental Records to Identify
    and Analyze Hydrocarbon Contaminated Sites, *Ronald D.
    Miller* 501

30. Development of Remediation Endpoints for Gasoline in
    Soils and Groundwater Using Risk Assessment, Transport,
    and Dispersion Models, *Kenneth Thomas, Ruth Custance,
    and Michael J. Sullivan* 517

31. The Effects of Toxicological Uncertainties in Setting
    Cleanup Levels for Hazardous Waste Sites, *Mike Alberts
    and Fred Weyman* 527

Glossary of Acronyms 537

List of Contributors 541

Index 547

Hydrocarbon Contaminated Soils and Groundwater

Volume 2

PART I

Perspectives on Hydrocarbon Contamination

CHAPTER 1

Solvable Problems of Hydrocarbon Contaminated Soils: An Industry Perspective

Evan C. Henry, Environmental Services, Bank of America, Newport Beach, California

INTRODUCTION

As a lender, I'm in the middle of the industry-consultant-regulator triangle—on any given day I can be any one of them. This book, however, is about contaminated soil. Soil is sometimes colloquially referred to as dirt, which is also a frequently used term for real estate. And that's what really is the subject of all the contamination talk—the real estate. This chapter presents my perspective (with an acknowledged lender bias) on what's happening in the development of hydrocarbon contaminated real estate.

WHAT DRIVES ENVIRONMENTAL CLEANUP OF REAL ESTATE?

The most obvious driver of environmental cleanup is regulatory requirements relating to the protection of natural resources and/or public health. Second, but related to regulatory pressures in the environmental concerns which affect real estate development, is the fact that to develop or redevelop a contaminated property, one needs to clean up the property for one or a combination of the following reasons:

- to make it safe (protect natural resources or public health)
- to satisfy a regulatory requirement
- to limit future liability.

All of these reasons to clean up a property apply to hydrocarbon contaminated soils.

When talking about hydrocarbon contaminated soils, we're talking about a wide range of properties, but all of them linked to the oil industry. In California we have the entire range of oil industry:

- oil and gas fields
- refineries
- terminals
- gas stations
- pipelines

Many of the largest commercial developments on former oil-related properties are by oil companies themselves (i.e., Chevron Land and Development Corporation and Unocal Land and Development Corporation). Projects on oil company lands in southern California include Chevron's Manhattan Village project, Unocal's Torrance Tank Farm, Texaco's sale of its tank farm in Cypress to a housing developer, redevelopment of the Golden Eagle Refinery in Lomita as well as the relocation of terminals in Los Angeles and Long Beach harbors.

HOW ARE THESE PROPERTIES RELATED TO
THE REAL ESTATE DEVELOPMENT PROCESS?

First, oil fields in urban areas such as Los Angeles now represent the single largest source of open land for development. Virtually all of the oil fields along the Newport–Inglewood fault are now undergoing commercial or residential redevelopment. These include:

Cheviot Hills
Baldwin Hills
Dominguez Hills
Signal Hill
Huntington Beach Bluffs
Newport Beach Bluffs

Other oil field areas are also being developed in the Los Angeles basin in Santa Fe Springs, LaHabra, and Brea.

Many of these oil fields are near the end of their productive life. Extraction methods can now co-exist more easily because of methods such as directional drilling. Land values also make it more attractive to spend more on extraction that co-exists with real estate uses, whether the use is commercial, industrial, or residential. Unlike old landfill properties which are restricted to commercial (if not golf course use), oil well properties can be remediated to allow just about any use if you try hard enough.

Second, there is increasing pressure to relocate oil refineries and terminals. Air quality regulations make it harder to do business as a refinery in the Los Angeles Basin. This, combined with factors such as more lucrative uses (auto import lots or shipping) of harbor terminal properties tends to push oil-industry related properties toward redevelopment. Port facilities probably include some of the most concentrated areas of hydrocarbon contamination of shallow soil and groundwater.

Third, gas station properties are constantly undergoing redevelopment. Consolidation in the industry has resulted in one or two stations per intersection instead of four. This allows for redevelopment of valuable properties at intersections. Some developers specialize in redevelopment of gas station sites into strip malls or convenience outlets.

Each one of these oil-industry-related properties leaves legacies that the enterprising real estate developer must deal with.

HOW DID WE HANDLE THIS IN THE PAST?

In ancient times, five or so years ago, two basic approaches were used:

1. Cleanups were done haphazardly and to widely varying standards, if at all.
2. Properties lay fallow due to the inability to clean up or uncertainty of cleanup.

Neither of these approaches were very comforting from either the development or environmental perspective, but I think we have come a long way in the past five years. The outlook on developing hydrocarbon contaminated properties is much different now, and better.

WHY DID IT CHANGE?

I think the single most important force driving changes in the way hydrocarbon contaminated real estate is addressed is the attention given to underground storage tanks. Underground storage tank regulations brought the impact of hydrocarbons on the environment into the limelight. Underground storage tank regulations are part of the overall increase in environmental concern and resultant regulatory controls in the United States. This is especially so in California, where regulations were ahead of most of the rest of the country.

But the bottom line effect is that we have had to learn more about the fate and transport, toxicity, and cleanup of hydrocarbons. We are now capable of dealing with many problems that were beyond our capabilities five years ago.

In addition, the ever-rising value of land has provided economic incentives to tackle previously unmanageable cleanup problems. This may seem ironic when real estate values are now declining, but it is a relative thing. The old gas station property on the intersection is more valuable than the property at the edge of town. Real estate development in tough times tends to focus on redevelopment in established areas where infrastructure is in place and access to existing markets, jobs, etc., is more favorable. This promotes dealing with that corner gas station problem.

HOW ARE WE NOW DIFFERENT?

First, regulatory guidelines have been better defined, if not institutionalized. Documents such as the California Leaking Underground Fuel Tank Manual provide needed guidance. According to an article in last year's conference proceedings, there are only eight states that do not have some sort of cleanup standards for hydrocarbons.

Second, cleanup technologies have been invented, acquired a track record, and become off-the-shelf. "Catalog engineering" is a term I heard the other day from a consultant doing bioremediation. Call in the appropriate bioremediation specialists, such as the bacteria and nutrient specialist, or call the vacuum extractor, the organic vapor treater, etc. The availability of proven technology results in better estimation of cleanup costs necessary for the development process.

Third, real estate developers are smarter. On real estate deals, 90% of developers come to the lenders with Phase I site assessments in hand. They know the environmental issues need to be addressed.

Fourth, lenders are smarter. We can lend where we can quantify the risks. There are ways to do business (even if it's complicated) if you can know the risks and account for them.

WHAT'S THE IMPACT OF THIS KNOWLEDGE?

Given a preference of contaminants to deal with, most developers will choose hydrocarbons. I know as a lender, I'm more comfortable with hydrocarbons. The work that goes into quantifying and cleaning up hydrocarbons is becoming more and more routine. This relieves a lot of pressure on cities with high redevelopment potentials, and assists in necessary emphasis on "infill" rather than sprawl.

WHAT MORE NEEDS TO BE DONE?

First, there is always room for a better mousetrap, or in this case a better bio-degradation, vapor extraction, catalytic burner, or other system. Especially needed are space-efficient and time-efficient methods. It's tough to effectively bioremediate a corner gas station because of the space limitations. Further research and development of effective and efficient soil and groundwater cleanup methods are necessary.

In addition, we need better definition of "How clean is clean?" There is still confusion over the standards for cleanup vs characterization as a hazardous waste. And there is too much confusion between regulatory agencies. For example, what consultant in the room can say definitely what analyses should be done on oily soil on an oil well drilling property in Long Beach that will enable the Long Beach Health Department, the California Regional Water Quality Control Board, and the State Department of Health Services to all agree on a reasonable petroleum hydrocarbon concentration that can be left onsite? And where must the soil go if development requires offsite disposal to a Class I, II, or III landfill or to an asphalt plant? How many fat-headed minnows do you have to kill? If you can answer that, I may have a lucrative contract for you.

From a "certainty" standpoint, there are distinct benefits to rigid soil and groundwater quality standards which allow developers to develop and lenders to lend. There is a need to be conservative, but not too conservative. We cannot operate under the extreme of "zero" tolerance, as was proposed for pesticides under the now-history Big Green Initiative of last year in California. However, conservatively set standards may not always make sense from a risk assessment standpoint. Significant cost savings (efficiency with no increase in environmental impact) could result from a risk-based approach.

I suggest that standards should be set where applicable, such as 100/ppm total petroleum hydrocarbons for gasoline in soil. There is general recognition that such a level protects the environment and public health. The set levels should be fallback position; conservative enough not to be questioned in most circumstances by an agency or private party.

However, where set standards may not be appropriate, regulatory agencies should allow for raising of set standards based on risk assessment. This would be beneficial where the cleanup project is large, and higher allowable concentrations would still protect while significantly decreasing costs. Future site use should be a major parameter in the risk assessment. But risk assessment standards and procedures set by state agencies need to trickle down to the local level—the fire department or county health department, especially those agencies dealing with smaller hydrocarbon problems. The combination of set standards and risk assessment would allow flexibility to get the job done, protect the environment, and get the necessary "sign-off" to get the redevelopment done.

Similar to the trickle-down from state to local agencies, there needs to be better coordination between state regulatory agencies. In concept, I favor a California EPA. I look forward to seeing more from Governor Wilson on his proposal. Perhaps the biggest quandary I've been in is being restricted from doing timely cleanup of soil and groundwater because of short-sighted air quality concerns by an air quality management district.

But all this talk about technology and regulatory improvements ignores an issue central to the redevelopment process; where projects succeed and cleanups are accomplished, liability is not an issue, but where cleanup is not successful as part of the development, someone can get left holding the bag. There is still a need to place liability where it belongs. It is in the best interest of all parties to environmental cleanup and rehabilitation of hydrocarbon contaminated properties to place the responsibility for cleanup on either the party that contaminated the property or the successor/owner of the contaminated site who now stands to profit from the cleanup and redevelopment.

However, right now there is incredible pressure on lenders, due to Superfund precedents which assign lenders liability for foreclosed real estate collateral. In the day-to-day world, the petroleum exclusion from Superfund has little meaning when it comes to liability. Lenders must push for regulatory agency sign-off prior to construction lending or any other type of lending secured by commercial real estate to avoid the potential of being left with an incomplete development on contaminated land. Most lenders require that properties must be cleaned up before lending can be done comfortably.

Many properties need lender financing, not only for the construction, but for cleanup as part of construction. Under the present lender liability scenario, as long as there is known contamination of the real estate collateral, the development of contaminated properties will be severely curtailed due to lack of financing. Many of the benefits of more research and development and more regulatory definition may be lost without freeing up access to funds from lenders.

What needs to be done is to clarify on both federal and state levels, through legislation, that lenders will not be held liable for the cleanup of properties owned by their borrowers. The present lending structure does not cover the costs of the borrower's collateral. The risk of not being repaid is enough, without the additional risk of cleanup costs to the lender.

The developer stands to make the profit on the real estate development and should take the responsibility for environmental conditions as part of his risk/reward ratio. A lender is a facilitator of the solution, just as the consultant and regulatory agencies are. Lending is a service industry. If faced with liability for a project on which his client goes belly up, the consultant would close down his activities and the regulatory agency would, too. So why shouldn't lenders?

Cleanup and redevelopment of hydrocarbon contaminated sites is good. It creates jobs, puts fallow properties into productive use, and cleans up the environment all at the same time. This benefits all parties to the development of contaminated properties in the long run. But it requires fairness to all parties involved, rather than putting inordinate risks on the lending community.

I encourage you to support changing lender liability laws to allow lenders to be part of the team, rather than scaring the lenders into taking their ball and going home.

In summary, the assessment and cleanup of hydrocarbon contamination of soil and groundwater has come a long way, but it is far from being "old hat," and we shouldn't be complacent as we go forward. The most effective way to move forward is to work as a team: scientists, consultants, regulators, developers, and lenders. By increasing our technology, by defining our cleanup goals and standards, and by placing responsibility on the appropriate shoulders, I am optimistic about a future where hydrocarbon contamination will no longer be the monkey wrench in the works of real estate development that it has been so often in the past.

CHAPTER 2

A Cost-Effective Approach to Regulating Contaminated Soil: Set De Minimis Concentrations for Eight Different Exposure Scenarios

James D. Jernigan* and **Rená Bass,** ChemRisk Division, McLaren/Hart Environmental Engineering, Springfield, Missouri [*Currently with Amoco Corporation, Chicago, Illinois]
Dennis J. Paustenbach, ChemRisk Division, McLaren/Hart Environmental Engineering, Alameda, California

INTRODUCTION

With over 2,000 different potential soil contaminants at hundreds of sites throughout the United States, both regulators and the regulated community face an almost overwhelming task of deciding "how clean is clean." There are only three or four federal Environmental Protection Agency (EPA) guidelines for chemicals in soil, including PCBs,[1] lead,[2] and 2,3,7,8-tetrachlorodibenzo-p-dioxin (TCDD).[3] Although these cleanup guidelines are intended to protect human health, they are by their very nature nonsite-specific. Many state regulatory agencies have published guidelines for cleanup levels of some soil contaminants,[4-6] particularly petroleum contaminated soils. The stated rationale for many of the state guidelines is to prevent contamination of groundwater and/or to protect human health, although in many instances the scientific underpinnings for some state criteria are elusive. For example, the so-called "Type B" cleanup criterion for benzene in soils in the State of Michigan was derived[7] by multiplying the groundwater criterion (1 ppb) by a factor of 20. In many states, informal or formal guidelines

11

specify that remediation is required when soil concentrations of total petroleum hydrocarbons (TPH) exceed 100 ppm, while others require cleanup to background or analytical detection limits.[4-6] The problems associated with the multitude of guidelines and criteria that are currently being developed or proposed in the United States are that: (1) the basis or rationale for some of the cleanup levels is not clearly presented, (2) approaches to setting cleanup levels vary between states, (3) cleanup objectives often vary among agencies within the same state, and (4) the cleanup levels are usually set in a haphazard way, rather than in a systematic and scientifically defensible manner.[6] The cost and inefficiencies of this situation are not trivial for the regulated community and society as a whole. For example, even a three-fold difference in cleanup levels (which is often considered of negligible regulatory significance) can result in over a 10-fold difference in the cleanup costs. Even in the case of relatively nonvolatile, nonmobile chemicals such as TCDD, the difference in costs to remediate soils to 1 ppb, 10 ppb, or 50 ppb can be dramatic. Thus, if a generally accepted and scientifically defensible methodology could be developed to accurately identify cleanup levels for contaminated soils, remediation monies could be efficiently and properly allocated.

One goal of the Council for the Health and Environmental Safety of Soils (CHESS) is to establish a generic method for deriving soil cleanup levels.[8] CHESS represents an ambitious effort to optimize the scientific process and has brought together as many as 30 of the nation's experts on contaminated soil. Since the group's inception in 1986, these experts have exhaustively evaluated all known techniques for setting soil cleanup levels, and have written a critique of those approaches.[9] The group is in the process of giving advice for developing a generic computer model which will account for all of the factors which should go into setting site-specific and chemical-specific criteria. This model will probably be completed by mid-1992. The approach discussed here is not in conflict with the goals of CHESS, but rather, it represents a way for agencies to quickly take advantage of the formulas and calculations suggested by CHESS.

The objective of this chapter is to propose that the most effective way to quickly regulate contaminated soil is to set eight "types" of soil cleanup guidelines for each chemical. This chapter presents the rationale and describes a systematic approach for setting each guideline. For the sake of simplicity, this chapter describes the techniques and the factors that were considered. Two examples of how the approach can be applied are presented. Each type of guideline would apply to a particular land-use scenario or would satisfy a need to protect a particular environmental receptor (Table 2.1). In short, based on our nearly 20 years of experience, we believe that eight guidelines, rather than one or two, may be necessary for states to properly regulate contaminated soils. Several of the types of soil cleanup guidelines (Table 2.1) address exposure scenarios which are of significant concern to the public and to regulatory agencies. For example, residential, recreational, agricultural, and industrial exposure scenarios are settings evaluated in many risk assessments.[10,11] However, other issues and scenarios should also be considered. For example, protection of groundwater has been an important

Table 2.1.   The Eight Most Common and Important "Types" of Exposure Scenarios Which Might Require a Cleanup Guideline

- Residential
- Recreational
- Agricultural
- Industrial
- Protection of Wildlife
- Protection of Groundwater
- Runoff Hazard
- "de minimis" Quantity

environmental issue for nearly two decades, and is more dependent on site-specific considerations, such as the geology of the site and depth to groundwater, rather than classic exposure pathways. Another reason for cleaning soil is to control the runoff hazard and to protect wildlife. During the past four years, these have become important issues, and will continue to be so in the coming years (Table 2.2).[12]

The final factor for deciding whether to remediate soil is an evaluation of the severity of the contamination. A so-called "de minimis" quantity represents an amount of a chemical simply too small to be worthy of concern. The rationale for identifying de minimis amounts of contamination is two-fold: (a) to encourage persons to perform a mass calculation of the total amount of a particular substance (subtracting background) that is in the environment, and (b) to ensure that money is not wasted on writing a risk assessment on a site that intuitively could pose no significant risks to humans or the environment.

The guidelines presented here were not intended to be site-specific. Rather, these guidelines represent "screening-level" soil concentrations. The objective was to identify concentrations for eight common exposure scenarios for each contaminant for which, when encountered, no further evaluation would need to be conducted. We believe that this approach can help agencies respond more efficiently since they could rapidly dismiss those sites which pose a negligible hazard. When soil concentrations exceed the screening-level, de minimis, or safe concentration, a site-specific assessment would be conducted. Although it may appear cumbersome to set eight soil cleanup levels for each compound, this seems likely to represent an improvement over the 30–50 separate state guidelines that attempt to address only two or three different scenarios.

DESCRIPTION AND RATIONALE
FOR EIGHT SOIL CLEANUP CRITERIA

Residential Scenario (Nonvolatiles)

When setting a soil cleanup guideline for residential areas, incidental soil ingestion by children is nearly always the most important factor for nonvolatile

Table 2.2. Factors that "Drive" Soil Cleanup Levels

Residential	Recreational	Industrial
• soil/dust ingestion	• soil ingestion	• soil ingestion
• dermal uptake	• dermal contact	• dermal uptake
• garden vegetables	• runoff	• inhalation of fugitive dust
• dermatitis hazard		• groundwater/runoff

Agricultural	Wildlife
• uptake via crops	• uptake by birds
• uptake by grazing animals	• adverse effects on predator food chain
• runoff (aquatic)	• effects on development and reproduction
• groundwater hazard	
• fugitive dust	

chemicals and, frequently, for the volatiles. This has been shown in several assessments of contaminated soils.[10-15] Particularly with the nonvolatile chemicals, where contamination may be limited to the top six to twelve inches of soil, incidental soil and dust ingestion will often represent from 60% to 85% of the risk.[16] Dermal contact with contaminated soil is usually the second most important source of uptake in residential scenarios, followed by the ingestion of garden vegetables grown in contaminated soil. Inhalation of suspended particulates rarely contributes more than 1% to 2% of the total uptake and risk.[10,11,17-20]

Residential Scenario (Volatiles)

For the volatile chemicals, inhalation due to vaporization from soil and ingestion of soil will constitute the primary sources of exposure (risk), particularly if the contamination is recent. Occasionally, even when the contamination is below ground surface (as is often the case with leaking underground storage tanks), inhalation can pose an acute health hazard or an explosion hazard, especially in basements located near the contaminated soil. The second most important route of human exposure is usually dermal contact. Due to their relatively short environmental half-life, volatile compounds rarely enter the food chain. In general, the primary reason for remediating sites with volatile chemicals is the threat of contamination of shallow groundwater aquifers or the threat to aquatic species, rather than direct human exposure.

Industrial or Occupational Scenario

The most significant routes of exposure at industrial sites are normally dermal contact with soil and incidental soil ingestion by adults.[17,20] Dermal contact tends to be the predominant pathway because adults do not intentionally ingest soil; however, due to hand-to-mouth contact, they may ingest 0–10 mg/day or more.[14,21] Because many industrial sites are contaminated with volatile organic chemicals, which tend to have high mobility in soils, protection of groundwater is usually the driving force behind establishing cleanup levels. Thus far, it has become commonplace for most agencies to set two cleanup guidelines for soil; one for residential and the other for industrial/occupational settings.[1,10,22]

Recreational Scenario

Sources of soil contamination at recreational areas can be numerous. In many instances, recreational areas, such as parks, baseball fields, playgrounds, etc., are former industrial properties which have been abandoned and deeded to the city. Often, recreational areas were former municipal or industrial landfills which have been capped and considered safe. In other instances, the use of sludge from municipal wastes, paper and pulp processing, and other sources have been used as soil amendment materials on recreational lands because they contain nutrients required for plant growth such as nitrogen, phosphorus, calcium, and minor trace elements.[18,23] Sometimes, the use of pesticides and herbicides on recreational lands to control poison ivy, poison oak, etc., has been a source of concern. Like the residential scenario, health risks at recreational sites are driven primarily by incidental soil ingestion and dermal contact with soil by children.

In areas where there has been a widespread application of sludge as fill or soil amendment (fertilizer), runoff may be the biggest potential environmental and human health concern for recreational lands. The potential problem is that runoff of contaminants on topsoil via suspended particulates can be transported to rivers, lakes, streams, and harbors and can pose a threat to both wildlife and people.[15,24,25]

Agricultural Scenario

Although it might be anticipated that the ingestion of vegetables or cereals grown in contaminated soil would be the primary concern in an agricultural exposure scenario, usually the most significant route of human exposure involves uptake of soil by grazing animals.[26,27] This is considered an indirect route of uptake of the contaminant. To the surprise of many persons, although animals will ingest potentially contaminated crops and vegetation, the primary problem is that grazing cattle ingest an average of 0.9 kilograms of soil per day.[27,28] The ingestion of soil can represent 80% of the cow's uptake, since weathering processes usually will remove the majority of the particles which deposit onto crops. For the non-volatile chemicals, the human hazard can be greater than expected because persistent, lipophilic chemicals accumulate in beef tissue and fat, resulting in a potentially significant dose to humans who eat the meat and drink the milk.

Runoff of contaminated soils into streams, creeks, and lakes is the second most important concern when evaluating an agricultural scenario. This is not surprising, given the millions of acres treated with a myriad of pesticide products and the proximity of most farmland to streams, lakes, and other waterways. For example, the U.S. Department of Agriculture (USDA) estimated that from 1968 to 1977, 11 million acres of land were treated annually with mirex bait,[29] yet it was a relatively small volume pesticide. Thousands of tons of many chemicals, especially herbicides, are applied annually to agricultural soils. As before, compounds that are lipophilic and environmentally persistent (e.g., stable) pose the most significant hazard because of their propensity to bind strongly to soil, which becomes sediment and then accumulates in the tissues of aquatic organisms.[15]

Protection of Wildlife Scenario

Concern over the potential risks to wildlife has become a major issue at contaminated sites in recent years and is likely to become even more so in the next decade.[30,31] Uptake of contaminated soil by fish, mollusks, game birds, song birds, deer, and endangered species tend to be the most important considerations in setting cleanup levels based on protection of wildlife.[32] Issues such as the migratory patterns of birds, changes in biological diversity, increased susceptibility to predators, etc., tend to make these assessments very complex. Adverse effects on the predator food chain can also be significant.[30]

Protection of Groundwater

The primary concern at sites contaminated with small to moderate molecular weight chemicals is the contamination of groundwater. Clearly, the number of pounds of contaminant in the soil, the water solubility of the contaminant, the potability of water from the aquifer, and the toxicity of the chemical dictate the level of concern.[15,33] During the 1980s the threat of contamination of groundwater by soil was the primary reason for addressing nonresidential sites contaminated with high levels of chemicals. Recent evidence suggests that it may not be technologically possible to remediate groundwater aquifers which have become contaminated, and therefore this may not be as important a rationale in the 1990s.[34]

Runoff Hazard

The possibility that persistent chemicals can be present for decades in soils and subsequently be transported through runoff to streams will drive the cleanup at a greater number of sites in the 1990s. The recent emphasis on PCBs, dioxins, furans, metals, and PAHs is evidence that this will be one of the major issues in the coming years.[35]

De Minimis Level of Contamination

Often, the amount of contaminant lost to the environment due to a leaking underground tank, an accidental release, or some other event is known (or can be approximated). In some cases, the quantity is so small as to be unworthy of further consideration.

KEY EXPOSURE PARAMETERS

To identify the most accurate soil cleanup guideline or standard, the most appropriate exposure factors need to be selected. This has proved to be a significant task since high quality information on exposure is often not available. On the other hand, the range of likely or reasonable values can often be identified.[36] The EPA has suggested numerous exposure factors in a number of guidance documents,[37-39] but these are often quite conservative. These conservative factors, if combined together, can produce unrealistic estimates of exposure.[15] A more reasonable alternative is the use of best-estimate values and to account for the range of plausible values using Monte Carlo methodologies.[14,40,41] This approach should accurately predict the range of likely values for the human uptake of chemicals with more confidence than point estimates.[14,41,42] Typically, when developing health-based cleanup levels, one or two key parameters tend to "drive" the risk assessment at sites having contaminated soil (Table 2.2.).[15] That is, one or two exposure parameters tend to account for a vast majority (75% to 90%) of the risk. The following section discusses the eight different scenarios that we believe deserve specific guidance. The key parameters, or "drivers," that dictate the soil cleanup levels are explained.

Soil Ingestion Rate

Foremost among the exposure factors is the rate of soil ingestion, particularly among children. Incidental soil ingestion occurs at all ages as a result of hand-to-mouth activities. However, it is widely accepted that children approximately 2 to 6 years old are the primary age group that consumes a potentially significant amount of soil.[12,14,43,44] Several literature surveys have been conducted which attempt to characterize what is known about soil ingestion rates among children. The majority of published estimates of soil ingestion rates for adults and children are presented in Table 2.3.

The most thorough and rigorous studies to date are those by Calabrese and co-workers[44] and van Wijnen and colleagues.[45] In the Calabrese study, eight different tracer elements were quantitatively evaluated in the stools of 65 school children, aged one to four years old. Although this study originally reported that the three most reliable tracers were Al, Si, and Y, as validated by a supplemental adult study,[46] a more recent analysis by the same group[47,48] indicated that Zr and Ti are probably the only reliable tracers, and therefore the bulk of the published work probably overestimates the true value. A value of 20–50 mg/day probably represents a conservative, yet realistic, soil ingestion rate for children.

In the van Wijnen study,[45] two tracer elements (Al and Ti) were quantitatively measured in the stools of children ages one to five. For children one to two years

Table 2.3.   The Range of Estimates for Soil Ingestion by Humans for Various Age Groups Which Have Been Published Over the Past 20 years (mg/day)

Researcher	AGE GROUP			
	Infants[a]	Toddlers[b]	Children[c]	Adults[d]
Baltrop (1973)[106]		100		
Lepow et al. (1974)[51]		100		
Day (1975)[107]		100		
Duggan & Williams (1977)[108]		50		
NRC (1980)[28]		40		
Kimbrough et al. (1984)[10]	0–1000	1000–10,000	100–1000	100
Hawley (1985)[109]	0	90	21	57
Binder et al. (1986)[49]		180		
Bryce-Smith (1986)[110]		33		
Schaum (1986)[55]		100–5000		
USEPA (1986)[111]		100		
Clausing et al. (1987)[112]		56		
LaGoy (1987)[113]		100		
Paustenbach (1987)[21]		25–50		0–10
Lipsky (1988)[114]		240		
Sedman (1988)[115]		50–200		
USEPA (1989[e])[38]		200	100	100
Calabrese et al. (1987,[116] 1989,[44] 1990,[46] 1991[47,48])		9–40	1–10	0–31
Davis et al. (1990)[50]		25–80	25–80	
van Wijnen et al. (1990)[45]		0–90		
Sheehan et al. (1991)[11]	0	33	10	10

[a]Infants have usually been defined as children aged 0–2 years and who receive significant adult supervision.
[b]Toddlers have usually been defined as children aged 2–4 or 2–6 years.
[c]Children are usually defined as 4–12 years of age.
[d]Adults are often considered 13–75 years of age.
[e]The current USEPA position on soil ingestion.

of age, the mean ingestion rate ranged from 0–90 mg/day. For children aged three to five, the mean ingestion rate ranged from 0–60 mg/day.

The validity of soil ingestion estimates for children derived from soil tracer methodologies has recently been questioned.[47,48] By calculating the soil recovery variances and soil detection limits for each tracer, it was shown that recent studies[44,45,49,50] provide little convincing evidence upon which to derive a precise quantitative estimate of soil ingestion. For the three or four major studies, only the soil ingestion rates (16 and 55 mg/day) reported by Calabrese et al.[46] for two out of eight tracers were within acceptable statistical norms.[47] Thus, an average soil uptake of about 40–50 mg/day appears to be credible. While this soil ingestion rate is less than the 200 mg/day rate currently recommended by the EPA for the typical child,[13,38] it should be noted that the EPA value predates the work of Calabrese et al. and van Wijnen et al.

The majority of soil ingested by adults is thought to occur via hand-to-mouth contact (such as during smoking) and soil on produce.[16] As such, adults are expected to ingest quantities of soil which are perhaps 2- to 10-fold less than children.[12,41] Although there is inadequate data to characterize the distribution of ingested soil by adults, a value of 10 mg of soil per day is expected to be representative.

Dermal Contact Rate

As mentioned previously, the second major driver for most soil cleanup levels is dermal contact. While there are several factors to consider, one of the more important ones is the dermal contact rate or soil loading factor. The soil contact rate is an estimate of the amount of soil that adheres to a given area of human skin over a specific period of time. As presented in Table 2.4, several studies have estimated or evaluated the amount of soil or dust that is likely to be in contact with skin. Lepow et al. found that an average of 0.5 mg soil/cm^2 skin could be removed from the hands of children aged 2 to 6 years old.[51,52] Soil samples were collected from 10 children's hands by pressing a preweighed self-adhesive label on the hand surface. Roels et al. determined the amount of lead on the hands of 11-year-old rural school children by pouring dilute nitric acid over the palm and fingers of the dominant hand.[53] Although the amount of soil on the hand was not determined by the authors, others have estimated this value by dividing the amount of lead on the hand by the concentration of lead in the soil. Using this method, the South Coast Air Quality Management District (SCAQMD) estimated about 120 mg soil on the hand.[54] Assuming that the surface area of one hand is 342 cm^2 and that about 60% of the hand was sampled,[54] they estimated a soil loading rate of about 0.6 mg/cm^2.

Table 2.4. **Summary of Soil Loading Rates on the Palms of Hands of Human Test Subjects**

Soil Loading Rate (mg/cm^2)	Type of Soil	Age and Sex	Reference Source
1.45	Commercial potting soil	Male adult	USEPA, 1989a[37]
2.77	Kaolin clay	Male adult	USEPA, 1986a[111]
1.48	Playground dirt and dust	Males 10 to 13 years old Rural Areas	Modified from Roels et al., 1980[53]
1.77	Playground dirt and dust	Males 10 to 13 years old Urban Areas	Modified from Roels et al., 1980[53]
0.6	Unsieved	Adult male	Driver et al., 1989[58]
0.5	Not available	Children, –4 years old	Lepow et al., 1975[52]
0.33	House dust	Children, 1–3 years old	Gallacher et al., 1984[117]
0.11	House dust	Adult woman	Gallacher et al., 1984[117]
0.2	Not available	Adults	Que Hee et al., 1985[56]

Schaum suggested[55] using a range of 0.5 to 1.5 mg/cm^2 (applicable to the entire exposed body) for adults and children based on the results of the Lepow et al. and Roels et al., respectively.[52,53] The estimate of the upper end of the range of Roels et al. is higher than the SCAQMD figure, due primarily to differences in the estimates of the hand surface area sampled in the study.

Que Hee et al. examined the retention of soil on the hands of a small adult.[56] Based on this study and several assumptions regarding soil particle size and adherence to the skin, the California Department of Health Services (CDHS) estimated that an average of 31.2 mg soil adhered to the palm of a small adult.[57] Using 168 cm^2 as the surface area of the palm of a small adult, CDHS estimated a soil contact rate of 0.2 mg soil/cm^2 skin.[57] Driver et al. surveyed various soils of different organic content for their ability to adhere to human hands (adult male).[58] It was determined that the average amount of unsieved soil retained on human hands was 0.6 mg/cm^2.

The data from each of these studies can be expected to overestimate the amount of soil that will adhere to the skin surface of other body parts (e.g., forearms, neck, and head) because it is often just the palms that are in direct contact with soil surfaces. This assumption has been generally confirmed in studies of agricultural workers.[59] In addition, these soil loading rates are almost certain to overestimate the actual average exposure because: (1) not everyone works in community gardens; (2) many persons wear gloves when working intimately with dirt; (3) gardeners work directly with soil primarily during planting and weeding only; (4) most people do not garden each day; and (5) the number of days of precipitation during the gardening season further diminishes the frequency of exposure.[21] Thus, a reasonable, yet conservative, estimate for soil contact rate appears to be 0.5 mg soil/cm^2 skin.

Inhalation of Suspended Particulates

Contrary to what most citizens believe, inhalation of suspended particulates will rarely, if ever, constitute a route of entry of sufficient magnitude to dictate cleaning contaminated soils.[10,11,17,18] To calculate the inhalation load, one has to estimate the suspension of soil/dust into the air. Usually, only a fraction of the total suspended particulates (TSP) is assumed to be due to contaminated soil. For example, in the dioxin risk assessment conducted by the Centers for Disease Control (CDC) on Times Beach, an average air concentration of TSP of 0.14 mg/m^3 was assumed,[10] and most of it was thought to be from the contaminated site. When actual field data from Missouri and elsewhere were considered, TSP measurements were often much less than this value. For example, the EPA data showed[10,60] that TSP for a rural area was about 0.070 mg/m^3. For other geographical locations, concentrations of 0.15 mg/m^3 for urban environments and 0.10 mg/m^3 for rural environments have been suggested.[12] These estimates for TSP appear appropriate for establishing soil cleanup levels in the absence of site-specific data.

Fraction of Soil from the Contaminated Source

Typically, only a fraction of soil ingested by a person in any one day will be from a contaminated source. In addition, for any particular site, usually only a fraction of the surface soil is contaminated. In setting de minimis or screening-level soil guidelines, choosing one fraction over another is problematic because that fraction is site-specific. However, it is reasonable to assume that the fraction of ingested, inhaled, or contacted soil that originates from the contaminated source is that fraction of a day spent at the source itself, whether it is a residence, a workplace, or a theme park. Data from national time-use studies performed at the University of Michigan and analyzed by the U.S. EPA are useful for assessing exposure at impacted residential areas. These data indicate that adults spend approximately 40% of their waking hours at home,[38] while children and youths are expected to spend approximately 47% of their waking hours at the residence.[38] Workers can conservatively be expected to be exposed to soils at the workplace for 8 hours per 16 awake hours per day (50%).

Exposure Duration

People generally spend only a fraction of their lifetime in any one location. U.S. census data indicate that the average residence time is 9 years in any one house.[13] That same report states that the 90th percentile of the population remains in one residence for 30 years.[13] We suggest that this value be used to establish screening-level soil cleanup criteria at residential sites. For an industrial scenario, a 25-year exposure duration is considered an upper-bound value[39] and would appear to be adequately conservative for purposes of risk assessment.

Bioavailability

Environmental contaminants are able to cross biological barriers with varying degrees of efficiency. When chemicals are bound to soil, the efficiency of uptake decreases even more. The bioavailability, i.e., the percentage of a chemical in soil that is absorbed by humans, is governed primarily by (1) the physicochemical properties of the contaminant, (2) the environmental matrix in which it is present, and (3) the nature of the biological membrane. Chemicals in soil are usually absorbed to a lesser degree than the chemicals in pure form.[16,61,62] Although this chapter addresses screening-level soil cleanup guidelines, due to the simplicity of the concept, there is no reason to ignore chemical-specific bioavailability in setting those guidelines if the appropriate data are available.

In the absence of chemical-specific data, an approach similar to that used by the State of Michigan could be adopted.[63] For purposes of risk assessment, oral bioavailability is assumed to be 100% for volatile organic compounds and 10% for semivolatiles and metals bound to soils. In the absence of chemical-specific data, dermal bioavailability is assumed to be 10% for volatile compounds and 1% for semivolatiles and metals.[63] Although a gross and conservative approach to the problem, it seems adequate until more data are available. A sophisticated model for dermal bioavailability which should be superior to other estimation techniques was recently proposed.[62]

Environmental Fate and Transport

Although not an exposure factor per se, understanding a chemical's environmental fate and transport is essential to properly characterize the hazards of contaminated soil. The first step in understanding the environmental hazard is to examine the physical and chemical properties of that compound.[12] In general, chemicals introduced into the environment may adsorb to soils, dissolve into bodies of water, leach from soil into water, volatilize from soil and water into the atmosphere, or be taken up from soil by vegetation.[33,64] These are known as "transport" processes. A chemical may also undergo photo- or microbial degradation. These are called "fate" processes.

Environmental fate and transport are dictated largely by a compound's physicochemical properties.[33,64-69] These characteristics govern the ability of a chemical to move from one matrix to another. In short, transport and distribution of all these compounds in the environment is determined to a large extent by the following physicochemical properties:

- water solubility
- organic carbon coefficient (K_{OC})
- vapor pressure
- Henry's law constant
- octanol/water coefficient (K_{OW})

The physicochemical properties of concern that govern transport in soil are water solubility and organic carbon coefficient (K_{OC}). The K_{OC} is a measure of relative sorption potential, and indicates the tendency of an organic chemical to be adsorbed to soil or sediment. It provides an important measure of a chemical's mobility in soil because it is largely independent of soil properties.[70] Typically, K_{OC} values can range from one to ten million, with higher values indicative of greater sorption (binding) potential. A low K_{OC} indicates a greater potential for leaching from soil into groundwater, followed by rapid transport through an aquifer. In contrast, a higher K_{OC} suggests a relatively lower potential for leaching into groundwater over a period of time.

K_{OC} values representative of chemicals that are immobile to very mobile in soil are shown in Table 2.5.

Vapor pressure and Henry's law constant are two measures of chemical volatility that are important in estimating releases to ambient air. Vapor pressure is an important determinant of the rate of vaporization from contaminated soil, but other factors, including temperature, wind speed, degree of adsorption to soil, water solubility and soil conditions, are also important.

Table 2.5.   Mobility Range of K_{oc} Values

Mobility	K_{oc} Coefficient Range (dimensionless)
Immobile	$K_{oc} > 2,000$
Low	$K_{oc} > 500$ and $< 2,000$
Intermediate	$K_{oc} > 150$ and < 500
Mobile	$K_{oc} > 50$ and < 150
Very mobile	$K_{oc} < 50$

Henry's law constant combines vapor pressure, water solubility, and molecular weight to characterize a chemical's ability to evaporate from aqueous media, including moist soils.[70] Typical values for Henry's constant (H_a) for compounds ranging from low to high volatility are as shown in Table 2.6.

Table 2.6.   Volatility Values for Henry's Constant Range

Volatility	Range of Henry's Constant (H_a) (atm-cu.m/mol)
Low	$H_a < 3 \times 10^{-7}$
Intermediate	$H_a > 3 \times 10^{-7}$ and $< 10^{-5}$
Moderate	$H_a > 1 \times 10^{-5}$ and $< 10^{-3}$
High	$H_a > 1 \times 10^{-3}$

The n-octanol-water partition coefficient (K_{ow}) describes the tendency of a nonionized organic chemical to accumulate in lipid tissue and to sorb onto soil particles or onto the surface of organisms or other particulate matter coated with organic material.[14] Although a powerful tool for understanding organic chemicals, it is not as good a predictor for inorganic chemicals, for metal organic complexes, or for dissociating and ionic organic compounds.[71] No published scale for interpretation of K_{ow} data is currently available. However, high K_{ow} values reflect lower water solubilities and higher bioconcentration factors.

ACCEPTABLE LEVELS OF RISK

The level of risk considered acceptable will ultimately dictate the soil cleanup level. In setting generic cleanup concentrations, the levels of acceptable risk are usually dictated, in part, by whether the risk is voluntary or involuntary. Traditionally, the residential scenario, and most others, are considered involuntary settings while industrial sites are often considered voluntary. For noncarcinogens, the EPA and others have usually used the Reference Dose (RfD) for a specific chemical as the basis for establishing its soil cleanup level.[37]

Noncarcinogens

For the purpose of setting de minimis screening-level soil guidelines, perhaps allowing one-tenth the RfD as the maximum allowable dose to be absorbed due to contaminated soil would be appropriate. Given the conservatism of many of the uncertainty factors used to establish RfDs, this value should adequately protect human health. In contrast, setting acceptable contaminant levels to protect the environment or wildlife involves the evaluation of several endpoints. To protect wildlife against the adverse effects of a noncarcinogen, perhaps 1/100th the animal no-observed-adverse- effect-level (NOAEL) would be appropriate. Similarly, 1/100th the aquatic NOAEL could be used as a basis for establishing soil cleanup concentrations when protection of aquatic species is the primary concern. An alternative (or complementary) level of acceptable risk could involve setting the soil level so that the steady-state chemical concentration in fish would ensure that no more than one-tenth the human RfD could be due to the contaminated site. The theoretical human receptor could be represented by the 80th percentile fish-eater. This level would most likely account for the possibility of additivity of chemicals, as well as other factors. The contamination of sediment poses an additional environmental concern. One approach could involve ensuring that the steady-state concentration of the chemical in a bottom-feeding fish (e.g., catfish) not exceed two times the background or result in a human uptake of the chemical not exceeding 1/10th the ADI. For groundwater protection, the goal is to prevent the contaminant from reaching groundwater, with perhaps 50% of the Maximum Contaminant Level (MCL) as the endpoint to avoid.

Carcinogens

Establishing acceptable levels of risk for carcinogens has been a topic of considerable debate within the risk assessment community. Inasmuch as soil cleanup levels are often "driven" by the presence of carcinogens, it is important to understand the regulatory history of "acceptable risk." It is a common misperception within risk assessment that all occupational and environmental regulations have, as their goal, a theoretical maximum cancer risk of 1 in 100,000 (1×10^{-5}) or 1 in 1,000,000 (1×10^{-6}). When a risk falls above this level, the public and the media quite often misinterpret this as a serious threat to public health. In a 1987 journal article, the former commissioner of the Food and Drug Administration (FDA), Dr. Frank Young, discussed this misunderstanding.[73]

> The risk level of one in one million is often misunderstood by the public and the media. It is not an actual risk—i.e., we do not expect one out of every million people to get cancer if they drink decaffeinated coffee. Rather, it is a mathematical risk based on scientific assumptions used in risk assessment. FDA uses a conservative estimate to ensure that the risk is not understated. We interpret animal test results conservatively and we are extremely careful when we extrapolate risks to humans.

When FDA uses the risk level of one in one million, it is confident that the risk to humans is virtually nonexistent.

In short, the cancer risk levels assumed by some policy makers to represent trigger levels for regulatory action, actually represent levels of risk that are so small as to be of negligible concern.

Recent reviews[75,76] indicate that the theoretical cancer risks associated with currently enforced environmental regulations are in the vicinity of 1 in 100,000, not 1 in 1,000,000. In a retrospective review of the use of cancer risk estimates in 132 federal decisions, Travis et al. examined the level of cancer risk that triggered regulatory action.[75] The authors considered three measures of risk: individual risk (an upper-limit estimate of the probability that the most highly exposed individual in a population will develop cancer as a result of a lifetime exposure), size of the population exposed, and population risk (an upper-limit estimate of the number of additional cases of cancer in the exposed population). Travis et al. found[75] that for exposures resulting in a small-population risk, the level of risk above which agencies almost always acted to reduce risk was approximately 4×10^{-3}. For large-population risks (the entire U.S. population) agencies typically acted on risks of about 3×10^{-4}. For effects on small populations, regulatory action was never taken for individual risk levels below 1×10^{-4}. For large-population effects, the de minimis risk level dropped to 1×10^{-6}. Consequently, the level of acceptable individual risk is usually dictated by the size of the exposed population. Recently, final revisions to the National Contingency Plan[77] have set the acceptable risk range between 10^{-4} and 10^{-6} at hazardous waste sites regulated under the Comprehensive Environmental Response, Compensation, and Liability Act (CERCLA). In the recently promulgated Hazardous Waste Management System Toxicity Characteristics Revisions (55 FR 11798–11863), the U.S. EPA has stated that:

> For drinking water contaminants, EPA sets a reference risk range for carcinogens at 10^{-4} to 10^{-6} excess individual cancer risk from lifetime exposure. Most regulatory actions in a variety of EPA programs have generally targeted this range using conservative models which are not likely to underestimate the risk.

Interestingly, the U.S. EPA has selected and promulgated a single risk level of 1 in 100,000 (1×10^{-5}) in the Hazardous Waste Management System Toxicity Characteristics (TC) Revisions (55 FR 11798–11863). In their justification, the U.S. EPA cited the following rationale:

> The chosen risk level of 10^{-5} is at the midpoint of the reference risk range for carcinogens (10^{-4} to 10^{-6}) targeted in setting MCLs. This risk level also lies within the reference risk range (10^{-4} to 10^{-6}) generally used to evaluate CERCLA actions. Furthermore, by setting the risk level at 10^{-5} for TC carcinogens, EPA

believes that this is the highest risk level that is likely to be experienced, and most if not all risks will be below this level due to the generally conservative nature of the exposure scenario and the underlying health criteria. For these reasons, the Agency regards a 10^{-5} risk level for Group A, B, and C carcinogens as adequate to delineate, under the Toxicity Characteristics, wastes that clearly pose a hazard when mismanaged.

Few state regulatory agencies have adopted a one in one million (1×10^{-6}) risk criterion in making environmental and occupational decisions. The states of Virginia, California, Maryland, Minnesota, Ohio, and Wisconsin have often employed, or propose to use, the one in one hundred thousand (1×10^{-5}) level of risk in their risk management decisions (personal communications with state agencies, 1990). The State of Maine Department of Human Services (DHS) uses a lifetime risk of one in one hundred thousand as a reference for nonthreshold (carcinogenic) effects in its risk management decisions regarding exposures to environmental contaminants.[78] Similarly, a lifetime incremental cancer risk of one in one hundred thousand is used by the Commonwealth of Massachusetts as a cancer risk limit for exposures to substances in more than one medium at hazardous waste disposal sites.[79] This risk limit represents the total cancer risk at the site associated with exposure to multiple chemicals in all contaminated media. The State of California has also established[80] a level of risk of one in one hundred thousand for use in determining levels of chemicals and exposures that pose no significant risks of cancer under the Safe Drinking Water and Toxic Enforcement Act of 1986 (Proposition 65). Workplace air standards developed by the Occupational Safety and Health Administration (OSHA) typically reflect theoretical risks of about one in one thousand (1×10^{-3}) or greater.[81]

ESTABLISHING DE MINIMIS SOIL CLEANUP GUIDELINES: TWO EXAMPLES

The suggested approach to setting screening-level soil guidelines presented here can be best illustrated by examining two chemicals which have been of significant concern when found in soil: 2,3,7,8-tetrachlorodibenzo-p-dioxin (TCDD) and the volatiles benzene, toluene, and xylene (BTX) which are indicator chemicals for gasoline.

TCDD

Environmental Fate and Distribution

The chemicophysical properties of TCDD (Table 2.7) indicate that this compound is virtually water insoluble, has a relatively high lipid solubility, has a high soil-binding constant (K_{OC}), almost no vapor pressure, a low Henry's law constant, and a fairly substantial bioconcentration factor (BCF). Thus, TCDD

Table 2.7.   Chemical and Physical Properties of TCDD

Molecular Weight (g/mol)	322
Water Solubility (mg/L)	2.0×10^{-4}
Lipid Solubility (log K_{OW})	6.72
Soil Adsorption Constant (K_{OC}) (ml/g)	3,300,000
Vapor Pressure (mm Hg)	1.7×10^{-6}
Theoretical Bioconcentration Factor (BCF)	5,000–50,000

is relatively nonvolatile and essentially immobile in soil. Moreover, one would expect to find TCDD in sediment rather in the water column, and its bioaccumulation potential in the food chain (but not plants) would be very high.

Residential, Industrial, and Recreational

The method for deriving health-based cleanup levels for the residential, industrial, and recreational situations is very similar. In many ways, calculating these levels is a risk assessment in reverse. First, potential exposure routes are identified and exposure parameters established to quantitatively describe exposure potential. By examining the physicochemical properties of the compounds of interest, one can determine which exposure routes will be significant (e.g., ingestion and dermal contact could be significant for TCDD, whereas inhalation of vapors would not). Second, the relative contribution of each of the routes of exposure to the total dose is determined by assigning a hypothetical chemical concentration in soil (e.g., 1 ppm) and apportioning exposure according to standard exposure equations (see below). By doing this, the relative contribution of each exposure pathway can be used to calculate the amount of total dose that can safely be received via each route of exposure.[32] Parameter values for each route of exposure are presented in Tables 2.8–2.11.

Table 2.8.   Exposure Parameters Used in Equation 5 to Estimate the Uptake by Humans of Contaminated Soil in Suspended Particulates via Inhalation

	Scenario		
Parameter	**Residential**	**Occupational**	**Recreational**
Exposure Duration (ED)	30 years	25 years	30 years
Exposure Frequency (EF) (fraction of year)	0.962 days/days	0.343 days/days	0.011 days/days
Average Lifetime (LF)	70 years	70 years	70 years
Body Weight (BW)	70 kg	70 kg	70 kg
Breathing Rate (BR)	20 m^3/day	11 m^3/day	20 m^3/day
Conc. of Total Suspended Particulates (TSP)	0.25 mg/m^3	0.1 mg/m^3	0.1 mg/m^3
Conversion Factor (CF)	1×10^{-6} kg/mg	1×10^{-6} kg/mg	1×10^{-6} kg/mg

continued

Table 2.8. *continued*

Parameter	Scenario		
	Residential	Occupational	Recreational
Fraction of Soil Originating from Source (FC)	0.47	0.5	1
Fraction Deposited to Lung Tissue (PDa)	0.25	0.25	0.25
Percent of Time Spent Outdoors (TO)	19%	100%	100%
Percent of Time Indoors (TI)	81%	0%	0%
Indoor Dust Factor (IDF)	0.75	0.75	0.75

Table 2.9. **Exposure Parameters Used in this Evaluation to Estimate the Human Uptake of Chemicals Due to Ingestion of Contaminated Soil (Used in Equation 1)**

Parameter	Scenario		
	Residential	Occupational	Recreational
Exposure Duration (ED)	30 years	25 years	30 years
Exposure Frequency (EF) (fraction of year)	0.962 days/days	0.343 days/days	0.011 days/days
Average Lifetime (LF)	70 years	70 years	70 years
Body Weight (BW)	70 kg	70 kg	70 kg
Ingestion Rate (IR)	12 mg/day	10 mg/day	10 mg/day
Conversion Factor (CF)	1×10^{-6} kg/mg	1×10^{-6} kg/mg	1×10^{-6} kg/mg
Fraction of Soil Originating from Source (FC)	0.47	0.5	1

Table 2.10. **Exposure Parameters Used in this Evaluation to Estimate the Human Uptake via the Skin Following Contact with Contaminated Soil (Used in Equation 2)**

Parameter	Scenario		
	Residential	Occupational	Recreational
Exposure Duration (ED)	30 years	25 years	30 years
Exposure Frequency (EF) (fraction of year)	0.962 days/days	0.343 days/days	0.011 days/days
Average Lifetime (LF)	70 years	70 years	70 years
Body Weight (BW)	70 kg	70 kg	70 kg
Conversion Factor (CF)	1×10^{-6} kg/mg	1×10^{-6} kg/mg	1×10^{-6} kg/mg

continued

Table 2.10. *continued*

Parameter	Scenario		
	Residential	Occupational	Recreational
Percent of Time Spent Outdoors (TO)	19%	100%	100%
Percent of Time Indoors (TI)	81%	0%	0%
Indoor Dust Factor (IDF)	0.75	0.75	0.75
Skin Surface Area (SA)	1980 cm^2	1980 cm^2	1980 cm^2
Mass of Soil Adhering to Skin (M)	0.5 mg/cm^2	0.5 mg/cm^2	0.5 mg/cm^2
Fraction of Soil Originating from Source (FC)	0.5	1	0.5

Table 2.11.   Exposure Parameters Used in Equation 4 to Estimate the Uptake by Humans Due to the Inhalation of Vapors from Contaminated Soil (Used in Equation 4)

Parameter	Scenario		
	Residential	Occupational	Recreational
Exposure Duration (ED)	30 years	25 years	30 years
Exposure Frequency (EF) (fraction of year)	0.962 days/days	0.343 days/days	0.011 days/days
Average Lifetime (LF)	70 years	70 years	70 years
Body Weight (BW)	70 kg	70 kg	70 kg
Breathing Rate (BR)	20 m^3/day	11 m^3/day	20 m^3/day
Conversion Factor (CF)	1×10^{-6} kg/mg	1×10^{-6} kg/mg	1×10^{-6} kg/mg

The following formulas were used to calculate the lifetime average daily dose (LADD) for humans exposed to soil via ingestion (Eq. 1), dermal contact (Eq. 2), inhalation of vapors (Eq. 3), and inhalation of suspended particulates (Eq. 4).

$$LADD = ED \times EF \times IR \times CF \times FC / BW \times LF \qquad \text{(Eq. 1)}$$

$$LADD = ED \times EF \times CF \times FC \times TO \times TI \times IDF \times SA \times M / BW \times LF \qquad \text{(Eq. 2)}$$

$$LADD = ED \times EF \times BR \times CF / BW \times LF \qquad \text{(Eq. 3)}$$

$$LADD = ED \times EF \times BR \times TSP \times CF \times FC \times PDa \times TO \times TI \times IDF / BW \times LF \qquad \text{(Eq. 4)}$$

Where: LADD = Lifetime Average Daily Dose
        ED   = Exposure Duration (years)
        EF   = Exposure Frequency (fraction of a year)
        IR   = Ingestion Rate (mg/day)
        CF   = Conversion Factor (kg/mg)
        FC   = Fraction of Soil Originating from Source
        TO   = Percent of Time Spent Outdoors
        TI   = Percent of Time Spent Indoors
        IDF   = Indoor Dust Factor
        SA   = Skin Surface Area (cm^2)
        M   = Mass of Soil Adhering to Skin (mg/cm^2)
        BR   = Breathing Rate (m^3/day)
        TSP   = Concentration of Total Suspended Particulates (mg/m^3)
        PDa   = Fraction Deposited to Lung Tissue
        BW   = Body Weight (kg)
        LF   = Average Lifetime (years)
        B   = Bioavailability

The parameters to be used in these calculations for the residential, occupational and recreational scenarios are presented in Tables 2.8 to 2.11.

Finally, by substituting a dose that represents an acceptable level of risk (see above) in the standard exposure equations and solving the equation for soil concentration, a chemical concentration in soil that should result in no adverse health effects can be derived. For TCDD, the relative contribution of the oral, dermal, and inhalation (particulates) exposure under an industrial scenario was 16%, 81%, and 2%, respectively. The conceptual approach is discussed in significant detail in the EPA guidance handbook.[32] Multiplying these proportions by the Risk Specific Dose (RsD) at 10^{-5} risk (1.0×10^{-10} mg/kg-day), based on the recent re-evaluation of the Kociba et al. bioassay[82] of TCDD[83,84] results in apportioned RsDs as follows:

Oral 1.7×10^{-10} mg/kg-day
Dermal 8.4×10^{-10} mg/kg-day
Inhalation 2.3×10^{-11} mg/kg-day

These values can then be substituted for the LADD in the above equations and the soil concentration, Cs, can be calculated as follows:

$$Cs_{oral} = (\text{Apportioned RsD} \times BW \times LF)/(IR \times CF \times B \times FC \times EF \times ED) \qquad \text{(Eq. 5)}$$

$$= (1.7 \times 10^{-10} \text{ mg/kg-day})(70 \text{ kg})(70 \text{ yr})/ (10 \text{ mg/day})(10^{-6})(0.4)(0.5)(0.343)(25 \text{ yr})$$

$$= 0.05 \text{ mg/kg, or 50 ppb}$$

$$Cs_{dermal} = (Apportioned\ RsD \times BW \times LF)/(SCR \times SA \times$$
$$FC \times CF \times B \times EF \times ED) \qquad (Eq.\ 6)$$

$$= (8.4 \times 10^{-10}\ mg/kg\text{-}day)(70\ kg)(70\ yr)/$$
$$(0.5\ mg/cm^2)(1980\ cm^2)(10^{-6})(0.01)(0.343)(25\ yr)$$

$$= 0.05\ mg/kg,\ or\ 50\ ppb$$

$$Cs_{inhal} = (Apportioned\ RsD \times BW \times LF)/(BR \times PM_{10} \times$$
$$FC \times CF \times B \times EF \times ED) \qquad (Eq.\ 7)$$

$$= (2.3 \times 10^{-11}\ mg/kg\text{-}day)(70\ kg)(70\ yr)/(11\ m^3/day)$$
$$(25\ \mu g/m^3)(10^{-9})(1.0)(0.343)(25\ yr)$$

$$= 0.05\ mg/kg,\ or\ 50\ ppb$$

Thus using this approach, the screening-level soil cleanup guideline for TCDD at an industrial site would be approximately 50 ppb. This is similar to the value of 130 ppb recently proposed by Paustenbach et al.[85] Soil cleanup guidelines for TCDD for the residential and recreational scenarios are presented in Table 2.12. It is important to note that ingestion of homegrown vegetables was not included in the residential cleanup levels because several studies suggest that uptake and translocation of TCDD from soils to plants is negligible and does not present an appreciable health risk to humans.[27,86-88]

Table 2.12. Suggested Clean-up Levels for TCDD in Soil Based on the Methodology Proposed

Residential	= 26 ppb
Industrial	= 50 ppb
Recreational	= 200 – 1000 ppb
Agricultural	= 400 – 10,000 ppb
Groundwater	= 10,000 or greater
Runoff	= 200 ppb[a]
Wildlife	= 100 ppt – 10 ppb
"de minimis" Quantity	= 2–10 grams[b]

[a]Based on a 200 acre site.
[b]The amount of TCDD that could contaminate 100′ × 100′ square section of land to a depth of 6″ at a concentration of 10–50 ppb.

Agricultural

As mentioned above, agricultural cleanup levels tend to be driven by the ingestion of soil by grazing animals. The potential uptake of TCDD by grazing cattle and dairy cows has been examined extensively elsewhere.[27] Their examination suggested that soil cleanup levels for TCDD under an agricultural scenario could range from 400 to greater than 5,000 ppb, depending on the availability for forage as an animal feed source, the extent cattle are pastured, and the length of time the animals are held in feed lots prior to slaughter. For example, lactating dairy cows are rarely pastured, and some form of supplemental feeding is almost always employed.[27] The fattening period for nonlactating cows may be as long as 150 days, during which time animals can gain as much as 60% to 70% in body weight; such lipophilic chemicals as TCDD will therefore be diluted in the expanding body fat pool. Although such soil concentrations may not pose a hazard to the animal, or humans who eat them, the runoff hazard might be too great to be acceptable.

Runoff

Runoff of TCDD-contaminated soil into nearby streams or lakes can sometimes become the controlling factor for establishing soil cleanup levels. The potential for surface runoff is important because relatively small amounts of TCDD-contaminated soil may eventually contaminate nearby streams and produce potentially excessive levels of TCDD in fish.

Due to TCDD's low water solubility, low vapor pressure, and strong tendency to adsorb to organic material, it does not readily dissolve in runoff waters. Consequently, it is not easily transported across the soil surface in a dissolved phase.[89] However, since TCDD is bound tightly to the soil matrix, surface transport from TCDD contaminated areas may be possible through erosion of soil-bound TCDD. The most common method used to estimate the rate of surface runoff is the Universal Soil Loss Equation (USLE)[90] and sediment delivery ratios (SDR).[37] Although originally designed to estimate runoff of soils from agricultural lands, modifications and adaptations of this equation have been devised to accommodate a wide variety of chemical contamination scenarios.[13,91] The application of the SDR was intended to correct for the overestimation of soil loss produced by the USLE.[30,92] It was developed from empirical data that quantified sediment movement in streams. However, the equations did not differentiate between sediments contributed by bank or gully erosion and actual soil movement from the land surface to the stream.[92] The results of this equation, presented below, can be used to calculate the amount of soil-bound TCDD that might run off to a neighboring stream and accumulate in sediments:

$$SL = Rf \times Ef \times SLf \times Cf \times SPf$$

Where: SL = Site soil loss rate (kg/acre/yr)
        Rf = Rainfall/runoff factor (years)$^{-1}$
        Ef = Erodibility factor (kg/acre)
        SLf = Slope-Length factor
        Cf = Cover/management factor
        SPf = Support practice factor

Table 2.13.   Parameters Used to Estimate Soil Loading Due to Run-Off to a Stream Neighboring a Hypothetical Residential or Industrial Site

Rainfall/Runoff Factor	375 years^{-1}
Erodibility Factor	390 kg/acre
Slope-Length Factor	1.5
Cover/Management Factor	0.011
Support Practice Factor	1
Rate of Soil Loss from Site	2413 kg/acre/year
Area of Site	200 acres
Sediment Delivery Ratio for Site	0.77 (dimensionless)
Area of Watershed	9000 acres
Rate of Soil Loss from Watershed	2413 kg/acre/year
Sediment Delivery Ratio for Watershed	56.8

Parameters used to estimate soil loading to a stream neighboring a hypothetical site are based on U.S. EPA guidance[89] and are summarized in Table 2.13. The cumulative concentration of TCDD in sediment is then,

$$CSDC = \sum_{i=1}^{n} \text{ of SDCi}$$

$$SDCi = (SC \times SL \times SA \times SSD)/(WA \times WA \times WSD)$$

Where:
- CSDC = Cumulative Sediment Concentration (μg/kg)
- SDCi = Sediment Concentration in year i (μg/kg)
- SC = Soil Concentration (μg/kg)
- SL = Site Soil Loss Rate (kg/acre-year)
- SA = Site Area (acres)
- SSD = Site Sediment Delivery Ratio
- WA = Watershed Area (acres)
- WL = Watershed Soil Loss Rate (kg/acre-year)
- WSD = Watershed Sediment Delivery Ratio

Parameters used to estimate cumulative sediment concentration are presented in Table 2.13. Uptake of TCDD by fish was calculated using the following equation and the bioaccumulation index methodology described by Goeden and Smith[93] and Cook.[94]

$$FC = CSDC \times Lf \times BI \times 1/OCf$$

Where:
- FC = Fish Concentration (μg/kg)
- CSDC = Cumulative Sediment Concentration (μg/kg)
- Lf = Lipids Factor (0.03)
- BI = Bioavailability Index (0.15)
- OCf = Organic Carbon Factor (0.1)

The catfish was selected in this analysis as a representative bottom-feeding fish that is commercially and recreationally desirable by fishermen and consumers. The potential for exposure to TCDD-contaminated soil through the consumption of fish from a neighboring stream was calculated using the following equation.

$$LADD = FC \times CR \times CKf \times EDa \times EDy / BW \times LF$$

Where: LADD = Lifetime Average Daily Dose (μg/kg-day)
      FC   = Fish Concentration (μg/kg)
      CR   = Fish Consumption Rate (kg/day)
      CKf = Cooking Loss Factor
      BW = Body Weight (kg)
      EDy = Exposure Duration (years)
      LF   = Lifetime (days)

Parameters used to estimate exposure through ingestion of catfish caught at a stream neighboring a hypothetical residential or industrial site are summarized in Table 2.14.

Table 2.14. Exposure Parameters Used to Estimate Uptake of TCDD by Adults Due to the Ingestion of Fish Caught at a Stream Neighboring a Hypothetical Residential or Industrial Site

Parameter	Value
Fish Consumption Rate	1.48 g/day
Proportion of TCDD Remaining after Cooking	0.5
Exposure Duration	58 years
Body Weight	68.5 kg

Using the assumptions and equations described above, together with the RsD at 10^{-5} risk, an acceptable or de minimis soil concentration of TCDD for areas where runoff is a concern would be approximately 200 ppb.

Groundwater, Wildlife, and "De Minimis"

To set soil guidelines for TCDD to protect groundwater is usually unnecessary. The chemicophysical properties of TCDD show that it is essentially immobile in soil and virtually not water-soluble, so groundwater is unlikely to be threatened by TCDD. The only exception might be in those situations where the TCDD is present (in solution) in solvents in soil. In such cases, TCDD has been shown to move through soil in the solvent (e.g., co-elution).

Because of the diversity of wildlife populations, it is difficult to determine a nonspecies-specific soil cleanup level for TCDD. The health risks to birds and deer exposed to TCDD in soil was recently evaluated for sludges contaminated with TCDD which were applied to soil.[95] Primary food sources (e.g., insects and earthworms) for the receptor species need to be identified in these assessments and bioconcentration factors of TCDD for those food sources and absorption coefficients estimated. As shown in that paper,[95] the calculations can be quite complex and are outside the scope of this chapter. However, the conclusions of other assessments is that soil concentrations of approximately 100 ppt to 10 ppb have been shown conservative and health-protective for most species.

As mentioned earlier, the purpose of setting a "de minimis" quantity is to quickly determine when the loss of the chemical is too small to warrant further evaluation. Often, with chemicals like dioxin it is not too difficult to calculate the mass available for ultimate release in the environment. Based on several analyses, a total mass of two grams of TCDD into the environment probably, under any reasonable conditions, would be a de minimis concern.

In summary, the eight screening-level soil cleanup guidelines for TCDD can range from 0.01 ppb to about 10,000 ppb (Table 2.12). Clearly, there will be instances where, for example, recreational or industrial sites will also have a runoff problem. In those cases, it may be appropriate to default to the more conservative cleanup guideline.

Benzene, Toluene, Xylene

Unlike TCDD, the chemicophysical properties of benzene, toluene, and xylene (BTX) (Table 2.15) indicate that they are relatively mobile in soil and have a relatively high potential for volatilizing into ambient air. Soil binding is much less tenacious for these three chemicals and they have a low potential for bioconcentration or biomagnification.[12]

Table 2.15.   Select Chemical and Physical Properties of Benzene, Toluene, and Xylene (BTX) Which Are Often Needed to Evaluate the Environmental and Human Health Risks

	B	T	X
Water Solubility (mg/L)	1000	515	160
Lipid Solubility	2.0	2.8	3.0
Soil Adsorption Constant (K_{oc})	70	2.5	—
Vapor Pressure (mm Hg)	98	28	6.7
Theoretical BCF	24	50	2.2
Surface Soil Half-life (days)	20	5	5
Odor Threshold	3	2.5	1.0
Flash Point (°C)	− 11	10	29

The calculation of soil cleanup guidelines for chemicals which are volatile is somewhat different than the procedure used to assess TCDD. Acceptable ambient air concentrations of BTX can be estimated using the equations below, where Ca is the acceptable air concentration.

$$Ca_{nc} = BW/(\text{Apportioned RfD} \times BR)$$
$$(\text{for noncarcinogens})$$

$$Ca_c = (BW \times LF)/(\text{Apportioned RsD} \times BR \times ED)$$
$$(\text{for carcinogens})$$

Where BW is body weight, RfD is the U.S. EPA-derived Reference Dose, BR is breathing rate (m³/24 hour day), LF is average lifetime (years), and RsD is the Risk- Specific Dose (mg/kg-day) for benzene. The second step is to develop a "Box Model" so as to estimate an emission flux rate. Finally, this flux rate is incorporated into an air dispersion model such as the Behavior Assessment Model (BAM)[96] to determine a corresponding soil concentration. BAM accounts for all three transport mechanisms in soil (vaporization, capillary action, liquid-phase diffusion). BAM is often considered appropriate for these situations because:

- The vapor emission rates predicted by the model are based on several conservative loss pathways, such as transport of a chemical subject to volatilization at the soil surface and leaching in the soil column via evapotranspiration,
- The model always conserves mass, and
- The model takes into account the time-varying depletion of the contamination in soil, since only a finite amount of chemical is initially present.

Applying Equations 1–7 and the exposure factors presented in Tables 2.8 to 2.11, together with the approach described for TCDD, some approximate health-based soil screening levels for benzene, toluene, and xylene for residential, industrial, and recreational sites are presented in Table 2.16. The soil concentrations presented in Table 2.16 assume that two other factors do not come into play: odor and/or significant degradation of groundwater. The odor threshold for BTX has been studied by several investigators;[97] the weight of evidence suggests the BTX is detectable at approximately 30 ppm in air to the unacclimated nose.[97] Acceptable soil concentrations for BTX developed for the protection of groundwater must be developed on a site-specific basis (Table 2.16).

Table 2.16.   Screening Level Soil Concentrations for BTX for Eight Typical Exposure Scenarios

	Benzene	Toluene	Xylene
Residential	2 ppm	1600 ppm	16,000 ppm
Industrial	5 ppm	2700 ppm	26,000 ppm
Recreational	40 ppm	6000 ppm	30,000 ppm
Agricultural[a]	400 ppm	2000 ppm	1000 ppm
Groundwater	Site-specific	Site-specific	Site-specific
Runoff	Not applicable	Not applicable	Not applicable
Wildlife	Site-specific	Site-specific	Site-specific
"de minimis" Quantity[b]	50–100 gal.	200–500 gal.	200–1000 gal.

[a]These concentrations could be phytotoxic.
[b]The amount of gasoline that if spilled over a 100¹ × 100¹ section of land could pose a vapor hazard or explosive hazard.

Soil cleanup levels for protection of wildlife, like protection of groundwater, will be very site-specific for BTX because of the variety of species that could potentially be involved. The setting of an agricultural and runoff guideline is much less difficult with BTX, compared to TCDD, because (1) the potential for plant uptake is low, and (2) the volatility and short half-life of BTX in soil usually will preclude significant soil runoff. Consequently, concentrations in excess of a few hundred ppm will often present no significant problem (except phytotoxicity).

Setting a de minimis quantity for BTX would be useful to regulatory agencies in light of the number of leaking underground storage tanks that have been found in the United States. Although this is dependent on site-specific issues, a leak of 50 to 100 gallons of gasoline, depending on the potential for explosion or fire in the basement of nearby homes, probably would pose no appreciable hazard.

COMPARISON TO CURRENT STATE GUIDELINES

The soil cleanup guidelines presented here are higher than typically observed in most sites in the United States. Although there are no formal or official (i.e., regulatory-sanctioned) health-based cleanup levels for TCDD in soil, cleanup levels of 1 ppb for residential areas and 20 ppb for industrial areas were adopted at several sites throughout the United States during the 1980s.[85,98] These informal cleanup levels are primarily a result of a paper published in 1984 by the U.S. Centers for Disease Control (CDC)[10] and a subsequent counter-paper by Paustenbach et al.[17] The CDC paper, which discussed an approach to setting an acceptable soil concentration of TCDD in residential settings, used a number of conservative exposure assumptions in the calculations. The paper concluded that "one part per billion of 2,3,7,8-TCDD in soil is a reasonable level at which to begin consideration of an action to limit human exposure to contaminated soil."[10] This recommendation formed the basis for the cleanup criteria of 1 ppb TCDD which has been applied for the cleanup of residential sites.

During the past eight years, several authors have identified health-based cleanup levels for TCDD in soil using risk assessment methods.[10,17,18,27,83,99-102] The most frequently cited studies are the risk assessments for Times Beach, Missouri, by the Centers for Disease Control (CDC),[10] the U.S. Environmental Protection Agency,[90] and Paustenbach et al.[17] In an effort to address public health concerns in residential settings like Times Beach, Kimbrough et al. concluded that 1 part per billion (ppb) of 2,3,7,8-TCDD in soil was a reasonable level at which to begin consideration of action to limit human exposure.[10] Several limitations and uncertainties have been identified in the Times Beach risk assessment.[17,103-105] Notable among these was the failure to consider site-specific conditions and the application of overly conservative exposure estimates for direct exposure pathways such as soil ingestion and dermal contact.[17,93,105] Other risk assessments, such as that by Paustenbach et al.,[17] concluded that TCDD soil levels as high as 10^{-100} ppb and 1,000 ppb might be acceptable for most residential and industrial sites, respectively, as long as surface soil runoff was negligible.

In a recent paper by Paustenbach et al.,[85] allowable levels of TCDD in residential and industrial soils were determined using risk assessment methodologies and the most recent data for estimating human uptake of TCDD through dermal contact, soil ingestion, inhalation of fugitive dust, and the consumption of contaminated fish. For a residential site, concentrations of TCDD in soil as high as 19 ppb did not exceed a lifetime incremental cancer risk of 1 in 100,000 for the typically exposed individual. Using quantitative uncertainty analysis of the key exposure parameters used in the risk assessment, TCDD soil concentration for the 75th and 95th percentile person were 12 ppb and 7.5 ppb (10^{-5} risk), respectively. At an industrial site where exposures to workers were limited to dermal contact, soil ingestion, and inhalation of fugitive dust, and consumption of fish by an offsite receptor was considered, TCDD concentrations in soil could range between 131 and 579 ppb (10^{-5} risk), depending on the amount of time a worker spends outdoors in direct contact with soil. The acceptable range of TCDD concentrations in industrial soils was not significantly reduced when the consumption of fish from a neighboring waterway by offsite receptors was considered. Industrial soil concentrations of approximately 108 ppb and 46 ppb could pose a 10^{-5} risk for the 75th and 95th percentile worker, respectively. In light of the histopathology re-evaluation of TCDD, these values could probably be much higher yet pose no appreciable health risk.[83,84]

The variety of regulatory cleanup levels for BTX in soils is much larger[4-6] than those for TCDD, primarily because BTX contamination is more frequently encountered. In general, state guidelines range from 10 to 100 ppm Total Petroleum Hydrocarbons, with cleanup levels for benzene much lower, usually in the 10 to 60 ppb range.[7] Although the rationale for many of these state guidelines is seldom made clear, it appears that protection of groundwater is the primary concern. The two key advantages of the methodology we propose in this chapter is that the rationale for the soil guideline is clearly defined and a quantitative approach for determining acceptability is clearly presented.

With the exception of New Jersey, few states consider different exposure scenarios when setting soil guidelines. The New Jersey Department of Environmental Protection has recently issued a set of draft guidelines[22] that attempts to use a logical and scientifically based approach to setting soil cleanup levels for both residential and industrial settings, and includes draft standards for the cleanup of the interior surfaces of buildings.[22]

DISCUSSION

The concept of setting eight different soil guidelines for each chemical contaminant is one based on more than 15 years of experience at evaluating how local, state, and federal agencies have struggled with the "how clean is clean" issue. The process of setting soil cleanup standards has clearly been an arduous one for the United States because, after nearly 20 years, less than five nationally recognized soil cleanup levels are available. The intent of our approach is to establish a rapid and efficient method for identifying soil contaminant concentrations (or quantities) below which it is unlikely that health hazard could be present. Having identified acceptable or ''de minimis'' concentrations for the eight most routinely encountered exposure scenarios, then agencies or companies could efficiently dismiss those settings which are not worthy of further evaluation and certainly don't deserve a thorough quantitative risk assessment. The approach is not unlike that currently used by industrial hygienists who rely upon the American Conference of Governmental Industrial Hygienists (ACGIH) Threshold Limit Values (TLV) to determine whether the airborne concentration of a chemical in the workplace is likely to pose a health hazard to the typical employee. Like the TLVs, some degree of professional judgment would need to be used when relying upon these eight guidelines to make the final determination that the site or incident poses no significant risk, but the benefits of developing such a set of guidelines would surely outweigh the problems that are inherent in the current inconsistent and redundant ad hoc guidelines which exist throughout the United States.

This approach could be the next logical step in the program which has been initiated by CHESS. Currently, it is anticipated that they will have developed a software package which will rapidly calculate site-specific soil cleanup levels based on the best environmental fate and transport models, human exposure parameters, and toxicology data. This model, which would clearly represent the state-of-the-art in exposure assessment, could serve as an invaluable tool for calculating the eight guidelines advocated here. Having used the model to calculate soil cleanup guidelines for perhaps as many as 100 chemicals for the eight exposure scenarios, a booklet much like the one published by the ACGIH (which contains the TLVs) could be published and distributed. Like the TLVs, the guidelines could be revised, updated, or expanded annually based on new information or the efforts of the group.

Although the computer program being derived by CHESS would allow state or federal agencies to conduct their own site-specific risk analyses in a relatively rapid and efficient manner, we believe that many of these professionals will simply not have the time to learn how to use the software package or the time to use it in their daily practice. It is also possible that as the CHESS software package is updated annually, not everyone will be able to stay abreast of the changes in the software or formulas. It seems to us that for the initial screening evaluation, most health professionals would welcome the simplicity and efficiency of being able to look up in a table the chemical and particular exposure scenario of interest to them so that they could reach a rapid decision as to whether they had a small, large, or negligible problem. We believe that if a booklet of values were developed and approved each year by CHESS, or a similar group of acknowledged professionals, this would provide an important mechanism for bringing order to what is rapidly becoming a morass of conflicting guidelines, rules, regulations, and recommendations.

REFERENCES

1. James, R. C., A. C. Nye, G. C. Millner, and S. M. Roberts. "Risks from Exposure to Polychlorinated Biphenyls (PCBs) in Soil: An Evaluation of the TSCA Soil Cleanup Guidelines for PCBs" *Regul. Toxicol. Pharm.* (1992) (in review).
2. "Interim Guidance for Cleanup of Lead in Soils," U.S. Environmental Protection Agency, Office of Emergency and Remedial Response. OSWER Directive 9355.4-02. Washington, D.C. September 7, 1989.
3. Johnson, B. R. Letter to David Wagoner, Director, Waste Management Division, EPA Region VII. 1987.
4. Bell, C. E., P. T. Kostecki, and E. J. Calabrese. "State of Research and Regulatory Approach of State Agencies for Cleanup of Petroleum Contaminated Soils," in P. T. Kostecki and E. J. Calabrese, Eds., *Petroleum Contaminated Soils.* Vol. 2 (Chelsea, MI: Lewis Publishers, Inc., 1989), pp. 73–94.
5. Bell, C. E., P. T. Kostecki, and E. J. Calabrese. "An Update on a National Survey of State Regulatory Policy: Cleanup Standards," in P. T. Kostecki and E. J. Calabrese, Eds., *Petroleum Contaminated Soils.* Vol. 3. (Chelsea, MI: Lewis Publishers, Inc., 1990), pp. 49–72.
6. Bell, C. E., P. T. Kostecki, and E. J. Calabrese. "Review of State Cleanup Levels for Hydrocarbon Contamination," in P. T. Kostecki and E. J. Calabrese, Eds., *Hydrocarbon Contaminated Soils and Groundwater.* Vol. 1 (Chelsea, MI: Lewis Publishers, Inc., 1991), pp. 77–89.
7. "Selected Type B Cleanup Criteria." Michigan Department of Natural Resources, Waste Management Division, Lansing, MI. 1990.
8. Kostecki, P. T., and E. J. Calabrese. *Hydrocarbon Contaminated Soils and Groundwater.* Vol. 1. (Chelsea, MI: Lewis Publishers, Inc., 1991).
9. Kostecki, P. T., and E. J. Calabrese. "Council for Health and Environmental Safety of Soils," in P. T. Kostecki and E. J. Calabrese, Eds., *Petroleum Contaminated Soils.* Vol. 2 (Chelsea, MI: Lewis Publishers, Inc., 1989), pp. 485–495.
10. Kimbrough, R., H. Falk, P. Stehr, and G. Fries. "Health Implications of 2,3,7,8-tetrachlorodibenzo-p-dioxin (TCDD) Contamination of Residential Soil," *Risk Anal.* 5:289–302 (1984).
11. Sheehan, P. J., D. M. Meyer, M. M. Sauer, and D. J. Paustenbach. "Assessment of the Human Health Risks Posed by Exposure to Chromium-Contaminated Soils," *J. Toxicol. Environ. Health* 32:161–201 (1991).
12. Paustenbach, D. J. "A Methodology for Evaluating the Environmental and Public Health Risks of Contaminated Soil," in P. T. Kostecki and E. J. Calabrese, Eds., *Petroleum Contaminated Soils.* Vol. 1. (Chelsea, MI: Lewis Publishers, Inc., 1989a), pp. 225–261.
13. "Superfund Exposure Assessment Manual," U.S. Environmental Protection Agency. Washington D.C. (1988).
14. Paustenbach, D. J. "Important Recent Advances in the Practice of Health Risk Assessment: Implications for the 1990s," *Regul. Toxicol. Pharmacol.* 10:204–243 (1989).
15. Paustenbach, D. J. "A Survey of Health Risk Assessment," in *The Risk Assessment of Environmental and Human Health Hazards: A Textbook of Case Studies,* D. J. Paustenbach, Ed. (New York: John Wiley & Sons, 1989).

16. Shu, H. P., D. J. Paustenbach, J. Murray, L. Marple, B. Brunck, D. Dei Rossi, and P. Tietelbaum "Bioavailability of Soil-Bound TCDD: Oral Bioavailability in the Rat." *Fund. Appl. Toxicol.* 10:648–654 (1988).

17. Paustenbach, D. J., H. P. Shu, and F. J. Murray. "A Critical Analysis of Risk Assessments of TCDD Contaminated Soil," *Regul. Toxicol. Pharmacol.* 6:284–307 (1986).

18. Eschenroeder, A., R. J. Jaeger, J. J. Ospital, and C. P. Doyle. "Health Risk Assessment of Human Exposures to Soil Amended with Sewage Sludge Contaminated with Polychlorinated Dibenzodioxins and Dibenzofurans," *Vet. Hum. Toxicol.* 28:435–442 (1986).

19. Motto, H. L., R. H. Daines, D. M. Chilko, and C. K. Motto. "Lead in Soils and Plants: Its Relationship to Traffic Volumes and Proximity to Highways," *Environ. Sci. Technol.* 4:231–237 (1970).

20. Paustenbach, D. J., T. T. Sarlos, V. Lau, B. L. Finley, D. A. Jeffrey, and M. J. Ungs. "The Potential Inhalation Hazard Posed by Dioxin-Contaminated Soil," *J. Air Waste Manage. Assoc.* 41:1334–1340 (1991).

21. Paustenbach, D. J. "Assessing the Potential Environment and Human Health Risks of Contaminated Soil," *Comments Toxicol.* 1:185–220 (1987).

22. "Preliminary Draft Cleanup Standards," New Jersey Department of Environmental Protection, Division of Hazardous Waste Management, Trenton, NJ (1991).

23. Keenan, R. E., M. M. Sauer, F. H. Lawrence, E. R. Rand, and D. W. Crawford. "Examination of Potential Risks from Exposure to Dioxin in Sludge Used to Reclaim Abandoned Strip Mines," in D. J. Paustenbach, Ed., *The Risk Assessment of Environmental and Human Health Hazards.* (New York: John Wiley & Sons, 1989), pp. 935–998.

24. Rand, G. M. and S. R. Petrocelli. *Fundamentals of Aquatic Toxicology.* (New York: McGraw-Hill, 1985).

25. Paustenbach, D. J. "A Comprehensive Methodology for Assessing the Risks to Humans and Wildlife Posed by Contaminated Soils: A Case Study Involving Dioxin," in *The Risk Assessment of Environmental and Human Health Hazards,* D. J. Paustenbach, Ed. (New York: John Wiley & Sons, 1989).

26. Fries, G. F. "Potential Polychlorinated Biphenyl Residues in Animal Products from Application of Contaminated Sewage Sludge to Land," *J. Environ. Qual.* 11:14–20 (1982).

27. Fries, G. F., and D. J. Paustenbach. "Evaluation of Potential Transmission of 2,3,7,8-Tetrachlorodibenzo-p-Dioxin-Contaminated Incinerator Emissions To Humans via Foods," *J. Toxicol. Environ. Health* 29:1–43 (1990).

28. "Lead in the Human Environment." National Research Council, Washington, DC (1980).

29. Hayes, W. J., and E. R. Laws, Jr. *Handbook of Pesticide Toxicology.* (New York: Academic Press, Inc., 1991).

30. "Reducing Risk: Setting Priorities and Strategies for Environmental Protection," U.S. Environmental Protection Agency, Science Advisory Board, Washington, DC (1990).

31. Charters, D. "Current Uses and Abuses of Environmental Effect and Fate Data," *Proceedings of the Annual Meeting of ASTM.* (Atlantic City: American Society of Testing and Materials, 1991).

32. "Guidance for Establishing Target Cleanup Levels for Soils at Hazardous Waste Sites," U.S. Environmental Protection Agency, Exposure Assessment Group, Office of Health and Environmental Assessment, Washington, DC (1989).
33. Conway, R. A. *Environmental Risk Analysis of Chemicals.* (New York: Van Nostrand-Reinhold, 1982).
34. "Throwing Good Money at Bad Water Yields Scant Improvements," *Wall Street Journal,* May 15, 1991.
35. Bopp, R. "Dioxins in Newark Bay," Annual Report, Lamont-Doherty Geological Observation of Columbia University (1990).
36. Harris, M., K. Conner, V. Lau, and D. J. Paustenbach. "The Range of Reasonable Exposure Factors for Use in Risk Assessment," *Risk Anal.* (in review) (1992).
37. "Risk Assessment Guidance for Superfund. Volume 1. Human Health Evaluation Manual (Part A), U.S. Environmental Protection Agency, Office of Emergency and Remedial Response. Washington, DC (1989).
38. "Exposure Factors Handbook," U.S. Environmental Protection Agency, Exposure Assessment Group, Office of Health and Environmental Assessment, Washington, DC (1989).
39. "Human Health Evaluation Manual, Supplemental Guidance: Standard Default Exposure Factors," U.S. Environmental Protection Agency, Office of Soil Waste and Emergency Response, Washington, DC (1991).
40. Finkel, A. M. *Confronting Uncertainty in Risk Management: A Guide for Decision-Makers,* (Washington, D.C.: Center for Risk Management, Resources for the Future, 1990).
41. Paustenbach, D. J., D. M. Meyer, P. J. Sheehan, and V. Lau. "An Assessment and Quantitative Uncertainty Analysis of the Health Risks to Workers Exposed to Chromium Contaminated Soils," *Toxicol. Ind. Health* (1991).
42. Harris, M., and D. J. Paustenbach. "An Assessment of a Pentachlorophenol-Contaminated Site," *Toxicol. Environ. Health* (in review) (1992).
43. Cooper, M. *Pica.* (Springfield, IL: Thomas, 1957), pp. 65–66.
44. Calabrese, E. J., R. Barnes, E. J. Stanek, H. Pastides, C. E. Gilbert, P. Veneman, X. Wang, A. Lasztity and P. T. Kostecki. "How Much Soil Do Young Children Ingest: An Epidemiologic Study." *Reg. Toxicol. Pharm.* 10:123–137 (1989).
45. van Wijnen, J. H., P. Clausing, and B. B. Runedreff. "Estimated Soil Ingestion by Children," *Environ. Res.* 51:147–157 (1990).
46. Calabrese, E. J., E. J. Stanek, C. E. Gilbert, and R. M. Barnes. "Preliminary Adult Soil Ingestion Estimates: Results of a Pilot Study," *Reg. Tox. Pharm.* 12:88–95 (1990).
47. Calabrese, E. J. and E. J. Stanek. "A Guide to Interpreting Soil Ingestion Studies." *Regul. Toxicol. Pharm.* 13:278–292 (1991).
48. Calabrese, E. J., and E. J. Stanek. "A Guide to Interpreting Soil Ingestion Studies. II. Qualitative and Quantitative Evidence of Soil Ingestion," Workshop Presentation at the Second Annual West Coast Conference on Hydrocarbon Contaminated Soil and Groundwater, Newport Beach, CA, March 4–7, 1991.
49. Binder, S., Sokal, D. and Maughan, D. "Estimating Soil Ingestion: The Use of Tracer Elements in Estimating the Amount of Soil Ingested by Young Children," *Arch. Environ. Health* 41(6):341–345 (1986).
50. Davis, S., P. Waller, R. Buschbom, J. Ballou and P. White. "Quantitative Estimates of Soil Ingestion in Normal Children Between the Ages of 2 and 7 Years:

Population-Based Estimates Using Aluminum, Silicon, and Titanium as Soil Tracer Elements," *Arch. of Environ. Health* 45(2):112–122 (1990).

51. Lepow, M. L., L. Bruckman, R. A. Robino, S. Markowitz, M. Gillette, and J. Kapish. "Role of Airborne Lead in Increased Body Burden of Lead in Hartford Children," *Environ. Health Persp.* 99–102 (1974).

52. Lepow, M. L., L. Bruckman, M. Gillette, S. Markowitz, R. Robino, and J. Kapish. "Investigations into Sources of Lead in the Environment of Urban Children," *Environ. Res.* 10:415–426 (1975).

53. Roels, H. A., J. P. Buchet, R. R. Lauwerys, P. Braux, F. Claeys-Thoreau, A. Lafontaine, and G. Verduyn. "Exposure to Lead by the Oral and Pulmonary Routes of Children Living in the Vicinity of a Primary Lead Smelter," *Environ. Res.* 22:81–94 (1980).

54. *Multi-Pathway Health Risk Assessment Input Parameters Guidance Document.* South Coast Air Quality Management District. Prepared by Clement Associates, Inc., June, 1988.

55. Schaum, J. "Risk Analysis of TCDD Contaminated Soil," U.S. Environmental Protection Agency, Office of Health and Environmental Assessment, Office of Research and Development, Washington, DC (1984).

56. Que Hee, S. S., B. Peace, C. S. Scott, J. R. Boyle, R. L. Bornschein, and P. B. Hammond. "Evolution of Efficient Methods to Sample Lead Sources, Such as House Dust and Hand Dust, in the Homes of Children," *Environ. Res.* 38:77–95 (1985).

57. "The Development of Applied Action Levels for Soil Contact," Final Draft. California Department of Health Services 1987 (cited in Reference No. 54).

58. Driver, J. H., Konz, J. J., and Whitmyre, G. K. "Soil Adherence to Human Skin," *Bull. Environ. Contam. Toxicol.* 43:814–820 (1989).

59. Knaak, S. B., I. Yutaka, and K. T. Maddy. "The Worker Hazard Posed by Re-Entry into Pesticide-Treated Foliage: Development of Safe Reentry Times, with Emphasis on Chlorthiophos and Carbosulfan," in D. J. Paustenbach, Ed., *The Risk Assessment of Environmental and Human Health Hazards.* (New York: John Wiley & Sons, 1989), pp. 797–844.

60. Trijonis, J., J. Eldon, J. Gins, and G. Berglund. "Analysis of the St. Louis RAMS Ambient Particulate Data," EPA Report 450/4-80-006a. Produced by Technology Service Corporation under EPA Contract 68-02-2931 for the Office of Air, Noise, and Radiation of the U.S. Environmental Protection Agency, Washington, DC (1980).

61. Goon, D., N. S. Hatoum, J. D. Jernigan, S. L. Schmitt, and P. J. Garvin. "Pharmacokinetics and Oral Bioavailability of Soil-Adsorbed Benozo[a]pyrene (BaP) in rats," *Toxicologist* 10:218 (1990) and Goon, D., N. S. Hatoum, M. J. Klan, J. D. Jernigan, and R. G. Farmer. "Oral Bioavailability of "Aged" Soil-Adsorbed Benzo[a]pyrene in Rats," *Toxicologist* 11:345 (1991).

62. McKone, T. E. "Dermal Uptake of Organic Chemicals From a Soil Matrix," *Risk Analysis* 10:407–419 (1990).

63. "State of Michigan Draft Risk Assessment Guidelines," Michigan Council on Environmental Quality. Committee on Risk Assessment. Lansing, MI, May 17, 1990.

64. Haque, R., Ed. *Dynamics, Exposure and Hazard Assessment of Toxic Chemicals.* (Ann Arbor, MI: Ann Arbor Science, 1980).

65. Thibodeaux, L. J. *Chemodynamics: Environmental Movement of Chemicals in Air, Water, and Soil.* (New York: John Wiley & Sons, 1979).

66. Beck, L. W., A. W. Maki, N. R. Artman, and E. R. Wilson. "Outline and Criteria for Evaluating the Safety of New Chemicals," *Regul. Toxicol. Pharmacol.* 1:19–58 (1981).

67. Mackay, D., and S. Paterson. "Calculating Fugacity," *Environ. Sci. Technol.* 15: 1006–1014 (1982).

68. Bergmann, H. L., R. A. Kimmerle, and A. W. Maki. *Environmental Hazard Assessment of Effluents.* (New York: Pergamon Press, Inc., 1986).

69. Woltering, D., and W. Bishop. "Assessing the Environmental Risks of Detergent Chemicals," in D. J. Paustenbach, Ed., *The Risk Assessment of Environmental and Human Health Hazards.* (New York: John Wiley & Sons, 1989), pp. 345–390.

70. Lyman, W. J., W. F. Reehl, and D. H. Rosenblatt. *Handbook of Chemical Property Estimation Methods. Environmental Behavior of Organic Compounds.* (New York: McGraw-Hill Book Company, 1982.)

71. "Environmental Assessment Technical Handbook," U.S. Food and Drug Administration, Center for Food Safety and Applied Nutrition and the Center for Veterinary Medicine, Washington, DC (1984).

72. Dragun, J. *The Soil Chemistry of Hazardous Materials.* (Silver Spring, MD: Hazardous Materials Control Research Institute, 1988).

73. Young, F. A. "Risk Assessment: The Convergence of Science and Law," *Regul. Toxicol. Pharmacol.* 7:179–184 (1987).

74. Dragun, J. "The Fate of Hazardous Materials in Soil (What Every Geologist and Hydrologist Should Know): Part 2," *Hazardous Materials Control* 1(3):41–65. (1988).

75. Travis, C. C., S. A. Richter, E. A. Crouch, R. Wilson, and E. Wilson. "Cancer Risk Management. A Review of 132 Federal Regulatory Decisions," *Environ. Sci. Technol.* 21(5):415–420 (1987).

76. Travis, C. C., and Hattemer-Frey, H. A. "Human Exposure to 2,3,7,8-TCDD," *Chemosphere.* 16(10–12):2331–2342 (1987).

77. "National Oil and Hazardous Substances Pollution Contingency Plan," 40 CFR Part 300. Environmental Protection Agency. Washington, DC (1990).

78. "Policy for Identifying and Assessing the Health Risks of Toxic Substances," Maine Department of Human Services, Environmental Toxicology Program, Division of Disease Control, Bureau of Health. February 1988.

79. "Draft Interim Guidance for Disposal Site Risk Characterization–In Support of the Massachusetts Contingency Plan," Massachusetts Department of Environmental Quality Engineering. Office of Research and Standards. October 3, 1988.

80. "Safe Drinking Water and Toxic Enforcement Act of 1986 (Proposition 65)," Health and Welfare Agency, Office of the Secretary, Sacramento, California, 1986.

81. Rodricks, J. V., S. M. Brett, and G. C. Wrenn. "Significant Risk Decisions in Federal Regulatory Agencies," *Regul. Toxicol. Pharmacol.* 7:307–320 (1987).

82. Kociba, R. J., D. Keyes, J. Beyer, R. M. Carreon, C. E. Wade, D. A. Dittenber, R. P. Kalnins, L. E. Frauson, C. N. Park, S. D. Barnard, R. A. Hummel, and C. G. Huminston. "Results of a Two-Year Chronic Toxicity and Oncogenicity Study of 2,3,7,8-Tetrachlorodibenzo-p-dioxin in Rats," *Toxicol. Appl. Pharmacol.* 46:279–303 (1978).

83. Paustenbach, D. J., M. W. Layard, R. J. Wenning, and R. E. Keenan. "Risk Assessment of 2,3,7,8-TCDD Using a Biologically-Based Cancer Model: A Reevaluation of the Kociba et al. Bioassay Using 1978 and 1990 Hissopathology Criteria." *Toxicol. Environ. Health* 34:11–26 (1991).

84. Keenan, R. E., D. J. Paustenbach, R. J. Wenning, and A. H. Parsons. "Pathological Re-Evaluation of the Kociba et al. Bioassay of 2,3,7,8-TCDD; Implications for Risk Assessment," *J. Toxicol. Environ. Health* (in press) (1991).

85. Paustenbach, D. J., R. J. Wenning, V. Lau, N. W. Harrington, D. K. Rennix and A. H. Parsons. "Recent Developments on the Hazards Posed by 2,3,7,8-Tetrachlordibenzo-p- dioxin in Soil: Implications for Setting Risk-Based Cleanup Levels at Residential and Industrial Sites," *Toxicol. Environ. Health* (in review) 1991.

86. Wipf, H. K., E. Homberger, N. Neuner, U. B. Ranalder, W. Vetter, and J. P. Vuilleumier. "TCDD Levels in Soil and Plant Samples from the Seveso Area," in *Chlorinated Dioxins and Related Compounds*, O. Hutzinger et al., Eds. (New York: Pergamon Press, Inc., 1982), pp. 115-126.

87. Jensen, D. J., M. E. Getzendaner, R. A. Hummel, and J. Turley. "Residue Studies for (2,4,5-tri-chlorophenoxy)acetic Acid and 2,3,7,8-tetrachlorodibenzo-p-dioxins in Grass and Rice," *J. Agric. Food Chem.* 31:118-122 (1983).

88. Stevens, J. B., and E. N. Gerbec. "Dioxin in the Agricultural Food Chain," *Risk Anal.* 8:329-335 (1988).

89. "Assessment of Risks from Exposure of Humans, Terrestrial and Avian Wildlife, and Aquatic Life to Dioxins and Furans from Disposal and Use of Sludge from Bleached Kraft and Sulfite Pulp and Paper Mills," U.S. Environmental Protection Agency. Office of Solid Waste and Emergency Response, Washington, DC (1990).

90. "Risk Analysis of TCDD Contaminated Soil," U.S. Environmental Protection Agency, Office of Solid Waste and Emergency Response, Washington, DC (1984).

91. "Dioxin Transport from Contaminated Site to Exposure Locations: A Methodology for Calculation of Conversion Factors," U.S. Environmental Protection Agency, Office of Research and Development (1985).

92. Knighton, M. D., and R. M. Solomon. "Applications and Research in Sediment Delivery and Routing Models in the USDA Forest Service," in S.S.Y. Want, Ed., *Sediment Transport Modeling.* (New York: American Society of Civil Engineers, 1989), pp. 344-350.

93. Goeden, H. E., and A. H. Smith. "Estimation of Human Exposure from Fish Contaminated with Dioxins and Furans Emitted by a Resource-Recovery Facility," *Risk Anal.* 9(3):377-383 (1989).

94. Cook, P. M., A. R. Batterman, B. C., Butterworth, K. B. Lodge, and S. W. Kohlbry. "Laboratory Study of TCDD Bioaccumulation by Lake Trout for Lake Ontario Sediments, Food Chain and Water," Draft Copy, Chapter 6. Environmental Reasearch Laboratory, U.S. Environmental Protection Agency, Duluth, MN.

95. Keenan, R. E., J. W. Knight, E. R. Rand, and M. M. Sauer. "Assessing Potential Risks to Wildlife and Sportsmen from Exposure to Dioxin in Pulp and Paper Mill Sludge Spread on Managed Woodlands," *Chemosphere* 20:1763-1769 (1990a).

96. Jury, W. A., W. F. Spencer, and W. J. Farmer. Behavior Assessment Model for Trace Organics in Soil: I. Model Description," *J. Environ. Qual.* 12(4):558-564 (1983).

97. *Complilation of Odor and Taste Threshold Values Data.* F. A. Fazzalari, Ed. International Business Machines. American Society for Testing and Materials Philadelphia, PA. (1978).

98. Gough, M. "Human Exposures from Dioxin in Soil—A Meeting Report," *J. Toxicol. Environ. Health* 32:205-245 (1991).

99. Nauman, C. H. and J. L. Schaum. "Human Exposure Estimation for 2,3,7,8-TCDD." *Chemosphere* 16: 1851-1856 (1987).

100. Birmingham, B., A. Gillman, D. Grant, J. Salminen, M. Boddington, B. Thorpe, I. Wile, P. Toft, and V. Armstrong. "PCDD/PCDF Multimedia Exposure Analysis for the Canadian Population: Detailed Exposure Estimation," *Chemosphere* (1989).

101. Heida, H., M. van den Berg, and K. Olie. "Risk Assessment and Selection of Remedial Alternative, the Volgemeerpoider Case Study," *Chemosphere* 19:615–622 (1989).

102. di Domenico, A. "Guidelines for the Definition of Environmental Action Alert Thresholds for Polychlorodibenzodioxins and Polychlorodibenzofurans," *Reg. Toxicol. Pharm.* 11:118–23.

103. Houk, V. "Uncertainties in Dioxin Risk Assessment," *Chemosphere* 15:1875–1881 (1985).

104. Fishbein, L. "Health-Risk Estimates for 2,3,7,8-tetrachlorodibenzo-dioxin: An Overview," *Toxicol. Ind. Health* 3:91–134 (1987).

105. Kimbrough, R. D. "Estimation of Amount of Soil Ingested, Inhaled, or Available for Dermal Contact," USEPA Workshop on Health Hazards of Contaminated Soil. Andover, MA (May 6, 1986).

106. Baltrop, D. "Sources and Significance of Environmental Lead in Children." Proceedings Int. Environ. Health Aspects of Lead, Commission of European Communities. Center for Information and Documentation, Luxembourg (1973).

107. Day, J. P., J. E. Fergusson and T. M. Chee. "Solubility and Potential Toxicity of Lead in Urban Street Dust," *Bull. Environ. Contam. Toxicol.* 23:497–502 (1975).

108. Duggan, M. J., and S. Willams. "Lead-in-Dust in City Streets," *Sci. Tot. Environ.* p. 791–797 (1977).

109. Hawley, J. K. "Assessment of Health Risk from Exposure to Contaminated Soil," *Risk Anal.* 5(4):289–302 (1985).

110. Bryce-Smith, D. "Lead Absorption in Children," *Phys. Bull.* 25:178–181 (1974).

111. "Superfund Public Health Manual," U.S. Environmental Protection Agency. Office of Emergency and Remedial Response, Washington, DC. EPA/540/1-86/060. October, 1986.

112. Clausing, P., B. Brunekreef, and J. H. van Wijnen. "A Method for Estimating Soil Ingestion by Children," *Int. Arch. Occup. Environ. Health* 59:73–82 (1987).

113. LaGoy, P. "Estimated Soil Ingestion Rates for Use in Risk Assessment," *Risk Anal.* 7:355–359 (1987).

114. Lipsky, D. "Assessment of Potential Health Hazards Associated with PCDD and PCDF Emissions from a Municipal Waste Combustor," in D. J. Paustenbach, Ed. *The Risk Assessment of Environmental and Human Health Hazards.* (New York: John Wiley & Sons, 1988), pp. 631–686.

115. Sedman, R. M. "The Development of Applied Action Levels for Soil Conact: A Scenario for the Exposure of Humans to Soil in a Residential Setting," *Environ. Health Perspect.* 79:291–313 (1988).

116. Calabrese, E. J., P. T. Kostecki and C. E. Gilbert. "How Much Soil do Children Eat? An Emerging Consideration for Environmental Health Risk Assessment," *Comments Toxicol.* 1(3):229–241 (1987).

117. Gallacher, J. E., P. C. Elwood, K. M. Phillips, B. E. Davies, and D. T. Jones. "Relationship Between Pica and Blood Lead in Areas of Differing Lead Exposure," *Arch. Dis. Childhood* 59:40–44 (1984).

PART II

Regulatory

CHAPTER 3

Canadian Approach to Establishing Cleanup Levels for Contaminated Sites

Connie Gaudet, Amanda Brady, Mark Bonnell and **Michael Wong,** Environmental Quality Guidelines Division, Water Quality Branch, Environment Canada, Ottawa, Ontario

INTRODUCTION

"For the first time I met a people whose job was not to manage what already existed but to develop it without stint. No one thought about limits, no one knew where the frontier was . . ."

Jean Monnet, founder of the European Economic Community as an eighteen-year-old salesman traveling in Canada in 1906.

This "legacy of the past"—captured so prophetically in Monnet's words—has become the challenge of the present. It has been increasingly apparent over the last number of years that Canada, like many other countries, is beginning to suffer from past mismanagement of hazardous materials and toxic waste. Accidents, inadequate environmental controls, and simply the age of some storage facilities have been putting at risk both the health of Canadians and of our environment. Soil and water contamination has occurred from sources such as coal tar pits, mine tailing wastes, landfill sites, and abandoned subsurface leaking waste drums and tanks. Responsible parties and government have addressed the contaminated sites issue on a number of fronts and in a number of ways in Canada; consequently, there has been a lack of national consistency in cleanup responses.[1]

One difficult but necessary aspect of contaminated site remediation is assessing the significance of contamination and determining the extent of cleanup required. In Canada, the derivation of cleanup levels for contaminated sites varies from province to province. Some provinces have established provincial guidelines or criteria for the cleanup (remediation) of contaminated sites in the context of the current or intended future land use.[2-5] There is, however, no national consistency in approach.

THE NATIONAL CONTAMINATED SITES REMEDIATION PROGRAM (NCSRP)

In recognition of the potential magnitude of the problem of contaminated sites in Canada and the lack of a consistent national approach to deal with it, this issue was placed on the agenda of the Canadian Council of Ministers of the Environment (CCME). [The CCME is comprised of federal, provincial, and territorial environment ministers from across Canada.]

In October of 1989, the CCME announced plans to initiate a program to clean up contaminated sites in Canada—the National Contaminated Sites Remediation Program (NCSRP). The program has been established (1) to promote a coordinated, nationally consistent approach to the identification, assessment, and remediation (cleanup) of contaminated sites in Canada which impact or have the potential to impact on human health or the environment; (2) to provide the necessary government funds to remediate high risk "orphan" sites for which the responsible party cannot be identified or is financially unable to carry out the work; (3) and to work with industry to stimulate the development and demonstration of new and innovative remediation technologies. The first five years of the program will provide the necessary funds to initiate the cleanup of high risk "orphan" contaminated sites.

To ensure effective implementation, the NCSRP is based on the following general principles:

- jurisdictions will have the necessary laws, regulations, and programs in place to ensure the remediation of all high-risk contaminated land sites where the responsible party can be held accountable *consistent with the polluter-pays principle;* and
- common assessment and remediation (cleanup) criteria/guidelines will be used in the management of contaminated sites.

In order that site assessment and remediation could be initiated as soon as possible, the development of common assessment and remediation tools was identified as an urgent priority in the NCSRP.

Public Consultation

The NCSRP is dedicated to a consultative process to ensure that the concerns and interests of all stakeholders (industry, government, ENGOs) are addressed in the development of the common tools. In April and November of 1990, the CCME held multi-stakeholder workshops to discuss the key factors in setting priorities for the remediation of contaminated sites, and the development of assessment and remediation criteria. Key recommendations from these workshops[6,7] included the need for a simple classification system (high, medium, low) to identify priority sites; a "two-tiered" approach (i.e., generic and site-specific) to the development of assessment and remediation criteria; and equal consideration of human health and the environment in the development of all common tools for use in the NCSRP.

APPROACHES TO ESTABLISHING CLEANUP LEVELS

Absolute vs. Relative Approaches: Irreconcilable Differences?

There has been a great deal of controversy over whether the absolute or relative approach is best suited to setting cleanup levels for contaminated sites. Clearly, each approach has certain advantages and limitations. Absolute approaches based on generic criteria are relatively simple and objective, but do not consider site-specific circumstances. This can lead to situations where cleanup levels (and therefore ensuing costs) are established beyond that required to protect human health and the environment. A relative (i.e., site-specific) approach, based on an assessment of risk to human or other receptors, enables cleanup levels to be tailored to a site and may maximize efficiency of expenditures. However, this approach usually requires extensive and expensive site characterization and risk assessment, and does not provide a consistent basis for decisionmaking, especially during early stages of site assessment and investigation.

In a review of international approaches to setting cleanup levels[8] it was concluded that a combined approach (incorporating both absolute and relative criteria) was the most appropriate, and that such an approach should consider land use and impacts to humans and the environment. Further, it was recommended that a systematic classification system is needed to screen and classify sites according to their potential hazard, and that soil and groundwater criteria are needed to facilitate initial site assessment and establishment of preliminary cleanup goals. Siegrist concluded that for sites considered high hazard, a site-specific risk assessment would be needed to verify cleanup goals, whereas for low hazard sites, use of soil and groundwater criteria alone may suffice.[8]

Marrying the Options: The NCSRP Framework for Contaminated Site Assessment and Remediation

The NCSRP approach outlined below encompasses a combined approach, marrying the strengths of absolute and relative criteria in order to target cleanup efforts as efficiently and effectively as possible in the protection of human health and the environment.

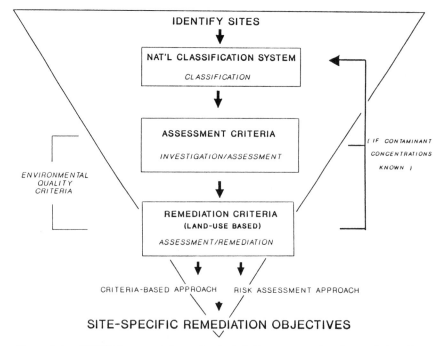

Figure 3.1.   NCSRP framework for contaminated site assessment and remediation. Note: Where contaminant concentrations are known for a site, the Assessment and Remediation Criteria (i.e., Environmental Quality Criteria) can be used in site classification.

Based on workshop recommendations and on an evaluation of existing approaches as outlined above, a general framework was created for the development of common tools to promote consistent identification, assessment, and remediation of contaminated sites in Canada. This framework, illustrated in Figure 3.1, is based on a series of screening and assessment tools intended to ensure effective decisionmaking in the development of remediation objectives (cleanup goals). The triangle underlying the figure represents the intent to focus efforts and funds through a tiered approach. The approach encompasses:

- National Classification System: a screening tool to assess relative risk/hazard at a site and to evaluate the need for further action;
- National Assessment Criteria: benchmarks for preliminary assessment which, if exceeded, indicate that further action/investigation is required;
- National Remediation Criteria: benchmarks to provide guidance on the need for remediation/detailed site assessment, or as the basis for site-specific remediation objectives. In the Canadian system, these criteria are developed for three distinct land-uses;

• Remediation Objectives: site-specific cleanup goals. Dependent on site-specific considerations and on socioeconomic factors, remediation objectives may be based directly on the generic National Criteria (''criteria-based approach'') or on a detailed site risk assessment (''risk assessment approach'').

The classification system and environmental quality criteria (assessment and remediation criteria have collectively been called environmental quality criteria) have been developed in the first year of the NCSRP (Phase 1). Phase 2 of the program will focus on the next component of this framework—establishment of a consistent process for the development of site-specific remediation objectives.

NATIONAL CLASSIFICATION SYSTEM

The CCME National Classification System for Contaminated Sites is a method for evaluating contaminated sites according to their current or potential adverse impact(s) on human health or the environment. This National Classification System was developed to establish a rational and scientifically-defensible tool for comparable assessment of contaminated sites across Canada, and has been developed to contribute national consistency in the overall implementation of the NCSRP.[9]

Participants of the consultative workshops and members of the CCME Subcommittee charged with developing such a classification system determined that, for the purposes of the NCSRP, the National Classification System should be simple (though sufficiently comprehensive to be scientifically sound); be risk-based; use existing or generally available site information (i.e., not require extensive onsite assessment); be flexible enough to accommodate professional judgement; place equal emphasis on human and environmental health; classify sites into general categories (i.e., sites will not be ranked); be a screening process; be traceable in terms of decisionmaking; and provide a nationally consistent method to evaluate sites.[6]

Review and Evaluation

To review existing methods of classifying contaminated sites and to recommend the format and content for a Canadian classification method for use under the NCSRP, a consulting firm was contracted by Environment Canada for the CCME Subcommittee on the Classification of Contaminated Sites.[10] The contractor's study methods included a review of existing systems from all Canadian provinces, several countries across Europe, and the United States EPA Hazard Ranking System (HRS). Each system was analyzed, and their strengths, weaknesses, and applicability to the Canadian situation were assessed. In addition, a questionnaire to determine the availability of contaminated site information in Canada was distributed to all provinces and territories, and the results evaluated.

The review of contaminated site classification systems from Britain, France, Germany, Holland, the U.S. and all Canadian provinces revealed that there are three general types of classification approaches: (a) an additive factorial approach, where a variety of site characteristics (factors) are considered and a score for these is added to reach an overall site score; (b) a risk/hazard assessment approach where in-depth investigation and assessment is undertaken and probabilities are considered; and (c) an ad hoc approach where professional judgment only is considered.[10]

Recommendations

Based upon their review, the consultants proposed to the CCME a National Classification System which used assessment factors developed in the Canadian provincial systems and added new environmental factors considered by the revised U.S. EPA HRS. The factors were organized into a 'risk' pathway approach, and the scoring was made consistent and defensible. An additive model was proposed because this type is a widely used method of multi-attribute decisionmaking, is simple to use and understand, and has been used by most Canadian provinces. In addition, the scores can be rationalized and documented for review by funding agencies (e.g., NCSRP) if required. Though the system is simple, a defined amount of site characterization information is required for a site to be classified using the system.

System Overview

The National Classification System is organized in such a way that the user assesses an "event" (i.e., a contaminated site) through evaluation of the hazard of that event, the potential for release of contamination, potential pathways, and potential receptors. The framework of the classification system reflects this hazard pathway (Table 3.1).

Table 3.1. Sample Hazard Pathway

Event	Hazard	Release	Pathway	Receptor
waste storage at Site A →	quantity and nature of waste stored on-site →	leakage from storage facility →	movement of contaminants off-site in groundwater →	well water user drinks contaminated water

The evaluation factors within the system are grouped into categories of:

1. Contaminant Characteristics (including hazard, quantity, and physical state of contaminants)
2. Exposure Pathways (including groundwater, surface water, direct contact and air)
3. Receptors (human and environmental; through drinking water, other water uses, land uses, etc.)

The use of the National Classification System is an evaluation exercise to classify a site, using a score intended to represent its hazard potential (maximum score equals 100). A sample page from the National Classification System is shown in Figure 3.2. Accompanying the system is a comprehensive "User's Guide" which explains the rationale for including each evaluation factor, as well as definitions and interpretations to assist in scoring a site. A worksheet is also provided in order that rationales for assigning scores to each factor can be documented. The system permits systematic, rational, and consistent evaluation of contaminated sites across Canada using existing or generally available site information.

Once the total site score has been determined, sites are grouped into one of four general categories according to the total site score (Table 3.2). The definitions of these classes of contaminated sites are purposely vague to reduce the likelihood of abuse of this classification system, which has been designed only as a tool to *aid* in the evaluation of contaminated sites. Its purpose is to provide scientific and technical assistance in the identification of sites which may be considered 'high', 'medium', or 'low' risk. It is beyond the scope of this system to address specific factors such as those of a technological, socioeconomic, political, or legal nature. Additional investigations will usually be required before remedial designs or regulatory requirements can be finalized.

Table 3.2. Site Classification Categories

Class	Site Score	Risk Potential/Action Required
Class 1	≥ 70–100	Risk potential is considered high, and action is required (although action may consist of further investigation or site assessment)
Class 2	≥ 50–< 70	Risk potential is considered medium, action is likely required
Class 3	≥ 37–< 50	Risk potential is considered medium low, action may be required
Class N	< 37	Risk potential is considered low, action not likely required

NATIONAL ENVIRONMENTAL QUALITY CRITERIA FOR CONTAMINATED SITES

Based largely on workshop recommendations as well as on the example of the National Guidelines for Decommissioning Industrial Sites,[11] a two-tiered approach was recommended for the development of NCSRP Environmental Quality Criteria for Contaminated Sites. The first tier consists of generic (i.e., absolute) criteria with national applicability. The second tier consists of criteria formulated for use on a site-specific (i.e., relative) basis. The workshop participants further recommended that the national criteria could be based initially on appropriate existing standards, criteria, or guidelines currently issued by regulatory agencies in Canada, the U.S., and Europe following a critical evaluation of the rationales behind these existing criteria.

DETAILED EVALUATION FORM (Cont'd.)

II

A

EXPOSURE PATHWAYS (Maximum Total Score is 33)

Groundwater (Maximum Score is 11)

Complete **Section 1 (Known)** or **2 (Potential)**

If answer is an estimate, circle the question mark (?) beside your score; if **not** an estimate circle the checkmark (✓).

Factors		Scoring Guideline	Site Score	Totals
1	**Known Contamination of Groundwater at or beyond the Property Boundary** (measured contamination of, or known contact with, groundwater (max. 11)			
	• Groundwater significantly exceeds CDWG (by >2x) or known contact of contaminants with groundwater;	11		
	• Between 1 and 2x CDWG or strongly suspected contact with groundwater	6		
	• Meets Canadian Drinking Water Guidelines	0	___ ? ✓	Section 1 **max. 11**
	If impact on groundwater is not known, complete **2**			
2	**Potential for Groundwater Contamination (max. 8)**			
	a) Engineered subsurface containment (max. 4)			
	• No containment	4		
	• Partial containment	2		
	• Full containment	0.5	___ ? ✓	

continued

2   Potential for Groundwater Contamination (max. 8) *continued*

b) Thickness of confining layer over aquifer (max. 1.5)
- Unconfined (3 m or less)   1.5
- Semi-confined (3 to 10 m)   1   — ? ✓
- Confined (>10 m)   0

c) Hydraulic conductivity of the confining layer (max. 1.5)
- $>10^{-4}$ cm/sec   1.5
- 10^{-4} to 10^{-6} cm/sec   1   — ? ✓
- $<10^{-6}$ cm/sec   0.5

d) Annual rainfall (max. 1)
- $>1,000$ mm   1
- 600 mm   0.6
- 400 mm   0.4   — ? ✓
- 200 mm   0.2

e) Hydraulic conductivity of aquifer (max. 3)
- $>10^{-2}$ cm/sec   3
- 10^{-2} to 10^{-4} cm/sec   1.5   — ? ✓
- $<10^{-4}$ cm/sec   0.5

Section 2
max. 11

Figure 3.2. Sample page from the National Classification System for Contaminated Sites.

Review and Evaluation

Several characteristics that were considered desirable for a national set of environmental quality criteria were identified and served as the basis of the criteria review and evaluation (Table 3.3). Based on this set of characteristics, Angus Environmental Limited (AEL), under the direction of the CCME Subcommittee for Environmental Quality Criteria for Contaminated Sites, carried out a comprehensive review of remediation criteria and approaches from regulatory agencies in various parts of the world.

Table 3.3. Desired Characteristics of NCSRP Criteria

- be applicable to a wide range of sites, site conditions, and contaminants
- consider all environmental media or compartments
- consider various exposure pathways and associated risks
- adapt to missing data
- consider present and future land use(s)
- place equal emphasis on the environment and human health
- consider aesthetics and phytotoxicity
- consider background or ambient concentrations of contaminants
- consider analytical detection limits

Twenty-one regulatory agencies were reviewed by AEL from across Canada, the U.S., Netherlands, United Kingdom, Australia, West Germany, and France.[12] Of these, 10 agencies were selected as potential sources for NCSRP criteria. Table 3.4 lists the 10 agencies reviewed that were of primary interest to the NCSRP and summarizes the extent to which these agencies meet the NCSRP desired characteristics. The review of existing criteria also revealed that one key difference between regulatory agencies was the number of 'categories' of criteria, and their implications for action (Figure 3.3). From this review, the consultants recommended a set of interim criteria for consideration by the CCME Subcommittee.

Table 3.4. Results of Evaluation of Existing Criteria According to NCSRP Desired Characteristics (adapted from AEL, 1991)

Characteristics Considered	AGENCY									
	Alta.	B.C.	Calf	CCME PAH	CCME PCB	Neth	N.J.	Ont.	Que.	U.K.
Widely Applicable	L	Y	Y	N	N	Y	L	L	Y	L
All Media	N	L	L	L	Y	L	N	N	L	N
All Routes of Exposure	N	L	Y	N	Y	N	N	N	N	L
Various Receptors	N	L	Y	N	Y	N	L	N	L	L

continued

Table 3.4. ***continued***

Characteristics Considered	AGENCY Alta.	B.C.	Calf	CCME PAH	CCME PCB	Neth	N.J.	Ont.	Que.	U.K.
Missing Data	L	L	L	N	N	N	N	N	L	N
Various Land Uses	L	Y	Y	L	Y	Y	L	Y	Y	Y
Neighboring Property Use	N	N	N	N	N	N	N	N	N	N
Environmental = Human Health	L	L	Y	N	N	?	?	L	?	L
Aesthetics	?	?	N	N	N	?	N	L	?	L
Phytotoxicity	N	?	N	N	N	N	N	Y	N	Y
Background	Y	Y	N	L	L	Y	Y	Y	Y	Y
Detection Limits	Y	Y	N	N	N	Y	N	N	Y	N

Notes: Y = Yes
N = No
L = Limited
? = Uncertain
See Appendix A for full reference to agencies listed in this table

APPROACH	JURISDICTION	CATEGORIES
ONE NUMBER	ALBERTA	TIER 1 GUIDELINES
	NEW JERSEY	ISAL
TWO NUMBER	UNITED KINGDOM	ACTION TRIGGER THRESHOLD TRIGGER
	ONTARIO	C/I GUIDELINE R/P GUIDELINE
THREE NUMBER	BRITISH COLUMBIA NETHERLANDS QUEBEC CCME	ABC CRITERIA VALUE LEVEL
FOUR NUMBER	FRANCE	URGENCY THRESHOLD TREATMENT THRESHOLD INVESTIG. THRESHOLD ANOMALY THRESHOLD

Figure 3.3. Comparison of approaches used by various agencies to establish cleanup levels (adapted from AEL, 1991).

Recommendations

Though most agencies fulfilled several of the NCSRP desired characteristics, no one agency considered all of the factors desired. It was therefore concluded that no existing set of criteria could be directly adopted as NCSRP criteria. However, given the urgent need to begin remediation, it was recommended that a set of interim criteria be selected based on existing criteria with the best (i.e., most scientifically-defensible) supporting rationale. It was also recommended that a new approach be developed to validate or modify these interim criteria on an ongoing basis. These recommendations were supported at the second multi-stakeholder workshop in November 1990.[7]

Interim Environmental Quality Criteria for Contaminated Sites

As no one set of existing criteria was considered directly adoptable, the interim set of criteria recommended for the NCSRP was derived from several agencies.[12] Based on the review and evaluation, it was decided that specific criteria from the provinces of British Columbia, Alberta, and Ontario, in addition to the CCME interim guidelines for PAHs, Canadian Water Quality Guidelines and Canadian Drinking Water Guidelines, would provide the basis for a working set of interim criteria for the NCSRP.

The interim criteria (called Canadian Environmental Quality Criteria for Contaminated Sites) are numerical limits or narrative statements intended to protect, maintain, or improve current and future use(s) of soil and water at contaminated sites.[13] The Interim Canadian Environmental Quality Criteria[13] encompass numerical criteria for both the *assessment* and *remediation* of soil and water in the context of agricultural, residential/parkland, and commercial/industrial land uses (Table 3.5). The criteria also include the Canadian Water Quality Guidelines[14] and Guidelines for Canadian Drinking Water Quality[15] for specified uses of water likely of concern at contaminated sites.

Table 3.5. **Canadian Environmental Quality Criteria for Contaminated Sites**

Assessment Criteria	Remediation Criteria (Soil)	Remediation Criteria (Water)
Soil	Agriculture (AG)	Freshwater Aquatic Life
Water	Residential/Parkland (R/P)	Irrigation
	Commercial/Industrial (C/I)	Livestock Watering
		Drinking Water

The interim assessment criteria are approximate background concentrations *or* approximate analytical detection limits for contaminants in soil or water. These are numerical values that, when exceeded, indicate that investigative action should be considered to assess the extent of contamination and the nature of any hazards at a site, and to determine the scale and urgency of any required further action. When contaminant concentrations in soil and water do not exceed the level of the assessment criteria, no further action is required.

The interim remediation criteria are intended to be nationally applicable and do not address site-specific conditions. They are considered protective of specified uses of soil and water at contaminated sites, based on the experience and professional judgment of experts in jurisdictions where these criteria are in use. The criteria may be used in a number of ways, including as indicators of the environmental quality of a site, guidance for determining when further site investigation is necessary, guidance for determining when site remediation or risk assessment is necessary, and to verify the adequacy of site cleanup. The criteria can also serve as a common basis for the establishment of site-specific cleanup levels. Once applied to a particular site, the criteria are known as *site-specific remediation objectives*. When remediation objectives are exceeded, remedial action is required to reduce the contamination to below the level set as the remediation objective for the intended soil or water use. Remediation is considered to be complete once contaminant levels have been reduced below the level of the remediation objective. Remediation criteria (or objectives) may also serve as the basis for the development of legally enforceable standards.

NCSRP REMEDIATION OBJECTIVES (CLEANUP LEVELS)

Before remediation of a contaminated site is initiated, remediation objectives must be established for that specific site. There are two basic approaches to this (Figure 3.4). The first approach, known as the *criteria-based approach,* involves the direct adoption or adaption of the existing interim environmental quality criteria in light of site-specific conditions (e.g., environmental, socioeconomic, and/or technological considerations). The second approach uses *site-specific risk assessment* to characterize potential risks, hazards, and exposures of receptors to contaminants at a contaminated site (for example, where no criteria exists for a contaminant or in the case of contaminant mixtures). Only the criteria-based approach uses the national environmental quality criteria directly.

WHERE DO WE GO FROM HERE?—PHASE 2 OF THE NCSRP

In order to ensure effective and consistent application of the common scientific tools developed under Phase 1 of the national program, Phase 2 of the NCSRP will focus on scientific validation of the interim environmental quality criteria and the development of guidance documents for establishing site-specific objectives and protocols for human health and environmental risk assessment.

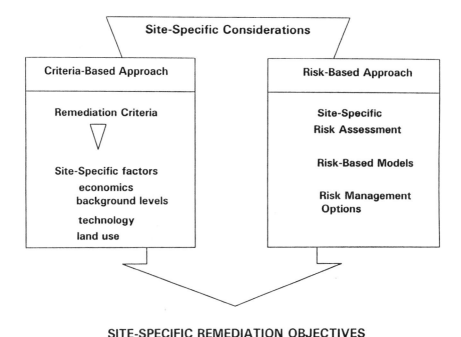

SITE-SPECIFIC REMEDIATION OBJECTIVES

Figure 3.4.   Approaches to establishing site-specific objectives in the NCSRP.

Validation of Interim Environmental Quality Criteria

At the November 1990 consultation workshop,[7] it was generally recognized that, although necessary to meet the urgent need for site remediation under the NCSRP, the interim criteria did not embody many of the characteristics desired in a national set of criteria. It was recommended that validation of the interim numbers to establish scientifically-based national environmental quality criteria was required to promote consistent, effective, and scientifically sound assessment and remediation of contaminated sites in Canada. It was also recommended that data requirements and an approach for validation and/or modification of the interim numbers be established as soon as possible. Some of the types of information that will be considered when validating interim criteria are listed below:

- physical-chemical data
- contaminant behavior in the Canadian environment
- ecotoxicological, phytotoxicological, and epidemiological data
- aesthetic information
- pathway information for receptors of interest

- guidelines/criteria for other media
- environmental fate modeling

Currently, minimum data requirements for validating criteria are being identified. Once a protocol for validation and development of criteria has been established, it is anticipated that approximately five contaminants (based on established priorities) will be evaluated per year.

Guidance Documents for Establishing Site-Specific Objectives

It was also recommended at the November 1990 consultation workshop that a scientifically sound method is needed in the NCSRP for consistent site-specific application of the national criteria.[7] As discussed above, two complementary but distinct approaches have been identified at the site-specific level. The criteria-based approach and the risk assessment approach will each be considered for establishment of remediation objectives at a site, dependent on site-specific circumstances. Both approaches may be seen as a single overall strategy or framework for the establishment of remediation objectives incorporating both human health and environmental considerations.

SUMMARY

The National Contaminated Sites Remediation Program (NCSRP) has recently been established by the Canadian Council of Ministers of the Environment (CCME) to promote consistent assessment and remediation of high priority contaminated sites in Canada. The first five years of the program will provide the necessary funds to initiate the cleanup of "orphan" contaminated sites. To promote consistency in the overall management of this program, the CCME has requested the development of common scientific tools for classification, assessment, and remediation of contaminated sites in Canada. Based on consultation with stakeholders and evaluation of existing approaches world-wide, a tiered approach to the assessment and remediation of contaminated sites has been established. A National Classification System and National Environmental Quality Criteria (assessment and remediation) have been developed under the first phase of the NCSRP. These tools will serve as a common scientific basis for contaminated site assessment, as well as for the development of remediation objectives. Development of these key tools is described in this chapter.

Phase 2 work in the NCSRP will focus on scientific validation of the Interim Canadian Environmental Quality Criteria for Contaminated Sites and the development of a nationally consistent approach to setting site-specific objectives in the NCSRP. This approach will consider the development of site-specific objectives both from the national generic criteria and from human-health and ecological risk assessment methods.

ACKNOWLEDGMENTS

This work has been carried out by Environment Canada in collaboration with the CCME Contaminated Sites Task Group, the CCME Subcommittee on Classification of Contaminated Sites, and the CCME Subcommittee for Environmental Quality Criteria for Contaminated Sites. Health and Welfare Canada also provided scientific and secretarial support to the subcommittees in development of these common scientific tools.

We would like to acknowledge the chairmen of the subcommittees, Mr. Walter Ceroici (Alberta) and Dr. John Ward (British Columbia) for their expert advice and guidance.

APPENDIX

Alta	= Alberta Draft Tier 1 Guidelines (Reference No. 5)
B.C.	= British Columbia Criteria for Managing Contaminated Sites (Reference No. 2)
Calf	= California Technical Standard (Reference No. 16)
CCME PAH	= CCME Interim PAH Guidelines (Reference No. 17)
CCME PCB	= CCME Interim PCB Guidelines (Reference No. 19)
Neth	= Netherlands Soil Guidelines (Reference No. 18)
N.J.	= New Jersey Interim Soil Action Levels (Reference No. 20)
Ont.	= Ontario Soil Clean-up Guidelines (Reference No. 4)
Que.	= Contaminated Sites Rehabilitation Policy (Ministere de l'Environmennement du Quebec (Reference No. 3)
U.K.	= United Kingdom Trigger Concentrations (Reference No. 21)

REFERENCES

1. Foote, T. *Proceedings from the Annual Conference of the Chemical Institute of Canada.* Halifax, Nova Scotia. July 17–18, 1990.
2. British Columbia Ministry of the Environment (BC MOE), 1989. *Criteria for Managing Contaminated Sites in British Columbia.* Draft Report. Waste Management Program.
3. Ministère de l'Environnement du Québec (MENVIQ), 1988. *Contaminated Sites Rehabilitation Policy.* Prepared by the Hazardous Substances Branch. January.
4. Ontario Ministry of the Environment (MOE), 1989. *Guidelines for Decommissioning and Clean up of Sites in Ontario.* February.
5. Alberta Environment, 1990. *Alberta Tier 1 Criteria for Contaminated Soil Assessment and Remediation.* Draft Report. Wastes and Chemicals Division, Soil Protection Branch.
6. Energy Pathways Inc., 1990. *Final Report on the April 1990 National Contaminated Sites Consultation Workshop.* Prepared for the Contaminated Sites Consultation Steering Committee, Canadian Council of Minsters of the Environment (CCME). November.

7. Energy Pathways Inc., 1991. *Final Report on the November* 1990 *National Contaminated Sites Consultation Workshop.* Prepared for the Contaminated Sites Consultation Steering Committee, Canadian Council of Minsters of the Environment (CCME). March.

8. Siegrist, R. L. 1989. *International Review of Approaches for Establishing Clean up Goals for Hazardous Waste Contaminated Land.* Institute for Georesources and Pollution Research, Norway.

9. Canadian Council of Ministers of the Environment (CCME), 1991. *National Classification System for Contaminated Sites.* Draft Report. Prepared for the CCME National Contaminated Sites Task Group. March.

10. Trow, Dames, and Moore (TDM), 1991. *Review of Contaminated Site Classification Methods and Recommendations for a National Classification System.* Prepared for the CCME Subcommittee on Classification of Contaminated Sites. Mississauga, Ontario. June.

11. Monenco Consultants Limited, 1989. *National Guidelines for Decommissioning Industrial Sites, Volume* 1. Draft Report. Prepared for Environment Canada, Conservation and Protection.

12. Angus Environmental Limited (AEL), 1991. Review and Recommendations for Canadian Interim Environmental Quality Criteria for Contaminated Sites. Prepared for the CCME Subcommittee on Environmental Quality Criteria for Contaminated Sites. February.

13. Canadian Council of Ministers of the Environment (CCME), 1991. *Interim Canadian Environmental Quality Criteria for Contaminated Sites.* Prepared for the CCME National Contaminated Sites Task Group. September.

14. Canadian Council of Resource and Environment Ministers (CCREM), 1987. *Canadian Water Quality Guidelines.* Prepared by the Task Force on Water Quality Guidelines of the Canadian Council of Resource and Environment Ministers.

15. Health and Welfare Canada (HWC), 1989. *Guidelines for Canadian Drinking Water Quality.* Environmental Health Directorate. Ottawa, Ontario. 4th ed.

16. California Department of Health and Safety (DHS), 1990. *Technical Standard for Determination of Soil Remediation Levels.* Draft Report. Prepared by the Toxic Substances Control Program. August.

17. Canadian Council of Ministers of the Environment (CCME), 1988. *Proposed Interim Guidelines for PAH Contamination at Abandoned Coal Tar Sites.* Prepared for the Waste Management Committee, Toxic Substances Advisory Committee by the Ad Hoc Federal-Provincial Working Group on Interim PAH Guidelines.

18. Moen, J. E. T., 1988. "Soil Protection in the Netherlands." *Contaminated Soil '88* pp. 1495-1503. Kluver Academic Publishers.

19. Canadian Council of Ministers of the Environment (CCME), 1987. *Interim Guidelines for PCBs in Soil and Sediment.* A report to the Waste Management Committee of Canadian Council of Ministers of the Environment.

20. New Jersey Department of Environmental Protection (NJDEP), 1990. *Basis for NJDEP Interim Soil Action Levels.* Division of Hazardous Site Mitigation, Bureau of Environmental Evaluation and Risk Assessment. February.

21. United Kingdom Department of the Environment (U.K.), 1990. *Contaminated Land: The Government's Response to the First Report from the House of Commons Select Committee on the Environment.* July.

CHAPTER **4**

Innovative Hydrocarbon Contaminated
Soil Treatment Techniques in Nevada:
A Regulatory Perspective

Jim Najima and **Allen Biaggi,** Nevada Division of Environmental Protection, Bureau
of Waste Management, Underground Storage Tank Branch, Carson City, Nevada

INTRODUCTION

Hydrocarbon contaminated soils are generated a number of ways. The most common are from leaking underground storage tanks; however, aboveground releases and other spillage also contribute a significant volume. The state of Nevada has a soil action and remediation standard of 100 mg/kg total petroleum hydrocarbons by method 8015 modified. It is estimated that in the state of Nevada over 80,000 tons per year of this waste stream is generated. An additional 10,000 to 15,000 tons are imported from out of state, of which the majority originate in California.

Traditionally, Nevada has allowed contaminated media to be placed into Class I sanitary landfills (serving a population over 2,000 user equivalents). While Nevada has perhaps the most desirable climatic and geologic conditions for land disposal in the country, it is realized that such disposal was not an environmentally sound, long-term method of waste disposal.

In 1987, the Division of Environmental Protection began negotiations with a private firm for the thermal treatment of hydrocarbon contaminated soils as an alternative to landfilling. Based upon the success of this operation, the state has encouraged the development and operation of similar facilities which utilize various treatment technologies.

The state of Nevada currently has three fully permitted and operating soil treatment facilities, one experimental facility in technical review and development, and two additional facilities in the permitting process. It is the goal of the state, in cooperation with private industry, to provide cost-effective and environmentally sound options for contaminated soil treatment, and to reduce or eliminate land disposal or other less desirable disposal practices.

TREATMENT TECHNOLOGY OVERVIEW

Thermal Treatment

Two facilities in Nevada use thermal treatment for hydrocarbon contaminated soils. Both utilize rotating drum kiln technology.

Soil is initially screened on the soil storage pad to remove debris and oversized material. The soil is then stockpiled prior to introduction into the kiln. Debris is disposed of at a municipal landfill while stones, boulders, and cobbles are steam cleaned in a specially designed containment unit and used for riprap in erosion control and landscaping projects.

Both plants are fired by propane and operate at temperatures between 1,300° and 1,800°F. The soil input rate and retention time is computer controlled and based upon contaminant concentration, soil type, and product type. Typical throughput rates are 10 to 50 tons per hour.

Analytical soil testing is conducted one to three times per day for total petroleum hydrocarbons (TPH). The treatment goal for both facilities is 50 mg/kg TPH. Soils not achieving this standard are blended with untreated soil and reburned.

Secondary burners are used to reduce volatile emissions and baghouses are used to control particulate air emissions. Baghouse dust is slurried with water and added to the treated soil after the kiln to reduce dust.

Treated soils are stockpiled for reuse or disposal. At one facility near Las Vegas, soils are blended with raw materials and used as roadbase, while a facility near Reno uses the treated soils for fill. This facility has established an in-house treatment goal of less than 10 mg/kg TPH. Both facilities provide the waste generator with certificates of destruction after successful treatment.

A recent study of the thermally treated soils from the northern Nevada facility has shown that the volatile fractions are removed via the thermal process. However, some polynuclear aromatics are present in the treated soils. Leachability testing has indicated, however, that these compounds are tightly bound to the soil particles and are thought to have a low leaching potential. State and local regulatory agencies are still examining available data in regard to the unrestricted reuse of treated soil.

The cost of thermal destruction at these facilities is $40 to $80 per ton.

Biological Treatment

Nevada's biological treatment facility is located in southern Nevada near Las Vegas. The treatment units are sited near a landfill serving the Las Vegas metropolitan area. Ten treatment impoundments are present at the facility which are used to store and treat the contaminated soil.

Soils received at the facility are screened to remove oversized material and debris and are placed in the treatment impoundment at a nominal depth. The soils are then inoculated with a commercially available strain of bacteria capable of utilizing the petroleum hydrocarbons as a food source. Water and nutrients (nitrates and phosphates) are added to maintain and enhance the biota. Sample bacterial cultures and nutrient concentrations are periodically obtained and reinoculations of bacteria and nutrients are conducted on an as-needed basis.

Soils are frequently turned via a tractor-driven disk, and watering is conducted on a daily basis using a commercial farm sprinkler irrigation system. In the hot, dry summer months, sprinkler irrigation is replaced by a water truck which is driven directly onto the treatment pad.

The average time for treatment to a concentration below 100 mg/kg TPH is 60 days. Treated soils are removed from the impoundments and are used for cover material at the sanitary landfill. After treatment is completed, the facility does provide the waste generator with a certificate of destruction for liability purposes.

The cost of soil treatment at this facility is $30 to $40 per ton.

Chemical Treatment

The state of Nevada and Carson City recently authorized the operation of a pilot plant for the chemical treatment of hydrocarbon contaminated soils. If successful, this technology may be utilized at the Carson City landfill and other landfills throughout the state to reduce the negative environmental impacts of hydrocarbon contaminated soils.

Contaminated soils are placed in a lined treatment/storage impoundment in 4 inch lifts. Soils are tested for buffering capacity and the pH is adjusted as necessary. Catalyzed hydrogen peroxide (Fenton's reagent) is then added to the soils. Homogeneity is ensured by mixing the soil/solution with a backhoe. Successive lifts are added on top of the initial lift, along with additional chemical treatment.

Post-treatment sampling indicated a reduction of TPH concentrations to below the 100 mg/kg criteria. pH adjustment may be required after treatment and prior to soil reuse. Soils chemically treated may not be suitable for reuse but may be utilized as landfill cover material.

Based on initial trials, the cost of chemical treatment using Fenton's reagent is competitive with thermal and biological treatments.

Results of this and previous trials are promising. Chemical treatment would appear to be a cost-effective means of soil treatment.

Facility Permitting

The state of Nevada, through the Department of Conservation and Natural Resources, Division of Environmental Protection, requires the permitting of soil treatment facilities. Required permits include a National Pollutant Discharge Elimination System (NPDES) permit, an Air Pollution Control Permit, and all applicable county and local permits.

The NPDES permit addresses the storage of soils prior to treatment and the operational requirements of the facility. The permit places a ''zero discharge standard'' on the operation which prohibits the discharge of any pollutants to any waters of the state, including groundwater. Permit conditions mandate the installation of at least two groundwater monitoring wells immediately downgradient of the soil storage impoundment. These wells must be sampled quarterly for total petroleum hydrocarbons, benzene, toluene, xylene and ethylbenzene, as well as any other pollutant unique to the operation. Additionally, the nearest domestic, municipal, or irrigation well to the facility must be monitored on a quarterly basis.

The soil storage impoundment must have a synthetic liner at least 40 mL thick which is chemically resistant to compounds typically found in petroleum products. Liner seam quality and integrity must be demonstrated via prearranged QA/QC procedures. Liner underbedding material must meet manufacturers' specifications, and leak detection systems (soil moisture blocks, observation tubes) must be included underlying the impoundment. Design requirements mandate containment of the 100 year, 24 hour duration storm event when the impoundment is at maximum storage capacity.

Due to the need to operate heavy machinery on the liner, the Division requires a minimum one foot sand layer along with an indicator material, (i.e., orange lining material, gravel, colored sand), to ensure the liner is not damaged during operation.

All operational irregularities must be reported and all analytical data generated by permit requirements must be submitted to the Division on a quarterly basis.

Air quality permitting requirements mandate compliance with the applicable air quality attainment standards for the region. Airborne toxic emissions are also examined and permitted if necessary.

County and local permitting requirements vary, but generally include local business licenses, fire protection and notification, building permits, and special use (zoning) issues.

Operational Requirements

The operation of treatment facilities must reduce the potential of public exposure to contaminated media within the current regulatory framework. To ensure reduced exposure, the Division has imposed strict operational limitations and requirements associated with permitting, and requires the development of a comprehensive management plan for facility operation. The management plan is made an attachment of the NPDES permit, and is a binding condition.

Transportation

Facility management plans are required to address the issue of transportation of contaminated media to and from the site. The plans are required to address emergency response in case of accident, and most importantly, decontamination of the vehicles prior to departure from the facility.

Waste Screening

The soil treatment facility is designed and regulated to handle only nonhazardous waste. Because of this, it is imperative that the waste generator certify their waste as nonhazardous by current state and federal definitions. Additionally, the waste must not contain any constituents which may pose a risk to the storage units, treatment process, or facility employees. The Division has mandated that each waste stream be characterized with regard to hazardous and toxic compounds. The Division has required the following minimum testing requirements be conducted to ensure the presence of nonhazardous wastes:

- Representative soil sampling for Total Petroleum Hydrocarbons by method 8015 modified;
- Representative soil sampling and analysis for halogenated organics by EPA Method 8240 for volatile organics;
- Pre-acceptance screening (prior to the facility entrance) for halogens using the copper wire test or equivalent;
- Screening for other contaminants which may be expected to be present in the waste stream (i.e., dielectric fluids or mineral oil—PCBs, waste oils—heavy metals).

The Division considers waivers for certain tests if it can be conclusively determined what the waste source is and the lack of potential for waste mixing or contamination with other regulated materials.

Because of the final disposition of the treated soils, some treatment operations require additional testing to protect the end user. Testing beyond the requirements of the Division are recognized and are discretionary on the part of the treatment facility.

Waste loads that fail any of the above tests are denied treatment and the Division is notified. The waste load is returned to the generator and notification is made to the proper state and local agencies to ensure the proper management of the waste material.

TCLP Requirements

The federal Toxicity Characteristic Leachate Procedure (TCLP) became effective in September 1990. These rules added 25 organic chemicals which must be evaluated in a hazardous waste determination. One of the compounds added to this listing is benzene, which is a major constituent in gasoline and diesel fuels. The TCLP rule currently defers Subtitle I regulated underground storage tanks and associated contaminated media from regulation. While the majority of the soils accepted by the treatment facilities originate from Subtitle I tanks, some soil is from other sources. Contaminated material from these other sources must be evaluated for hazard via the TCLP requirements prior to facility acceptance unless other knowledge can be applied which documents the nonhazardous nature of the waste. Treatment facilities accepting TCLP hazardous wastes are subject to enforcement actions for permit and hazardous waste violations.

Acceptance of Out-of-State Waste

Nevada has long been looked to for the disposal of waste material. This is due in part to Nevada's vast open spaces, abundance of federal land, low population, and lenient environmental regulations. The state's proximity to California has, in particular, placed extreme pressure on the state in regard to waste importation from that state.

Nevada regulations currently require that if a waste is considered hazardous in its state of origin, that waste must also be considered hazardous in Nevada.

In some states, and specifically in California, the presence of hydrocarbon contamination may classify the waste as hazardous. Waste determined to be hazardous may not be treated at a Nevada certified soil treatment facility.

The waste generator does have options for importation. Acceptance at a permitted facility is allowed:

1. If a letter from the regulatory entity governing hazardous waste management within the state of origin attests to the nonhazardous nature of the waste, or;
2. In the case of California, hydrocarbon contaminated waste may be declassified via the procedures outlined in Article 11, Title 22 of the California Administrative Code. The declassification is recognized only if:

   • The sampling protocol is certified as representative by a California registered professional engineer or geologist;
   • All tests required under the title 22 requirements for self-declassification are below specified critical levels;
   • All samples, protocols, test methods, certifications, and results are reviewed by a Nevada Registered Professional Engineer knowledgeable in California and Nevada waste regulations. This review must

include a recommendation to the Division on adherence to Title 22
requirements.
- All testing and screening procedures required by the Division and the
  facility are satisfactorily met;
- A waste importation fee of $3.00 per ton is paid to the Division and/or
  the local solid waste authority.

The Division reserves the right to terminate waste importation at any time.

Conclusions

The state of Nevada has recognized potential problems in the proper treatment
and disposal of hydrocarbon contaminated soils. For the past four years the state,
in association with private industry, encouraged the research,development, and
operation of facilities capable of treating petroleum contaminated media. Progress
has been made in biological, chemical, and thermal treatment technologies. Our
goal is to increase the efficiency and safety of treatment facilities and to provide
an incentive for corrective action. Environmentally sound, cost-effective treatment of petroleum contaminated soil will help safeguard Nevada's precious
groundwater.

CHAPTER 5

Regulatory Complexities Associated with the Cleanup of Petroleum Contaminated Sites

Michael R. Anderson, Oregon Department of Environmental Quality, Portland, Oregon

INTRODUCTION

As with most environmental work, the investigation and cleanup of petroleum contamination requires the knowledge of principles and the application of procedures from a wide variety of technical disciplines. By the time that a project has reached the corrective action phase, it may have required input from hydrogeologists, geologists, chemists, biologists, and engineers. Since such remedial work must usually be carried out within a regulatory framework, project managers must also have a clear understanding of federal, state, and local regulations covering these activities. In many cases, the timely completion of a project may depend as much on the knowledge of and ability to operate within the regulatory environment as it does on the technical complexities of the site. However, even the most knowledgeable manager will have projects slowed by the routine requirements of regulatory agencies. Such delays are common to state-run projects as well as to those being directed by consultants working for private environmental firms.

With individual states developing their own Underground Storage Tank (UST) Cleanup Programs, there will understandably be differences from one state to the next. However, because of the nature of regulatory agencies, there are a number of factors affecting the efficiency of petroleum cleanup activities which will probably be common to many states. Some of these factors are:

1. Insufficient Staff

This is certainly not surprising. There are probably very few, if any, government programs which would claim to have sufficient staff to handle its assigned workload within a reasonable period of time. This problem is particularly acute in many state UST programs. The Environmental Protection Agency (EPA) estimates that there are approximately two million regulated USTs within the United States and that 84% are made of bare steel and are highly susceptible to corrosion.[1] Even if a leak does not occur, problems still arise. Since it was considered an acceptable practice in the past to allow tank overfills to run into the backfill around the tanks, and since federal regulations now require the reporting of all below-ground releases of petroleum products from regulated tanks regardless of the volume of the release (40 CFR 280 Subpart F), almost every tank decommissioning or upgrade will result in a reportable release that must be dealt with by a state regulatory agency.

2. Nonexistent or Changing Cleanup Goals

Federal regulations related to UST systems were first passed in 1984 as part of the Hazardous and Solid Waste Amendments (HSWA) to the Resource Conservation and Recovery Act (RCRA). To implement the new regulations, the EPA decided to franchise the program to the individual states. It was hoped that such an approach would allow for consistent enforcement of the regulations while still providing for flexibility to meet the needs of individual states. Although some states began to develop their UST programs quite quickly, others have only recently done so. Because of the relative newness of many state petroleum cleanup programs, project managers may find themselves in the unenviable position of starting a project without having specified procedures or firm cleanup goals in mind. Furthermore, regulatory conditions may change during the course of a project. Sites which began work when "no odor or sheen" was the cleanup goal may now find themselves sampling and working toward specific numeric soil cleanup levels. Before cleanup is complete, these projects may be subject to new rules establishing additional and possibly more stringent cleanup procedures and parameters.

3. Conflict with Pre-Existing Programs

Newly established cleanup programs often get bogged down in pre-existing bureaucracy. Historically, state environmental programs have been in the business of permitting releases; that is, controlling the release of industrial discharges through programs like the National Pollutant Discharge Elimination System (NPDES). Such programs generally set limits for the amount of pollutants that may be released from industrial sources, and monitor for compliance with the permit limits by requiring sampling for specified parameters. Although such programs work well in their originally-intended context, their requirements are often inclusive enough to demand compliance at even the smallest cleanup site where they may not only be inappropriate, but where they may halt cleanup for months and allow contamination to continue to spread while the required permits are being obtained.

With the types of difficulties listed above in mind, this chapter will discuss the development of Oregon's UST Cleanup Program over the past three years and examine some of its attempts to improve the efficiency of petroleum cleanup activities. A case history will be presented to illustrate some of the regulatory complications that may be encountered at a moderately complex petroleum cleanup site. This will be accompanied by a discussion of some of the changes that have been made in Oregon's UST Cleanup Program as well as changes that are currently being considered to further improve the staff's ability to handle the cleanup of petroleum contaminated properties in a timely fashion.

CASE HISTORY

In April 1988, the Oregon Department of Environmental Quality (DEQ) received a call from the owner of a Recreational Vehicle (RV) Park who complained of gasoline in his irrigation well. At the time the site was reported, the DEQ was in the process of organizing a new division, the Environmental Cleanup Division (ECD), to handle a broad range of cleanup activities within the state (Figure 5.1). One section within that division, the UST Cleanup Section (USTCS), was designated to have specific responsibility for cleaning up releases of petroleum products from USTs that are regulated under the EPA UST statutes. At the time that this release was reported, a manager had been hired for the USTCS, but technical staff had not yet been hired and rules had not yet been adopted.

When a DEQ regional office manager traveled to the RV Park site to investigate the complaint, he discovered about 10-12 inches of a petroleum product floating on the water in the well casing. Tests run on a sample of the product showed that it consisted of hydrocarbons in the C6 through C35 range. These chromatographic results were indicative of a mixture of gasoline and fuel #2 (diesel). The laboratory also indicated that the gasoline fraction showed little signs of weathering.

By surveying the surrounding area and interviewing local residents about possible sources of petroleum contamination, six different potentially responsible parties (PRPs) were identified in the vicinity of the RV Park. These potential sources were (Figure 5.2):

- A livestock yard located northwest of the RV Park,
- A bulk fuel distribution and card lock station located west of the RV Park,
- An abandoned gasoline station located west of the cardlock station,
- Railroad property located south of the abandoned gas station,
- Bureau of Land Management (BLM) property located northwest of the abandoned gas station, and
- An agrichemical company located on the west side of the abandoned gas station.

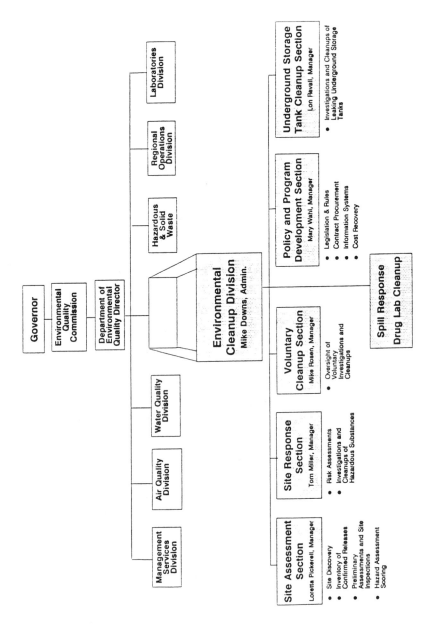

Figure 5.1.   Environmental Cleanup Division organizational chart.

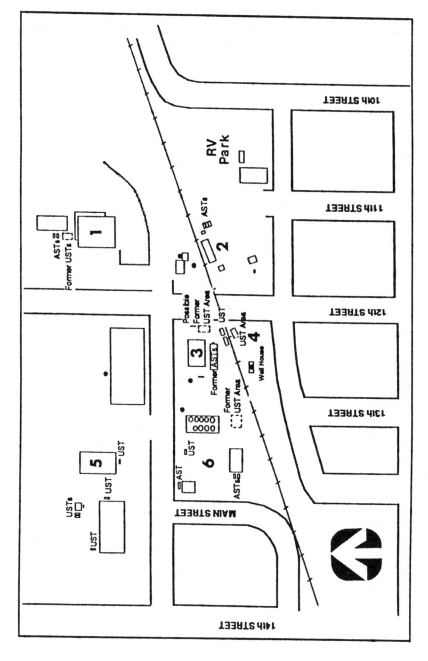

Figure 5.2. Map of the RV Park site showing the location of the six potential sources of petroleum contamination.

The RV Park itself had no known current or former storage tanks of any kind. A residential area was located to the east of the RV Park and a river was located beyond the houses about 500 meters east of the park. Although data were not yet available, the assumed direction of local groundwater flow was east toward the river.

After identifying possible sources, the department formally notified all six of the parties about the petroleum contamination in the area, informed them that they were considered PRPs for the contamination found at the RV Park, and requested information about tank inventories, tank tightness tests, and possible known leaks or spills at their sites. Because of poor record-keeping practices, many of the PRPs were not able to provide much of the requested information. The information that was received proved to be inconclusive.

By this time, the department had adopted two sets of rules regarding activities overseen by the ECD. The Environmental Cleanup Rules (OAR 340-122-010 through 110)[2] were developed to deal with hazardous substance contamination and were geared to high priority ''state superfund'' sites. The Cleanup Rules for Leaking Petroleum UST Systems (OAR 340-122-205 through 260)[3] were meant to handle the cleanup of petroleum releases from regulated UST systems. Because of the relative newness of the program, the department decided to meet with the PRPs, describe what was known about the site and explain the requirements of recently adopted statutes and rules regulating environmental cleanup. The parties were informed that they could take responsibility for the investigation and cleanup either individually or jointly. If not, the department would perform the investigation, make a determination of responsibility, and then charge the responsible party or parties (RPs) for all costs incurred by the state. The RPs would also then be ordered to perform the cleanup of the contamination. It was also explained that the advantage to the PRPs in performing their own investigation was that they would have better control over the scope of the investigation and therefore the costs. Despite the meeting, all parties continued to claim that there were no releases of petroleum products on their properties and therefore they felt that they were not responsible for performing the proposed investigations.

Because of the unwillingness of the PRPs to investigate the source of the release, the department had to consider other strategies for moving the site cleanup forward. It was not clear that there was sufficient evidence at this point to attempt a formal enforcement action against any of these parties. While investigation at the site was at a temporary impasse, the ECD was completing contract negotiations with several consulting firms for the purpose of using these firms to prepare plans and carry out investigations and remedial actions as directed by division project managers. When the contracts were finally signed, the decision was made to prepare a task order and statement of work for the initial stages of investigation at the RV Park. The first phase of the investigation was carried out to:

- Identify drinking water wells in the vicinity of the RV Park and collect water samples from several homes whose residents had recently complained of contaminated water,

- Perform a reconnaissance of the area to make sure that all potential sources had been identified, and
- Complete a soil gas survey to assist in finding the possible sources of the petroleum contamination.

As a result of the initial investigation, four homes were found to have contaminated drinking water supplies. Analyses of the water samples showed not only contamination that may be related to petroleum products (total petroleum hydrocarbons identified as both gasoline and diesel fractions, benzene and 1,2-dichloroethane), but also contamination that may be related to agricultural chemicals [1,2-dichloropropane (1,2-DCP) and chlorobenzene]. All of these homes obtained their water from private backyard wells completed in the shallow unconfined aquifer at depths of only 25 to 35 feet. Because of the presence of petroleum-related contamination in these water supplies, the department was able to use federal LUST trust funds to provide bottled water to the affected families.

The tank survey turned up evidence of at least 27 current and former storage tanks at the 6 previously identified facilities (Table 5.1). The results of the soil gas survey showed the highest levels of soil hydrocarbon vapors at the abandoned gas station (PRP 3), the card lock station (PRP 2), and a residence located north of the card lock station. Lower levels of contamination were found on the BLM property (PRP 5), the railroad property (PRP 4) and the agrichemical company property (PRP 6). No significant levels of hydrocarbon vapors were detected on the livestock yard property (PRP 1).

A summary of the first phase of the investigation was prepared and sent to the PRPs. The department stated that it did not consider the soil gas results to be positive proof of contamination, but that these results would be used to direct the next phase of the investigation. The parties were given the opportunity to take over the investigation of the contamination on their individual properties. If they chose to do this, the PRPs were to coordinate their work with the department, have all workplans approved prior to implementation, and submit copies of all reports of their work. If the PRPs chose not to proceed, the DEQ would continue the work with the same goal of determining the RPs and then recovering the state's costs.

With the results of the soil gas survey in hand, the department received a considerably different response from the PRPs. After further discussions, the six sites were handled as follows:

1. The livestock yard was no longer considered a PRP.
2. The card lock station agreed to hire a firm to perform the next phase of the investigation. Their goal was to determine if there was any petroleum contamination on their property and, if so, if that contamination originated from their facility and was connected to the plume affecting the RV Park.
3. The owner of the abandoned gas station had declared bankruptcy and moved out of state. He continued to claim that a leak never occurred

Table 5.1.   Tank Inventory for Potential Sources of the Petroleum Contamination at the Recreational Vehicle Park

Location	Tank Type	Capacity (gal.)	Fuel Type	Approximate Age (yr.)	Comments
PRP 1	AST[a]	500	Diesel	Unknown	In use
	AST	500	Gasoline	Unknown	In use
	UST[b]	500	Unknown	> 11	Removed about 1975
	UST	500	Unknown	> 11	Removed about 1975
	UST	3000	Gasoline	> 11	Disposition unknown
PRP 2	AST	19500	Gasoline	Unknown	In use
	AST	19500	Diesel	Unknown	In use
	AST	9500/	Unleaded/	Unknown	Both compartments
		9500	Stove Oil		in use
PRP 3[c]	AST	15000	Gasoline	> 10	Removed in 1987
	AST	15000	Diesel	> 10	Removed in 1987
	AST	15000	Unleaded	> 10	Removed in 1987
	AST	15000	Unknown	> 10	Removed in 1987
PRP 4	UST	300	Gasoline	> 20	Last used in 1981
	UST	300	Gasoline	> 20	Last used in 1981
	UST	250	Lube Oil	> 20	Last used in 1981
PRP 5	UST	300	Waste Oil	3	In use
	UST	2500	Gasoline	19	Not in use
	UST	1500	Diesel	19	Not in use
	UST	500	Heating Oil	Unknown	Heating Oil #2
	UST	1000	Heating Oil	Unknown	Heating Oil #2
PRP 6	AST	285	Gasoline	1	In use
	AST	285	Diesel	1	In use
	UST	300	Heating Oil	14	In use
	UST	1000	Gasoline	Unknown	Removed in 1984
	UST	500	Unleaded	Unknown	Removed in 1984
	UST	500	Diesel	Unknown	Removed in 1984
	AST	1000	Waste Oil	> 20	In use

[a]Aboveground Storage Tank.
[b]Underground Storage Tank.
[c]Conflicting reports were also received about possible USTs located on this property.

while he owned the station and that any contamination on the property must be from the previous owner. The department decided to task one of the state contractors to perform the next phase of the work at this site. That phase would involve not only an investigation into the magnitude and extent of contamination, but also into previous owners and other PRPs for the abandoned gas station property.

4. The railroad no longer needed their tanks and agreed to decommission them and perform an investigation at that time.

5. A release of fuel from one of the diesel tanks on the BLM site occurred after the soil gas investigation. The BLM agreed that their investigation and cleanup of this release would include information and data to confirm or refute their potential connection to the RV Park plume.

6. The agrichemical company agreed to perform the next phase of the investigation on their property. Because some of the drinking water samples from downgradient homes showed the presence of agricultural chemicals, the department required the investigation to include analyses for 1,2-dichloropropane and other agricultural chemicals as well as for contamination from petroleum products.

Project Difficulties

As work on this site continued, problems were encountered. First of all, the work did not proceed very efficiently since the rapidly increasing number of reported releases quickly reduced the ability of the staff to respond in a timely fashion. The Oregon UST Cleanup program is coordinated through the DEQ main office in Portland. However, the release reports and field activities are carried out primarily in five regional offices around the state. Although all five regional offices eventually hired a UST Cleanup staff person, the region in which this site is located was one of the last to hire their staff. Also, with only one UST Cleanup staff person initially allocated to each region, some staff members were responsible for over 200 reported releases. It is not surprising that reports submitted to the department often sat for many months before they could be reviewed and the next stages of work approved.

Because of the varied sources of contamination, not all of the sites around the RV Park could be handled with the same set of cleanup rules. Since the department's Cleanup Rules for Leaking Petroleum UST Systems were based on comparable federal rules, they only covered releases from regulated petroleum tanks. Therefore, although a gasoline or diesel release from a UST would be covered by these rules, the same product released from an aboveground storage tank (AST) would not be covered. Releases from ASTs would require cleanup under the more rigorous Environmental Cleanup Rules or under an older and less clearly defined set of Spill Response Rules. Cleanup was further complicated by the presence of the 1,2-DCP in the shallow groundwater, and a subsequent RCRA inspection of the agrichemical facility initiated by the department's Hazardous and Solid Waste (HSW) Division.

Even at those locations where the tanks are regulated and work has begun under a well-defined set of rules, progress can be delayed by the requirements of other programs. Emissions to the air from any planned vapor extraction systems or air stripping towers are regulated by the Air Quality (AQ) Division. AQ Rules require the submission of a Notice of Construction for such systems. Routine procedures for this notice would begin with the RP submitting a completed application to AQ in Portland. The notice is then forwarded to the appropriate regional office for review and approval. The regional office then returns it to AQ in Portland. The turnaround time for this process is typically several weeks. This does not include the time required for the RP to prepare the application and accompanying plans.

Discharges to waters of the state are covered by the Water Quality (WQ) Division and also require a permit. For sites at which the groundwater has been contaminated, proposed discharges may be associated with pump-and-treat systems installed for the purpose of long-term remediation. In many cases, however, the only discharge necessary from a UST site may be for the purpose of pumping relatively clean water from a tank excavation so that soil samples can be collected or new tanks can be installed. As with AQ permits, WQ permits also require significant paper work and time to obtain. An NPDES permit application may require as long as three months to process. Depending upon the volume of water, the level of contamination in the water, and the location of the site, the RP may be able to avoid the NPDES permit by having the water pumped out and hauled to a fuel recycler, or by getting a permit from a publicly owned treatment works to discharge to a sanitary sewer.

Almost all activities at UST sites, from routine tank removals to complex remediations, will involve excavation of contaminated soils. The disposal of such soils is controlled by HSW. In the past these soils were routinely taken to local landfills. However, as landfill operators became more aware of their potential liabilities for accepting petroleum contaminated soils, fewer landfills agreed to take them. Also, a new rule which went into effect in Oregon on January 1, 1991, further restricted the disposal of such soils. As a result, many gas stations around the state now find themselves stuck with piles of petroleum-contaminated soils after other cleanup activities have been completed.

Disposal problems have been further complicated by the adoption of EPA's new Toxicity Characteristic Leaching Procedure (TCLP) requirements which went into effect on September 25, 1990. In implementing the new requirements, the EPA has granted an interim exemption for the cleanup of petroleum contamination due to releases from tanks regulated under the 1984 amendments to RCRA. Therefore, petroleum-contaminated soil or water due to a release from a regulated UST would not have to meet the TCLP requirements. However, identically contaminated soil or water due to a release from an AST would be required to meet the TCLP requirements or be handled as a hazardous waste. This is a classic example of a regulatory decision that has no technical basis nor merit, and which does nothing to improve the regulated community's impression of bureaucracy. Since the exemption for regulated tanks is only an interim measure while the EPA considers the impact, it is possible that they may decide to require all petroleum contamination to meet the new TCLP requirements. If that should happen, most petroleum cleanups would be severely hampered and considerably more expensive.

Program Improvements

Because the delays discussed above were restricting the ability to efficiently handle petroleum cleanups without providing added protection to public health, safety and welfare and the environment of the state, the Oregon UST program looked for ways to streamline procedures. The first improvement was the development and adoption of Numeric Soil Cleanup Levels for Motor Fuel and Heating Oil (OAR 340-122-305 through 360, the ''soil matrix'').[4] The soil matrix rules established specific soil cleanup levels for petroleum contaminants. The levels are based on both the characteristics of the site and the nature of the product. Establishment of soil cleanup levels was an important goal since it allowed cleanup contractors to make decisions about their sites without having to consult the DEQ.

Although cleanup levels are an important goal, the soil matrix rules have an even more important purpose: they are designed to reduce the amount of staff time spent overseeing cleanup activities where petroleum contamination is restricted to the soils at the site. The UST cleanup rules originally adopted by the department were based on the federal regulations and required initial response, abatement measures and investigation followed by preparation of a corrective action plan (CAP). Staff time was required to fill out the initial release report, review the prescribed 20-day and 45-day site reports, and review and approve CAPs, progress reports, and final reports. Although this may be a practical approach for moderate to large releases, this is overkill for contamination found at most routine tank decommissionings and other minor releases where contamination is restricted to onsite soils. The soil matrix provides a ''fast track'' approach in which the rules define exactly how a site must be evaluated and the cleanup level determined, designates where and how samples are to be collected and analyzed, and specifies how results are to be evaluated and reported. Figure 5.3 shows where the matrix rules fit in with the previously adopted UST cleanup rules. For most soil matrix sites, staff involvement has been reduced to filling out the initial release report, reviewing the matrix report and issuing the final closeout letter.

In addition to the soil matrix rules, the department has initiated a licensing program for contractors engaged in tank-related activities. Contractors are required to take and pass qualifying examinations and obtain a license in order to perform tank installation or retrofitting, tank decommissioning, cathodic protection installation, and tank tightness testing. To this list the department has recently added a licensing requirement for contractors performing soil matrix cleanups. It is hoped that having licensed soil matrix cleanup contractors will further improve the quality of the work done at these sites and offset the department's reduced involvement. Oregon is also working with the other states in Region 10 in order to provide licensing reciprocity for comparable activities throughout the region.

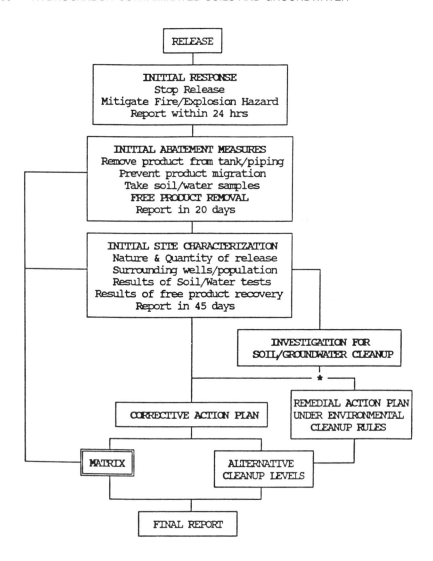

* Based on the magnitude or complexity of the release, the Director may
require the investigation and cleanup to follow OAR 340-122-010 through
110 (the Environmental Cleanup Rules).

Figure 5.3.   Flowchart of the Oregon UST Cleanup Rules.

With respect to the delays resulting from permit requirements of other state
programs, the UST Cleanup staff have begun to coordinate efforts with the other
applicable divisions (WQ, AQ, and HSW) to explore ways to expedite petroleum
cleanup activities while still ensuring that public health, safety and welfare, and
the environment are protected. In most cases, the need for permits for certain
activities associated with petroleum cleanups is not a point of dispute. All parties

recognize the usefulness of permits for specifying acceptable discharge limits, and, even more importantly, for providing the department with the power to enforce violations of the permit. The main topic of debate is trying to agree on ways to modify rules and procedures so that permits can be issued rapidly and cleanup activities can proceed with little delay and reduce the further spread of contamination.

Although the permit coordination effort has only begun, some progress has already been made. Two alternatives to the NPDES permits are currently being used for UST cleanup activities. A special permit is available for discharges that will be completed within less than 60 days. These are typically used for pumping accumulated water from tank pits or other short-term discharges. Special permits can generally be issued in about 10 days. Longer term discharges can now be permitted under a general permit developed specifically for UST cleanup work. General permits may take up to three weeks to process. Both permits have associated fees.

Disposal of petroleum-contaminated soil is still a major problem in Oregon. To provide some assistance, the department has established a permit addendum which can be issued to a regulated tank/facility permit holder and used for allowing a one-time onsite or offsite soil treatment program. This is typically used for onsite aeration of gasoline-contaminated soils. Unfortunately, such permits are not yet available for contaminated soils associated with unregulated tanks. More help may be available in the future from private industry. Several firms in the state are in the process of developing and obtaining permits for mobile thermal treatment units. These will be used to reduce or eliminate petroleum levels in soils so that the soils may be reused rather than disposed in landfills.

Future Proposals

Though recent progress is encouraging, the program still has a long way to go in streamlining the UST cleanup process. Toward this end, work is continuing in the following areas:

1. Groundwater cleanup rules for petroleum-related contaminants are being prepared. As mentioned earlier, the soil matrix rules have been very beneficial since they provide RPs with specific target cleanup levels. The regulated community has commended the department on its efforts in this area and several other state UST programs have expressed an interest in developing a similar approach. It is hoped that the development of rules for assessing petroleum-related groundwater contamination and providing numeric groundwater cleanup levels will further streamline petroleum cleanup activities in the state.

2. More progress needs to be made on the matter of necessary permits. The department's Northwest Regional Office UST staff have recently prepared a thorough overview of those areas where UST cleanup activities overlap with existing WQ, AQ, and HSW programs. They have

also made specific suggestions for improving the efficiency of UST cleanup permitting. One goal is to eliminate much of the time-wasting paper shuffling by allowing regional project managers to issue certain types of permits for UST cleanup activities. Because of the possible need for a number of rule amendments, such changes will not occur quickly. However, the cooperative efforts between the divisions are continuing and there is no reason to suspect that more progress cannot be made in this area.

3. The USTCS is considering a proposal to broaden the scope of its authority to include all petroleum cleanups. If this can be accomplished, a consistent approach would be used for handling petroleum contamination regardless of whether the source was a regulated UST, an unregulated AST or simply a spill from an overturned tank truck or railcar. This makes sense from a technical point of view and acknowledges the fact that UST cleanup staff are already having to deal with an increasing number of unregulated tanks as more bulk storage facilities are being closed down and cleaned up.

4. In addition to petroleum contamination, the USTCS must also deal with incidental nonpetroleum contamination. Such contaminants are commonly found during the cleanup of releases from waste oil tanks. This effort will be made by coordination with the division's new Voluntary Cleanup Section (VCS), which was recently created to handle voluntary hazardous substance investigations and cleanups. Part of the work of VCS will be to develop soil cleanup standards and rules similar to the UST cleanup soil matrix. These standards will be for chlorinated hydrocarbons, heavy metals, and other common environmental contaminants. When such standards are adopted, the USTCS will be arranging with VCS to apply the same standards for the incidental nonpetroleum contamination found at USTC sites.

5. Finally, it is recognized that not all changes that help the program come in the form of major rule packages or agreements with other divisions. The department has worked with EPA to flowchart tank decommissioning and petroleum cleanup activities in order to look for ways to streamline some of the routine activities. The department also schedules regular meetings at which all headquarters and regional office staff get together and discuss problem areas and share solutions. Such meetings also serve the purpose of providing more consistency in how the program is run in the different regional offices.

The development and implementation of these ideas in conjunction with previously made program changes should result in considerable improvement in the ability of the Oregon UST cleanup staff to handle an increasing number of petroleum contaminated sites in a manner that is both efficient and technically defensible while still providing the necessary protection to both the public and the environment.

REFERENCES

1. United States Environmental Protection Agency, Office of Underground Storage Tanks, *Tank Tour,* March 1989.
2 Oregon Administrative Rules 340-122-010 through 340-122-110, Environmental Cleanup Rules.
3. Oregon Administrative Rules 340-122-205 through 340-122-260, Cleanup Rules for Leaking Petroleum UST Systems.
4. Oregon Administrative Rules 340-122-305 through 340-122-360, Numeric Soil Cleanup Levels for Motor Fuel and Heating Oil.

CHAPTER 6

Evaluation of the Use of the Toxicity Characteristic in Cleanup Standard Selection

J. David Thomas, Radian Corporation, Washington, D.C.

One of the most contentious issues in site remediation today is the choice of cleanup goals and standards. For some sites, the choice of one remedial technology versus another can result in increased expenditures of orders of magnitude. In such a high stakes situation, it is in the best interests of all parties involved to fully understand the derivation of, and implications of, various potential cleanup level candidates. Two relatively new schemes that are, and increasingly will be, used to identify contaminant concentrations defining remediation goals and standards are the recently promulgated Resource Conservation and Recovery Act hazardous waste Toxicity Characteristic and the proposed Corrective Action, Action Levels.

This chapter will briefly compare and contrast the basic philosophies of the two principal, national, remediation programs in place today—the Comprehensive Environmental Response, Compensation, and Liability Act (CERCLA or Superfund) program and the Resource Conservation and Recovery Act (RCRA) Corrective Action program. Next, the background and basis for the RCRA Toxicity Characteristic regulation, promulgated approximately one year ago will be discussed. Third, I will review the background and basis for the Corrective Action, Action Levels proposed in July of 1990. Finally, I will discuss the efficacy of using the Toxicity Characteristic and the Corrective Action, Action Levels as the basis for identifying cleanup goals for the remediation of waste sites under the two programs.

THE CERCLA AND RCRA REMEDIATION PROGRAMS

Today, two major environmental regulatory programs deal with site remediation: the Superfund program and the RCRA Corrective Action program.

CERCLA

The Superfund program, enacted in 1980 and extensively revised in 1986, principally addresses remediation of abandoned or inactive sites containing hazardous substances. Candidate sites are identified for the program and scored for their potential hazard using a Hazard Ranking System (HRS). Some 25,000 candidate sites have been identified thus far in the program. Of these, approximately 2,000 sites have scored sufficiently high in the ranking system to be placed on the National Priority List (NPL) designating the site as posing a significant potential hazard to human health and the environment.

Originally, CERCLA did not specifically address cleanup standards, except to direct the Environmental Protection Agency (EPA) to select remedial actions that (1) complied with the National Contingency Plan, at the time unpromulgated, and that (2) were cost-effective and balanced the need for protection of public health and the environment against the availability of money from the Superfund Trust Fund to respond to other sites. Unhappy with both the pace and the level of remediation in the first years of the program, Congress in 1986 added Section 121 to CERCLA, establishing requirements for selecting case-by-case cleanup goals and standards for the remediation of NPL sites. Under Section 121, each site is to be remediated so that contaminants of concern are reduced to levels meeting "legally applicable or relevant and appropriate requirements" (ARARs). Legally applicable standards are those federal or state standards that directly apply to the site or the remedial activity, e.g., land disposal restriction standards for waste to be re-land disposed. Relevant and appropriate requirements are those that are not legally applicable to the site or remedial activity, but which establish "safe" concentrations in applicable media, e.g., state water quality criteria. The statute specifically lists the following federal statutes that might contain applicable, or relevant and appropriate, standards: Toxic Substances Control Act (TSCA), Safe Drinking Water Act, Clean Air Act (CAA), Clean Water Act (CWA), Marine Protection, Research, and Sanctuaries Act, and RCRA. In addition, cleanup decisions are required to consider:

- the goals and requirements of RCRA;
- the uncertainties of land disposal;
- the persistence, mobility, and tendency to bioaccumulate of the substances of concern;
- the short- and long-term threats to human health;
- the long-term maintenance costs;

- the potential for future cleanup costs if the remedy were to fail; and
- the potential threat to health and the environment associated with excavation, transportation, redisposal, or containment.

CERCLA contains a strong preference for permanent remedies that destroy or remove the contaminants of concern as opposed to those remedies that merely contain the contamination, going so far as to require periodic re-evaluation of nonpermanent remedies. Because most NPL sites are inactive or abandoned, the clear public policy is to require, where feasible, remediations that result in allowing full, unrestricted use of the property without the need for continuing custodial care.

Finally, Section 121 requires that remedial actions at least attain any directly applicable groundwater or surface water Safe Drinking Water Act Maximum Contaminant Level Goals and CWA water quality criteria.

RCRA

RCRA Corrective Action addresses releases of hazardous constituents from RCRA Subtitle C interim status or permitted facilities. Thus, unlike CERCLA, which principally deals with inactive or abandoned sites, RCRA deals with sites that have continuing operations. An important limitation to Corrective Action is that it does not reach to ongoing manufacturing activities at the facility. Prior to enactment of the 1984 Hazardous and Solid Waste Amendments to RCRA, EPA's corrective action authority at hazardous waste facilities was limited to:

- Section 7003 orders to correct imminent and substantial endangerment situations;
- Section 3013 authority to compel investigations where the presence of hazardous wastes may pose a substantial hazard; and
- 40 CFR Part 264F requiring permitted units to address known releases of hazard wastes and constituents as a condition of the permit.

The 1984 Amendments substantially expanded the Corrective Action program by adding:

- Section 3004(u) requiring that any hazardous waste permit issued address corrective action for releases of hazardous wastes or constituents from any active or inactive solid waste management unit at the facility;
- Section 3004(v) authorizing EPA to require corrective action beyond the boundary of the facility; and
- Section 3008(h) authorizing EPA to require corrective action at interim status facilities.

The EPA proposed regulations to implement the 1984 Amendments corrective action program on July 27, 1990 (see 55 FR 30798).

This proposal, together with the National Contingency Plan (see 55 FR 8666, March 8, 1990) define the Agency's overall approach to the cleanup of environmental contamination resulting from the management of solid and hazardous waste. The EPA estimates that some 5,700 facilities, with some 80,000 solid waste management units, are currently in the RCRA Subtitle C program and thus potentially subject to corrective action.

There are certain overlaps between the CERCLA and RCRA programs. The EPA states in the July preamble that "Substantive provisions of the (corrective action) rule, when promulgated, generally will be applicable to response actions under CERCLA. . . . These provisions may also be 'relevant and appropriate' to other CERCLA response actions (55 FR 30802)." There are also important distinctions. For example, the agency further sets the basic premise of the corrective action program by recognizing that "because of the wide variety of sites likely to be subject to corrective action, EPA believes that a flexible approach based on site-specific analyses, is necessary. . . . Therefore the program has to allow significant latitude to the decisionmaker in structuring the process, selecting the remedy, and setting cleanup standards appropriate to the specifics of the situation (55 FR 30802)."

In further recognition of the distinction between CERCLA NPL (uncontrolled) sites and corrective action (ongoing activity) sites, the agency expressly states that cleanup standards should be based on "all actual and reasonably expected uses" (55 FR 30803) of the site. Conditional remedies are also sanctioned in contrast to the CERCLA preference for permanent remedies.

The remediation goal for corrective action "is, to the extent practicable, to eliminate significant releases from solid waste management units that pose threats to human health and the environment, and to clean up contaminated media to a level consistent with reasonably expected, as well as current, uses. The timing for reaching this goal will depend on a variety of factors, such as the complexity of the action, the immediacy of the threat, the facility's priority for corrective action, and the financial viability of the owner/operator (55 FR 30804)."

The Corrective Action program, like the CERCLA program, must address the issue of cleanup goals. Instead of the ARAR approach of CERCLA, the EPA has proposed that "different clean-up levels will be appropriate in different situations, and that the levels are best established as part of the remedy selection process. . . . To be protective of human health, EPA believes that clean-up levels for carcinogens must be equal to or below an upperbound excess lifetime cancer risk level of 1 in 10,000 (1×10^{-5}). . . . For non-carcinogens, clean-up levels would be set at a level at which adverse effects would not be expected to occur (55 FR 30804)."

Consistent with this cleanup level philosophy, de facto cleanup levels are established for the program in the form of Corrective Action, Action levels. These Action Levels are media specific, constituent specific concentrations. Interestingly, the proposal places the Action Level in an appendix, not in the rule itself. This should not be construed to assume the Action Levels do not have regulatory effect, similar to Appendix VIII or 40 CFR Part 261. Sites that have concentrations exceeding the Action Levels must generally be remediated; sites with concentrations below the Action Levels are "safe." Action Levels thus serve as a rebuttable presumption of cleanup levels. It is important to note that once a determination is made that remediation is required, generally because an Action Level is exceeded, remedial action does not stop once the Action Level is met. Instead, all remedial action alternatives that meet or exceed the Action Level are evaluated, and selection is based on a number of factors, including cost, feasibility, etc.

The basis for establishing the Corrective Action, Action Levels is discussed in more detail later.

Once a site is listed on the NPL pursuant to CERCLA, or is determined to potentially require corrective action pursuant to RCRA, the processes of site investigation, remedial alternative evaluation, and ultimate remediation are similar. (However, while the process may be similar, the ultimate remedy selected may be different, given the differing bases and goals of the two programs). Under CERCLA, EPA or an identified potentially responsible party (PRP) must conduct a Remedial Investigation/Feasibility Study (RI/FS). Under RCRA the process is termed a Facility Remedial Assessment/Facility Remedial Investigation/ Corrective Measures Study. In either case the process is intended to:

1. Characterize the source of contamination by completing a:
   - waste characterization
   - unit characterization
   - magnitude of any known releases, and
   - timing of any known releases;

2. Characterize the environmental setting by completing a:
   - soil characterization
   - groundwater characterization
   - air characterization, and
   - identification of potentially exposed populations;

3. Characterize any releases by designing and implementing monitoring and identification programs;

4. Identify remediation cleanup goals and standards consistent with applicable regulatory requirements;

5. Characterize potential remedial actions by:
   - selecting potential remedial alternatives that meet or exceed the cleanup goals and standards
   - studying the feasibility of selected potential alternatives, and
   - designing the selected remediation alternative;

6. Implementation of the selected remedial actions by:
   - installing or constructing the selected alternative
   - operating the selected alternative
   - monitoring the progress of the alternative in abating releases and removing contaminants of concern to the remediation levels required.

THE TOXICITY CHARACTERISTIC

RCRA, Section 3001, authorizes the Administrator of the Environmental Protection Agency to identify by listing of particular wastes or by characteristic those solid wastes that may:

"pose a substantial present or potential hazard to human health and the environment when improperly treated, stored, transported, disposed, or otherwise managed." (Section 1004(5) definition of hazardous waste).

The hazardous waste characteristics promulgated by the EPA define broad classes of wastes that are regulated as hazardous due to inherent properties. In the May 19, 1980 rules originally establishing hazardous waste regulations, the EPA announced two basic criteria for identifying hazardous waste characteristics: (1) the characteristic should be capable of being defined in terms of physical, chemical, or other properties which cause the waste to meet the statutory definition; and (2) the properties defining the characteristic must be measurable by standardized testing protocols or reasonably detected by generators through their knowledge of the waste (see 40 CFR 260.10).

The approach adopted by the EPA is thus to establish properties of wastes that result in harm to human health and the environment if the waste is mismanaged. The agency then establishes test methods and regulatory levels for each identified characteristic property. Solid wastes that exceed the regulatory level for any characteristic property are hazardous wastes.

The regulatory levels for characteristics are intended to provide a high degree of certainty that wastes exceeding those levels would pose an unacceptable hazard if improperly managed. The EPA thus views the characteristics as a first line of hazardous waste definition.

The EPA initially promulgated four characteristics: ignitability, corrosivity, reactivity, and the Extraction Procedure Toxicity Characteristic (EPTC), the predecessor to the Toxicity Characteristic. The EPTC established regulatory levels for 14 individual constituents in leachate. The EPTC assumed codisposal of the waste of concern in a municipal landfill. Regulatory levels were established at 100 times a recognized chronic health based number, e.g., the then National Interim Primary Drinking Water Standard for the eight identified metals. The 100 times factor was an estimated dilution and attenuation factor that estimated the dilution and attenuation of the toxic constituents in the wastes as they travel through the subsurface from the point of leachate generation (i.e., the landfill) to the point of human or environmental exposure (i.e., a drinking water well). In the final rule, the EPA noted that the 100 times factor was somewhat arbitrary in that there were few empirical data on which to base the factor (see 45 FR 33084, May 19, 1980). The EPA stated at that time that it would adjust the factor if future studies (undesignated) indicated that another factor was more appropriate.

In the early 1980s, Congress and others were unhappy with what they perceived as the slow pace of hazardous waste regulation by the EPA. This displeasure manifested itself in the 1984 Hazardous and Solid Waste Amendments to RCRA. The Amendments required the EPA to undertake a major overhaul of the entire RCRA regulatory program. Among other things, the amendments added a new Section 3001(g) specifically requiring the EPA to re-examine the Extraction Procedure as an accurate predictor of the leaching potential of wastes and to make appropriate revisions. (The legislative history makes it clear that Congress intended for the EPA to promulgate a more aggressive leaching procedure and one specifically designed with organic constituents in mind).

New Section 3001(h) required the EPA to promulgate additional characteristics, including toxic levels of organic constituents.

In response, the EPA on June 13, 1986 (see 51 FR 21648), proposed to expand the EPTC to 52 compounds, including the original 14 substances plus 38 new organic constituents. The June 13 proposal used a subsurface fate and transport model to determine compound-specific dilution and attenuation factors. The 38 new toxicants included all the Appendix VIII hazardous constituents for which appropriate drinking water chronic toxicity reference levels were available and for which there existed adequate fate and transport data to establish a compound specific factor. Published reference doses for noncarcinogens and risk-specific doses for carcinogens were used.

Between the June 13 proposal and the March final rule, the EPA published a total of five additional notices relating to the Toxicity Characteristic rulemaking. (See 51 FR 40572, November 7, 1986; 52 FR 18583, May 18, 1987; 53 FR 18024, May 19, 1988; 53 FR 18792, May 24, 1988; and 53 FR 28892, August 1, 1988).

The final rule replaces the EPTC and adds 25 of the original 38 organic constituents proposed. The Toxicity Characteristic uses a new leaching procedure, the Toxicity Characteristic Leaching Procedure (TCLP) in place of the Extraction Procedure. It retains the generic dilution and attenuation factor of 100 in lieu of the originally proposed constituent specific factors. Thus the regulatory levels for the original 14 EPTC constituents remain the same under the Toxicity Characteristic. However, because the TCLP replaces the Extraction Procedure, wastes that were not EPTC hazardous may now be regulated, and conversely, wastes that exceeded regulatory levels when leached by the Extraction Procedure may not be hazardous when the TCLP is substituted.

In summary, hazardous wastes are that subset of solid wastes that pose an unacceptable risk to human health and the environment when improperly managed. The Toxicity Characteristic establishes constituent concentration levels for eight metals, four pesticides, two herbicides, and 28 organics. Wastes which leach one

or more of these constituents above the established levels must be managed as
hazardous wastes. The levels are established by multiplying the drinking water
chronic toxicity level by 100. EPA intends the Toxicity Characteristic to be a
first order hazardous waste definition, capturing those wastes that clearly should
be in the system.

THE CORRECTIVE ACTION, ACTION LEVELS

As discussed previously, the Corrective Action, Action Levels are a series of
media-specific, constituent-specific concentrations established largely to define
those Solid Waste Management Units requiring remediation. The Action Levels
are set at concentrations determined to be fully protective of human health and
the environment. Thus, they are sometimes termed "eatable, drinkable standards,"
in reference to the fact that they are set at levels consistent with ingestion or inhalation without adverse effect. Where possible, Action Levels are based on
promulgated standards. Such specific promulgated standards include:

- For groundwater—Maximum Contaminant Levels (MCLs) pursuant to
  the Safe Drinking Water Act (See 40 CFR Part 141B);
- For surface water—Numerical State or Federal Water Quality Standards
  pursuant to the Clean Water Act (See 40 CFR Part 131); or MCLs.

For air and soils, no promulgated standards are available. For these media and
in other cases, Action Levels are established on the basis of the following general criteria:

1. The concentration is derived in a manner consistent with the principles
   and procedures set forth in agency guidelines for assessing the health
   risks of environmental pollutants (see 51 FR 33992, 34006, 34014,
   34028, September 24, 1986);
2. Toxicology studies used to derive action levels are scientifically valid
   and conducted in accordance with the Good Laboratory Practice Standards (40 CFR 792);
3. Concentrations used as action levels, for carcinogens, must be associated with a 1×10^{-6} upperbound excess cancer risk for Class A and
   B carcinogens, and a 1×10^{-5} upperbound excess cancer risk for Class
   C carcinogens (e.g., the Unit Cancer Risk level); and
4. For systemic toxicants, the action level must be a concentration to which
   the human population, including sensitive subgroups, could be exposed
   on a daily basis without appreciable risk of adverse effects during a lifetime. (e.g., the Reference Dose or RfD).

A complete list of Action Levels is proposed as Appendix A of the July 1990
proposal.

WHEN DO I USE WHICH ONE AND WHY?

The Toxicity Characteristic establishes a concentration threshold in leachate above which solid wastes are regulated as hazardous wastes and, conversely, below which there is protection of human health and the environment even if the material is plausibly mismanaged. However, it is important to remember that the Toxicity Characteristic concentrations are derived by multiplying "eatable, drinkable" numbers times a standard dilution and attenuation factor of 100. The dilution and attenuation factor de facto assumes that there is no direct contact with the material. The proposed Corrective Action, Action Levels, on the other hand, establish a concentration threshold below which there is no anticipated human health or environmental concern regardless of the disposition of the material. In fact, it is argued that the material is safe for lifetime direct contact, ingestion, or inhalation. Thus, clearly, the proposed Corrective Action, Action Levels afford a substantially higher level of protection than that afforded by the Toxicity Characteristic. In practice, for most constituents included in both regulations, the Toxicity Characteristic concentration is 100 times that of the Groundwater Action Level.

For CERCLA NPL sites, the Toxicity Characteristic is not generally useful as a potential cleanup standard. It is not legally applicable as a standard, in that it merely defines under which regulatory structure the material must be managed [i.e., Subtitle C (hazardous waste) or Subtitle D (solid waste) of RCRA]. However, once defined as a hazardous waste, a number of standards may be legally applicable, such as Best Demonstrated Available Technology Land Disposal Restriction levels, Incineration Destruction and Removal Efficiency requirements, and Groundwater Protection Standards. Even if the material being managed does not meet either a hazardous waste listing or a characteristic, such derivative RCRA regulatory requirements may still be relevant and appropriate requirements and standards.

For the same reasons, the Toxicity Characteristic is not useful for establishing cleanup goals and standards as a "relevant and appropriate requirement."

For RCRA Corrective Action sites, the Toxicity Characteristic is again not useful as a cleanup standard or goal. The Corrective Action program addresses releases from Solid Waste Management Units. If the release itself, or the source material resulting in the release, exceeds the concentrations set in the Toxicity Characteristic, the situation ceases to be subject to the Corrective Action portion of the permit program, and instead, is subject to the full panoply of the RCRA hazardous waste regulatory scheme.

The Corrective Action, Action Levels are proposed, but have not yet been finalized. (They are included in the EPA's existing corrective action guidance documents.) As such, they do not currently have any regulatory force, and therefore, cannot be either legally applicable standards or relevant and appropriate requirements pursuant to CERCLA. They may be considered in establishing cleanup goals and standards under the auspices of information "to be considered" in remediations. Once the Corrective Action Program regulations are promulgated, however, the Action Levels will clearly be "relevant and appropriate requirements." Because the agency intends to publish the Action Levels as an Appendix and not as formal standards, they will likely not be legally applicable standards.

Of course, for the Corrective Action Program, the Action Levels, if the EPA finalizes the July 1990 proposal, will generally become the basis for determining whether a particular unit requires remediation. As discussed previously, they will not be cleanup levels per se. Instead, they will likely become minimum cleanup goals. In general, remediation to at least the Action Levels will be required in the absence of demonstration that attaining such levels is not feasible.

In summary, the Toxicity Characteristic is one basis for defining wastes as hazardous wastes. It is a sweeping new regulation that greatly expands hazardous waste characteristics and which furthers the process of establishing constituent concentrations as a basis for hazardous waste regulation. The Toxicity Characteristic does not, however, add to the list of potential cleanup goals or standards for CERCLA NPL sites or RCRA Corrective Action sites. The Corrective Action, Action Levels, on the other hand, will, when promulgated, have an immediate and pervasive impact on both CERCLA and RCRA remediations. If Action Levels are established for all constituents in all media, they may well become de facto threshold, national cleanup standards for both programs.

PART III

Environmental Fate and Modeling

The Influence of Methanol in Gasoline Fuels on the Formation of Dissolved Plumes, and on the Fate and Natural Remediation of Methanol and BTEX Dissolved in Groundwater

J. F. Barker, C. E. Hubbard, L. A. Lemon, and **K. A. Vooro,** Waterloo Centre for Groundwater Research, University of Waterloo, Ontario, Canada

INTRODUCTION

The monoaromatic hydrocarbons in gasoline represent a significant potential source of groundwater contamination. They include benzene, toluene, ethylbenzene, and the three xylenes, m-, o- and p-, as a group termed BTEX. These constitute the most soluble, most mobile fraction of gasoline, and benzene especially has a very low standard in drinking water of 5 μg/L. Currently, many oxygen-bearing organics (oxygenates) such as alcohols and ethers are added to gasoline to boost octane and to reduce air pollution from combustion. Methanol, in particular a mixture of 85% methanol and 15% gasoline termed M85, has been proposed as a fuel in response to concern about urban air quality. Little is known about the impacts of these oxygenates on the groundwater environment.

Based on current experience, significant groundwater impacts can result from the spillage, leakage, and improper disposal of methanol fuels. The impact of methanol on groundwater quality is of considerable concern, given that methanol is miscible with water and so will likely occur at a high concentration in impacted groundwater. Another concern is that the methanol will exacerbate the impacts of BTEX. Three aspects of BTEX behavior in the subsurface are of particular concern with respect to the use of oxygenates in gasolines:

- the possible enhanced aqueous solubility of BTEX
- the possible enhanced mobility of BTEX dissolved in groundwater
- the possibility that the presence of methanol will decrease the natural bio-degradation of BTEX and will thus enhance the persistence of BTEX in groundwaters.

This chapter deals with the dissolution of gasoline hydrocarbons from methanol-containing fuels into groundwater and the subsequent mobility and persistence of methanol and BTEX in groundwaters. Laboratory and field experiments have been undertaken to address these concerns. Preliminary results are briefly summarized and discussed in this chapter. More detailed presentation and discussions are to be found in Barker et al.,[1] Poulsen et al.,[2] Hubbard (in prep.), and Vooro.[3]

Enhanced Aqueous Solubility of BTEX Due to Methanol in Gasoline

Stumm and Morgan,[4] among others, suggested that the aqueous solubility of a particular component of gasoline can be reliably predicted from the aqueous solubility of that pure component and the mole fraction of that component in the mixture, as:

$$S_i = Sp_i \times MF_i \qquad (1)$$

where the aqueous concentration of the ith component, S_i, of the organic phase is the product of the solubility of the ith component from a phase of pure i, Sp_i, times the mole fraction of the ith component in the gasoline mixture, MF_i. This approximation is probably reliable to within a factor of 2 where the solubility of the ith component in water is low.[5,6] Even the most soluble of the aromatic compounds—benzene—has a solubility of only about 1800 mg/L when present in pure form. Since benzene makes up less than 5% of the gasoline phase, the maximum aqueous benzene concentration associate with gasoline is less than 60 mg/L.

However, this solubility can be dramatically increased if a water-soluble cosolvent is present in the gasoline. Oxygenates are examples of potential cosolvents. Their water solubilities range from a few % (e.g., ethers) to complete miscibility with water (e.g., ethanol and methanol). These alcohols and other oxygen-containing organic compounds, particularly methyl-tert-butyl-ether (MTBE), are becoming more common in gasolines. This raises concerns that groundwater impacted by such gasolines could contain high BTEX concentrations due to a cosolvent effect.

One aspect that must be included in any model of enhanced solubility by cosolvents is the requirement to conserve mass. This becomes critical when discussing the solubility of hydrocarbons when the ratio of the hydrocarbon phase to water becomes less than about 1:20. When the water phase becomes relatively large, there is insufficient organic component to partition between the two phases and attain the maximum predicted aqueous concentration. Such a situation could result from the release of small volumes of gasoline, for example, into a large mass of water with rapid mass transfer of components from the gasoline to the water.

Therefore, we feel that a useful approach to the problem of estimating aqueous concentrations of components such as BTEX from complex mixtures such as gasoline, M85, and MTBE-containing gasoline is that presented by Maijanen et al.[7] and Shiu et al.[8] It treats the dissolution of the components of an organic mixture as attaining an equilibrium partitioning between aqueous and organic phases of specified volumes or volume ratios.

For fuels with high oxygenate content the BTEX contents are likely to be much lower. Applying Equation 1, the decrease in the BTEX content of the gasoline would tend to decrease the concentrations of BTEX in impacted groundwaters. However, the cosolvency effect of the oxygenate would alter the pure phase solubility term so as to make the BTEX more soluble in the impacted groundwaters (Figure 7.1, for a benzene- methanol-water system, for example). Poulsen et al.[2] demonstrate that the actual BTEX concentrations in the groundwaters impacted by oxygenate fuel spills can only be predicted when both the proportions of BTEX and oxygenates in the gasoline are known and the ratio of gasoline and impacted groundwater can be specified.

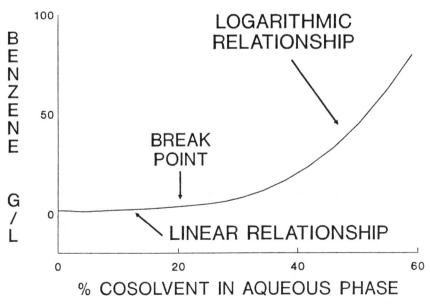

Figure 7.1.   The aqueous concentration of benzene as a function of the methanol content of the aqueous phase. Based on experimental data where benzene was equilibrated with various water-methanol mixtures (data from Poulsen et al.[2]).

For methanol, a cosolvent effect is evident, especially when the water phase contains more than 20% methanol. Figure 7.2 shows the aqueous benzene concentration as a function of methanol content of the water for the water-methanol-gasoline system (from Poulsen et al.[2]).

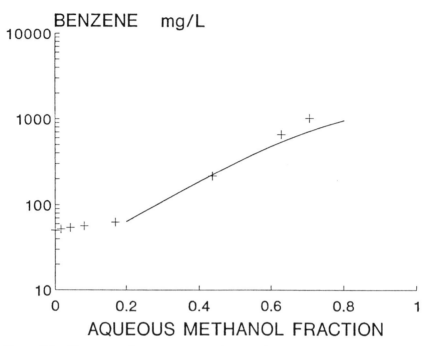

Figure 7.2.   The aqueous benzene concentrations as a function of aqueous methanol content. PS-6 gasoline was equilibrated with various water-methanol mixtures and the equilibrium concentrations reported. The line represents the calculated benzene concentrations in the logarithmic part of the curve (from Poulsen et al.[2]).

How can high aqueous methanol contents be created in impacted groundwaters? Two conditions must be met. First, the fuel must have a high methanol content. Currently,one of the fuels being considered for distribution is M85, which contains 85% methanol and 15% gasoline. This fuel would meet this first criteria. The second requirement is that the ratio of water to fuel attaining equilibrium in the subsurface be low. The groundwater is likely to contact the spilled fuel along a narrow interface, whose effective, equilibrating water and fuel volumes are unknown. Whether the second requirement can come about is speculative at present.

A preliminary column experiment was conducted to see if a cosolvent effect was possible.[3] M85 (10 cm³) was injected via a syringe into the central portion of sand within 2 cm of the influent end of a water-saturated column packed with Borden sand. The column was 9.6 cm long, with an inside diameter of 6.4 cm. The pore volume of the column was about 105 cm^3, assuming the sand had a porosity of 0.34 which is typical of Borden sand columns. A groundwater flow (linear velocity of about 250 cm/day) was re-established immediately after the M85 was injected. Figure 7.3 shows the effluent breakthrough curves (concentration in effluent versus effluent volume) for methanol, benzene, and total xylenes. Flow was interrupted for about 12 hours after 715 cm^3 of water had eluted, producing the slight, temporary increase in the effluent concentrations. It appears that the methanol did enhance the solubility of BTEX, producing the initial high benzene concentration when the aqueous methanol concentration was also high. Subsequently, the effluent BTEX concentrations declined to levels commonly associated with normal gasoline impacts (benzene < 20 mg/L). Within the 9 pore volumes flushed from the column, about 95% of the injected methanol was recovered, while only 78% of the benzene and 20% of the xylenes were recovered. Clearly, a residual of essentially pure gasoline has remained in the column. That gasoline residual is considerably different chemically from the gasoline introduced in the M85. It has lost much of the most water-soluble fraction. This experiment does suggest that a plume of contaminated groundwater emanating from an M85 spill may have a leading segment with very high methanol levels and much higher BTEX levels than encountered in conventional gasoline spills.

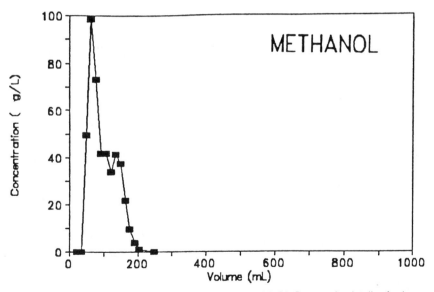

Figure 7.3a.   Effluent breakthrough curve for methanol (g/L). See text for details of column and its operation (from Vooro, in prep.).

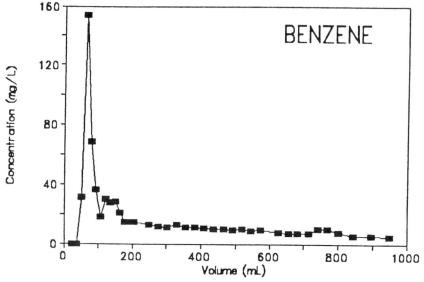

Figure 7.3b. Effluent breakthrough curve for benzene (mg/L). See text for details of column and its operation (from Vooro, in prep.).

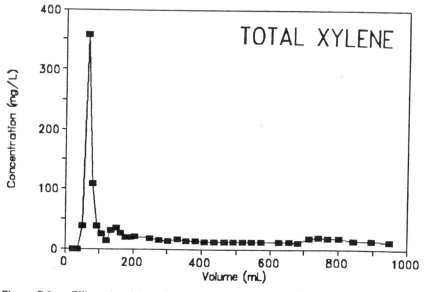

Figure 7.3c. Effluent breakthrough curve for total xylenes (mg/L). See text for details of column and its operation (from Vooro, in prep.).

The Mobility and Persistence of Methanol and BTEX

What is the subsequent fate of the dissolved gasoline components, especially methanol and BTEX? Methanol is miscible and so is anticipated to occur at higher concentrations than even BTEX in oxygenate/gasoline-impacted groundwaters. Highly soluble compounds have a low sorption potential, so methanol should be more mobile in groundwater than BTEX. Methanol, when present at high concentrations could decrease the sorptive retardation of BTEX via cosolvent effects which could lead to enhanced mobility of BTEX in groundwater. Methanol could be biodegraded in preference to BTEX and so could consume the oxygen required for BTEX biodegradation. Alternately, methanol could be toxic or inhibitory to BTEX-degraders.

A natural gradient injection experiment has been performed at CFB Borden to investigate the influence of methanol on the behavior of the dissolved BTEX in groundwater. The experimental zone is in a 5-meter lens of aerobic, uncontaminated groundwater which overlies a landfill-leachate plume. The site has been used for previous natural gradient experiments so the hydrogeology is well understood. The medium to fine sand has small-scale layering and heterogeneities. The groundwater flows essentially horizontally at an average velocity of 0.09 m/day.

Dissolved contaminants were injected below the water table to represent dissolved plumes which might emanate from subsurface contamination by normal gasoline (PS-6, supplied by API), gasoline with 15% MTBE, and M85. The three plumes were injected simultaneously, side-by-side, so that they would all encounter the same degree of aquifer heterogeneity, would flow at about the same rate and in the same biogeochemical environment. The MTBE plume will not be considered here. Injection took place on July 13, 1988 and groundwater monitoring continued for 476 days. Three long, thin plumes of contaminated groundwater with minimal lateral overlap were created (see Figure 7.4). The average injection concentrations in the gasoline and M85 plumes are given in Table 7.1.

Table 7.1. Average Concentrations (mg/L) of Injected Solutes

	PS-6	85% Methanol
Chloride	478	577
Methanol	—	7030
Benzene	7.2	7.7
Toluene	5.0	5.2
Ethylbenzene	0.8	0.8
p-Xylene	0.8	0.8
m-Xylene	2.0	2.1
0-Xylene	1.2	1.2
Total BTEX	17.0	17.8

The transport of the solutes in groundwater was monitored by sampling from a detailed piezometer network for the solutes and for dissolved oxygen. Six sampling "snapshots" were taken over the 476-day experiment with the collection and analysis of 850 to 1600 samples each time.

Figure 7.4 shows the distribution of chloride, methanol, benzene, and p-xylene at 6 and 476 days after injection. The contours are of vertically integrated concentration data and are shown only to indicate the position of the solute plumes. In detail the migration is more complex, reflecting the small-scale heterogeneities in the Borden aquifer. Clearly, solutes are migrating and dispersing, with longitudinal dispersion or spreading far exceeding transverse dispersion. Methanol is migrating at essentially the same rate as groundwater and the conservative solute, chloride. Benzene is somewhat retarded and p-xylene is more retarded, as anticipated. There does not appear to be any consistent variation of the apparent mobility of BTEX due to the presence of methanol. Perhaps much higher concentrations of methanol would have produced less retardation of BTEX.

The mass loss for each organic over time has been assessed but this analysis is being refined. Table 7.2 gives the preliminary estimates of total mass for each solute injected and in each plume at days 6, 42, 106, and 476 after injection. The chloride decline indicates either a poor approximation of the distribution or that some of the chloride mass lies outside the sampled network. In any event, the declines in the organics can still be appreciated. By day 476, chloride mass has declined to about 50% of that estimated to have been injected. All monoaromatics have declined to less that 20% of the initial mass; all except benzene have declined to less than 10% of their initial mass. Methanol essentially disappeared, with its mass having declined to about 0.02% of the initial mass.

Table 7.2.   Estimates of the Mass (g) of Each Solute Injected and in Each Plume at Selected Days After Injection

	Injected	Day 6	Day 42	Day 106	Day 476
METHANOL PLUME					
Chloride	1640	1200	790	1090	750
Methanol	19900	10900	7400	10300	3.1
Benzene	21.9	13.1	8.9	11.6	4.3
Toluene	14.6	8.5	6.3	6.3	0.08
m-Xylene	5.9	3.1	2.9	1.4	0.05
p-Xylene	2.3	1.3	1.1	0.64	0.29
PS-6 PLUME					
Chloride	1400	820	880	690	650
Benzene	20.4	13.8	13.4	8.1	2.1
Toluene	14.0	8.6	7.9	4.9	0.05
m-Xylene	5.8	3.5	3.7	1.6	0.004
p-Xylene	2.3	1.4	1.4	0.86	0.06

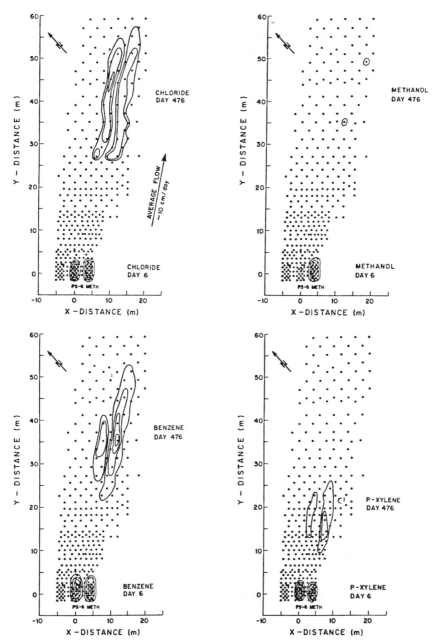

Figure 7.4. The vertically-averaged distribution of solutes in groundwater about 6 and 476 days after injection. Contours are shown only to indicate the approximate location of the bulk of the solute mass.

Although by day 476 most of the methanol had been biodegraded, the BTEX in the methanol plume were at higher relative concentrations than in the PS-6 plume. This suggests that methanol slowed the BTEX biodegradation rate compared to oxygenate-free groundwater. Laboratory experiments found no significant biodegradation of BTEX in the presence of 7000 mg/L methanol but some BTEX biodegradation was noted when the methanol concentration was only 1000 mg/L (Hubbard, in prep.). In the field experiment, dilution and different rates of migration bring about continuously lowering methanol concentration in waters containing BTEX. Thus, while initially methanol could be inhibiting BTEX biodegradation, lowering methanol concentrations could permit subsequent BTEX biodegradation. On the other hand, as methanol itself is biodegraded and oxygen is consumed, BTEX biodegradation could still be inhibited by lack of oxygen.

Biodegradation of only a small proportion of the 7000 mg/L of methanol present in the injected water would consume all the available dissolved oxygen, thus inhibiting aerobic biodegradation of BTEX. In fact, it appears that the removal of methanol aerobically would require much more oxygen than is available in the methanol-impacted groundwater. The possibility of anaerobic methanol biodegradation needs to be considered.

It is tentatively concluded that the mechanism for greater BTEX persistence with methanol being present is the initial inhibition of BTEX biodegradation by high methanol concentrations and then inhibition due to removal of oxygen by methanol biodegradation. An examination of the distribution of dissolved oxygen in the M85 plume is under way and should clarify this issue.

CONCLUSIONS

Overall, the impact of methanol in gasolines on groundwater contamination is serious and complex. The replacement of BTEX by methanol will provide a lower total load of these contaminants to the groundwater and so should lessen their impact. Enhanced solubility of BTEX will likely be found only with high methanol gasolines where the effective gasoline:water ratio in contact is relatively high. Once dissolved, methanol will be very mobile, but not necessarily persistent. Methanol appears to biodegrade, but in so doing it appears to enhance the persistence of BTEX. Those dealing with groundwater contamination will be faced with a much more complex groundwater problem than simple gasoline has produced in the past. Much more research and the rapid communication of the findings are required to prepare the appropriate response.

ACKNOWLEDGMENTS

This research has been supported by the American Petroleum Institute, the Canadian Petroleum Products Institute (formerly PACE), and the University Research Incentive Fund, Ontario Ministry of Colleges of Universities. The opinions expressed are those of the authors, however. The encouragement and advice of the members of both the Soil/Groundwater Technical Task Force (API) and the Groundwater Taskforce (CPPI) was welcomed. The efforts of numerous technicians, staff, and graduate students at Waterloo made this research possible and their assistance is gratefully acknowledged. In particular, Karen Berry-Spark, now of Beak Consultants, Guelph, Ontario, was responsible for initiating the field experiment. CFB Borden allowed the field research on their facility and provided logistic support.

REFERENCES

1. Barker, J. F., C. E. Hubbard, and L. A. Lemon. "The Influence of Methanol and MTBE on the Fate and Persistence of Monoaromatic Hydrocarbons in Groundwater," Proceedings of the National Water Well Association/American Petroleum Institute Conference: Petroleum Hydrocarbons and Organic Chemicals in Groundwater, Houston, TX, Oct. 31–Nov. 2, 1990, pp. 113–127.
2. Poulsen, M., L. Lemon, and J. F. Barker. "Solubility of BTEX from Gasoline-Oxygenate Mixtures," Report to API, Project GW-16, January, 1991.
3. Vooro, K. A. "M85 Migration in Groundwater: Investigation of BTEX Concentrations Using Laboratory Soil Column Experiments," B.Sc. thesis, University of Waterloo, 1991.
4. Stumm, W., and J. J. Morgan. *Aquatic Chemistry,* 2nd Ed., (Toronto: John Wiley & Sons, 1970).
5. Burris, D. R., and W. G. MacIntyre. "Solutions of Hydrocarbons in a Hydrocarbon-Water System with Changing Phase Compositions Due to Evaporation," *Environ. Sci. Technol.,* 20:296–299 (1986).
6. Leinonen P. J., and D. Mackay. "The Multicomponent Solubility of Hydrocarbons in Water," *Can. J. Chem. Eng.,* 51:230–233 (1973).
7. Maijanen, A., A. Ng, W. Y. Shiu, and D. Mackay. "The Preparation and Composition of Aqueous Solutions of Crude Oils and Petroleum Products," Undistributed Report, PACE, 1984.
8. Shiu, W. Y., A. Maijanen, A. L. Y. Ng, and D. Mackay. "Preparation of Aqueous Solutions of Sparingly Soluble Substances: II. Multicomponent Systems—Hydrocarbon Mixtures and Petroleum Products," *Environ. Sci. Technol.,* 7:125–137 (1988).

CHAPTER 8

Effects of Oxygenated Fuels on Groundwater Contamination: Equilibria and Transport Considerations

William G. Rixey and **Ira J. Dortch,** Shell Development Company, Houston, Texas

INTRODUCTION

There has been considerable interest in the use of oxygenated compounds in gasoline as a result of the phasing out of lead and, more recently, the Clean Air Act Amendments of 1990 designed to control the generation of CO and ozone.[1] Many technical issues regarding the operability of such fuels have already been addressed.[2] One which has received little attention until recently is the potential impact of oxygenates on groundwater contamination. Potential problems include contamination of groundwater by the oxygenates themselves and possible cosolubility effects; i.e., enhancement of the mobility of gasoline components such as benzene, toluene, ethylbenzene, and xylenes (BTEX) due to the presence of oxygenates.

In contrast to the issue of contamination by the oxygenate itself, the potential for enhancing BTEX concentrations in groundwater resulting from oxygenates, commonly referred to as the cosolubility effect, has been the subject of considerable speculation. A first step in assessing this impact is to look at the equilibrium partitioning behavior between an oxygenated fuel and an aqueous phase. This has been done by several researchers.[3-6] From these studies it can be concluded that, for fuels containing MTBE, there will be an insignificant enhancement in BTEX concentrations. Fuels containing oxygenates such as methanol and ethanol, however, may have an impact on BTEX contamination. One study showed, for example, that a concentration of 50% methanol in the aqueous phase increases the solubility of BTEX by a factor of seven over that for methanol-free water.[3]

Equilibrium studies alone, however (except for the MTBE case), do not give enough information to determine the impact on BTEX plumes. Transport effects must also be considered. One way to do this is to first make an assumption of how much water comes in contact with the fuel to determine the aqueous methanol concentrations and the resultant BTEX concentrations near the source. This information can then be used as input to a transport model to predict groundwater concentration profiles. This approach has been used recently and is a significant step in properly assessing the impact of oxygenates on BTEX.[3] However, the conclusions are very dependent on the choice one makes for the ratio of water to gasoline used to generate a source input. Moreover, this method does not address the potential for oxygenates to leave a trail of a residual free-phase of gasoline downstream of the spill zone. The approach used in this chapter is an attempt to develop an alternative method to answer the BTEX enhancement concerns as well as the residual free-phase hydrocarbon distribution.

PREDICTING AQUEOUS BTEX CONCENTRATIONS— EQUILIBRIA CONSIDERATIONS

We need to first set up a scenario to simulate a spill of an oxygenated fuel. The case chosen for this study is a spill of a fuel containing 85% methanol and 15% regular unleaded gasoline. A fuel of this composition is commonly referred to as M85, an alternative fuel that has been proposed for some sensitive areas such as the Los Angeles basin.

Let's first make the assumption that after a fuel spill is released in the vadose zone and reaches the groundwater table, it completely and instantaneously enters the groundwater. This is a worst case scenario to assess the impact of a given quantity of fuel. One way to simulate this case in the laboratory is with a column experiment like that suggested in Figure 8.1. It will be assumed that at time, t = 0, the M85 is introduced instantaneously into the saturated zone without any mixing with the groundwater. Then, at the edges of the M85 zone, which is surrounded by groundwater, a residual gasoline phase will be formed. At this boundary the concentrations of methanol and hydrocarbons will be determined by the phase equilibria. As the M85 plume moves downstream of the source, the methanol concentrations in the plume decrease due to dispersion, and more residual hydrocarbon comes out of solution. The methanol plume will ultimately reach a position where the concentrations of methanol are sufficiently low that free-phase residual gasoline no longer comes out of solution.

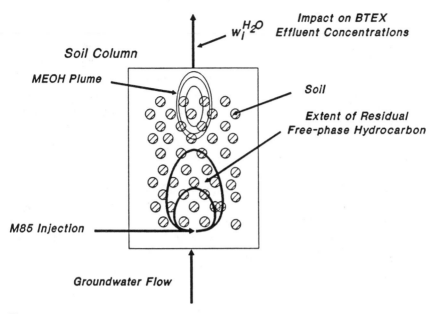

Figure 8.1.   Laboratory set-up for simulating a spill of an 85% methanol-15% gasoline (M85) mixture.

In Figure 8.1 the methanol plume has progressed sufficiently far from the spill zone so that the concentrations are low enough that a trail of free-phase residual gasoline is left behind. Two contours are shown for the residual hydrocarbon phase on the soil. These contours define levels of contamination on the soil. The area between the outer and inner contours represents the region of free-phase contamination. Note that one would predict no contamination within the inner contour, since this is a region which will have seen a high enough methanol concentration while the plume moves downstream of the source, so that a separate phase is not formed. A semiquantitative description of the formation of this separate gasoline phase is presented later in the chapter.

We would like to model this case mathematically as well as experimentally in one-dimensional column studies. A rigorous solution to this problem would involve a finite difference model which incorporates the effects of axial and radial dispersion and fluid-phase equilibria. This involves performing equilibrium flash calculations at each node and is beyond the scope of this present study. As an approximation to this case, we will describe an equilibrium-based method which approximates the effects of dispersion and equilibrium. This method will be used to approximate the multidimensional as well as the one-dimensional case.

Modeling Fluid Phase Equilibria

The first step in the mathematical development is to be able to predict the fluid-fluid equilibria over the entire concentration range which will be encountered. For the case of a spill of M85, this requires modeling the fluid-fluid equilibria over the region shown in the diagram in Figure 8.2. In Figure 8.2 a line is drawn between pure water and the initial 85% methanol-15% gasoline mixture. As the M85 mixture is diluted with water, eventually the point b will be reached. Point b represents the concentrations of methanol, gasoline, and water that will be in the aqueous phase at the point that a second hydrocarbon phase, in equilibrium with the aqueous phase, begins to form. The composition of the second hydrocarbon phase is defined by the point b′. Aqueous and hydrocarbon phases that are in equilibrium are connected by tie-lines. As the methanol-water-gasoline mixture is diluted further, the equilibrium aqueous concentrations follow the curve shown by points b′-f′. Again the corresponding hydrocarbon equilibrium concentrations are b′-f′.

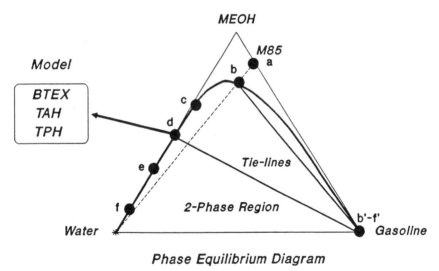

Phase Equilibrium Diagram

Figure 8.2.   Predicting equilibrium BTEX aqueous concentrations from a spill of M85 (85% methanol-15% gasoline).

We wish to determine the aqueous phase concentrations from point a all the way to point f. The approach for modeling the phase equilibrium is described below.

Fluid-Phase Equilibria—Mathematical Development

When the two-phase region is entered, there are an aqueous phase and a gasoline phase present. At equilibrium we can write for each component:

$$\gamma_i^{aq} \, x_i^{aq} \; = \; \gamma_i^{HC} \, x_i^{HC} \tag{1}$$

where γ_i^{aq} and γ_i^{HC} are the activity coefficients, in the Raoult's law sense, for a particular component i in the aqueous and hydrocarbon phases, respectively; and x_i^{aq} and x_i^{HC} are the mole fractions of component i in the aqueous and hydrocarbon phases, respectively.

Fortunately, for this case of a high methanol content fuel, the modeling can be greatly simplified by accounting for nonidealities in the aqueous phase only. Starting with M85, the hydrocarbon phase in the two-phase region will always have a very low methanol and water content. From the data of Letcher et al., the maximum methanol and water concentrations in the hydrocarbon phase will be less than 10% and 1%, respectively, if M85 is used as the fuel.[6] Hence, in the subsequent equilibrium flash calculations it will be assumed that essentially all of the methanol partitions into the aqueous phase. It will also be assumed that the hydrocarbon phase is relatively ideal for the gasoline components. This means that the activity coefficients will be assumed to be equal to unity.

The aqueous phase, on the other hand, is extremely nonideal. We chose to model the aqueous phases as a pseudo-three component system with the components being methanol, water, and solubilized gasoline. The composition of this gasoline component can be taken as the sum of the individual gasoline components. Let's take gasoline to be component 1, methanol to be component 2 and water to be component 3. Now the mole fraction of gasoline (component 1), x_1, can be written as the sum of the individual gasoline species as follows:

$$x_1 \; = \; \sum_{i=1}^{n} x_i^{aq} \tag{2}$$

where x_i^{aq} is the mole fraction of a given gasoline component in the aqueous phase and n is the total number of gasoline components.

The mole fractions of the gasoline components in the aqueous phase can be determined from Equation 1 if the activity coefficients are known. We have chosen to use the simplest activity coefficient model for a ternary system: the two-suffix Margules relationship. For the M85 case, this simple model predicts the equilibrium behavior quite well. Starting with lower concentrations of alcohols in fuels, such a model would be considerably less accurate, because as one gets closer to the plait point (point on the phase diagram when the concentrations in both phases are equal), the concentrations of water and methanol in the hydrocarbon phase are no longer insignificant.[6] Accounting for these nonidealities requires a more sophisticated approach to modeling liquid-liquid phase equilibria, such as one which uses the Wilson or NRTL equations.[7] Use of these equations also makes the equilibrium flash calculations to be discussed later more difficult.

Using the two-suffix Margules relation for a ternary system, the activity coefficient of a hydrocarbon solute in a solvent mixture of methanol and water (components 2 and 3) γ_i^{aq} can be expressed as follows:[7]

$$\ln\gamma_i{}^{aq} = A_{12}x_2^2 + A_{13}x_3^2 + (A_{12} + A_{13} - A_{23})\, x_2x_3 \tag{3}$$

where A_{12}, A_{23}, and A_{13} are constants that can be obtained from binary data. The components and compositions used to simulate a regular unleaded gasoline along with the values for A_{12} and A_{13} are shown in Table 8.1. These values are based on reported values of infinite dilution activity coefficients for various solutes in methanol and solubilities in water.[8-11]. The value of A_{23} was determined to be 0.43 from the infinite dilution activity coefficient of methanol in water.[7]

Table 8.1. Pseudo-Gasoline Composition and Margules Constants Used for Equilibrium Flash Calculations

Component	wt. %	A_{12}	A_{13}
Benzene	1.0	1.8	7.8
Toluene	5.0	2.1	9.2
Ethylbenzene	10.0	2.4	10.3
Butylbenzene	20.0	2.8	12.8
Nonane	64.0	3.6	17.5

Approximate Method for Simulating Equilibria and Transport

One way to simulate the spill process is to use an approach shown in Figure 8.3. We will call this process a differential flash process. M85 is contacted with a differential amount of water. The M85 will become diluted by water and remain a single phase until a concentration is reached where a hydrocarbon phase begins to form. From this point on, an introduction of a differential amount of water will result in a differential amount of hydrocarbon coming out of solution. The amount and composition of this differential amount will be determined by the equilibrium between the aqueous and the hydrocarbon phases. This process is a simplified attempt to simulate the effect of dispersion on the M85 introduced into the aquifer.

The hydrocarbon phase which begins to come out of solution is richer in the less methanol and water soluble compounds. Thus, nonane, for example, comes out of solution in a proportionately higher concentration than benzene. But a significant amount of benzene will also come out of solution. This process will continue until a point is reached when hydrocarbon no longer comes out of solution. Past this point all hydrocarbons disperse roughly in proportion to the dispersion of methanol. Our calculations are set up so that, as each increment of hydrocarbon comes out of solution, it is separated from the M85 aqueous fraction. This is an attempt to simulate the assumed transport situation in the subsurface: as the M85 plume moves through the aquifer, hydrocarbon comes out of solution and then becomes immobile in the pores of the formation, so that the plume no longer sees the hydrocarbon phase.

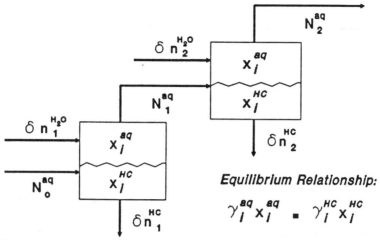

Equilibrium Relationship:

$$\gamma_i^{aq} x_i^{aq} = \gamma_i^{HC} x_i^{HC}$$

Differential Mole Balances:

Total:	$\delta N^{aq} = \delta n^{H_2O} - \delta n^{HC}$
Hydrocarbon Component:	$\delta (x_i^{aq} N^{aq}) = -(x_i^{HC} \delta n^{HC})$
Water Component:	$\delta (x_{H_2O}^{aq} N^{aq}) = \delta n^{H_2O}$
Methanol Component:	$\delta (x_{MEOH}^{aq} N^{aq}) = 0$

Figure 8.3. Two iterations of differential equilibrium flash calculations to approximate the dispersion-equilibrium process of an M85 spill in groundwater after the plume has become saturated with hydrocarbons.

Differential Equilibrium Flash Calculations— Mathematical Development

Assume that we start with a methanol-gasoline mixture with a composition given by point a in Figure 8.2. From point a to b only one phase is present, and the change in moles can be written as follows:

$$\Delta N^{aq} = \Delta n^{H_2O} \tag{4}$$

where N^{aq} is the total number of moles in the aqueous phase, and Δn^{H_2O} is an incremental number of moles of water added to the aqueous phase.

The concentrations of each of the components except water can be determined from the following equation:

$$\delta(x_i^{aq} N^{aq}) = 0 \tag{5}$$

where x_i^{aq} is the mole fraction of a given hydrocarbon component or methanol. The concentration of water is determined as follows:

$$\Delta(x_{H_2O}^{aq} \, N^{aq}) = \Delta n^{H_2O} \tag{6}$$

where $x_{H_2O}^{aq}$ is the mole fraction of water in the aqueous phase.

At point b, the second (hydrocarbon) phase begins to form. Equation 4 becomes:

$$\delta N^{aq} = \delta n^{H_2O} - \delta n^{HC} \tag{7}$$

where δn^{HC} is a differential number of moles of hydrocarbon that comes out of solution from the addition of a differential number of moles of water. The symbol δ has been used here rather than Δ to emphasize that differential elements will be taken in the two-phase region.

The hydrocarbon component mole balance equations become:

$$\delta(x_i^{aq} \, N^{aq}) = -(x_i^{HC} \, \delta n^{HC}) \tag{8}$$

where x_i^{HC} is the mole fraction of hydrocarbon component i. At each increment x_i^{HC} can be related to x_i^{aq} through the equilibrium relationship described by Equation 1. The activity coefficients in the hydrocarbon phase referred to in Equation 1 will be assumed equal to unity. The aqueous phase activity coefficients, γ_i^{aq}, are determined from Equation 3, where component 2 will be considered methanol and component 3 to be water, i.e., $x_2 = x_{Meoh}^{aq}$ and $x_3 = x_{H_2O}^{aq}$.

Similar to Equation 6, the mole fraction of water is determined from the following equation:

$$\delta(x_{H_2O}^{aq} \, N^{aq}) = \delta n^{H_2O} \tag{9}$$

It will be assumed that for these calculations an insignificant amount of methanol partitions into the hydrocarbon phase. (For the path taken in Figure 8.2, this is a reasonable assumption). The mole fraction of methanol in the aqueous phase is then determined from the following equation:

$$\delta(x_{Meoh}^{aq} \, N^{aq}) = 0 \tag{10}$$

One final equation is required to carry out these differential calculations:

$$\sum_{i=1}^{n} x_i^{HC} = 1 \tag{11}$$

For the two-phase region the equations above can be solved numerically using the following approach: for each increment of water, δn^{H_2O}, the differential amount of hydrocarbon that comes out of solution, δn^{HC}, is determined through iteration with sufficiently small increments until Equation 11 is satisfied.

Results of Differential Equilibrium Calculations

The results of numerical calculations are shown in Figure 8.4. Plotted are the predicted aqueous phase concentrations of benzene, BTE (the sum of benzene, toluene, and ethylbenzene), and TPH (the sum of all of the 5 components shown in Table 8.1). Initially, the concentration of the gasoline is 150,000 mg/kg (15 wt%) before any water has been introduced into the system. As the M85 is mixed with water, the hydrocarbon and methanol concentrations decrease as shown in the upper figure in Figure 8.4. When the methanol concentration reaches a wt. fraction of 0.77 (point b on both figures), a hydrocarbon phase begins to come out of the aqueous phase. From this point (point b) the concentration of methanol in the aqueous phase actually begins to increase slightly with further additions of water, because hydrocarbon is coming out of solution.

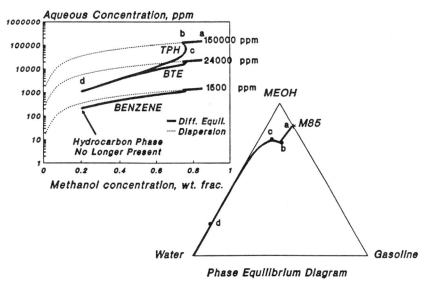

Figure 8.4. Aqueous hydrocarbon concentrations predicted from differential equilibrium flash calculations starting with an 85% methanol-15% gasoline mixture. See text for explanation of points on curves.

With further additions of water, the methanol concentration reaches a maximum (point c) and then begins to decrease. The hydrocarbon concentrations then begin to decrease exponentially with methanol concentration. Finally a point (point d) is reached when hydrocarbon no longer comes out of solution with further additions of water. This occurs when the methanol concentration reaches a weight fraction of 0.20. At this concentration the benzene concentration is predicted to be 200 ppm and the BTE concentration is predicted to be still over 1000 ppm. The butylbenzene and nonane concentrations are predicted to be less than 1 ppm at this point. Also shown in Figure 8.4 are curves assuming that each of the hydrocarbons are dispersed along with methanol, neglecting phase equilibria. (This assumes that the hydrocarbons have the same diffusivities as methanol.) These curves demonstrate the effect of equilibria on removing the more insoluble hydrocarbons from the aqueous phase. The assumption that all of the BTE components transport and disperse with methanol results in a concentration six times higher than that predicted from the equilibrium calculations.

For comparison with the differential equilibrium calculations, in Figure 8.5 a curve is shown for BTE concentrations based on batch equilibrium flash calculations. For these calculations only one equilibrium flash is carried out, but for different amounts of water added to the original M85 mixture. This is essentially the method others have used to assess the effect of oxygenates on BTE concentrations.[3-6] It predicts significantly lower values of BTE than that from the differential flash calculations. For example, for a methanol weight fraction of 0.2, the BTE concentration is 5 times lower than that predicted for the differential equilibrium calculation. (Below the methanol concentration of 0.20 wt. fraction, we have assumed that BTE concentrations decrease in proportion to methanol concentrations for the differential equilibrium case.)

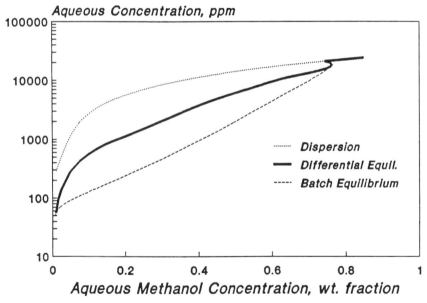

Figure 8.5.   Comparison of BTE (sum of benzene, toluene, and ethylbenzene) concentrations from differential flash equilibrium calculations, batch flash equilibrium calculations, and dispersion only.

In conclusion, in predicting benzene, toluene, and ethylbenzene concentrations, the equilibrium flash method shown here predicts concentrations intermediate between what is predicted by simple dilution and by batch equilibrium. Despite the added complexity of the model, we prefer the differential method because it also gives us information regarding the extent of residual hydrocarbon formation in the aquifer, something which the simpler methods cannot do.

The aqueous concentrations predicted from the differential equilibrium model will be compared with experimental effluent concentrations of column studies with injections of M85 described below.

SIMULATING AN M85 SPILL—COLUMN STUDIES

Column studies were conducted in order to understand the behavior of a spill of M85 into groundwater. As shown in Figure 8.1, we are interested in predicting the impact of methanol plumes on BTEX concentrations and the extent of the residual hydrocarbon phase contamination produced from the methanol plume.

Procedure

A mixture of methanol and a pseudo-gasoline (composition given in Table 8.1) was injected near the bottom of a vertical glass column. The column had a diameter of 1.0 in. and a height of 24 inches. The column was filled with 500 g of glass beads (105–150 microns in diameter). A steady upward flow of water (1.0 g/min) was established prior to injection of the methanol-gasoline mixture. A column porosity of 0.36 was determined from the initial packing of the glass beads in a water-filled column. Three separate experiments were run with injections of 2.5 g, 5 g, and 10 g of the methanol- gasoline mixture. Effluent samples were collected and analyzed with a Varion Model 3700 gas chromatograph equipped with a packed column and flame ionization detector. The gasoline mixture contained 0.5 wt% Sudan III (Diagnostic Systems, Inc.) in order to visualize the formation of a separate hydrocarbon phase, as the methanol-gasoline pulse passed through the column.

Methanol Effluent Concentrations

The effluent concentrations for methanol resulting from the 5 gram injection of M85 are shown in Figure 8.6. Also shown is an effluent breakthrough curve predicted with the following simple equation:[12]

$$C_i^{aq}(t) = \frac{Mw_{io}}{2A\theta \sqrt{\pi D_x t}} \exp\left[\frac{-(L-v_x t)^2}{4D_x t}\right] \qquad (12)$$

where C_i^{aq} is the concentration of methanol in the aqueous effluent at a given time (g/cm³), M is the total mass of M85 injected (grams), w_{io} is the weight fraction of methanol in M85, L is the length of column (cm), v_x is the interstitial velocity (cm/sec), A is the cross sectional area of the column (cm²), θ is the porosity (cm³ pore volume/cm³ total volume), and D_x is the effective axial dispersion coefficient (cm²/sec). Equation 12 assumes that there is an instantaneous

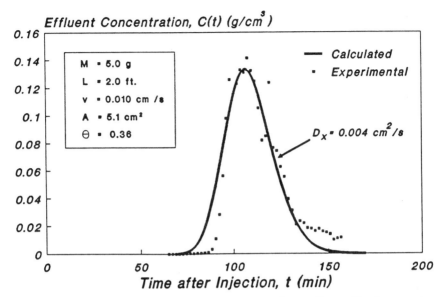

Figure 8.6.   Experimental effluent methanol concentration profile resulting from a pulse of an 85 wt% methanol-15 wt% gasoline (M85) mixture. The curve was calculated with Equation 12.

release of methanol uniformly over the cross-sectional area of the column at the point of injection, and that there is not a concentration gradient in the radial direction. Equation 12 also assumes no sorption of solute occurs. A value of 0.004 cm²/sec for D_x was used to fit the curve to the data. This value is nearly 10 times that for typical values given for packed beds.[13]

A possible explanation for the high value for D_x is that all of the methanol is not released instantaneously. A separate hydrocarbon phase does form near the inlet of the column. There should have been some partitioning of methanol into this hydrocarbon phase, resulting in methanol being retarded somewhat at the leading edge of the breakthrough curve. A small amount of methanol present in this hydrocarbon phase may also explain the inability of Equation 12 to fit the observed concentrations on the trailing portion of the curve.

BTE Concentrations

The corresponding effluent concentrations for benzene, toluene, and ethylbenzene are shown in Figure 8.7. For comparison the calculated curve for methanol from Figure 8.6 is also shown. The maximum observed concentration for benzene was 125 ppm while that for ethylbenzene was 350 ppm for a corresponding maximum methanol concentration of 0.13 g/cm³ (130,000 ppm). Note that the curves for the benzene, toluene, and ethylbenzene reach asymptotic values. These asymptotes reflect the relatively slow leaching process of BTE from the residual hydrocarbon that is left on the column.

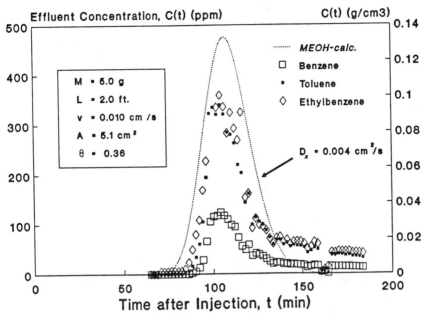

Figure 8.7.   Experimental benzene, toluene, and ethylbenzene effluent concentrations from a pulse of an 85 wt% methanol-15 wt% gasoline (M85) mixture. The calculated methanol curve of Figure 6 is shown for comparison.

The maximum concentrations of benzene, toluene, ethylbenzene, butylbenzene, and nonane observed in the column effluent from all three of the M85 injections are shown in Table 8.2. The maximum methanol concentrations were 7, 13, and 25 wt%, respectively. Also shown in Table 8.2 are values for aqueous hydrocarbon concentrations expected when gasoline containing no methanol is injected onto the column. (The gasoline should become trapped in the pores near the inlet of the column. This gasoline will serve as a continuous source of hydrocarbons in the effluent.[14]) The concentrations for the gasoline-only case in Table 8.2 were determined from the composition of the gasoline given in Table 8.1 and the partition coefficients of the pure components.[14]

Assuming that the extent of residual hydrocarbon contamination left near the source of the spill is the same for a regular fuel and an oxygenated fuel, Table 8.2 can be used to estimate the maximum enhancement in concentration of hydrocarbons due to the methanol plume. For the column experiment in which 2.5 grams of a 85% methanol-15% gasoline were injected, the maximum effluent concentration was 7 wt%. This produced a maximum benzene concentration of 70 ppm, which is only 2.5 times the pseudo steady-state concentration predicted to emanate from residual gasoline. However, the BTE concentration was enhanced 4.5 times over that from gasoline alone. When the M85 injection was increased to 10 grams, the maximum methanol concentration was 25 wt%, and the enhancements in benzene and BTE maximum concentrations were factors of 9 and 23, respectively.

Table 8.2.   Hydrocarbon Component Effluent Concentrations for Various Injections of a Methanol-Gasoline Mixture onto a Column of Glass Beads

	Gasoline Only[a]	M85 Injection		
M85 Injection, grams	0	2.5	5	10
Max. Methanol Effluent Concentration, wt%	0%	7%	13%	25%
Component	**Maximum Hydrocarbon Concentrations, ppm**			
Benzene	28	70	125	250
Toluene	35	170	340	710
Ethylbenzene	17	110	360	900
BTE	80	350	825	1860
Butylbenzene	3.8	40	125	640
Nonane	0.1	15	5	35
TPH	84	405	955	2535

[a]Calculated from partition coefficients of pure components and hydrocarbon phase concentrations given in Table 8.1.

Is there a simple relationship between the observed maximum hydrocarbon and methanol concentrations? In Figure 8.8, the results of the column experiments are compared with the differential equilibrium calculations of Figure 8.4. When plotted against the observed maximum methanol concentrations, the observed

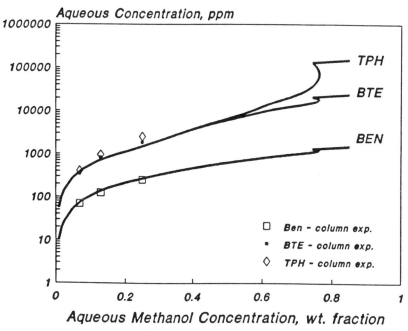

Figure 8.8.   Comparison of hydrocarbon concentrations from column studies with that from differential equilibrium flash calculations.

maximum benzene concentrations correspond with those calculated from the differential equilibrium model. The agreement between the observed and calculated concentrations for BTE is also quite good. For the higher methanol concentrations, however, the observed TPH values are significantly greater than those calculated. Recall that the differential equilibrium curves were extrapolated below methanol concentrations less than 20 wt% by assuming that the hydrocarbon concentrations decrease in proportion to the methanol concentration. The differential equilibrium calculations predict that a second hydrocarbon phase is no longer present once the methanol concentration reaches 20 wt%.

Overall, the differential equilibrium model agrees well with the results of our column studies. This agreement suggests a simple way to predict, semiquantitatively, the hydrocarbon concentrations which would accompany the solvent plume downstream of an oxygenated solvent spill. Of course, this assumes that there is no biodegradation of solvent or solubilized hydrocarbon components. It also assumes that there is no separation of solvent and hydrocarbon plumes due to sorption. The effect of sorption will tend to reduce the maximum observed concentrations. This will be discussed briefly later in this chapter.

SCALING LABORATORY COLUMN RESULTS TO FIELD CONDITIONS

How does one translate one-dimensional, laboratory scale data into predictions for a three-dimensional field case? Let's compare the size of a spill and the distance from the spill zone which achieves the same maximum concentration for the two cases.

For the one-dimensional case, the maximum concentration observed at a given position, L^{1D}, downstream of the source can be expressed as follows:

$$C_{max}^{1D} = \frac{M^{1D}w_{io}^{1D}}{2A\theta\sqrt{\pi D_x^{1D}\dfrac{L^{1D}}{v_x^{1D}}}} \tag{13}$$

where the variables are similar to those described for Equation 12.

For the three-dimensional case, if M85 is assumed to be spilled as a point source at the surface of the saturated zone, the maximum concentration observed at a given distance, L^{3D}, downstream of the source and a given depth, H^{3D}, from the surface can be determined from the following equation:[15]

$$C_{max}^{3D} = \frac{M^{3D}w_{io}^{1D}}{4\theta\left(\pi\dfrac{L^{3D}}{v_x^{3D}}\right)^{3/2}\sqrt{D_x^{3D}D_y^{3D}D_z^{3D}}}\exp\left[\frac{-(H^{3D})^2}{4\left(D_y\dfrac{L}{v_x}\right)^{3D}}\right] \tag{14}$$

The mass of the spill for the three-dimensional field case which yields the same maximum concentration as that for the one-dimensional column becomes:

$$M^{3D} = 2\pi\frac{M^{1D}}{A^{1D}} \sqrt{\frac{\left(D_x\frac{L}{v_x}\right)^{3D}\left(D_y\frac{L}{v_x}\right)^{3D}\left(D_z\frac{L}{v_x}\right)^{3D}}{\left(D_x\frac{L}{v_x}\right)^{1D}}} \ \exp\left(\frac{(H^{3D})^2}{4\left(D_y\frac{L}{v_x}\right)^{3D}}\right)$$

$$(15)$$

Using values of D_x^{3D}, D_y^{3D}, and D_z^{3D} of 1, 0.01, and 0.01 ft²/day, respectively, $L^{3D} = 500$ ft., $H^{3D} = 5$ ft., and $v_x^{3D} = 1$ ft/day; the 5 gram spill in the column experiment ($D_x^{1D} = 0.004$ cm²/sec, $L^{1D} = 61$ cm, and $v_x^{1D} = 0.01$ cm/sec) scales to a 30,000 lb (~ 5,000 gallon) spill in a large-scale case.

EFFECT OF M85 ON THE DISTRIBUTION
OF RESIDUAL FREE-PHASE HYDROCARBON

In addition to understanding the effect that fuels containing methanol may have on BTEX plumes, we would also like to know how far the methanol plume will carry a trail of residual free-phase hydrocarbon downstream of the spill zone. The results of the differential equilibrium flash calculations will be used to semi-quantitatively answer this question.

To estimate the extent of hydrocarbon contamination let's first determine the methanol concentrations in the column. We will assume that the hydrocarbon which comes out of solution is controlled by equilibria. Therefore, if we know the concentration history of methanol, then something can be said about the residual hydrocarbon phase.

In Figure 8.9 is shown a series of curves for various concentrations of methanol. Curves are shown for 77, 72, 52, and 20 wt% methanol. They represent the aqueous methanol concentrations seen by the column as a methanol plume progresses through the column. The curve identified by a methanol concentration of 77 wt%, for example, was generated by taking the locus of 77 wt% methanol contours for several times after injection of 10 grams of M85. The volume bounded by a surface generated from this curve represents a region of the column for which the concentration of methanol is always greater than or equal to 77 wt%. Since this is the concentration above which a separate hydrocarbon phase is not formed, no hydrocarbon should be present within this volume.

As the methanol concentration goes below 77 wt%, hydrocarbon readily comes out of solution so that when the methanol concentration reaches 72 wt%, nearly 75% of the gasoline present in the M85 is removed from the aqueous phase. Let's assume that as the hydrocarbon comes out of solution, it becomes trapped in the pores and does not move along with the methanol plume. If the 75% removal of gasoline is averaged over the incremental volume bounded by the 77 and 72 wt% methanol surfaces, the level of contamination of hydrocarbon on the soil

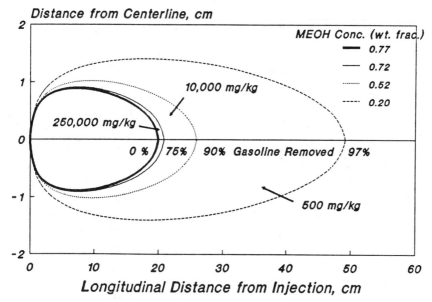

Figure 8.9. Estimated gasoline formation following a 10 gram injection of an 85% methanol-15% gasoline (M85) mixture.

would be 250,000 mg/kg by the equilibrium calculations. (This level of hydrocarbon corresponds to complete saturation of the pores with gasoline; it is more likely that the residual saturation in the presence of water will be 20–30% of this value.) When the methanol concentration is reduced to 52 wt%, an additional 15% of the gasoline is removed. Over the corresponding incremental volume, the average hydrocarbon level is calculated to be 10,000 mg/kg on the column. When a methanol concentration of 20 wt% is reached, no more hydrocarbon comes out of solution. The level of hydrocarbon on the soil averaged over this final increment is estimated at 500 mg/kg.

Our equilibrium calculations predict that a total of 97% of the gasoline will come out as a separate phase—only 3% of the total is solubilized and is transported further downstream with the methanol plume. This 3% is comprised of essentially only benzene, toluene, and ethylbenzene. It is predicted that 52% of the benzene, 29% of the toluene, 8% of the ethylbenzene, and an insignificant fraction of the butylbenzene and nonane in the initial M85 injection are solubilized in the aqueous phase and transported along with the methanol plume.

For the M85 case, the volume occupied by the residual gasoline will be assumed to be equal to the volume bounded by a surface corresponding to a methanol concentration of 20 wt%. This volume, determined through integration of an equation for the methanol concentration contours, is described mathematically as follows:

$$V = \frac{3}{4} \pi v_x D_y \left(\frac{Mw_{io}}{4C\theta\pi^{3/2}\sqrt{D_x D_y D_z}} \right)^{4/3} \qquad (16)$$

where D_y and D_z are the dispersion coefficients in the y and z directions, respectively (cm²/sec); C is the concentration of methanol for a given surface whose cross-sectional area is shown in Figure 8.9 (g/cm³); V is the volume bounded by the surface generated by rotating the appropriate curve in Figure 8.9 about the x-axis (cm³); and all other variables have been defined previously. (It was assumed that $D_y = D_z$ in the integration to determine the volume given by Equation 16.)

Equation 16 provides a semiquantitative tool for estimating the distribution of a gasoline phase formed from a spill of an oxygenated fuel. Recall that dyes were used to follow the gasoline phase in the column experiments discussed earlier. The distribution of the dyed gasoline (not shown) after breakthrough of methanol for the 10 gram injection of M85 is in qualitative agreement with the hydrocarbon distribution predicted in Figure 8.9. Recall also that the concentrations of benzene and toluene predicted from the column experiments agreed well with the equilibrium predictions. However, the concentrations of ethylbenzene, and especially butylbenzene and nonane, were significantly higher than the calculated concentrations. This does suggest that some of the hydrocarbon which comes out of solution can be transported with the methanol. Significant mobilization of this second phase can result in higher downstream hydrocarbon concentrations than that reported here. This potential for mobilization of the hydrocarbon phase in the presence of methanol is the subject of current research.[16]

EFFECT OF RETARDATION ON HYDROCARBON CONCENTRATIONS FROM A SPILL OF AN OXYGENATED SOLVENT— SOME PRELIMINARY THOUGHTS

In order to assess the potential impact of retardation, the following one-dimensional transport equation was used:[17]

$$\frac{\partial C_i^{aq}}{\partial t} + \frac{\rho_b}{\theta} \frac{\partial K_{d,i} C_i^{aq}}{\partial t} + D_x \frac{\partial^2 C_i^{aq}}{\partial x^2} - v_x \frac{\partial C_i^{aq}}{\partial x} \qquad (17)$$

where $K_{d,i}$ is the distribution coefficient of component i between the aqueous and soil phases (cm³ solution/g soil) and ρ_b is the bulk density of the soil (g soil/cm³ soil). Other quantities were defined previously with Equation 12. The distribution coefficient is variable since it is dependent on the methanol composition. Therefore, the component transport equations for methanol and the hydrocarbon components must be solved simultaneously.

The results of a numerical solution for the column effluent concentrations following a 10-gram injection of M85 are shown in Figure 8.10. Shown in the figure are curves for benzene, toluene, and methanol. For these simulations, we have assumed that all of the hydrocarbon from the injection enters the aqueous phase.

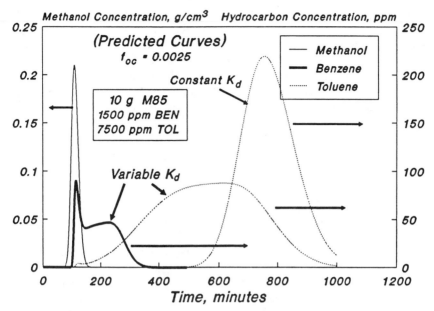

Figure 8.10. Effect of retardation on benzene and toluene from an injection of M85. L = 61 cm, D_x = 0.004 cm³/sec, v_x = 0.01 cm/sec. θ = 0.36, and ρ = 1.6 g/cm³.

With this assumption, the maximum concentration for benzene is 90 ppm compared with a concentration of 250 ppm observed in the column experiment with 10 grams injection. The maximum toluene concentration is 85 ppm compared with a maximum value of 710 observed in the column experiments. Note that the concentration of toluene at the point at which the methanol concentration is a maximum is > 5 ppm. Also shown in Figure 8.10 is the effluent curve predicted for toluene, assuming that K_d is not affected by methanol. (The value for the distribution coefficient between soil and water was used for this case.)[14] Comparison of the two curves for toluene shows the predicted effect of methanol on retardation. These calculations suggest that, even for soils with moderate sorptive capacity (characterized by a fraction of organic content, f_{oc}, of 0.0025), a spill of M85 may result in earlier detection of significant levels of BTEX at a given point downstream of the spill zone. The effect, of course, will be very dependent on the size of the methanol spill.

The quantity of M85 used in the simulation for Figure 8.10 was chosen to be consistent with that of the column experiments. This case demonstrates the potential effect of an extremely large spill at a point far downstream of a spill (or the effect on concentrations at a short distance from the source of a small spill). If Equation 15 is used, the 10 gram M85 spill in the small-scale, one-dimensional case would correspond roughly to a 17,000 gallon spill for a large-scale case (assuming values for D_x, D_y, and D_z of 3, 0.01, and 0.01 ft²/day, respectively, L = 500 ft, H = 5 ft, and v_x = 1 ft/day). Values of D_x, L, and v_x were chosen, so that $D_x/(Lv_x)$ would be the same for the small-scale and large-scale cases. This is necessary for the curves shown in Figure 8.10 to simulate the effect of retardation in the large-scale case. In addition, the time, t, in Figure 8.10 should be scaled by L/v_x to represent a large-scale case.

Additional calculations can be carried out with the one-dimensional numerical model for lower effluent methanol concentrations to assess the potential effect of retardation on BTEX concentrations for smaller spills of M85.

SUMMARY AND CONCLUSIONS

A mathematical framework, which accounts for the combined effects of phase equilibria and transport, has been presented. This framework should be useful in assessing the potential impact of oxygenated fuels on groundwater contamination.

If assumptions are made that (1) the oxygenated fuel is spilled instantaneously and completely into the groundwater, (2) the effects of biodegradation and sorption can be neglected, (3) phase equilibria initially control the concentration profiles, and (4) hydrocarbon which comes out solution because of equilibria remains in the pores and is not transported downstream with the plume; then it is possible, with the approach presented here, to relate the expected BTEX concentrations to the oxygenated solvent concentration at some point downstream of a spill.

The gasoline component concentrations expected from a spill of a 85% methanol-15% gasoline (M85) mixture, using such an approach, were summarized in Figure 8.8. (These calculations agreed well with one-dimensional column experiments.) The distribution of the residual gasoline which comes out of solution was also estimated, as shown in Figure 8.9—again for a M85 fuel.

The results presented in this chapter apply specifically to a M85 fuel. The mathematical framework for the equilibrium calculations can be adapted, however, to any gasoline-oxygenate fuel provided the gasoline-oxygenate-water phase equilibria are known. The appropriateness of the assumptions listed above, however, should be carefully considered when interpreting results of the method presented here.

NOTATION

A   = cross sectional area of laboratory column, cm^2

A_{12} = binary interaction parameter for hydrocarbon (1) and methanol (2)

A_{13} = binary interaction parameter for hydrocarbon (1) and water (2)

A_{23} = binary interaction parameter for methanol (2) and water (3)

C_i^{aq} = solute concentration in aqueous effluent, g/cm^3

C_{max}^{1D} = maximum oxygenate concentration at a given position downstream of the source for the one-dimensional case, g/cm^3

C_{max}^{3D} = maximum oxygenate concentration at a given position downstream of the source for the three-dimensional case, g/cm^3

D_x^{1D} = axial dispersion coefficient for one-dimensional case, cm^2/sec

D_x^{3D} = dispersion coefficient in the x direction for the three-dimensional case, cm^2/sec

D_y^{3D} = dispersion coefficient in the y direction for the three-dimensional case, cm^2/sec

D_z^{3D} = dispersion coefficient in the z direction for the three-dimensional case, cm^2/sec

H^{3D} = vertical distance from source of spill, cm

L = length of the laboratory column, cm

L^{1D} = distance downstream of the source in the one-dimensional case, cm

L^{3D} = distance downstream of the source in the three-dimensional case, cm

M = total mass of fuel injected at the column inlet, grams

M^{1D} = total mass of fuel spilled for the one-dimensional case, grams

M^{3D} = total mass of fuel spilled for the three-dimensional case, grams

N^{aq} = total number of moles in the aqueous phase

n^{H_2O} = total number of moles of water added to the oxygenated fuel

n^{HC} = total number of moles of hydrocarbon removed from the aqueous phase

t = time after injection of fuel, sec

V = volume occupied by residual gasoline at long times after oxygenate spill, cm^3

v_x^{1D} = groundwater velocity in the x direction for the one-dimensional case, cm/sec

v_x^{3D} = groundwater velocity in the x direction for the three-dimensional case, cm/sec

w_{io} = weight fraction of oxygenate in the original fuel mixture

x_1 = mole fraction of gasoline in the aqueous phase

x_2 = mole fraction of methanol in the aqueous phase

x_3 = mole fraction of water in the aqueous phase

x_i^{aq} = mole fraction of gasoline component i in the aqueous phase

x_i^{HC} = mole fraction of gasoline component i in the hydrocarbon phase

Δ = change in number of moles for equilibrium calculations

δ = differential change in number of moles for equilibrium calculations

γ_i^{aq} = activity coefficient of gasoline component i in the aqueous phase

γ_i^{HC} = activity coefficient of gasoline component i in the hydrocarbon phase

θ = porosity of the porous media, cm^3 pores/cm^3 total volume

ρ_b = bulk density of the media, g soil/cm^3 total volume

REFERENCES

1. Lieder, C. A., "The Impact on Fuels of the 1990 Clean Air Act Amendments," NPRA National Fuels and Lubricants Meeting, Houston, Texas, November 1–2, 1990.

2. American Petroleum Institute, "Alcohols and Ethers: A Technical Assessment of Their Application as Fuels and Fuel Components," API Publication 4261, 2nd ed., Washington D.C. (1988).

3. American Petroleum Institute, "Chemical Fate and Impact of Oxygenates in Groundwater: Solubility of BTEX from Gasoline-Oxygenate Mixtures," API Publication, in press (1991).

4. Groves, F. R., Jr., "Effect of Cosolvents on the Solubility of Hydrocarbons in Water," *Environ. Sci. Technol.*, 22:282 (1988).

5. Cline, P. V., J. J. Delfino, and P. Suresh C. Rao, "Partitioning of Aromatic Constituents into Water from Gasoline and Other Complex Solvent Mixtures," *Environ. Sci. Technol.*, 25:914 (1991).

6. Letcher, T. M., C. Heyward, S. Wootton, and B. Shuttleworth, "Ternary Phase Diagrams for Gasoline-Water-Alcohol Mixtures," *Fuel*, 65:891 (1986).

7. Reid, R. C., J. M. Prausnitz, and T. K. Sherwood, *The Properties of Gases and Liquids*, (New York, NY: McGraw-Hill, 1977).

8. Verschueren, K., *Handbook of Environmental Data on Organic Chemicals*, (New York, NY: Van Nostrand Reinhold, 1983).

9. Polak, J., and B. C.-Y. Lu, "Mutual Solubilities of Hydrocarbons and Water at 0 and 25°C," *Can. J. Chem.*, 51:4018 (1973).

10. Petrovic, S. M., S. Somic, and I. Sefer, "Utilization of the Functional Group Contribution Concept in Liquid Chromatography on Chemically Bonded Reversed Phases," *J. Chromatog.* 247:49 (1985).

11. Deal, C. H., and E. L. Derr, "Selectivity and Solvency in Aromatics Recovery," *Ind. Eng. Chem. Process Des. Dev.*, 3:394 (1964).

12. Carslaw, H. S., and J. C. Jaeger, *Conduction of Heat in Solids* (New York, NY: Oxford University Press, 1959).

13. Sherwood, T. K., R. L. Pigford, and C. R. Wilke, *Mass Transfer* (New York, NY: McGraw-Hill, 1975).

14. Rixey, W. G., P. C. Johnson, G. M. Deeley, D. L. Byers, and I. J. Dortch, "Mechanisms for the Removal of Residual Hydrocarbons from Soils by Water, Solvent, and Surfactant Flushing," *Hydrocarbon Contaminated Soils*, Volume 1, E. J. Calabrese and P. T. Kostecki, Eds., (Chelsea, MI: Lewis Publishers, Inc., 1991).

15. Baetslé, L. H., "Migration of Radionuclides in Porous Media," *Progress in Nuclear Energy, Series XII, Health Physics*, A. M. F. Duhamel, ed., (Elmsford, NY: Pergamon Press, 1969), p. 707.

16. Barker, J. F., and R. C. Donaldson, University of Waterloo, personal communication, September, 1990.

17. Bear, J., *Dynamics of Fluids in Porous Media* (New York, NY: American Elsevier, 1972).

CHAPTER 9

Estimating the Multimedia Partitioning of Hydrocarbons: The Effective Solubility Approach

Donald Mackay and **Wan Ying Shiu,** Department of Chemical Engineering and Applied Chemistry, University of Toronto, Ontario, Canada

INTRODUCTION

In this chapter our primary purpose is to suggest and illustrate a novel approach for estimating the approximate prevailing concentrations of hydrocarbons in the vicinity of a contaminated soil site. The basic concept is that hydrocarbons tend to migrate throughout such a system, striving to establish a common, prevailing chemical potential or fugacity in all compartments or media. Of course, precisely equal values are not likely to be achieved because of transport and transformation processes which result in displacement from equilibrium, but we argue that it is illuminating to explore the likely magnitudes of the concentrations which are being approached, and given sufficient time, could be achieved. In some cases this leads to an upper estimated limit on concentration. One objective of this approach is to enable "back of the envelope" concentrations to be estimated, which are useful for checking the reasonableness of the results of more complex calculations. These more complex calculations may be done by computer solution of the partial differential equations describing the hydrocarbon's fate in space and time in soil, air, and water phases and any resident biota. It is first useful to document sources of physical-chemical property data and assemble an illustrative set of data. There are several sources of data on the physical-chemical properties of hydrocarbons, including the important effect of temperature dependence.

137

A convenient recent review is that of Eastcott et al.[1] who compiled data for 99 hydrocarbons and derived simple correlations for vapor pressure, water solubility and octanol-water partition coefficient (K_{OW}), Henry's law constant and molar volume. Mackay and Shiu[2] and Miller et al.[3] have compiled physical-chemical data and their partitioning behaviors of some hydrophobic organic chemicals of environmental interest. Vapor pressures may be mainly obtained from Zwolinski and Wilhoit[4] and Sonnefeld et al.[5] Aqueous solubilities were obtained from McAuliffe,[6] Mackay and Shiu,[7] and the IUPAC Solubility Data Series[8,9]; octanol-water partition coefficients were obtained from Hansch and Leo.[10] Molar volume was calculated using the additive LeBas method.[11] Other handbooks such as the *CRC Handbook of Chemistry and Physics* (1983–84),[12] Verschueren,[13] and an EPA report on environmentally relevant chemicals[14] provide valuable information.

Table 9.1 gives data for selected hydrocarbons. This shows the striking variation in magnitude of these properties. Water solubilities vary by a factor of 10^6, vapor pressure by 10^8, and K_{OW} by 10^6. It is thus clear that widely differing fates and behavior are to be expected, as dictated by these properties.

THEORETICAL BASIS

There are several approaches for determining the concentration of hydrocarbon in one phase when contacted with a known concentration in another phase, under conditions such that equilibrium is assured.

The simplest is the partition coefficient or distribution coefficient approach, which is based on the Nernst distribution law

$$C_1/C_2 = K_{12}$$

Here C_1 and C_2 are concentrations in phases 1 and 2, and K_{12} is the partition coefficient. There are several options for selecting units of concentrations (C). Most commonly, C is expressed in mass/volume, e.g., mg/L or g/m^3, but for solid phases, mg/kg may be preferred. K_{12} may thus be dimensionless, or it may have dimensions such as L/kg. The disadvantage of the partition coefficient approach is that K_{12} depends on properties of the hydrocarbon in two phases; thus it is not immediately clear which role each phase plays. For example, a soil/water partition coefficient may be 4 for benzene and 20 for toluene. Is this difference because the toluene is five times more attracted to soil, or is benzene five times more attracted to water, or is it some combination of these effects? There is a compelling incentive to understand the basic determinants of these partition coefficients, because this knowledge enables hydrocarbon-to-hydrocarbon "structure-activity-property" relationships to be established. Such relationships are essential given the number of hydrocarbons of potential concern.

Second is the chemical potential approach, which is most satisfying to the thermodynamacist because it rigorously treats phase and reaction equilibria.

Regrettably, it is a fairly complex formalism. It is not easy to relate chemical potential to commonly measured quantities such as concentrations in soil. Chemical potential is logarithmically related to concentration; thus, it cannot be used in transport equations derived from Fick's law containing concentration differences.

Third is the fugacity approach which has been advocated by Mackay and Paterson[15] and Mackay.[16] Fugacity is linearly (or nearly linearly) related to concentration; thus, fugacity differences have the same inherent meaning as concentration differences. The fugacity (f, Pa) is related to concentration C (mol/m³) by a capacity term or Z value (mol/m³.Pa) as

$$C = Zf$$

At equilibrium between two phases f_1 equals f_2; thus,

$$C_1/C_2 = Z_1f_1/Z_2f_2 = Z_1/Z_2 = K_{12}$$

Fugacity is essentially partial pressure and can be viewed as an "escaping tendency." The Z terms reflect the capacity of a phase for the hydrocarbon in the same sense as solubility implies a maximum capacity of the phase for the solute. A Z value is thus "half" a partition coefficient. All Z values must have units of mol/m³.Pa; thus, all ratios of two Z values are dimensionless partition coefficients.

Z values are calculated first in the atmosphere, then other Z values can be deduced for the hydrocarbon in other media from the appropriate partition coefficients.

The problem with this approach is that the new concept of fugacity must be learned, with new terminology and strange dimensions. The advantages of the approach are that it leads to more elegant partitioning expressions, and it separates phases from each other, avoiding the "two-phase" problems of partition coefficients.

We believe that a fourth approach may prove useful. We present and discuss it here for consideration. It is essentially a "de-mystified" version of the fugacity approach in which fugacity is replaced by an "effective solubility." This has been used for describing metal distribution in lakes by Mackay and Diamond[17] who suggested the use of the similar concept of "aquivalence" as an equilibrium criterion which can replace fugacity, especially in situations when nonvolatile chemicals are being treated.

For each phase, designated by subscript i, a specific hydrocarbon at a specific temperature has an "effective solubility," S_i g/m³. This is the maximum concentration which the hydrocarbon can establish in that phase. Higher concentrations would result in "precipitation" or "phase separation" of pure hydrocarbon; i.e., the phase cannot accept any more hydrocarbon.

For water, S_W is simply the solubility in water. For example, S_W for benzene is 1780 g/m³ at 25°C. There are difficulties in using this approach to determine

Table 9.1. Physical-Chemical Properties of Selected Organic Chemicals at 25°C

Chemical	Molecular Mass g/mol	mp °C	bp °C	Density g/cm³	Molar Volume LeBas cm³/mol	Solubility g/m³	Vapor Pressure Pa	log K_{ow}
Benzene	78.10	5.53	80.0	0.879	96.0	1780	12700	2.13
Toluene	92.10	−95.00	111.0	0.867	118.0	515	3800	2.69
Ethylbenzene	106.20	−95.00	136.2	0.867	140.4	152	1270	3.13
p-Xylene	106.20	13.20	138.0	0.860	140.4	185	1170	3.15
1,3,5-Trimethylbenzene	120.20	−44.70	164.7	0.865	184.8	48	325	3.58
n-Propylbenzene	120.20	−101.60	159.2	0.862	170.0	55	449	3.69
Isopropylbenzene	120.20	−96.60	154.2	0.862	170.0	50	611	3.66
n-Butylbenzene	134.20	−88.00	183.0	0.860	184.8	13.8	137	4.28
1,2,4,5-Tetramethylbenzene	134.20	79.20	196.8	0.838	184.8	3.48	66	4.00
Pentylbenzene	148.25	−75.00	205.4	0.859	207.0	10.5	44	4.90
Hexylbenzene	162.28	−61.00	226.0	0.861	229.2	1.02	13.6	5.22
n-Pentane	72.15	−129.70	36.1	0.614	118.0	38.50	68400	3.62
n-Hexane	86.20	−95.00	68.0	0.660	140.6	9.50	20200	4.11
n-Heptane	100.20	−90.60	98.4	0.670	162.8	2.93	6110	4.66
n-Octane	114.20	−56.20	125.7	0.700	185.0	0.66	1880	5.18
n-Nonane	128.30	−54.00	151.0	0.720	207.2	0.122	571	5.65

continued

Table 9.1. *continued*

Chemical	Molecular Mass g/mol	mp °C	bp °C	Density g/cm³	Molar Volume LeBas cm³/mol	Solubility g/m³	Vapor Pressure Pa	log K_{ow}
n-Decane	148.28	– 9.60	174.2	0.730	229.4	0.052	175	6.25
n-Dodecane	170.33	– 9.60	216.3	0.766	273.8	0.0034	15.7	6.80
n-Tetradecane	198.38	5.86	253.7	0.763	318.2	0.000655	1.27	8.00
Cyclopentane	70.14	–93.90	49.3	0.799	100.0	156	42400	3.00
Cyclohexane	84.20	6.55	80.7	0.779	118.0	55	12700	3.44
Methylcyclohexane	98.19	–126.60	100.9	0.770	140.4	14	6180	2.82
Naphthalene	128.20	80.20	218.0	1.025	147.6	31.7	10.4	3.35
1-Methylnaphthalene	142.20	–22.00	244.6	1.030	199.7	28.4	7.9	3.87
Biphenyl	154.20	71.00	277.5	0.992	185.0	7.48	1.2	4.03
Fluorene	166.20	116.00	295.0		188.0	1.84	0.008	4.18
Anthracene	178.20	216.20	340.0	1.283	199.0	0.041	0.0008	4.63
Phenanthrene	178.20	101.00	339.0	0.980	199.0	1.29	0.016	4.57
Fluoranthene	202.30	111.00	383.0		217.0	0.263	0.00121	5.22
Pyrene	202.30	156.00	393.0	1.271	214.0	0.135	0.0006	5.22
Chrysene	228.30	255.00	441.0		251.0	0.002	0.00000385	5.79
Benzo[a]pyrene	252.30	175.00	496.0		263.0	0.0038	0.00000073	6.04

S for chemicals which are miscible with water, e.g., ethanol, but these special situations are treated later.

In air S_A is the concentration corresponding to the saturation vapor pressure P^S (Pa). For benzene at 25°C, P^S is 12700 Pa; thus, the concentration can be deduced from the gas law as n/V or P/RT as P^S/RT mol/m^3 where R is the gas constant (8.314 Pa.m^3/mol K) and T is absolute temperature (K). To convert to g/m^3 involves multiplication by the molecular mass W g/mol which is 78 for benzene. It follows that S_A for benzene at 25°C or 298 K is

$$12700 \times 78/(8.314 \times 298) = 400 \text{ g/m}^3$$

This quantity is the "vapor density" commonly used in agrochemical terminology.

Obviously the air/water partition coefficient K_{AW} is S_A/S_W, which is here 0.225. For water-miscible or even fairly water-soluble substances it is preferable to determine S_A then calculate S_W as S_A/K_{AW} because the water solubility (if measurable) is not of the pure chemical, but is of the chemical saturated with water. S_W can also be estimated in such situations from activity coefficient correlations as discussed by Reid et al.[11]

The solubility in a pure solute chemical phase S_P is simply the density (g/m^3) of the liquid hydrocarbon. For example, benzene with a density of 0.879 g/cm^3 has a S_P of 879,000 g/m^3. Another source of S_P which may be useful is an estimated molar volume using, for example, the LeBas correlation for molar volume at the normal boiling point as described by Reid et al.[11] That correlation suggests a molar volume of 96 cm^3/mol which corresponds to a density of 78/96 g/cm^3 or S_P of 813,000 g/m^3. It is noteworthy that this is the highest conceivable solubility because it represents pure undiluted chemical.

For organic solvents such as oils or n-octanol, the solubility S_O is generally somewhat lower than S_P. If the hydrocarbon behaves ideally, i.e., activity coefficients are 1.0, then S_O will equal S_P. Few oil-water partition coefficients are available, so the more widely available octanol-water partition coefficient may be used to estimate S_O from S_W since we can postulate

$$S_O/S_W = K_{OW}$$

For example, for benzene K_{OW} is 135; thus, S_O is 240000, a factor of 3.7 lower than S_P. Much lower solubility will be encountered for solid hydrocarbons because of the lower ideal solubility.[3] It should be noted that this solubility is not necessarily a real quantity; it is often hypothetical, especially if the corresponding concentration is high.

In soils, sorption to organic matter is an important partitioning process. The effective solubility in organic matter can be estimated from the water solubility and K_{OC}, the organic-carbon to water partition coefficient.

If no K_{OC} value is available, a satisfactory estimate is that K_{OC} is 0.41 K_{OW} where K_{OW} is the octanol-water partition coefficient.[18] Since K_{OC} is reported in units of L/kg it must be multiplied by the solid phase density ρ_S g/cm^3 or kg/L to render it dimensionless. Soils typically have solid densities of 2.5 g/cm^3.

The simplest approach is to estimate the solubility in the organic carbon as

$$S_{OC} = K_{OC}\rho_{OC}S_W$$

or the solubility in the whole soil solids as

$$S_S = yK_{OC}\rho_S S_W$$

where y is the mass fraction organic carbon in the soil and is typically 0.01 to 0.02. The use of this latter approach implies that sorption to mineral matter is negligible. For benzene S_{OC} is about 98400 g/m^3; thus, in a 2% organic carbon soil of density 2.5 g/cm^3, S_S is 4920 g/m^3.

For some purposes it will be preferable to calculate a solubility in organic matter as distinct from organic carbon. Assuming organic matter to be 56% by mass organic carbon, the solubility in organic matter S_{OM} is lower, i.e., it is 0.56 S_{OC}, or in the case of benzene, 55100 g/m^3. It may be useful to estimate or measure solubilities in water-solvent mixtures; for example, solutions of MTBE (methyl tert-butyl ether) or alcohols which may be encountered in spill situations. We treat these solutions later.

Table 9.2 lists the estimated effective solubilities of the series of selected hydrocarbons from Table 9.1. The solubilities are, of course temperature-dependent, especially the air solubility or vapor pressure.

Having assembled these data, certain simple calculations become feasible. These are essentially mass balance partitioning calculations.

MASS BALANCE CALCULATIONS

Distribution of a Pure Hydrocarbon in Soil

If we have a soil, which contains water and air phases, contaminated under saturation conditions with a hydrocarbon, it may be useful to calculate the likely concentrations in each phase. The volume of each phase V_i is estimated, and the "effective solubilities" S_i deduced. The maximum amount in each phase (corresponding to the solubility) is then V_iS_i g and the total amount in solution M_S is ΣV_iS_i.

If the amount actually present M equals M_S then all phases are fully saturated and no pure hydrocarbon phase exists.

If the amount present M exceeds M_S, then a pure hydrocarbon phase is likely to be present which will contain approximately $(M-M_S)$ g, and have a volume of $(M-M_S)/S_P$ m^3.

If the amount present M is less than M_S then subsaturation conditions exist. It is then reasonable to estimate that the concentration in each phase is proportionally

Table 9.2.   Effective Solubilities (g/cm³) of Selected Hydrocarbons at 25°C

Chemical	Density MW/V (liquid)	Air S_A	Water S_W	Pure Phase S_P	Octanol S_O	Organic Carbon S_{OC}	Organic Matter S_{OM}
Benzene	0.814	400	1780	8.14×10^5	2.40×10^5	9.84×10^4	5.51×10^4
Toluene	0.781	141	515	7.81×10^5	2.52×10^5	1.03×10^5	5.79×10^4
Ethylbenzene	0.756	54.4	152	7.56×10^5	2.05×10^5	8.41×10^4	4.71×10^4
p-Xylene	0.756	50.2	185	7.56×10^5	2.61×10^5	1.07×10^5	6.00×10^4
1,3,5-Trimethylbenzene	0.739	15.8	48	7.39×10^5	1.82×10^5	7.48×10^4	4.19×10^4
n-Propylbenzene	0.707	21.8	55	7.07×10^5	2.69×10^5	1.10×10^5	6.18×10^4
Isopropylbenzene	0.707	29.6	50	7.07×10^5	2.29×10^5	9.37×10^4	5.25×10^4
n-Butylbenzene	0.726	7.42	13.80	7.26×10^5	2.63×10^5	1.08×10^5	6.04×10^4
1,2,4,5-Tetramethylbenzene (s)	0.726	3.57	3.48	7.26×10^5	3.48×10^4	1.43×10^4	7.99×10^3
Pentylbenzene	0.716	2.63	10.50	7.16×10^5	8.34×10^5	3.42×10^5	1.91×10^5
Hexylbenzene	0.708	0.891	1.02	7.08×10^5	3.38×10^5	1.38×10^5	7.75×10^4
n-Pentane	0.611	1990	38.50	6.11×10^5	1.60×10^5	6.58×10^4	3.68×10^4
n-Hexane	0.613	703	9.50	6.13×10^5	1.22×10^5	5.02×10^4	2.81×10^4
n-Heptane	0.615	247	2.93	6.15×10^5	1.34×10^5	5.49×10^4	3.07×10^4
n-Octane	0.617	86.7	0.66	6.17×10^5	9.99×10^4	4.10×10^4	2.29×10^4
n-Nonane	0.619	29.6	0.122	6.19×10^5	5.45×10^4	2.23×10^4	1.25×10^4

continued

Table 9.2. *continued*

Chemical	Density MW/V (liquid)	Air S_A	Water S_W	Pure Phase S_P	Octanol S_O	Organic Carbon S_{OC}	Organic Matter S_{OM}
n-Decane	0.646	10.5	0.052	6.46×10^5	9.25×10^4	3.79×10^4	2.12×10^4
n-Dodecane	0.622	1.08	0.0034	6.22×10^5	2.15×10^4	8.80×10^3	4.93×10^3
n-Tetradecane	0.623	0.102	0.00065	6.23×10^5	6.55×10^4	2.69×10^4	1.50×10^4
Cyclopentane	0.701	1200	156	7.01×10^5	1.56×10^5	6.40×10^4	3.58×10^4
Cyclohexane	0.714	432	55	7.14×10^5	1.51×10^5	6.21×10^4	3.48×10^4
Methylcyclohexane	0.699	245	14	6.99×10^5	9.25×10^5	3.79×10^3	2.12×10^3
Naphthalene (s)	0.869	0.538	31.70	8.69×10^5	7.10×10^4	2.91×10^4	1.63×10^4
1-Methylnaphthalene	0.836	0.453	28.40	8.36×10^5	2.11×10^5	8.63×10^4	4.83×10^4
Biphenyl (s)	0.834	7.47×10^{-2}	7.48	8.34×10^5	8.01×10^4	3.29×10^4	1.84×10^4
Fluorene (s)	0.884	5.37×10^{-4}	1.84	8.84×10^5	2.78×10^4	1.14×10^4	6.39×10^3
Anthracene (s)	0.895	5.75×10^{-5}	0.041	8.95×10^5	1.75×10^3	7.17×10^2	4.02×10^2
Phenanthrene (s)	0.895	1.16×10^{-3}	1.290	8.95×10^5	4.79×10^4	1.97×10^4	1.10×10^4
Fluoranthene (s)	0.932	9.88×10^{-5}	0.263	9.32×10^5	4.36×10^4	1.79×10^4	1.00×10^4
Pyrene (s)	0.945	4.90×10^{-5}	0.135	9.45×10^5	2.24×10^4	9.19×10^3	5.14×10^3
Chrysene (s)	0.910	3.55×10^{-7}	0.002	9.10×10^5	1.23×10^3	5.06×10^2	2.83×10^2
Benzo[a]pyrene (s)	0.959	7.13×10^{-8}	0.0038	9.59×10^5	4.17×10^3	1.71×10^3	9.57×10^2

Solids are denoted (s).

reduced from M_S, i.e., it is $S_i M/M_S$. The ratio M/M_S is the "fraction of saturation" which we can designate F. The concentrations in each medium are then FS_i g/m³ and the amounts are V_iFS_i g.

Example

We consider a soil, air, water system as shown in Table 9.3 which is contaminated with benzene. The effective solubilities are as shown. The sum of the V_iS_i groups is 1578000 g or 1578 kg. If 1578 kg is present, conditions will be approximately as shown in the "Saturation" line 5. If 2000 kg is present, the "Supersaturation" line 6 will apply, with a pure benzene phase of 422 kg or 480 L present with a small volume of air being displaced. If only 1000 kg is present, the "Subsaturation" line 7 will apply, F being 0.634. The concentrations can then be calculated as shown.

Table 9.3. Calculation of Distribution of Benzene in a Soil

Soil, containing benzene, 1 m deep, 0.1 ha (1000 m²) in area of volume 1000 m³, consisting of 30% air, 20% water, 48% mineral matter and 2% organic matter, i.e., 1.12% organic carbon.

			Medium			
	Benzene	Air	Water	Organic Matter	Mineral Matter	Total
Volume (V, m³)	0	300	200	20	480	1000
Volume fraction	0	0.3	0.2	0.02	0.48	1
Solubility (S, g/m³)	879000	400	1780	55100	0	—
Mass, VS (g)	0	120000	356000	1102000	0	1578000
Saturation (kg)	0	120	356	1102	0	1578
Supersaturation (kg)	422	120	356	1102	0	2000
Subsaturation (kg)	0	76	225	699	0	1000
Subsaturation concentration (g/m³)	0	253	1125	35000	0	—

Distribution of a Hydrocarbon with an Oil Present

Clearly, from the above example, the concentration of hydrocarbon in a hydrocarbon or oil phase greatly exceeds those in air or water and even in soil. If an oil is present which contains (for example) benzene, it is likely that most of the benzene will be present in the oil; thus, the other phases will tend to adopt concentrations controlled by, or buffered by, that in the oil.

If the actual concentration in the oil is known and is, for example, 4.8 g/L or 4800 g/m³, then since S_O is 240000 g/m³, the benzene is present at 4800/240000 or 1/50th of saturation. F is then 1/50 or 0.02. The other phases will also tend to approach this fraction of saturation. For example, the concentration in water will be 0.02 × 1780 or 36 g/m³. Table 9.4 shows such a calculation for a

Table 9.4. Calculations of Subsaturation Conditions (F = 0.02) in the Presence of an Oil Phase Containing Benzene

	Medium					
	Oil	Air	Water	Organic Matter	Mineral Matter	Total
Solubility (S, g/m^3)	24000	400	1780	55100	0	—
Volume (V, m^3	50	250	200	20	480	1000
Benzene concentration (g/m^3)	4800	8	36	1102	0	
Benzene mass (kg)	240	2	7	22	0	271

soil containing a substantial quantity (5% by volume) of a petroleum product. The useful feature of this calculation is that it gives the maximum concentration of benzene which is likely to be present in each phase. Essentially, this is a modified Raoult's law calculation in which it is assumed that all activity coefficients are constant at their "infinite dilution" values.

The actual, prevailing concentrations are likely to be lower than is indicated above, but at least the method gives an order-of-magnitude estimate. This calculation also shows which hydrocarbons are likely to be encountered in each phase. For example, it is probably futile to search for n decane in water, unless there is appreciable emulsification, or a strong co-solvent is present.

If the amount of oil present is smaller, then the disposition of benzene between air, water, soil, and the oil phase can also be calculated.

DEPLETION BY TRANSPORT

In an actual contaminated soil situation, there is usually transport of hydrocarbon from the site by advective flow in air and water. It is useful to estimate this rate of loss, especially as a means of deducing the time for depletion. In Table 9.5, it is assumed that air is leaving the site at a rate of 200 m^3/h. This could be by deliberate soil venting, or by natural diffusion. In the diffusion case this volumetric rate can be estimated as

$$kA \text{ or } DA/Y \text{ m}^3/\text{h}$$

where k is a mass transfer coefficient (m/h), A is the area, D is a diffusivity in air m^2/h and Y is a diffusion path length.

Similarly, water is flowing at a rate of 0.1 m^3/h either by natural flow or induced by pumping.

The rates of loss of benzene in these flows can be readily deduced by multiplying the flowrate (m^3/h) by the concentrations (g/m^3) as shown. In this case, it is clear that evaporation is much more important as a loss mechanism, but dissolved benzene may, of course, cause severe "downstream" contamination problems in wells or aquatic systems.

Table 9.5.   Transport Rates of Benzene from the Soil Described in Table 9.4

Soil area (A)	1000 m^2
Mass transfer coefficient (k)	0.2 m/h
Volumetric loss of air (kA or G_A)	200 m^3/h
Concentration of benzene in air (C_A)	8 g/m^3
Rate of loss of benzene (GC$_A$)	1600 g/h or 1.6 kg/h
Water flow rate (W)	0.1 m^3/h
Concentration of benzene in water (C_W)	36 g/m^3
Rate of loss of benzene (WC$_W$)	3.6 g/h or 0.0036 kg/h
Half-life = (0.693 × 271) kg/(1.6 + 0.0036) kg/h =	117 h = 4.9 days
Half-life for evaporation	0.693 × 271/1.6 = 117 h
Half-life for leaching	0.693 × 271/0.0036 = 52000 h
Half-life for reaction	200 h (assumed)
Overall half-life = 1/(1/117 + 1/52000 + 1/200) =	74 h
Relative process rates are proportional	
to 1/half-lives, i.e.,	
evaporation	0.00855 or 63%
leaching	0.00002 or 0.1%
reaction	0.00500 or 37%
total	0.01357 or 100%

In this case, the rate of loss is 1.6 kg/h from a total amount of 250 kg, i.e., a fraction of 1.6/250 or 0.0064 or 0.64% is lost each hour or 15% each day. As the benzene is depleted, the concentrations will drop and a steady 15% of the amount present (not the initial amount) will be lost each day. It is interesting to calculate a "half-life" which can be shown to be 0.693/0.15 or 4.5 days. Often it will be apparent that this time is so long, i.e., many years, that transport is insignificant as a loss process. Reaction or degradation losses may then be dominant and there is little further incentive to deduce transport rates. If a reaction half-life is known it can be used to deduce a rate of loss and hence an overall half-life. The relative contributions of each process can be deduced as shown.

This approach can also be used to explore the feasibility of certain remedial measures such as soil venting or solvent extraction.

Concentrations in Nearby Atmospheres

The concentration of benzene in air in intimate contact with the oil-contaminated soil in Table 9.5 is 8 g/m^3 and the rate of loss by evaporation is 1600 g/h in an air flow of 200 m^3/h. This vapor will be appreciably diluted by dispersion when it reaches the atmosphere a meter or two above the soil surface. The dilution factor is essentially the ratio of the air ventilation rate above the soil to the air flow rate from the soil. If the spill site in Table 9.5 is 30 m wide and the wind velocity is 5 km/h or 5000 m/h, this gives a volumetric ventilation flow to a height of, say, 3 m (10 ft) of 5000 × 30 × 3 or 450000 m^3/h. A dilution factor of 450000/200 is thus expected, i.e., 2250. Prevailing air concentrations of about 8/2250 or 0.0036 g/m^3 or 3.6 mg/m^3 are expected.

This is a fairly adventurous calculation, but it may be sufficient to show that the concentration of hydrocarbon in air is definitely, or definitely not, of concern. The obvious concentration for comparison is a Threshold Limit Value (TLV) which in the case of benzene is 30 mg/m^3.[19]

This dilution factor is controlled by the ratio of the wind velocity (which is typically 5000 m/h) and the evaporation mass transfer coefficient (which is typically 0.2 m/h). This coefficient is the actual effective velocity with which air is migrating out of (and into) the soil surface layer. The ratio is thus about 25000 but may vary from this value by a factor of 3. It is then multiplied by the ratio of the atmosphere cross section (here 90 m^2) to the soil area (1000 m^2) which is typically of magnitude 0.1. It follows that concentrations in the air may be about 2500 times lower than the air in the soil. A compelling case can be made that when monitoring the site, it is better to analyze soil gas with its higher, more stable hydrocarbon concentration, than the atmosphere with its inherently greater variability, and much lower concentration.

It would be useful if measurements of air and soil air concentrations could be made at contaminated sites and reported to build up a database of actual ratios encountered in practice.

A similar approach is feasible in principle for dilution in water.

Concentrations in Biota

There is a fair knowledge of partitioning tendencies for chemicals from water and air into biota. For example, fish-water bioconcentration ratios are well correlated with K_{OW}.[20] Leaf-air concentration ratios have been developed recently.[21] There is a formidable literature on partitioning in the pharmacokinetic setting of mammals, especially rodents and humans. In most cases the primary sites of accumulation are lipids, waxes, or nonpolar organic phases. An attractively simple approach is to develop correlations based on the supposition that the animal or plant behaves as if it contains a fraction y of octanol in which all accumulation occurs. For fish, y is typically 0.05. The "solubility" in these biota are then S_B given by

$$S_B = yK_{OW}S_W$$

For plant foliage y appears to be in the range 0.005 to 0.05. There is even a possibility of applying this approach to burrowing animals which may be intimately exposed to contaminated soils.

The concentrations in these biota can then be estimated by treating them as yet another phase with a concentration FS_B. A useful feature of this approach is that it may be easier to collect and analyze biotic samples. Further, the resulting concentrations can be compared with body burdens known to cause adverse effects, or with tolerable concentrations in foodstuffs such as vegetables. Biota have the advantage of accumulating the "available" hydrocarbon, thus giving an estimate of exposure as distinct from concentration.

For vegetation the value of F which applies in the air phase will be much lower than that in the soil, thus the value in the plant will depend on whether the tissue is root, stem, fruit, or foliage and the magnitude of rates of transport in the plant. Talmage and Walton[22] have recently provided an excellent review of the use of small mammals as monitors of environmental contamination of soils by metals, radionuclides and organic chemicals.

If metabolism occurs within the organism, the hydrocarbon concentration will be reduced. In some cases there may be kinetic limitations to uptake rate resulting in much lower concentrations than the ultimate equilibrium value. Finally, for some organisms there may be biomagnification, i.e., higher than equilibrium concentrations can occur.

COSOLVENTS

Fortunately, hydrocarbons are generally sparingly soluble in water, thus groundwater contamination is mitigated by the low saturation levels which can be achieved. There is concern that the presence of cosolvents such as alcohols or ethers may worsen contamination by virtue of the enhanced hydrocarbon solubility in such systems.

The hydrocarbon solubility increases as solvent is added, generally in a near log-linear fashion, so that if the hydrocarbon solubility in water is S_W and that in the pure cosolvent is S_C, the solubility in a mixture S_M is approximately[23]

$$\log S_M = v_w \log S_W + v_C \log S_C$$

where v_W and v_C are the volume fractions of water and cosolvent. Generally, a fairly large volume fraction v_C is needed to enhance solubility appreciably. For example, if S_W is 1 g/m^3 and S_C is 10^5 g/m^3, a 1% cosolvent mixture ($v_C = 0.01$) will have a solubility of only 1.12 g/m^3. A 6% mixture is needed to double the solubility to 2 g/m^3.

This logarithmic-linear relationship must be used with caution because there is evidence that substantial deviations may occur.

A note of caution is appropriate when measuring or interpreting data for the solubility of hydrocarbons in cosolvent mixtures. Inevitably the cosolvent will migrate into (be extracted by) the liquid hydrocarbon; thus, the measured or observed solubility is not that of the liquid hydrocarbon, it is of the hydrocarbon-cosolvent phase.

Sorption to Emulsions or Organic Colloids

If water contains emulsified oil, surfactant micelles, or organic colloids, there may be an apparent "solubilization" which in reality is partitioning into these second phases at enhanced concentrations. If the solubility in these media is known, or can be estimated from the "solubilities" (S_O) in octanol, oil, or organic matter, then the solubility in the mixture (S_M) will be

$$S_M = v_w S_w + v_o S_O$$

where v_w and v_o are volume fractions of water and the second phase. This linear equation contrasts with the log-linear equation for single phase cosolvents, especially in the low concentration region.

If S_w is 1 g/m^3 and S_O is 10^5 g/m^3 as before, then a 1% mixture ($v_o = 0.01$) will have an apparent solubility of 1000 g/m^3, compared with 1.12 g/m^3 in the cosolvent case.

The conclusion is that such "two-phase" systems have a much greater capacity to dissolve and transport hydrocarbons than homogeneous cosolvent systems. The corollary is that, if in practice, significantly (e.g., factor of over 2) higher than saturation concentrations are encountered, it is probably the result of a "second phase" effect. Colloids and emulsions and surfactant micelles are clearly much more effective than cosolvents in promoting hydrocarbon transport in groundwater.

SOIL CLEANUP CRITERIA

An issue of considerable current interest is defining soil cleanup criteria or acceptable concentrations of various hydrocarbons. These criteria may be based on various adverse effects including soil ingestion, phyto-toxicity or groundwater contamination as discussed by Bell et al.[24] and Wong et al.[25] The similar problem which exists when defining sediment quality criteria can be resolved by the simple expedient of decreeing that the acceptable sediment concentration is the product of the equilibrium sediment-water partition coefficient and an acceptable water quality criterion. In essence the water quality criterion is applied to the sediment pore water. Alternatively the fugacity or fraction of the effective solubility applicable to water is also applied to sediment.

In the case of soils, air, water, or soil solids, criteria can be gathered as concentrations, then expressed as fractions of effective solubility for the chemical in question (see illustration in Table 9.6). The lowest, most demanding, fraction is selected and applied to all media. This will ensure that all media equilibrating with the soil will be at acceptably low concentrations. There is a certain elegance to defining these multimedia criteria as a single quantity. Further, soils with greater sorptive capacities will be assigned a higher acceptable concentration because the hydrocarbon is less "available." Undoubtedly there are disadvantages to this approach, but we suggest that it deserves assessment.

Table 9.6.   Illustration of Multimedia Criteria Setting by Applying a Single F of 10^{-3} to a Hypothetical Chemical

Medium	Effective Solubility	Criterion	Basis	Fraction F	Suggested Criterion (F = 10^{-3})
Air	100 g/m^3	0.1 g/m^3	Inhalation	10^{-3}	0.1 g/m^3
Water	5 g/m^3	0.01 g/m^3	Drinking water	2×10^{-3}	0.005 g/m^3
Soil solids	5000 mg/kg	25 mg/kg	Ingestion	5×10^{-3}	5 mg/kg

DISCUSSION AND CONCLUSIONS

The calculations illustrated here should not be regarded as accurate deductions of hydrocarbon concentrations or fluxes. Rather, they are order-of-magnitude estimations of concentrations which will prevail, or are approached, under near-equilibrium conditions. From experience, it is usually possible to estimate a typical time necessary for equilibrium to be achieved. For example, water contacted with oil generally reaches equilibrium within a day, provided that the volumes are not too large, which results in delays from diffusion or mixing. It is thus possible to identify phases which will be close to equilibrium (e.g., soil and soil water) and those which will be far from equilibrium (e.g., soil and air above). In the latter cases a dilution factor approach may be useful.

The ability to conduct rapid, approximate, back-of-the-envelope calculations as described here may be useful in several situations such as:

- checking the reasonableness of the results of more complex computer simulations
- deciding on the feasibility of analytical or monitoring programs
- undertaking calculations during emergency conditions
- determining a "worst case" situation with regard to human exposure
- presenting simple order-of-magnitude calculations for scrutiny during litigation where there is a need to convince nonscientists of the reasonableness of the assertions.

It is hoped that these ideas will contribute, in their limited way, to assessments and remedial actions of situations involving hydrocarbon contaminated soils.

REFERENCES

1. Eastcott, L., W. Y. Shiu, and D. Mackay. "Environmentally Relevant Physical-Chemical Properties of Hydrocarbons: A Review of Data and Development of Simple Correlations," *Oil & Chemical Pollut.* 4:191–216 (1988).
2. Mackay, D., and W. Y. Shiu. "A Critical Review of Henry's Law Constants for Chemicals of Environmental Interest," *J. Phys. Chem. Ref. Data* 10:1175–1199 (1981).

3. Miller, M. M., S. P. Wasik, G. L. Huang, W. Y. Shiu, and D. Mackay. "Relationships Between Octanol-Water Partition Coefficient and Aqueous Solubility," *Environ. Sci. Technol.* 19:522–529 (1985).

4. Zwolinski, B. J., and R. C. Wilhoit. *Handbook of Vapor Pressures and Heats of Vaporization of Hydrocarbons and Related Compounds.* API-44 TRC Publication No. 101, Texas A & M University, Evans Press, Fort Worth, Texas, 1971.

5. Sonnefeld, W. J., W. H. Zoller, and W. E. May. "Dynamic Coupled-Column Liquid Chromatographic Determination of Ambient Temperature Vapor Pressure of Polynuclear Aromatic Hydrocarbons," *Anal. Chem.* 55:275–280 (1983).

6. McAuliffe, C. "Solubility in Water of Paraffin, Cycloparaffin, Olefin, Acetylene, Cycloolefin and Aromatic Hydrocarbons," *J. Phys. Chem.* 70:1267–1275 (1966).

7. Mackay, D., and W. Y. Shiu. "Aqueous Solubility of Polynuclear Aromatic Hydrocarbons," *J. Chem. Eng. Data* 22:399–402 (1977).

8. *IUPAC Solubility Data Series, Volume 37: Hydrocarbons (C_6-C_7) with Water and Seawater.* Shaw, D. G., Ed. (Oxford, England: Pergamon Press, 1989).

9. *IUPAC Solubility Data Series, Volume 38: Hydrocarbons (C_8-C_{36}) with Water and Seawater.* Shaw, D. G., Ed. (Oxford, England: Pergamon Press, 1989).

10. Hansch, C., and A. Leo. *Substituent Constants for Correlation Analysis in Chemistry and Biology* (New York, NY: John Wiley & Sons, Inc., 1979).

11. Reid, R. C., J. M. Prausnitz, and B. E. Polling. *The Properties of Gases and Liquids.* 4th ed. (New York, NY: McGraw-Hill Book Company, 1987).

12. Weast, R. *Handbook of Chemistry and Physics,* 64th ed. (Boca Raton, FL: CRC Press, Inc., 1983–1984).

13. Verschueren, K. *Handbook of Environmental Data on Organic Chemicals,* 2nd ed. (New York, NY: Van Nostrand Reinhold, 1983).

14. Callahan, M. A., M. W. Slimak, N. W. Gabel, I. P. May, C. F. Fowler, J. R. Freed, P. Jennings, R. L. Durfee, F. C. Whitmore, B. Maestri, W. R. Mabey, B. R. Holt, and C. Gould. "Water-Related Environmental Fate of 129 Priority Pollutants. Vol. II, Halogenated Aliphatic Hydrocarbons, Halogenated Ethers, Monocyclic Aromatics, Phthalate Esters, Polycyclic Aromatic Hydrocarbons, Nitrosamines and Miscellaneous Compounds," EPA-440/4-79-029b, 1979.

15. Mackay, D., and S. Paterson. "Fugacity Revisited," *Environ. Sci. Technol.* 16:654A–660A (1982).

16. Mackay, D. *Multimedia Environmental Models: The Fugacity Approach.* (Chelsea, MI: Lewis Publishers Inc., 1991).

17. Mackay, D., and M. Diamond. "Application of the QWASI (Quantitative Water Air Sediment Interaction), Fugacity Model to the Dynamics of Organic and Inorganic Chemicals in Lakes," *Chemosphere* 18:1343–1365 (1989).

18. Karickhoff, S. W. "Semi-Empirical Estimation of Sorption of Hydrophobic Pollutants on Natural Sediments and Soil," *Chemosphere* 10:833–849 (1981).

19. Threshold Limit Values and Biological Exposure Indices for 1985–1986, American Conference of Governmental Industrial Hygienists, Cincinnati, OH, 1985.

20. Mackay, D. "Correlation of Bioconcentration Factors," *Environ. Sci. Technol.* 16:274–278 (1982).

21. Bacci, E., D. Calamari, C. Goggi, and M. Vighi. "Bioconcentration of Organic Chemical Vapors in Plant Leaves: Experimental Measurements and Correlations," *Environ. Sci. Technol.* 24:885–889 (1990).

22. Talmage, S. S., and B. T. Walton. "Small Mammals as Monitors of Environmental Contaminants," *Rev. Environ. Contam. Toxicol.* 119:47–145 (1991).

23. Morris, K. R., R. Abramowitz, R. Pinal, P. Davis, and S. H. Yalkowsky. "Solubility of Aromatic Pollutants in Mixed Solvents," *Chemosphere* 17:285–298 (1988).

24. Bell, C. E., P. T. Kostecki, and E. J. Calabrese. "Review of State Cleanup Levels for Hydrocarbons," in *Hydrocarbon Contaminated Soils and Groundwater.* Vol. I, P. T. Kostecki, and E. J. Calabrese, Eds., (Chelsea, MI: Lewis Publishers Inc., 1991), pp. 77–89.

25. Wong, J. J., G. M. Schum, E. G. Butler, and R. A. Becker. "Looking Past Soil Cleanup Numbers," in *Hydrocarbon Contaminated Soils and Groundwater.* Vol. I, P. T. Kostecki, and E. J. Calabrese, Eds., (Chelsea, MI: Lewis Publishers Inc., 1991), pp. 1–21.

PART IV

Sampling and Site Assessment

Application of Aerial Infrared Survey Techniques in Detailing Soil Contamination

J. M. Fernandez, Horizon Helicopters, Rancho Murieta, California

DEFINITION

All matter can be viewed and analyzed by sensing the many wavelength classifications along the spectral chart. Of particular interest to this chapter is the sensing of wavelengths in the 8 to 14 micron region. This band is referred to as long wave infrared. Matter can be viewed in this spectral region through the complete diurnal cycle, through dust, smoke, and other climatic systems. Electronic imagers, video systems, and photographic systems sense energy emissions from matter, using differential correlations to produce detailed imagery in video and photographic media forms.

HISTORY

Infrared technology was developed in the early 1950s for military applications. In 1960, limited infrared imagery was available to a select few in the civilian sector. In 1968, the technology was released to the general public. Currently, the systems available on the open market are equivalent in imagery and articulation to the most advanced military systems.

APPLICATIONS

Infrared imagery technology can be applied to a number of industrial applications. Typical applications are:

- Electrical Energy Transmission System Audits
- Pipe Line Integrity Audits
- Furnace Efficiency Audits
- Roof Moisture Surveys (structural)
- Building Energy Conservation Surveys
- Mechanical Systems Audits (bearings, valves)
- Circuit Board Quality Control
- Road and Bridge Deck Structural Analysis
- Environmental Audits

Electrical Energy Transmission System Audits

During the electrical transmission process, corroded or loosely applied connections will cause electrical arcing. The arcing creates higher temperatures and energy levels of electrical components. The resultant contrasting energy is detectable by infrared imagers.

Pipe Line Integrity Audits

Surface and subsurface pipe line systems must be coated and lined with insulation to prevent corrosion from occurring in the system. Damaged insulation and corrosion can be detected by infrared imagers. Product leakage (oils, liquid gas, vapor gases, etc.) will display abnormal temperatures and unique flow patterns.

Flue Efficiency Audits

Industrial furnaces work at optimum efficiency when the combustion patterns are correctly oriented and applying the proper amount of heat. Plugged jets create serious combustion problems. Combustion patterns and intensity can be imaged and measured by viewing the combustion chambers and kiln areas with infrared imagers.

Roof Moisture Surveys (Structural)

Roofing systems (outer protective layers, insulation, and support decks) are subjected to harsh environmental activities such as wind, rain, sun, and air pollution. When the protective layer is breached and moisture enters the insulation layer, the insulation efficiency drops, the weight of the added water increases roof loading (reduces structural integrity), and electrical systems can be damaged.

The water, with the aid of the insulation, retains solar energy during the day. At night, the moist areas emit the energy at a slow rate that is detectable by infrared imagers.

Building Energy Conservation Surveys

Building structures will shift after some period of time, exposing open joints, leaking windows and doors, and displaced insulation. The problem areas will allow internal heat and cold air (from environmental control systems) to escape the building. Outside and inside infrared imaging will detect the leak areas.

Mechanical Systems Audits (Bearings, Valves)

Excessive energy (heat) from mechanical wear on bearings, shafts, and other moving parts can reduce the useful life of operating machinery. Abnormal energy (heat patterns) can be detected by infrared imagers. Mechanical repairs can show immediate results in reduced energy loss.

Circuit Board Quality Control

Modern circuit boards contain many small (very small) electrical circuits layered in plastics. These individual circuits can be checked for integrity by energizing each circuit and viewing it with an infrared imager. If the circuit appears cold or too hot, the circuit may have a defect. The infrared system sits on an assembly line as circuit boards move along.

Road and Bridge Deck Structural Surveys

Road surfaces and bridge decks are susceptible to erosion, wear, and delamination. Water contamination of road beds can create voids below the surface causing a pumping action with sand that will cause serious pavement deterioration. Bridge decks are subjected to similar problems, The bridge decks are also susceptible to corrosion of structural members. Corrosion has a very distinctive energy level compared to normal metal, and can be detected by infrared imagers. The subsurface voids and moisture can be detected by noting cool areas in the pavement with the infrared imager.

Environmental Assessments

The modern infrared imagers can successfully identify subsurface voids, combustion energy activity, and rising energy levels by contrast and analysis. Environmental assessments include surface and subsurface pipe and tank systems, chemical spills, waste disposal areas, microbiological energy evolution, and wetlands identification.

Other Applications

Infrared imagers are used in the medical industry (tumor location, muscle and nerve damage, etc.), communications industry (voice and data), aircraft navigation, and general night vision activity.

Infrared Imagers

Infrared imagers can be designed to sense energy in short wave, long wave, or specific regions. The infrared detector/sensor design, scan cavity, and filters determine the wavelength region and flexibility.

The energy detected can be converted to various types of data signals. The media form varies from plotting data to video imagery. The data can be recorded on wire recording decks, computer disks, video decks, or sent via microwave and VHF transmission to receiving stations.

The discussion of system architecture will be structured to address an infrared imager with detection into 8 to 12 micron region and real time video output.

System Architecture

A typical infrared imaging system incorporates a series of infrared detectors, high-speed scan assembly, a gas or close cycle cooling system, optics for standard image and zoom, associated electronics for video output, and an articulation system for system movement. Only aerial mounted systems utilize this articulation system.

Incident radiation as a function of wavelength is measured by radiation entering an optical area (lens of the system), a diffraction grating system to spread the infrared radiation (like visible light through a prism), into a bandpass filter system, and then into the detector. A high speed oscillating mirror and one rotating facet mirror are used to scan the detector for infrared information. The optical systems are usually constructed of germanium or Irtran.

The primary detector material used for infrared scanners is mercury cadmium telluride (HgCdTe) with a spectral response extending from 2 to 14 microns. Most systems are usually limited to a range of 8 to 12 microns. The detector is organized in a 2×4 TDI array.

The typical system has two fields of view: wide field of view (WFOV) and narrow field of view (NFOV). The WFOV has a horizontal scan of 28 degrees and a vertical scan of 15 degrees. If the narrow field of view represents a magnification of four (4) times, the horizontal scan would be 7 degrees and the vertical scan would be 3.25 degrees.

The minimum resolvable temperature difference (MRTD) is 0.2°C for wide field of view and 0.25°C for narrow field view.

The instantaneous field of view (IFOV) for wide field of view is 1.4 milliradians and 0.35 milliradians for narrow field of view.

The video formats include EIA, RS170, and CCIR.

The cooling system maintains system integrity by insulating the scan area and detectors and maintaining temperatures of approximately $-195°C$ during operations. This ensures that only infrared radiation (energy) associated with a selected target will be processed.

Other Technical Considerations

The infrared imager in aerial format is portable and can be employed in the field without disruption to vegetation and soil. Typically, these systems are mounted to the underside of a helicopter or the wing of a fixed wing airplane.

The infrared imagers are nonintrusive systems. The infrared wave form (radiation/energy) travels to the imager. The imager does not alter sample material or invade space.

Limitations in Use

The infrared imager receives wave forms from all objects in the target area. In order to precisely define, in video form, all objects in the target area, a thermal differential among objects must be present. The differential could be as small as 0.10 to 0.20°C. Since all objects work within the parameters of thermal equilibrium, thermal differentials are almost always present.

Thermal influence from surrounding objects can be a factor. Overhead structures can mask the thermal characteristics of targeted objects. Energy emissions from surrounding objects can blend in with the targeted object, thus altering its thermal value. The impact of these limitations can be overcome by altering imaging times, imaging angles, and altitudes above the target zone.

Weather can disrupt the quality of imagery by defusing the infrared radiation or counteracting the energy releases from objects.

Wind can cause the infrared radiation to spread and reduce the quality of the thermal definition of an object. In video form, the image appears to lose edge definition.

Rain can reduce energy emissions to one value for all objects. The droplets act as a heat sink, absorbing energy from all objects at an equal rate.

The images produced by infrared systems usually appear in grey tones or color. The quality of the image is equal to a normal video camera image. The grey tones provide a good media for the human eye and computer systems to perceive very small rates of change.

The imagery can be very subtle and require skill and knowledge to analyze. Interpretation of the infrared images requires a certain degree of skill and knowledge about the infrared system and the specific application. Education and experience are key factors for proper interpretation of video infrared images.

Characteristics of Waste, Applicable to Infrared Imagery

The infrared imager receives wave forms from all objects in three different forms. The forms are:

- Emission
- Absorption
- Reflectivity

The emissive process can be generated by waste materials decomposing (high energy levels and combustion), microbiological activity (oil-eating bacteria generating heat), and other catalytic events (chemicals combining with nutrients in soil or two chemicals combining in a spill).

The absorption process can be generated by objects absorbing energy from surrounding objects and displaying a lower temperature and energy level.

The reflective process can be generated by energy from sunlight or emissive energy from surrounding objects.

Application of the emissive process would include imaging waste disposal sites, locating subsurface pipe systems transporting volatile fluids, improper leachates, and monitoring oil recovery processes.

Application of the absorption process would include locating subsurface vaults, trenches, pipe lines, tanks, and water systems (wetlands).

Application of the reflective process would include locating chemical spills in daylight hours, locating metals, and location of rock deposits for vernal pool sources.

Application of Infrared Imaging in Environmental Assessments

The application of aerial infrared imaging in site screening will reduce the overall cost of the screening and provide potentially more information about the site. Unauthorized disposal, drainage patterns, buried tanks and pipes, and chemical spills will be imaged if present on the site. The imagery gives a view of the extent of second phase tasks. The infrared imagery may yield a more accurate method for boring locations and soil sampling.

The area of coverage is complete and efficient. This could yield substantial cost savings to clients. A 3,000 acre site may require as little as two hours of imagery time on site.

Case Study

The client has granted permission to discuss the following study, under conditions that the company name, site locations, product, and development names not be used.

Three aerial infrared surveys were conducted to detect the presence of hydrocarbon products at and below the surface.

Site one housed a storage facility for hydrocarbon products, and is currently being converted to nonindustrial activity.

Site two was a large expanse of agricultural property designated for commercial use.

Site three was a storage facility for hydrocarbon products. The site is currently active as a residential community.

Sites one and three have been carefully analyzed and cataloged for presence of hydrocarbons.

SITE ONE

This site housed a containment facility for hydrocarbon products for a period of approximately 40 years. The facility became expendable and was reclassified as commercial/retail/residential property. Prior to any construction, the property was cleared of all containment facility material. During site excavation, large subsurface deposits of product were discovered at depths of surface to 60 feet. The product locations were localized. A plan was formulated to remove the product, meet air quality and water quality standards for the area, and facilitate a new topology for the site. Several methods were used to remove and transport the product from the surface and deep locations to holding area such as ponds and large berm systems. Soil blending was used when permitted to facilitate air and water quality standards.

Upon completion of the excavation, removal, and blending process, an aerial infrared survey was conducted. The survey monitored the holding areas, verified clean soil areas, and searched for areas in need of remediation.

The survey detected all known conditions on the site, several conditions offsite, and several areas in need of remediation.

The known conditions included subsurface pipe systems, holding ponds for recovered material, recovery mounds of product-saturated soil, recovery systems, and blended soil and product.

The offsite conditions included microbiological soil treatment moving onsite, potential drainage (surface and subsurface), product migration offsite, and an offsite waste disposal area.

The areas requiring remediation included a leaching recovery system, soil blending outside set limits, and one area untouched during the excavation process.

SITE TWO

This site was used for agricultural production from the 1930s to the late 1980s. The site activities included cattle and crop production. At times during the 50 year span, a small fuel pumping facility was located on or near the property for the purpose of dispensing fuel. Available written historical records were inconclusive in confirming the existence and location of this fuel pumping facility.

Confirmation of the fuel facility would be complete with the identification of sub-surface tanks, foundation systems, sumps, and/or pipe lines.

An aerial infrared survey was conducted over the entire property and just out-side the legal boundaries.

The survey detected subsurface foundations for a barn, farm house, well pump facility, and driveway. Natural drainage and a septic system were detected on the property. A subsurface metal pipe (approximately 2 inches in diameter) was discovered near the foundation. The pipe system turned at a right angle toward the barn. The head of the pipe system appeared to be located in the vicinity of a large freeway system. The pipe system was buried at a uniform depth of three feet.

SITE THREE

This site was similar in function to the Site One facility. The site was excavated at the surface and reclassified for use. The use activity included residential and recreational park facilities. Discovery of surfacing product required further evalu-ation and remediation. Since the site had been developed, excavation and removal could not be considered. New technology was implemented to draw the product to recovery systems.

Aerial infrared surveys were conducted before the remediation started and after the new technology system had been operating for approximately three months. The surveys were directed toward identifying any key areas for remediation, evalu-ating the location of test wells for placement accuracy, and monitoring the remedi-ation process of the new technology recovery system.

The survey confirmed the successful placement of the test wells, several areas of heavy product concentration, and movement from the product saturated areas to the recovery system.

CONSIDERATIONS

The aerial infrared survey process relies upon the presence of contrasting energy levels. The energy levels and resulting temperatures must be below or above am-bient air in order that infrared detection may occur. The process of thermal equilibrium aids in providing contrast in energy and corresponding temperatures. When hydrocarbons are induced into a soil system, energy contrast is accom-plished by several factors including microbiological activity and conductivity of solar energy.

The data in Table 10.1 were gathered from a test well at Site Three during the aerial infrared survey.

Table 10.1. Information Gathered at Site Three Test Well

Soil Temperatures			(8:30 a.m.)
Distance from Well (feet)	Depth (feet)	Temperature (°F)	Notes
10	1	64.5	Extrapolation Measured
10	3	71.2	"
10	5	76.7	"
10	7	81.0	"
20	5	69.6	"
40	5	67.6	"
150	3	55.8	Background data not in
150	5	57.1	area of remediation
150	7	59.4	"

The soil temperatures were taken at 8:30 a.m. The temperatures reflect the depth and distance from the well head. The surface temperature (52°F) was recorded at 8:30 a.m. Note the surface should be warmer than the one foot depth if subsurface thermal activity is not present.

CONCLUSIONS

The aerial infrared survey process provides an efficient method of discovery during Phase I assessments (a Phase I assessment void of any discoveries is as important as an assessment with discoveries). The survey process can be used to develop selection methods for boring locations, evaluate placement accuracy for current borings, verify remediation activity, search for surface and subsurface activity not identified during archive searches and walkthroughs, and develop information relevant to offsite activities that may affect the client's survey site. The survey process provides a nonintrusive method of analysis, and thus eliminates the need to disturb soil and vegetation. The helicopter platform provides fast search grids over large parcels and accessibility to areas not serviced by road systems.

GLOSSARY OF TERMS

Apparent Contrast: The target to background contrast seen by an observer or other sensor separated from the target scene by a contrasting-degrading medium such as the atmosphere.

Aspect Ratio (TV): The ratio of frame width to height (as defined by IEEE).

Bandwidth (Device): The range of frequencies over which a device is capable of operating within a specified performance limit.

Detection: The ability to distinguish that an artifact within the field of view is of interest. For thermal systems, detection is of two types: (1) MDT or "start" detection and (2) MRT or resolution detection.

Field of View: Maximum cone or fan of rays subtended by the entrance pupil that is transmitted by the instrument to form a usable image.

Gray Shades: A measure of the dynamic range of a display or system using a target with a set of stepped contrast levels. The most common target uses a square root of two brightness ratio between levels.

Instantaneous Field of View (IFOV): The width of the line spread function between the 50% intensity points measured at detection output.

Minimum Detectable Temperature Difference (MDT): The MDT of a thermal device is defined as the minimum temperature difference between a square (or circular) target and the background necessary for an observer to perceive the target source through the thermal imaging device. It is a function of target angular size and represents the threshold detection capability of the system.

Minimum Resolvable Temperature Difference (MRT): The central parameter in the modeling of infrared imaging hardware performance. It is both theoretically predictable and laboratory measurable. It is the minimum temperature difference required between a standard bar-type target pattern (four-bar, 7:1 aspect ratio) at which a trained observer with normal vision can distinguish the bar pattern as a four-bar pattern. The MRT is generally determined for a variety of different spatial frequency targets. It is plotted as a graph of minimum temperatures difference in Kelvins (1K = 1°C) normalized to 300K versus spatial frequency (cycles/milliradians) in object space.

Spatial Frequency: A term that describes the frequency of an evenly spaced bar type target pattern or a sinusoidal type pattern. It is the number of cycles of the pattern that occurs in a given distance. The distance is usually expressed in millimeters or the more common angular distance of milliradians (17.45 milliradian/degree). The milliradian dimension is most commonly used for system analysis because it projects the spatial frequency into object space.

Thermal Imaging: Pertaining to a class of devices that optically collect infrared radiation within a limited wavelength band; e.g., 3–5 microns or 8–12 microns, and convert the received energy into a ''thermal image'' of the scene which can be viewed by the human eye.

Thermal Signature: The mean target-background temperature difference referenced to a 300K (27°C) background temperature.

Visual Range: A measure of target detection range that depends only on the extinction of the atmosphere in the visual (0.4 to 0.7 microns) spectral region. Visual range is defined as the range at which visual atmosphere transmittance is 0.02 and can be found from the visual extinction coefficient (V) by V.R = 2.912/V.

Visual Radiation (*or Light*): Electromagnetic radiation in the wavelength range of approximately 0.4 to 0.7 micrometers, the wavelength to which the human eye is sensitive.

CHAPTER 11

The Use of a Portable Infrared Analyzer to Perform Onsite Total Petroleum Hydrocarbon Analysis

Roy A. Litzenberg, Richard H. Oliver, and **James J. Severns,** The Earth Technology Corporation, Commercial Waste Management Division, Long Beach, California

INTRODUCTION

The authors have found the portable field infrared analyzer (IR) to be a useful tool to obtain quick, reliable estimates of total petroleum hydrocarbon (TPH) concentrations for soil samples in the field. These estimates can prove the basis for decisionmaking, provided that one takes into consideration the limitations of the methodology. This chapter provides a discussion of the methodology.

The discussion begins with a description of applications and limitations of the methodology. Summaries of two actual projects are used to illustrate the methodology. Next, a discussion of the theory of Infrared Absorption Spectroscopy is presented with emphasis on its applicability to total petroleum hydrocarbon analysis. Following the discussion of theory, the chapter presents an overview of the field IR procedure used to obtain field TPH results. The final section of the chapter provides the authors' conclusions and recommendations.

APPLICATION

The Field Infrared Analyzer (Field IR) has proved to be a very useful instrument for providing timely onsite estimates of TPH concentrations in soil samples.

Correlations between field IR results for TPH and results obtained from California Certified Hazardous Waste Laboratories performing EPA Method 418.1 test vary, depending on the nature of the hydrocarbon materials in the soil.

The field IR is typically effective for detecting diesel fuels, lubricating oils, fuel oils, asphalts, tars, crude oils, grease, kerosene, and related substances. Hydrocarbons predominantly less than C-6, including solvents, gasoline, refrigerants, and the like, if detected by the field IR at all, will typically be detected at significantly lower concentrations than actually present in the soil.

Field IR results will typically be within one order magnitude of the corresponding laboratory EPA Method 418.1 findings. This limitation in accuracy is usually acceptable for the purpose of approximating the extent of contamination for investigations or remediations pending certified laboratory results. During the investigation phase, field IR results can be useful in preliminary estimates of the extent of contamination and selection of representative samples for certified laboratory analysis. Field identification of the apparent zone(s) of highest contamination can be useful in attempting to identify potential sources.

During remediation activity involving soils excavation, field IRs have been very effective tools for maximizing the effective use of expensive earth moving equipment. Soil exhibiting evidence of TPH contamination can be excavated until the field IR indicates the TPH contamination is close to cleanup goals (typically between 100 mg/kg and 1000 mg/kg). At that point, samples should be collected and analyzed by a certified laboratory to verify that goals have been met.

Tables 11.1 and 11.2 summarize the results from two projects with different hydrocarbon contamination present in the soil. By reviewing these two cases, the usefulness of the field IR becomes apparent.

Table 11.1.   Project I: Waste Hydraulic/Lube Oil in Silty Clay [results in mg/kg]

Field IR	Laboratory EPA Method 418.1
160	11
180	15
260	8.0
180	8.0
35,000	12,000
53,000	46,000
31,000	9,500
240	15
740	18
340	21
160	15
80	14

Table 11.2.   Project II: Diesel Fuel in Sand with Cobbles and Asphalt [results in mg/kg]

Field IR	Laboratory EPA Method 418.1
360	120
47	21
1,100	11,000
410	4,300
ND < 23	32
1,100	320
5,200	6,100
4,900	6,700
2,000	3,300
3,300	4,100

Project I

Project I was a remedial investigation of an industrial facility in which, over the years of operation, waste oil from hydraulic presses and cutting oils leaked into the soils underneath the foundation of a building. During the remedial investigation, various soil borings were placed in order to estimate the extent of contamination. The field IR was used to screen recovered soil samples onsite. The results of the onsite field IR results were used to determine final borehole depth, the need for additional borings, and to select samples for certified laboratory analyses. Table 11.1 lists field IR results along with the corresponding certified laboratory EPA Method 418.1 results for samples collected from the Project I site.

The field IR consistently read higher than the laboratory EPA Method 418.1 analysis, especially at low concentrations. However, there is a very clear rise in readings upon encountering higher concentrations of contaminants. The screening data provided a reliable indication of the zone of contamination during drilling, eliminating the need to remobilize a drill rig and crew after the results of the certified laboratory were obtained. The maximum downtime between holes due to this analysis was 10 minutes.

Project II

Project II involved the removal of diesel-contaminated soils related to former underground storage tanks. The field IR was used to determine the extent of contamination as the excavation proceeded. When it appeared that the excavation was near completion, soil samples were collected and analyzed using an onsite field IR and a stationary certified laboratory. The results are summarized in Table 11.2.

These results are less consistent than the previous project. This is not surprising, however, because of the soil type and location. The soil was a mix of fill soil, sand, and fairly large cobbles, with tiny to large chunks of asphalt mixed throughout. With this type of soil, it is very hard to get a representative sample in the field, or in the laboratory. Even with the mentioned difficulties, however, 7 of the 10 samples show a very high degree of correlation. Where the field IR readings indicated the soil contained less than 100 mg/kg TPH, it was verified by the laboratory, while all the field readings above 1,000 mg/kg, with one exception, were verified at over 1,000 mg/kg by the certified laboratory. The field IR was used in this situation to quickly determine the extent of diesel contamination, with a minimum of downtime for the excavating equipment.

THEORY

The field IR analysis detailed in this report is based on the principles of Infrared Absorption Spectroscopy. In this discussion we investigate the theory of infrared absorption in general, and then discuss how it applies specifically to the field screening procedure described below.

The instrument used in this procedure measures infrared radiation, commonly referred to as heat radiation. Infrared (IR) radiation is electromagnetic radiation with wavelengths ranging from 0.78 micrometers to 1,000 micrometers. The infrared spectrum is subdivided into three sections, the Near IR (0.78 micrometers–2.5 micrometers), the Middle IR (2.5 micrometers–50 micrometers), and the Far IR (50 micrometers–1,000 micrometers).[1]

The most useful part of the IR spectrum in analytical applications is the wavelength band of the Middle IR region between 2.5 micrometers and 15 micrometers. It is this range of wavelengths that corresponds to the vibrational-rotational frequency of molecular bonds.

The relative position of atoms in a molecule are not fixed, but in reality fluctuate constantly as a result of several different types of vibrations. These vibrations fall into two basic categories, stretching and bending. Stretching involves a continuous change in the interatomic distance along the axis of the bond between the atoms. Bending vibrations are characterized by a change in the angle between two bonds.

At the microscopic scale, IR radiation between 0.78 micrometers and 2.5 micrometers, when it encounters a molecule, can be absorbed by the molecule and increase the vibration of the atoms in that molecule. Only certain specific wavelengths of IR radiation are absorbed by a given type of bond (such as a carbon-hydrogen bond). Thus, for any given molecule absorption appears at discrete wavelengths.

Positive identification of molecules can be obtained by exposing molecules to the IR spectrum from 0.78 micrometers to 2.5 micrometers. Discrete wavelengths will be absorbed, creating absorption bands. Since interactions between atoms in a molecule effect their vibration, each type of molecule has a unique IR absorption "fingerprint." This fingerprint is used to identify compounds in various analytical applications.

The IR field screening technique utilizes a narrow band of the IR spectrum around 3.48 micrometers. This wavelength of IR radiation corresponds to the energy of absorption wavelength for carbon-hydrogen (C-H) bond stretching. When IR radiation at 3.48 micrometers encounters a C-H bond, it is absorbed. The more C-H bonds encountered, the more radiation will be absorbed. All hydrocarbons are characterized by a large number of C-H bonds. The number of C-H bonds in a given sample is thus generally directly proportional to the concentration of hydrocarbons in that sample. Therefore, by measuring the amount of IR radiation absorbed as it passes through a sample, the concentration of TPH in that sample can be determined.

In order to determine the TPH concentration in a given amount of soil, the TPH must be extracted from the soil into a reagent that does not absorb IR radiation in the vicinity of 3.48 micrometers. This reagent must be water-free because water's oxygen-hydrogen (O-H) bond absorbs in the same wavelength range as the C-H bonds. The reagent must also be a solvent for hydrocarbons.

A good reagent is trichlorotrifluoroethane (Freon). This compound is relatively nonreactive, dissolves hydrocarbons, is hydrophobic, and has no C-H bonds to interfere with the analysis.

Using Freon, the hydrocarbons can be extracted from the soil. Freon attracts the hydrocarbons from the soil particles and carries them into solution. The filtered Freon solution containing the dissolved hydrocarbons is then subjected to an IR beam. The amount of IR radiation absorbed is directly proportional to the level of TPH in the soils.

From Beer's law, Concentration is directly proportional to Absorbtion:

$$C \propto A \tag{1}$$

or

$$C = kA \tag{2}$$

with C being the concentration in mg TPH per kg Freon (mg TPH/kg Freon); A being the absorbance; and k being a constant associated with the cell pathlength, the absorbtion coefficient of the compounds involved and other variables. For a given set of circumstances:

$$k = C/A \tag{3}$$

By measuring the absorbance standards of known concentrations, k can be calculated. For example, if the concentration of the standard is 100 mg TPH/kg Freon, and the absorbance is 0.065, then k = 1,471 mg TPH/kg Freon. In the field, 100 mg/kg, 500 mg/kg, and 1,000 mg/kg standards are used to delineate k for a range of concentrations.

In order to calculate the concentration of TPH in a given amount of soil, the k value and the absorbance that is read from the IR device are substituted into Equation 2. For example, with k = 1,471 mg TPH/kg Freon and A = .010

$$C_{unk} = k \times A$$

$$= (1471 \text{ mg TPH/kg Freon}) \times 0.010$$

$$= 14.71 \text{ mg TPH/kg Freon}$$

This gives us the concentration of TPH relative to the amount of Freon used in the analysis. The concentration we need to know is the concentration of TPH to the weight of soil (mg TPH/kg soil). We therefore have to perform a substitution:

$$\text{Conc. TPH/Soil} = (\text{Conc. TPH/Freon}) \times \\ (\text{amt. Freon/amt. soil}) \times (\text{density of Freon}) \tag{4}$$

Substituting the concentration from above (14.71 mg TPH/kg Freon), plus the weight of soil (12 grams), and the volume of Freon (25 mL) used in the analysis:

$$C_{TPH} = (14.71 \text{ mg TPH/kg Freon}) \times (0.025 \text{ L Freon}/0.012 \text{ kg soil}) \\ \times (1.57 \text{ kg Freon/L Freon})$$

$$= (14.71 \text{ mg TPH/kg Freon}) \times (3.27 \text{ kg Freon/kg soil})$$

$$= 481.07 \text{ mg TPH/kg soil}$$

which, taking into account the accuracy of the test, can be expressed as:

$$= 480 \text{ mg/kg TPH in soil}$$

Since the amount of Freon and soil is the same for each analysis, Equations 2 and 4 can be combined to form:

$$C_{TPH} = A \times (k \times 3.27) \tag{5}$$

thus, by reading the absorbance from the IR device for a given sample and multiplying it by a constant, the TPH is calculated.

PROCEDURE

The following Standard Operating Procedure (SOP) is based on EPA Method 418.1 for Total Petroleum Hydrocarbons. Modifications have been made to adapt the 418.1 procedure to be used in field screening applications. It must be stressed that the modifications are substantial and this field procedure should be used as a SCREENING TEST only. Certified laboratory analysis should be used in all cases where the accuracy of the results is critical to decisionmaking or conclusions.

Soil samples are rarely homogeneous, so care should be taken to obtain a representative sample. Measuring spoons will not give consistent results; therefore, a small triple beam balance should be used to measure the soil. (Equipment for estimating TPH is listed in Table 11.3.)

The use of nylon mixing balls, sodium sulfate, and silica gel is imperative to the field procedure. The nylon balls aid in the mixing process; the sodium sulfate removes residual water from the Freon (water interferes with the IR measurements); and the silica gel removes polar compounds (i.e., water, soaps, alcohols, pesticides, fertilizers, etc.).

Table 11.3. Equipment for Obtaining Estimates of Total Petroleum Hydrocarbon Concentration

Item	Amount Required
Foxboro Miran-1A, CVF or FF IR Analyzer (or equivalent)	1 each
Sample cuvettes	2 minimum
Concentration standards (100, 500, 1000 mg/kg)	40 mL per each
Triple beam balance	1 each
Calculator	1 each
Watch to time extraction agitation	1 each
Generator	1 each
Extension cord (100 ft)	1 each
Stainless steel spatula	1 each
Teaspoon measuring spoons	1 each
Safety glasses	1 pair
Kimwipes™	1 box
125 mL Erlenmeyer flasks	4 flasks
Funnels	4 funnels
Exam gloves	5 pair/day
Recovery bottle for Freon	1 4-liter bottle
Quantitative measuring pump for Freon	1 pump, capacity: 25 mL
Freon (1,1,2-trichloro-1,2,2-trifluoroethane)	150 mL per extraction
Granular sodium sulfate (Na_2So_4)	0.5 tsp per extraction
Granular silica gel	1 tsp per extraction
40 mL VOA sample bottles	1 per extraction
Filter paper, qualitative 41, 9.0 cm diameter	1 per extraction
Disposable nylon ball bearings	1 per extraction

Procedure for Calibration to Establish k

The IR instrument is calibrated using three solutions containing known concentrations of hydrocarbons. The absorbance of each standard is measured, and then used in the calculation of the proportionality constant k. The concentration versus absorbance relationship for this method is nonlinear over the range of absorbance measured by the IR device. Therefore, a range of standards must be used to calibrate the IR equipment over the range of concentrations expected.

1. Add a known concentration standard (100 mg/kg, for example) to a sample cuvette that has been rinsed three times with clean Freon. Place the cuvette into the sample chamber of the IR. With the IR set to register in absorbance, read and record the absorbance.
2. Repeat this procedure for each standard.

From Equation 3, with a standard of 100 mg/kg and a recorded absorbance of 0.065:

$$k \quad = C/A \qquad\qquad (3)$$

$$k_{100} = 100/0.065$$

$$k_{100} = 1,538$$

When estimating an unknown sample concentration, the next highest standard absorbance calibration constant (k) is used. Thus, if the 100, 500, and 1,000 mg/kg standards had absorbances of 0.065, 0.32, and 0.64 respectively, sample readings for 0.000 to 0.065 would use the 100 mg/kg standard constant (k_{100}) to calculate the unknown sample concentration; sample absorbances from 0.066 to 0.32 would be calculated using the 500 mg/kg standard constant (k_{500}); and sample absorbance above 0.32 would be calculated using the 1,000 mg/kg standard constant ($k_{1,000}$). Once the k value has been determined, the procedures listed below can be used to measure TPH concentration in soil samples.

Procedure for Sample TPH Measurement

The following procedure is used to measure the TPH of sample soils:

1. Spread the soil in a clean stainless steel pan, and blend it thoroughly with the stainless steel spatula to get a representative sample.
2. From several locations in the mixed sample, collect 12 grams of soil (or approximately 2 teaspoons if a scale is not available) and place in a clean 40 mL VOA bottle.
3. Add a clean nylon ball to the VOA bottle.
4. Add 0.5 teaspoon of sodium sulfate.

5. Mix the soil and sodium sulfate together by closing the bottle and shaking, or if necessary, by using a clean stainless steel spatula.
6. Add 25 mL Freon to a VOA bottle; reclose the bottle.
7. Shake the VOA bottle briskly for a full 3 minutes.
8. Place a funnel on a 125 mL Erlenmeyer flask.
9. Fold filter paper and place into the funnel.
10. Add 1 teaspoon silica gel to the funnel and filter paper.
11. Decant the Freon mixture into the filter funnel. Avoid pouring soil and other solids from the 40 mL VOA bottle into the filter funnel.
12. After the Freon has filtered through the silica gel and filter paper, remove the funnel from the 125 mL flask and discard the filter paper and silica gel.
13. Rinse the sample cuvette twice with Freon.
14. Pour the filtered Freon mixture into the sample cuvette and record the absorbance reading.
15. Rinse the 125 mL flask and funnel with 20 mL of Freon; repeat.
16. Collect all excess Freon in the recovery bottle for subsequent recovery or proper disposal.

The concentration of TPH in a sample can now be calculated by using Formula 5:

$$C_{TPH} = A \times (k \times 3.27) \qquad (5)$$

where A is the recorded absorbance, k is the appropriate standard constant and 3.27 is the multiplier calculated above. For example, if an absorbance of 0.032 was read and recorded, with k = 1538:

$$C_{TPH} = 0.032 \times (1538 \times 3.27)$$

$$= 160.9$$

$$= 170 \text{ mg/kg TPH in soil}$$

Dilutions

If the filtered Freon extract does not appear colorless, or the IR reads off scale, dilution may be necessary. The following steps should be added between Steps 14 and 15 of the standard procedure:

14a. Pour the contents of the sample cuvette (4 mL) into a clean 125 mL flask.
14b. Add one or more cuvette volumes of Freon to the 125 mL flask.
14c. Carefully decant the mixed solution into the sample cuvette and reinsert into the IR.

14d. If the meter reads off scale again, repeat steps 14a through 14c.

14e. Record the dilution as follows:

Sample cuvette + 1 cuvette of Freon = 1:1 dilution

Sample cuvette + 2 cuvettes of Freon = 1:2 dilution, etc.

If a dilution was necessary, a multiplier must be applied when calculating the TPH concentration in soil. The dilution multiplier used is as follows:

1:0 dilution (no dilution) = 1

1:1 dilution = 2

1:2 dilution = 3

1:10 dilution = 11

etc.

In order to calculate the TPH in soil following dilution of the sample solution, Formula 5 becomes:

$$C_{TPH} = A \times (k \times 3.27 \times \text{dilution multiplier}) \tag{6}$$

Therefore, if the sample absorbance was 0.32 and the dilution was 1:7, with a $k = 1538$:

$$C_{unk} = 0.32 \times (1538 \times 3.27 \times 8)$$

$$= 12,874 \text{ mg/kg TPH in soil}$$

$$= 13,000 \text{ mg/kg TPH}$$

With a few hours of experience, the above procedure can be performed, including dilutions if necessary, and a result obtained within 10 minutes. The information thus gained can then be used immediately to make project management decisions.

CONCLUSION

Experience has demonstrated the utility of a portable IR for obtaining rapid, reliable estimates of the concentration of TPH in soil samples. The methodology discussed in this chapter results in analytical results which are generally within one order of magnitude of certified laboratory results. Factors which affect the correlation include:

1. The accuracy of the method used to measure the amount of soil and Freon in the field.

2. The consistency of the sample extraction/agitation processes used in the field.

3. Variations in concentration of the contaminant within the sample.
4. Physical/chemical reactions between the contaminant(s), the soil, and the Freon.
5. The potential for loss of Freon (evaporation) during the extraction, filtering, decanting, and cuvette handling steps.
6. Ambient weather conditions (temperature, light, etc).
7. The responsiveness of the instrument to the specific contaminants.
8. The instrument's accuracy.
9. Operator error.

The methodology discussed in this chapter is relatively simple to learn and compatible with the field environment. A greater level of accuracy and corresponding correlation with certified laboratory results can be obtained by addressing some or all of the above items. Care must be taken that modifications to increase the accuracy do not result in a methodology that is less portable, less rugged and/or more complex than the discussed methodology. Continued refinement of the methodology is expected to maximize the usefulness of the portable IR for field work.

REFERENCE

1. Skoog, D.A. *Principles of Instrumental Analysis, Third Edition,* (Philadelphia, PA: Saunders College Publishing, 1985), p. 316.

CHAPTER 12

Real-World Soil/Waste Standards: Solution to Differential Analytical Results Among Laboratories

David J. Carty, K. W. Brown & Associates, Inc., College Station, Texas
Janick F. Artiola, University of Arizona, Tucson, Arizona

Analytical results generated for submission to regulatory agencies often differ substantially, when split environmental soil/waste samples are analyzed by different laboratories. Differential results have been obtained from even the most straightforward chemical analyses performed on soils and soil/waste admixtures. Without submission of simultaneously analyzed real-world reference standard data to accompany environmental results, neither commercial clients, analytical laboratories, nor regulatory agencies have any firm basis for assessing whether or not analytical results are accurate. Furthermore, environmental managers cannot compare data from their facility to data from competing facilities, much less to the database from which pertinent regulatory criteria were initially established. An additional negative factor is that without established benchmark data for analytes from environmental reference standards, both environmental managers and regulatory agencies may find themselves susceptible to a full spectrum of undesirable litigation.

This chapter is written from the point of view of an experienced environmental consultant, and the director of a soil and water analytical laboratory at a major university. It is presented (1) to demonstrate the need for referenced soil and soil/waste standard performance samples; and (2) to propose a simple and relatively inexpensive method for providing and administrating the database of anticipated commonly available standards.

EVALUATION OF PROBLEM

Problem Statement

The problem of determining accuracy in environmental samples can be stated in two parts as follows:

1. Different analytical laboratories often generate substantially different analytical results from split soil and soil/waste samples.
2. In addition, the determination of accuracy for some parameters cannot be clearly established using current analytical protocols and procedures.

Predominant Causes of Differential Analytical Results

Soil and environmental scientists are aware that determination of the physical and chemical characteristics of environmental soil and soil/waste samples consistently challenges many aspects of analytical work. Specifically challenging are procedures involving sample preparation, laboratory analysis, and the reporting of data. Unlike manufactured products, native soils are continuously variable and complex mixtures of gases, liquids, solids, and biota. After industrial wastes are applied to soils, the resulting soil/waste mixtures contain synthetic or naturally occurring substances which may then be altered in the field by weathering, biodegradation, or other chemical and physical interactions. A review of some of these processes is given in Dragun et al.[1], among others.

It is clear that even in native soils, naturally occurring spatial variability is commonly the greatest single source of constituent concentration variability in field soil samples. Barth and Mason[2] estimate that sample to sample variation of field soils is often 30 to 40 times that of analytical precision. However, the true extent of variability introduced by laboratories can only be ascertained by comparing results from split samples, and such variability is often substantial.

Until split sample results from different laboratories are compared, environmental managers may sense an inappropriately high comfort level regarding the quality of data they are receiving. For example, when a given laboratory continues to report data which appear consistent and reasonable from a periodically monitored land area, the environmental manager may sense little reason for concern. Yet when split samples are sent to another competitive laboratory, he may be surprised to see the same level of precision at the new laboratory, but analytical results of a substantially different magnitude.

The basis of the problem lies in confusion about what constitutes accuracy, and how accuracy differs from precision. For the purpose of this discussion, precision is used to indicate the reproducibility of a constituent concentration from homogenized subsamples, using a given analytical procedure exactly as written, and using accepted analytical judgment where authorized. Precision, therefore, is a measure of the degree to which the same constituent concentration can be obtained repeatedly, as the experiment is performed time after time. Accuracy is a measure of the ability to generate the true central value which best represents the concentration of a constituent which should be obtained when a procedure is performed exactly as written, and using accepted analytical judgment. A laboratory which demonstrates excellent precision for an analyte may or may not be providing accurate results.

Fundamental Causes of Differential Analytical Results

After soil and soil/waste samples are collected, analytical results may differ as a result of (1) the splitting process, (2) alteration during, or as a result of, packaging or transport, (3) differential sample preparation at the receiving laboratory, (4) differential performance of the same cited procedures by different laboratories, and (5) differential, improper, or inaccurate calculations, documentation, QA/QC, and reporting.

With the exception of available nitrogen (N), the examples given later in this chapter have been chosen based on their relative stability in air-dried and ground samples. Therefore, differentiation as a result of the splitting, packaging, and transport processes, is considered to be negligible for these samples.

On a different level, several other factors contribute to the current disparity of analytical results among laboratories. First, based on misplaced confidence in good precision, laboratories which report inaccurate values may be consistently repeating some analytical idiosyncrasies. Second, currently acceptable analytical protocol is insufficient to resolve differences of opinion regarding which values are "accurate." Third, differential results are comfortably ignored as long as they are not discovered. Fourth, there is currently no centralized forum under which analytical laboratories and regulatory agencies can resolve analytical differences of opinion. Fifth, there are no commonly available soil or soil/waste reference standards of a matrix similar enough to real-world environmental samples to facilitate the resolution of the problem.

Potential Errors Discoverable With Appropriate Reference Standards

This section illustrates some of the ways in which an individual laboratory can introduce errors into the reported concentrations of constituent in soil and soil/waste samples. It is important to note that many of these errors cannot be detected without use of solid phase reference standards, preferably of a real-world matrix similar to that being analyzed.

Errors Generated During Sample Preparation: Samples may become contaminated in the laboratory environment with particulate or aerosols dispersed by air motion from the opening or shutting of doors or windows, or from fume hood activation. Volatiles commonly escape from reagent containers, or analytical devices, as well as other samples, during many phases of analytical work.

Samples in the laboratory environment may also lose some constituents prior to analysis. Selective constituent loss can occur when some organic gases, such as BTEX (benzene, toluene, ethylbenzene, and xylenes), inorganic gases, such as hydrogen sulfide, ammonia, nitrous oxide, carbon dioxide, and water vapor, and some metals volatilize. During preparation, it is almost impossible to avoid losing some very fine particulate. In this regard, the reactive surface area of very fine smectite may be greater than 650 m^2/g compared to less than 1×10^{-6} m^2/g for coarse quartz sand. Selective loss can also occur due to differences in specific gravities which range from less than 1.0 to greater than 5.2, and by various magnetic, static electrical, and adhesion losses to preparation equipment, surfaces, and materials.

Differentiation in analytical results can be a result of different mixing, drying, grinding, or splitting activities in laboratories. With regard to mixing, some procedures require "field-moist" samples, whereas air drying or oven drying is required for other procedures. Sealed samples often exhibit moisture condensation on container surfaces, indicating degassing of the samples in transport. Effective homogenization and subsampling of field moist soil materials is arduous even under the best conditions, especially when samples consist of hard peds, massive tight clays, or with segregations or striations of nodules, mottles, and various waste materials. Different laboratories may take far different approaches to homogenizing moist samples, and subsequent results may differ as well.

The duration and temperature of drying can also be critical for some soil materials. This is especially true for oxidation/reduction, solubility, adsorption, organic matter, and mineralogical changes. Different grinders in different laboratories may have very different effects on some materials, since some tend to crush or grind, whereas others function primarily by impact. The importance of sample splitting techniques has recently been addressed by Schumacher, et al.[3] They showed that closed-bin riffle splitting is more efficient, more effective, and had a lower loss of fines than open-bin riffle splitting, or cone and quartering techniques. Perhaps more importantly, they demonstrated that five passes were required for complete splitting.

Errors Generated During Sample Extraction, Digestion, and Analysis: The complexity of analytical chemistry for soil and soil/waste samples presents a challenge to any laboratory. One of the most critical aspects is the actual performance of the analytical procedure. To protect clients and laboratories from potential litigation, it is important to clearly establish guidelines for willful alterations from cited analytical procedures. The analyst may feel that "improvements" should be made to cited procedures when they are known to be ineffective for the analysis of some constituents. However, this may cause a disruption in applicability of new data to historical data, as well as create an incorrect impression that there has been a sudden increase in a waste constituent concentration. It may also put the client at an unfair disadvantage with regard to competitive industries using less adventuresome laboratories.

Attention to proper procedures also includes an awareness of which revision is currently in effect for a given regulatory program. Although formal revisions are often warranted and necessary, when they come into effect, they often create havoc with historically based statistical data. Such effects may include changes in average, variance, and skewness statistics in baseline and background data. For this reason, it may be advisable to have the laboratory also simultaneously analyze some subsamples by the previous version of the procedure, so that differences resulting solely from written procedural changes can be documented.

Analytical procedures can also be affected by inattention to various details during analysis. One potential problem involves use of properly prepared samples. Analysis of a soil/waste which has been dried to some extent should not be expected to produce the same results as field-moist material. Another problem area can be use of improper sample to reagent volumes and ratios. The ratios of sample to extractants, acids, or other reagents can also affect the outcome of laboratory analyses. Some extraction techniques are very dependent on time, temperature, and repetition cycle limits, during certain extraction or digestion phases of analysis. Whereas one laboratory may consistently consider the timing phase to begin or end at one point, another laboratory may be consistently utilizing the time frame concept very differently. As a result, both laboratories may show good precision, and believe their results to be accurate, but their reported constituent concentrations may be quite different. Even relatively uncomplicated situations such as reagent freshness may differ from laboratory to laboratory. Although use of standard curves will typically minimize this problem, there may still be some effect on the final results of environmental samples. Similarly to reagent freshness, each time reagents are prepared, it is possible to make a dilution or other matrix error. If data are not reviewed in a conscientious and timely manner, such errors can escape detection until holding times have been exceeded.

Differences in analytical results can also result from nonuniform equipment use and handling among laboratories. Use of nonuniform equipment can result in different detection limits, responses to electronic, or "white" noise, heat capacity and transfer properties, and various other effects. Even differential or improper use of simple equipment such as hot plates and ovens may have an effect on some parameters.

Errors Resulting from Analyst's/Laboratory's Performance: Soils and soil/waste mixtures differ from many materials subjected to analysis due to the vast range of properties which may be encountered. The mere handling of some of these materials requires experience which is not commonly obtained in non-soils laboratories. For example, soils may range from less than 1% to greater than 99% organic matter prior to exposure to wastes. Organic soils which are commonly found in wetlands, may be very hydrophobic, compared to typically hydrophilic mineral soils. These hydrophobic soils may be very difficult to "wet-up" in preparation for some procedures, and this situation can be very confusing to some analysts. Another example is difficulty dealing with a wide range of adhesive, viscosity, and plasticity characteristics. During transfer from one container to another, some soil and soil/waste materials may strongly adhere to the transferring implement. Soils which contain iron oxides can create localized hot spots, especially in microwave ovens, and improper handling of organic, amorphous, or expandable minerals may have a great effect on cation and anion exchange capacities.

Under the best of circumstances, it would be difficult to expect that analysts at different laboratories were trained similarly, and shared a uniform tendency regarding consistency of performance, and attentiveness to accuracy and precision. In addition to this fundamental problem, some special skills are required of soil and soil/waste analysts. Among some of the most critical and demanding skills are some of the least complex. As a result, the attention span and skill level of the analyst may waiver after many repetitions of seemingly simple exercises.

Some other soil-related skills which are easily misapplied include cation exchange capacity (CEC), completion of a Soxhlet extraction, and preparation and collection of a saturated paste extract. Other potentially deceptive skills include adjustment of simple weighing and wavelength verniers, performance of dilutions, and matrix adjustments. Another important skill for analysts is recognition of the point at which common instrumental deterioration must be attended to.

Laboratory management should assume responsibility for fundamental functions associated with laboratory operations. Such functions include guidelines for reasonable size and uniformity of sample batch sizes. Analytical errors may result when batches are either too small or too large to be managed equitably. One potential problem with large batches is that a small interruption can greatly prolong the time between an extraction and subsequent instrumental analysis.

Laboratory management should also maintain and orchestrate uniform use of procedural revisions and techniques. Periodic internal round-robin analyses should also be utilized to ensure that all technicians who perform a given analysis obtain similar results. Whereas laboratory management should expect each technician, individually, to achieve a high level of precision from split samples, they should also expect good precision when comparing results submitted by all technicians. Toward this end, laboratory management should strive to minimize analyst turnover, and encourage consistent availability.

Errors Resulting from Documentation, Calculations, QA/QC, and Reporting: Analytical errors which are due to lapses in documentation may be among the most difficult to detect without appropriate reference standards. For instance, a technician may simply forget to record and perform calculations based on an actual dilution of a sample extract. Similar problems may result when numbers are transposed or are attributed to the wrong sample. Simple format errors, such as misplacement of decimal points when reporting results in percents, can also occur. Use of real-world reference standards as a spot-check cannot entirely prevent such occurrences, but results from such standards may provide an estimate of the frequency of such problems, and thus overall laboratory performance.

Even conscientious performance of QA/QC protocols may not be able to reveal some analytical errors. Consistent failure to subtract significant reagent contaminants from blanks may go unnoticed, as well as adjustment of method detection limits where required. When duplicates are analyzed within the same batch, they only provide a limited estimate of overall analyst or laboratory precision. Therefore, it is important that replicate analyses are begun from original solid phase subsamples in addition to more limited replicates from extracts.

Likewise, matrix spikes added to an extract do not provide information about possible errors introduced during sample preparation or extraction procedures. Matrix spikes should be added in the mineral or organic form of materials expected in the environmental samples. For example, it is only of limited use to apply a matrix spike of a barium chloride ($BaCl_2$) standard solution to an environmental sample where the solid phase barium to be extracted is primarily in the form of barite ($BaSO_4$), which has a solubility in water of 1.1×10^{-10}. Except for possible precipitation and other matrix interferences, most laboratories should expect to get good barium recovery when analyzing barium from barium chloride, but such results only marginally address accuracy.

Laboratory reports should be "stand alone" documents. This includes citations of the procedures used, including revisions, instrumentation, and reasons for any deviations from written procedures. When unusual data are encountered, the laboratory should indicate that it recognizes this fact and has either verified its results, or has received instructions not to pursue the matter. A summary of QA/QC data should always be submitted with an analytical report. Significant figures should also neither grossly overestimate, nor underestimate confidence in analytical results, and confidence limits would also be useful.

Both the individual analyst and laboratory management are responsible for maintaining high quality analytical skill levels. Whereas the analyst should always be alert for improbable input/output data, the laboratory manager should always perform a critical evaluation of the final dataset before it leaves the laboratory. Errors which can be detected prior to forwarding reports to clients include checking for proper and consistent units, reasonable orders of magnitude, interrelated salt values, organic constituent ratios, and other typical soil and soil/waste properties. This should be done in addition to more complex QA/QC protocols.

Two Example Problems

Two examples are shown below to illustrate how reputable laboratories can generate differential analytical results from split samples. Permission could not be obtained to use a substantial amount of other data which would similarly illustrate the problem. Since laboratories rarely release results unless their QA/QC data are acceptable, it can be assumed that internal QA/QC checks, including matrix spikes, show good precision and, to the extent possible, accuracy.

Example #1: Data shown in Table 12.1 are from two industrially-affected soil samples which were air dried, ground, and riffle-split (i.e., each was homogenized) prior to being sent to four laboratories for simultaneous routine analysis. Each laboratory indicated use of the same analytical procedures. Due to space limitations, data from only 13 of the full suite of 50 parameters are shown.

Table 12.1.   Analytical Results Generated by Four Analytical Laboratories for Two Industrially-Affected Soil Samples.[a]

Parameter	Units	Laboratory			
		1	2	3	4
		FIRST SAMPLE			
pH	s.u.	5.4	3.9	5.2	5.0
1:1 EC	mmhos/cm	<0.1	0.02	0.04	0.06
1:1 SAR	ratio	2	1.80	0.7	2.78
Sol. Na	meq/100 g	0.1	0.18	0.1	0.59
CEC	meq/100 g	9	—	9.7	12.20
Exch. Na	meq/100 g	0.1	0.10	0.2	0.14
Total Cd	ppm	0.5	0.50	0.02	0.05
Total Cr	ppm	28	20.0	30.8	42.5
Total Pb	ppm	10	17.5	0.1	3.15
Total Se	ppm	0.13	0.03	0.2	0.33
Total Zn	ppm	25	25	21.6	24.5
		SECOND SAMPLE			
pH	s.u.	2.5	2.1	2.4	2.4
1:1 EC	mmhos/cm	7.7	3.63	5.14	6.85
Avail. N	ppm	44.7	6.51	0.6	—
SO_4	meq/L	143.4	53.8	45.3	158

[a]Analyses performed during 1990.

The data in Table 12.1 reveal several important facts. First, there is general agreement by all four laboratories for some analytes. Examples are exchangeable sodium (Exch. Na) and zinc (Zn) in the first sample, and pH in the second sample. Good agreement also was also reached for a number of the other 37 parameters which are not shown. The fact that all four laboratories obtained similar values for these three parameters would afford the environmental manager with a strong degree of confidence that in the first sample, exchangeable Na ranged from about 0.1 to 0.2 meq/100g, and that total Zn was about 24 ppm. He could also be confident that the pH in the second sample is about 2.4 standard units (s.u.).

Further evaluation shows that by dismissing the one outlier and utilizing the "closest three out of four" rule of thumb, the environmental manager could also reasonably assume that pH in the first sample is approximately 5.2, and soluble sodium (sol. Na) is somewhere between 0.1 and 0.2 meq/100g.

However, the environmental manager may not be able to reconcile, or base management decisions on some of the more diverse data, such as lead (Pb), available nitrogen (avail. N), or sulfate (SO_4). Although the lead data are all within typical background levels for some soils, the results differ by over two orders of magnitude, assuming each laboratory intended ppm to represent mg of lead per kg oven-dried equivalent weight soil. Reported concentrations of available nitrogen differ by more than one order of magnitude, which converts to differences in fertilizer recommendations of greater than 3.5. to 1. Data for sulfate are bimodal and suggest that there is equal division for two schools of thought regarding how to perform the same procedure for sulfate analysis.

Based on experience, within-laboratory precision for duplicate samples which have been similarly homogenized is usually much better than the diverse values shown here for lead, available nitrogen, and sulfate. If different laboratories were actually performing the same procedures in the same manner, the typically acceptable levels of intra-laboratory precision should also be expected for inter-laboratory precision.

In fairness to these laboratories, the true variance of these parameters within the split samples is not known. Without additional data to illustrate subsample variance, it may be unfair to suggest that laboratory error is responsible for some of the disparate results. For example, a combination of subsample variability, instrumental limitations, and other factors may account for seemingly diverse values reported for 1:1 electrical conductivity (EC), and a few of the other parameters shown.

Example #2: For several years during the latter half of the 1980s, a state regulatory limit for barium (Ba) in soil/drilling-mud-waste was 3,000 ppm (sic). Analysis by the Method 3050 acid extraction procedure in Test Methods for Evaluating Solid Waste (SW-846)[4] was required.

Four reputable commercial analytical laboratories which commonly performed similar work under the applicable state program were contacted about analyzing a soil/drilling-mud-waste for barium using Method 3050 as required by the state. These laboratories are identified below as Laboratories A, B, C, and D. These laboratory data have been adjusted by request, to preserve confidentiality, but still represent the situation encountered. Each of these laboratories had considerable experience in soil and environmental analyses and indicated frequent use of Method 3050. Ultimately, Laboratory D was not sent samples because, during a laboratory inspection, they could not find their copy of Method 3050, when asked to do so.

Five separate soil/drilling-mud-waste samples from related areas were split three ways and sent to Laboratories A, B, and C. For clarification, each laboratory was specifically told to report Ba results in mg/kg oven dry equivalent weight of sample. During an inspection, the analyst who was to perform the extraction at one laboratory had to be persuaded to use at least a water or sand bath, instead of simply setting the small refluxing beakers directly on an on/off-cycling hot plate, while monitoring the refluxing temperature at the specified temperature. Standard QA/QC data from Laboratories A, B, and C indicated acceptable precision and accuracy. The regulatory agency commonly accepted data from each of these laboratories without restrictions. Representative, but modified, data are shown in Table 12.2.

These data were initially intended for use as the basis of recommendations to be made by the environmental consultant. Recommendations to the industrial client and the state would be to categorize the materials sampled as roughly "in compliance" or "out of compliance." The environmental consultant was unsure of what to recommend because the regulatory limit was 3,000 ppm for barium. Four important questions were of immediate importance:

1. Is the material represented by the split samples in or out of compliance when the criteria is 3,000 ppm?
2. Which, if any, laboratory has provided accurate results?
3. When the regulatory criteria for barium in soil/waste were established, which technique variant of Method 3050 was used as the basis for the 3,000 ppm limit?
4. Which entity (i.e., industrial client, regulatory agency, and/or analytical laboratory A, B, and/or C) should bear the costs required to resolve the data and provide accurate results?

Table 12.2.   Representative Analytical Results of Barium from Mixed Soil/Drilling-Mud-Waste Samples Using Method 3050 from SW-846.[a]

| Sample No. | Laboratory | | | |
	A	B	C	D
	(mg Ba/kg oven dry equivalent weight of sample)			
1	10,500	450	4,370	Not Analyzed
2	13,450	730	5,570	Not Analyzed
3	11,790	680	4,820	Not Analyzed
4	9,860	595	7,340	Not Analyzed
5	12,330	870	8,660	Not Analyzed

[a]Cited procedure = acid extraction/refluxing of sediments, sludges, and soils; Revision 0 (September, 1986).

In defense of the analytical laboratories, it was, and is, common knowledge in the industry that for certain metals Method 3050 is a very tenuous procedure which requires great skill to achieve reasonable precision, even within the same laboratory. Nevertheless, these data clearly show fundamental and unacceptable differences in the performance of this procedure among the three laboratories. The following paraphrased review of Method 3050 shows some of the potential pitfalls of this procedure. The symbol (#) and underlined notations indicate steps which require the analyst's skill, constant attention, or subjective interpretations.

Heat 1 or 2 g field-moist soil + 10 mL 1:1 HNO_3 in conical beaker covered by a watchglass to 95°C and reflux for 10 to 15 minutes without boiling.

After cooling, similar steps are repeated with 5 mL additions of HNO_3 to assure complete oxidation of a final 5 mL solution.

Add 2 mL Type II water and 3 mL of 30% H_2O_2, cover beaker with watch glass, and return to hot plate to start peroxide reaction. Ensure no losses occur due to excessively vigorous effervescence.

Cool beaker <u>when effervescence subsides</u>.

Continue to add 30% H_2O_2 in 1-mL aliquots (10-mL max) <u>with warming until effervescence is minimal or until general sample appearance is unchanged</u>.

Add 5 mL concentrated HCl and 10 mL to Type II water.

Return to hot plate and reflux <u>without boiling</u> for 15 min. After cooling, dilute to 100 mL with Type II water.

<u>Remove particulates in the digestate that may clog the nebulizer by filtration, centrifugation, or gravity sedimentation</u>.

Analyze by flame or furnace atomic absorption, or ICP.

<u>The concentrations are to be reported on the basis of the actual weight of the sample</u>. If a dry weight analysis is desired, then the percent solids of the sample must also be provided. If percent of solids is desired, a separate determination of percent solids must be performed on a homogeneous aliquot of the sample.

POSSIBLE CONSEQUENCES WHEN REGULATORY AGENCIES ACCEPT UNCERTIFIED AND/OR DIFFERENTIAL DATA FROM DIFFERENT LABORATORIES

Possible Environmental Consequences

Consider the following hypothesis: The degree of environmental protection afforded by current or future regulations for soil and soil/waste mixtures cannot be known without commonly available real-world environmental reference standards. This hypothesis is based on evidence that laboratories often do not generate similar analytical results from split environmental samples. As a result, databases upon which regulatory criteria have been, and are being, formulated are partially dependent on the specific analytical laboratory which generated the results. When regulatory criteria are based on data generated from one laboratory, how should data from another laboratory be interpreted, when there are no commonly available benchmark standards for comparison?

A second aspect of this problem is the difficulty encountered when trying to determine environmental trends. Analytical laboratories often disagree on some results even when performed simultaneously, and it is to be expected that this problem will be exacerbated over time, since analytical procedures and accepted techniques are also frequently changing. With real-world soil and soil/waste reference standards, it may be possible to introduce correction factors into archival data to provide more accurate information regarding trends, as well as to relate proposed new criteria to archival databases.

A third aspect is that without commonly available environmental reference standards, it is improbable that even background soil constituent levels can be fairly documented. It is entirely possible that background levels of some soil constituents are highly dependent on which analytical laboratory generated the results. Based on experience, when an industrial client changes laboratories, they can expect to see new monitoring data which falsely indicates that there has been a simultaneous and permanent change in the levels of some constituents taken from the same background areas as on previous occasions. It is relatively easy to detect such laboratory induced artifacts in background data, because seasonal, tidal, and other predictable influences can be determined after several years of record-keeping. However, assessing the extent of laboratory induced effects on soil which has recently received waste is much more difficult, due to the limited size of the historical database.

Possible Consequences for Regulatory Agencies

Without real-world reference standards for environmental materials, the regulatory agency is faced with several concurrent problems. As stated above, existing environmental databases must be considered suspect, unless more than one laboratory has independently verified the data. For example, if the historical database of a given site was developed by one laboratory, and a second laboratory generated data after a spill, or purposeful waste application, it may be inappropriate to attempt to formulate conclusions about the effects of wastes on the environmental samples without agreement of background data.

It is possible that investigation of the effects of ''acid rain'' on soils over the last few decades has been limited by the lack of commonly available reference standards which could have been used to link historical data to data obtained in more recent years. In this circumstance, there have been major changes in analytical techniques and equipment, as well as in the analytes measured. To take a lesson from history, by initiating the production and distribution of real-world reference standards, now, it may be possible to correlate data obtained today with data which may be obtained decades into the future. This would also provide a firm foundation by which regulatory agencies and academicians could clearly establish long-term trends for some analytes.

The relationship of natural field constituent levels to regulatory criteria is difficult to establish in the absence of appropriate reference standards. Regulatory criteria are often based on a combination of assumptions, suppositions, possibilities, probabilities, and an array of correlated and uncorrelated environmental and other analytical data developed by specific laboratories. If the environmental analytical data upon which the criteria were based, were generated by a laboratory which was not in the ''mainstream'' with regard to the performance of critical analytical techniques, then the regulatory criteria developed from those data may likewise be based on misleading constituent levels. Therefore, subsequent monitoring or investigatory analytical data generated by other laboratories which are more in the mainstream of analytical technique, may be falsely judged as falling either within, or conversely, outside of, regulatory criteria.

Lack of complete confidence in the databases upon which regulatory perspectives and criteria have been, and are being established, could easily translate into reduced confidence in the ability of regulatory agencies to fulfill their mission. Industry may become much greater proponents of efforts to protect the environment when convinced that regulatory limits are based on data which have universal applicability over time, and among laboratories. Respect for regulatory agencies by industry may also suffer as environmental managers realize that they can "shop around" for laboratories which tend to generate more favorable analytical results than competitive laboratories.

Likewise, public confidence in regulatory agencies may also be partially dependent on their ability to prevail against technicality-based litigation efforts by industry. There may be numerous opportunities for technicality-based litigation, if initiated as a result of irregularities, ergo unfairness, in analytical data among different laboratories. As a result, without commonly available real-world environmental reference standards to show uniformity of performance among analytical laboratories, regulatory agencies may find it difficult to enforce regulations in a fair manner among competitive industries.

Possible Consequences for Analytical Laboratories

Potential consequences for analytical laboratories can be particularly damaging as a result of unfair competition among laboratories. Many commercial analytical laboratories compete for the environmental business of industry. It is obvious that when different analytical laboratories obtain different analytical results on split environmental samples, that there will be some fraction of those laboratories which will obtain results which are more favorable, compared to regulatory criteria, than data from other laboratories. As long as regulatory agencies fail to require different analytical laboratories to obtain the same results on blind environmental reference standards, industry may have no restrictions on seeking and retaining one of the laboratories which tend to give the most favorable results. In fact, there are already reports that some industrial firms may have selected subcontractor analytical laboratories based on their predisposition to obtain favorable analytical results. The unfairness of this situation could be avoided if regulatory agencies required submission of simultaneously analyzed blind real-world environmental reference standards with each dataset, and rejected field sample results accompanied by inaccurate or nonmainstream reference standard results.

Although it is possible that laboratories which tend to obtain the most favorable analytical results may be performing analytical procedures most "correctly," it is not possible to determine this without commonly available soil/waste reference standards of an appropriate matrix. As a result, when a laboratory manager realizes that a competitive laboratory is able to consistently achieve more favorable analytical results, and that potential industrial clients are selecting that laboratory for the same reason, difficult ethical questions must be confronted. Without evidence to the contrary, it is possible to hypothesize that each laboratory manager believes that the manner in which their laboratory performs analyses, is the "correct" way, in an academic sense. However, based on experience, different laboratories appear to devote different degrees of effort toward rigorous pursuit of accuracy. These differences convert into differences in overhead costs associated with the generation and substantiation of data. It may be that the laboratories which do not tend to achieve favorable analytical results are the same laboratories which are expending the greatest portion of their budget to achieve the most "correct" data. Under this scenario, regulatory agencies which do not require submission of simultaneously analyzed blind environmental reference standards, may be allowing industry to discriminate against analytical laboratories which are making the greatest effort to substantiate their analytical results. More bluntly, it is possible that regulatory agencies may be allowing the best analytical laboratories to either be forced out of business, or to consider compromising their ethical standards.

The respectability of analytical laboratories is also strained, because when different laboratories obtain differential analytical results, neither environmental managers, consultants, nor regulatory agencies know which laboratories are most trustworthy. As a result, data from all laboratories become suspect, and considerable additional expense must be devoted to examining data to find patterns indicating irregularity. If data are not consistently examined for irregularity, claims of negligence may be possible.

Without a regulated, or protective, forum through which analytical laboratories are strongly encouraged to work together to obtain similar analytical results on split environmental samples, laboratories are not likely to make themselves vulnerable to criticism by voluntarily sharing analytical data. Such a forum, based on the goal of achieving similar analytical results on real-world environmental reference standards, could be established and sponsored by regulatory agencies, or through laboratory certification programs.

When diverse analytical results from petroleum-contaminated split soil samples were reported by different commercial analytical laboratories participating in one state regulatory program, an interesting situation developed. Based in part on complaints from at least one consultant, a certain commercial laboratory volunteered to educate, largely at their own corporate expense, the less formally trained analysts from competing laboratories. Although not entirely unexpected, an amazing diversity of analytical results were obtained on homogenized split environmental samples for several different parameters. Although details of this situation cannot be published due to confidentiality limitations, it is known that there were some strongly contrasting opinions, associated with equally strongly held convictions regarding how cited analytical techniques should be performed. However, the willingness of the participating laboratories to undertake such self-examination is commendable.

This information shows that without use of split environmental samples or reference standards, analytical laboratories may be legitimately unaware that their data typically falls either on the periphery, or in the mainstream, of data generated by competing laboratories. Laboratories are thus not always in a position which clearly requires them to consider the possibility that they may have a problem with procedural technique or interpretation. In other words, there may be no incentive to seek improvements where there are no indications of a problem. Therefore, the ability of a laboratory to detect and correct analytical problems may be limited by the absence of commonly available environmental reference standards.

Possible Consequences for Industry

The two major problems that differential analytical results from different laboratories presents for industry is unfair competition, and unfair treatment by regulatory agencies. Those industrial firms which are able to discover and retain commercial laboratories which tend to generate favorable results may be able to limit expenses associated with unfavorable analytical data. For commercial waste disposal facilities, reduced expenses, relative to competitors, may result if favorable data suggest that they can receive and treat a greater quantity of waste using less expensive treatment inputs, including duration of time required for treatment. For generator facilities which have waste problems, expenses may be decreased if favorable analytical results show that waste constituents do not exceed regulatory criteria which otherwise might require expensive cleanup efforts, or if cleanup efforts have succeeded with minimal expenditure. Analytical expenses may also be less if the laboratory which generates the most favorable data also has a relatively low billing scale, in addition to a decreased need for repetitive analyses.

When different analytical laboratories generate differential analytical data, the conscientious environmental manager is faced with a dilemma similar to that faced by management of the laboratory which tends to obtain less favorable results. Even though the environmental manager may "sense" that the laboratory which tends to obtain less favorable results may be the more accurate laboratory, he may also feel pressure from upper management to limit costs and remain competitive with industrial firms which utilize another laboratory that obtains more favorable results. Pressure may be brought to bear on their laboratory of choice, to achieve equally favorable results. Thus, the more environmentally conscientious industrial firms and laboratories which resist such pressures, may be "penalized" by forfeiting such potential competitive advantages.

Differential analytical results from different laboratories also affords the opportunity for industrial firms to choose litigation over field cleanup efforts, as the preferred method to resolve environmental problems. It is not difficult to imagine that litigation could be based on a premise that the regulatory criteria are irrelevant to a situation because the historical data upon which it was based cannot be adequately linked to current data. Another argument might be based on unequal treatment of competitive industrial firms by regulatory agencies due to inconsistencies in data accepted by the regulatory agency. Pursued to the extreme, industry could attempt to discredit major portions of environmental databases used by regulatory agencies. Thus a continued lack of commonly available real-world soil/waste reference standards could conceivably result in the postponement of many environmental cleanup efforts.

PROBLEM SUMMARY

Analytical results of split soil and soil/waste admixture samples sent to different laboratories sometimes differ. There are numerous opportunities for analytical laboratories to generate differential data between the time a sample is received at the analytical laboratory, and submission of the final report. When regulatory agencies accept differential data from laboratories participating in a given regulatory program, it can have negative repercussions for environmental protection, the regulatory agency, the analytical laboratories, and industrial firms.

RESOLUTION OF PROBLEM

Resolution Statement

Problems associated with the generation of differential data from split environmental samples can be resolved if the following two conditions are met:

1. Real-world environmental soil and soil/waste reference standards are prepared and made commonly available.
2. Regulatory agencies require the results of simultaneously analyzed real-world soil and soil/waste reference standards of similar matrix with environmental data.

Determination of Accuracy

As noted above, based on training and experience, each analyst may believe that they are performing an analytical procedure "correctly." Even in academic circles there may be continuous debates about the proper way to perform certain established analytical procedures, and diverse "schools of thought" may develop as a result.

To avoid becoming mired in debates where one laboratory is set against another, the authors propose an interim modification to the concept of accuracy. The proposed interim concept of accuracy is termed "consensus accuracy." Consensus accuracy is defined here as the center-most cohesive range of analytical results for a given parameter generated from homogenized split soil or soil/waste reference standards by participating laboratories using identical cited procedures.

Thus, consensus accuracy would be the typical reproducible results which would be generated during the day-to-day operations by the majority of participating analytical laboratories.

The concept of consensus accuracy, or value, could be implemented in a manner similar to that used to certify four soil reference materials by the Canada Centre for Mineral and Energy Technology (CANMET), as described by Bowman et al.[5] Each participating laboratory would generate a data set with (n) number of replicate analyses for each parameter. Thus, suitable statistical procedures to reject outliers both within and among laboratory data sets would be used. Data from each laboratory could be examined for outliers using tests such as the Dixon or Grubbs tests.[6] Concurrently, tests such as the Youden and Cochran's tests[6] could be used to identify outliers among laboratory data sets. Although these tests would provide a statistical framework for rejection of outliers, subjective judgments would still have to be made in cases where statistics tests fall in a gray area.

The pooled data consisting of (N) laboratories, each with a set of (n) data points, could be evaluated for normality using suitable goodness-of fit tests, as described by Sokal and Rohlf[7] and others. Assuming a normally distributed homogeneous sample population, a parametric analysis of variance (one-way ANOVA) could be used to evaluate variances between and among data sets with the F-test at the 95% confidence level. Also, the 95% confidence limits could be computed using an appropriate $t_{0.975}$ value with the mean variances of all data sets (N). In this manner a consensus value with a 95% confidence interval would be computed and reported for each parameter. As a unifying measure of the overall quality of a reference material parameter, CANMET[5] also made use of critical value (CF) for each parameter measured. This was arrived at using formula relating the 95% confidence limit, the overall mean, and the mean coefficients of variance.[8] Accordingly, only parameters with a CF of less than 4 were deemed acceptable for final issuance of certification. However, this threshold criterion was originally developed for ores and may be too restrictive for soils and soil/waste materials.

The above described statistical methods could be used to accept and/or reject laboratory data and to compute a consensus value for each parameter. However, it is also recognized that analytical science should not be reduced to accuracy by popularity, as suggested by the term "consensus." Nevertheless, use of consensus accuracy to provide interim benchmarks for real-world matrices will allow industry, regulatory agencies, and analytical laboratories to cooperate in an atmosphere of fairness until the academic debates are concluded.

In addition, analytical laboratories which take issue with screening on a consensus basis, would still be permitted to provide data and arguments in defense of their position. It should be recognized that one of the outlier laboratories may actually be the best trained or most informed laboratory, and may be performing the procedure most correctly. Such a laboratory should be encouraged to defend its position. Similarly, continued academic debate would also still be encouraged.

However, most laboratories which submit outlier data would be expected to recognize that their analytical procedures for a given parameter in a given matrix, requires a purposeful review to discover sources of possible systematic errors. Conversely, laboratories that consistently submit results which fall within consensus guidelines would be assured that their techniques are in the mainstream.

During the process of developing the initial databases, all analytical data, plus a statistical review, would be made available to all participating analytical laboratories by a reference standard program administrator. The reference standard program administrator should be a neutral party, and not affiliated with any one laboratory. The administrator could also serve as a clearing house for the sharing of information relating to the performance of specific analytical procedures, and the development of a suitable database management format. Only data from the laboratory requesting the data should be identified by source, and data from other laboratories should only be identified through a protected code number.

It would be expected that after analyses of a few rounds of "known" reference standards, all laboratories participating in a given regulatory program would have modified their procedures to be able to generate consistently similar data. Correlations between measured additions of appropriate additives to base reference matrices and subsequent analytical results could also be generated during the initial testing period, to determine possible matrix interactions.

Once preliminary rounds of "known" real-world reference standards have been completed by participating laboratories, "unknown" standards could be distributed, when requested, for analysis with actual environmental data. The unknown standards would be developed, as above, from the same matrix base, but with unspecified additions of the correlated materials typically incorporated into industrial soil/waste admixtures. For example, one of several basic oilfield waste standards may receive additional precisely measured amounts of industrial grade sodium bentonite, chrome lignite, barite, zinc sulfonate, diesel oil, and salt mix.

An additional benefit to the use of environmental reference standards is that, where required, regulatory criteria could be readjusted to relate more closely to the typical day-to-day performance of applicable analytical procedures by participating laboratories.

Methods for evaluating dissimilar results from environmental samples analyzed at different laboratories have been proposed by Papp et al.,[9] and Kreamer and Stetzenbach,[10] among others. Consensus results from geologic materials have also been used to indicate the relative consistency of stress and fracture testing among several university rock mechanics laboratories.[11]

As of this writing, details for the preparation, distribution, and administration of real-world soil and soil/waste reference standards are being developed by several organizations, including K. W. Brown & Associates, Inc. Comments and expressions of interest are sought.

CONCLUSIONS

If required by regulatory agencies, submission of acceptable known and blind referenced performance standard results with field environmental data would substantiate analytical relevance and competence. Suitable standards would have applicable matrices and constituent-of-interest concentrations relative to the objectives of analysis. "Accuracy" for each parameter in standards could be evaluated based on reasonable allowable deviation from a central "consensus" value established by the distribution of data generated by participant laboratories. Regulatory criteria could be adjusted to relate to the empirically derived consensus database. Laboratories could detect and revise inappropriate, or nonstandard, procedural variations by using these standards and the resulting database. Conscientious environmental managers and laboratory managers would be less pressured to achieve inappropriately favorable analytical results, and regulatory agencies could more equitably enforce environmental protection laws.

REFERENCES

1. Dragun, J., J. Barkach, and S. A. Mason. "Misapplications of the EP-Tox, TCLP, and CAM-WET Tests to Derive Data on Migration Potential of Metals in Soil Systems," in P. T. Kostecki and E. J. Calabrese, Eds., *Petroleum Contaminated Soils, Volume* 3 (Chelsea, MI: Lewis Publishers, Inc., 1990).
2. Barth, D. S., and B. J. Mason. Soil Sampling Quality Assurance User's Guide. EPA-600/4-84-043, 1984.
3. Schumacher, B. A., K. C. Shines, J. V. Burton, and M. L. Papp. "Comparison of Three Methods for Soil Homogenization," *Soil Sci. Soc. Am. J.* 54:1187–1190 (1990).
4. Test Methods for Evaluating Solid Waste, U.S. Environmental Protection Agency (SW-846), 1986.
5. Bowman, W. S., G. H. Faye, R. Sutarno, J. A. McKeague, and H. Kodama. Soil Samples SO-1, SO-2, SO-3, and SO-4—Certified Reference Materials. Canada Centre for Mineral and Energy Technology (CANMET). Canadian Government Publishing Centre. Hull, Quebec, Canada, 1979.
6. Taylor, J. K. *Quality Assurance of Chemical Measurements.* (Chelsea, MI: Lewis Publishers, Inc., 1987).
7. Sokal, R. R., and J. Rohlf. *Biometry.* 2nd Ed. (New York, NY: W. H. Freeman and Company, 1981).
8. Sutarno, R., and G. H. Faye. "A Measurement for Assessing Certified Reference Ores and Related Materials," *Talanta* 22:676–681 (1975).
9. Papp, M. L., J. Miah, J. E. Teberg, R. D. Van Remortel, G. E. Byers, and B. A. Schumacher. "Estimating Uncertainty in Laboratory Measurements for Planning Optimization," 1988 Agronomy Abstracts, pp. 44–45.
10. Kreamer, K. K., and K. J. Stetzenbach. "Development of a Standard Pure-Compound Base Gasoline Mixture for Use as a Reference in Field and Laboratory Experiments," *Ground Water. Monit. Rev.* 10:135–145 (1990).
11. Magouirk, J. Personal communication. Research Specialist, Center for Tectonophysics, Texas A&M University, College Station, TX, 1990.

CHAPTER 13

Site Assessment, Remediation, and Closure
Under the Oregon UST Matrix

Robert A. Dixon, Sweet-Edwards/EMCON, Inc., Portland, Oregon

The Oregon Numeric Soil Cleanup Levels for Motor Fuel and Heating Oil (OAR 340-122-305 through 340-122-360), commonly called the Matrix, has been very useful in expediting site assessments, remediations, and closures since its implementation on July 21, 1989.[1] The Matrix has focused tremendous time and financial resources away from litigation and toward direct assessment and cleanup of underground storage tank (UST) sites with minimal regulatory oversight.

ECONOMIC AND POLITICAL CONTEXT

The Matrix was framed, as are all regulations, within the prevailing political and economic climate. Oregon's economy and population has undergone both rapid growth and extensive recession in the last 25 years, typical of the economies of other western states with natural resource bases. This historic pattern of boom and bust has produced a general job-protective political climate; one favoring jobs and the economy over the environment. At the same time, voters developed an opposition to taxation, which resisted funding to hire additional government employees to regulate job-supplying industries.

In this climate the Oregon Department of Environmental Quality (DEQ) and the UST Citizen Advisory Committee (USTCAC) were challenged to develop rules for implementing the federally mandated UST regulations that became effective in December 1988. These rules had to be straightforward enough for independent station owners to use when decommissioning their own tanks. Such rules would then have to be restricted to sites of a simple nature.

GENERAL TECHNICAL CONSIDERATIONS

The major technical consideration in forming the Matrix was the need to base cleanup levels on site-specific fate and transport risk assessments. Such assessments, however, are comprehensive and not economically feasible for the majority of UST assessment and cleanup jobs, which are small. Therefore, a simplified Matrix risk assessment was developed. Because of its simplicity, its applicability was restricted to simple sites: those whose groundwater has not been impacted and whose primary contaminants are gasoline or nongasoline petroleum hydrocarbons (usually diesel, heating oil, or lube oil). Such sites are typically the size of corner gas stations. The Matrix was designed to provide a straightforward protocol for assessing simple UST-removal sites, quickly determining site-specific cleanup levels, and effecting site closure.

SITE ASSESSMENT

Site assessment under the Matrix follows a simple protocol for soil sampling and analysis and for ranking a site.

Soil Sampling Protocol

The particulars of assessment soil sampling are given by OAR 340-122-340, Sample Number and Location:

> The collection and analysis of soil samples is required to verify that a site meets the requirements of these rules. These samples must represent the soils remaining at the site and shall be collected after contaminated soils have been removed or remediated. Each sample must represent a single location; composite samples are not allowed. The number of soil samples are to be collected as follows:
>
> (1) A minimum of two soil samples must be collected from the site:
>
>   (a) These samples must be taken from those areas where obviously stained or contaminated soils have been identified and removed or remediated.
>
>   (b) If there are two or more distinct areas of soil contamination, then a minimum of one sample must be collected from each of these areas.

(c) The samples must be taken from within the first foot of native soil directly beneath the areas where the contaminated soil has been removed, or from within the area where in-situ remediation has taken place.

(d) A field instrument sensitive to volatile organic compounds may be used to aid in identifying areas that should be sampled, but the field data may not be substituted for laboratory analyses of the soil samples.

(e) If there are no areas of obvious contamination, then samples must be collected from the locations specified in subsections (2) to (5) of this section which are most appropriate for the situation.

(f) If it is being proposed that a pocket of contamination be left in place pursuant to 340-122-355(4), then sufficient samples shall be collected from the site in order to estimate the extent, volume and level of contamination in this pocket.

(2) If water is NOT present in the tank pit:

(a) Soil samples must be collected from the native soils located no more than two feet beneath the tank pit in areas where contamination is most likely to be found.

(b) For the removal of an individual tank, samples must be collected from beneath both ends of the tank. For the removal of multiple tanks from the same pit, a minimum of one sample must be collected for each 150 square feet of area in the pit.

(3) In situations where leaks have been found in the piping, or in which released product has preferentially followed the fill around the piping, samples are to be collected from the native soils directly beneath the areas where obvious contamination has been removed. Samples should be collected at 20 lateral foot intervals.

(4) If water is present in the tank pit, regardless of whether obvious contamination is or is not present, the Department must be notified of this fact. The owner, permittee, or responsible person shall then either continue the investigation under OAR 340-122-240, or do the following:

(a) Purge the water from the tank pit and dispose of it in accordance with all currently applicable requirements. This may include obtaining appropriate permits from the Department or local jurisdictions.

(b) If the pit remains dry for 24 hours, testing and cleanup may proceed according to the applicable sections of these soil cleanup rules. If water returns to the pit in less than 24 hours, a determination must be made as to whether contamination is likely to have affected the groundwater outside of the confines of the pit as indicated below:

(A) For the removal of an individual tank, soil samples are to be collected from the walls of the excavation next to the ends of the tank at the original soil/water interface. For the removal of multiple tanks from the same pit, a soil sample is to be collected from each of the four walls of the excavation at the original soil/water interface.

(B) At least one sample must be taken of the water in the pit regardless of whether obvious contamination is or is not present. This sample shall be collected as required by 340-122-345(4).

(C) The soil samples must be analyzed for TPH and benzene, toluene, ethylbenzene and xylenes (BTEX), and the water sample must be analyzed for BTEX. These analyses must be made using the methods specified in 340-122-350. The results of these analyses must be submitted to the Department.

(D) The Department shall then determine how the cleanup shall proceed as specified in 340-122-355(3).

(5) In situations where tanks and lines are to remain in place in areas of suspected contamination, the owner, permittee or responsible person shall submit a specific soil sampling plan to the Department for its approval.

(6) In situations where TPH analysis indicates that contamination is present due to a release from a waste oil tank, at least one sample of the waste oil contaminated soils must be collected and analyzed for PCBs, volatile chlorinated solvents, volatile aromatic solvents, and leachable metals using the analytical methods specified in 340-122-350.

Site Ranking

To simplify the risk assessment process the Matrix establishes three levels of environmental sensitivity that should contain any given site. A site's level is determined by its score in an objective ranking system that assigns points for the following common site-specific parameters:

- depth to groundwater
- mean annual precipitation
- native soil type
- sensitivity of the uppermost aquifer
- potential receptors

The specific number of points assigned for each parameter are described in OAR 340-122-330, Evaluation Parameters:

The site-specific parameters are to be scored as specified in this section. If any of the parameters in 340-122-330(1)–(5) is unknown, that parameter shall be given a score of 10.

(1) Depth to Ground Water: This is the vertical distance (rounded to the nearest foot) from the surface of the ground to the highest seasonal elevation of the saturated zone.

The score for this parameter is:

>100 feet	1
50–100 feet	4
25–50 feet	7
<25 feet	10

(2) Mean Annual Precipitation: This measurement may be obtained from the nearest appropriate weather station.

The score for this parameter is:

<20 inches	1
20–45 inches	5
>45 inches	10

(3) Native Soil or Rock Type:

The score for this parameter is:

Low permeability materials such as clays, silty clays, compact tills, shales, and unfractured metamorphic and igneous rocks.   1

Moderate permeability materials such as sandy loams, loamy sands, and clay loams; moderately permeable limestones, dolomites and sandstone; and moderately fractured igneous and metamorphic rocks.   5

High permeability materials such as fine and silty sands, sands and gravels, highly fractured igneous and metamorphic rocks, permeable basalts and lavas, and karst limestones and dolomites.   10

(4) Sensitivity of the Uppermost Aquifer: Due to the uncertainties involved in the Matrix evaluation process, this factor is included to add an extra margin of safety in situations where critical aquifers have the potential to be affected.

The score for this parameter is:

Unusable aquifer, either due to water quality conditions such as salinity, etc.; or due to hydrologic conditions such as extremely low yield.   1

Potable aquifer not currently used for drinking water, but the quality is such that it could be used for drinking water.   4

Potable aquifer currently used for drinking water; alternate un-threatened sources of water readily available.   7

Sole source aquifer currently used for drinking water; there are no alternate unthreatened sources of water readily available.   10

(5) Potential Receptors: The score for potential receptors is based on both the distance to the nearest well and also the number of people at risk. Each of these two components is to be evaluated using the descriptors defined in this section.

(a) The distance to the nearest well is measured from the area of con-tamination to the nearest well that draws water from the aquifer of concern. If a closer well exists which is known to draw water from a deeper aquifer, but there is no evidence that the deeper aquifer is completely isolated from the contaminated aquifer, then the dis-tance must be measured to the closer, deeper well.

The distance descriptors are:

Near <1/2 mile
Medium 1/2–2 miles
Far >2 miles

(b) The number of people at risk is to include all people served by drink-ing water wells which are located within 2 miles of the contaminat-ed area. For public wells, count the number of users listed with the Oregon Health Division, Drinking Water Systems Section. For pri-vate wells, assume 3 residents per well. In lieu of a door-to-door sur-vey of private wells, it may be assumed that there is one well per residence.

The number descriptors are:

Many >3000
Medium 100–3000
Few <100

(c) The score for this parameter is taken from the combination of the two descriptors using the following grid:

	Many	Medium	Few
Near	10	10	5
Medium	10	5	1
Far	5	1	1

(6) The Matrix Score for a site is the sum of the five parameter scores in 340-122-335(1)–(5).

Based on its total number of points, the site is ranked Level 1 (greater than 40), 2 (from 25 to 40), or 3 (less than 25). Because each ranking level is presumed to indicate significantly different environmental sensitivity or potential to impact ground water, each carries with it different soil cleanup concentrations for the same general categories of petroleum hydrocarbons.

SITE-SPECIFIC SOIL CLEANUP CONCENTRATIONS

The Matrix soil cleanup concentrations were developed with the understanding that they only pertain to simple sites. Their primary intent is to protect local groundwater quality and minimize direct acute and chronic exposure to petroleum hydrocarbons that may affect human health or the environment.

In developing a conservative leaching model, the federal drinking water standard of 5 parts per billion (ppb) of benzene was taken as the minimum protective standard. Benzene has a clearly established federal standard and is considered the most toxic constituent (both acute and chronic) normally associated with refined petroleum hydrocarbon products. The percentage of benzene generally present in such products is predictable within a limited range. The DEQ chose benzene as the standard because, in the vast majority of cases, the risk posed by the other probable petroleum hydrocarbons present would probably be less.

The Matrix soil cleanup concentrations were integrated with the site-specific ranking to yield site-specific numeric soil cleanup concentrations (levels). The site-specific rules describing this integration are given in OAR 340-122-335, Numeric Soil Cleanup Standards:

(1) If the Matrix Score evaluated in 340-122-330 is:

    (a) Greater than 40, the site must be cleaned up to at least the Level 1 standards listed in 340-122-335(2).

    (b) From 25 to 40, inclusive, the site must be cleaned up to at least the Level 2 standards listed in 340-122-335(2).

    (c) Less than 25, the site must be cleaned up to at least the Level 3 standards listed in 340-122-335(2).

(2) The following table contains the required numeric soil cleanup standards based on the level of Total Petroleum Hydrocarbons (TPH) as measured by the analytical methods specified in 340-122-350.

	Level 1	Level 2	Level 3
TPH (Gasoline)	40 ppm	80 ppm	130 ppm
TPH (Diesel)	100 ppm	500 ppm	1000 ppm

(3) The Hydrocarbon Identification (HCID) test specified in 340-122-350(3) shall be used to identify the petroleum product contamination present at the site. The results of the HCID test shall be used to determine which analytical method or methods are required for verifying compliance with the Matrix cleanup levels. At locations where the soil is contaminated with both gasoline and diesel or other non-gasoline fraction hydrocarbons,

the gasoline contamination shall be shown to meet the appropriate gasoline cleanup standard and the diesel or other non-gasoline fraction contamination shall be shown to meet the appropriate diesel cleanup standard.

When the original Matrix was finalized in 1989, the DEQ believed that the United States Environmental Protection Agency (U.S. EPA) Method 418.1, which determined TPH for gasoline or nongasoline, often produced data that varied in its accuracy and precision. The DEQ and the USTCAC debated whether to use more expensive analytical gas chromatographic (GC) methods, but the absence of a standardized U.S. EPA Method utilizing GC further complicated the situation. The DEQ chose to make the Matrix cleanup concentrations conservative enough to offset the perceived lack of accuracy and precision of U.S. EPA Method 418.1.

This alternative was chosen by the DEQ for a 1-year trial. During this trial year, DEQ staff would participate in a U.S. EPA committee to address the need for a U.S. EPA Method for TPH by GC. Consequently, the Matrix cleanup levels were specific to the analytical method used.

Revisions to the Matrix effective March 1, 1991, implemented GC methods (TPH-G and TPH-D) for determining TPH concentrations in gasoline- or diesel-contaminated soil.[3] A modified version of U.S. EPA Method 418.1 was retained for determining TPH concentrations in soils contaminated with oils heavier than diesel. Matrix analytical methods are detailed in the DEQ's *Total Petroleum Hydrocarbons Analytical Methods, 11 December 90.*

REMEDIATION AND CLOSURE

Remediation under the Matrix can be expedient because the site can be ranked and assessed, and the target cleanup level in soil determined, without regulatory involvement (except for the initial telephone notification of a confirmed release to the DEQ). Once the cleanup level is determined, the most appropriate remediation technology can be selected.

Excavation and Landfilling

If this method is selected, soils are excavated until samples demonstrate that the Matrix cleanup levels have been attained. The excavation is then backfilled with clean material. Before the excavated soil is hauled to a landfill, a UST permit addendum must be filed with the DEQ. Processing the addendum generally takes from several days to several weeks, depending on the urgency of the project and the DEQ workload.

When the project is finished, a Matrix closure report describing site work, the volume of soil excavated, final disposition of excavated soil, and supporting analytical chemistry, is submitted to the DEQ. If the Matrix rules have been followed exactly, and the site meets the Matrix cleanup goals, the DEQ typically issues a No Further Action Required (NFAR) letter within 1 to 3 months. For projects with firm deadlines, such as the transfer of ownership of real property, the DEQ is usually responsive and sometimes has been able to issue an NFAR letter within 1 week, workload permitting.

Excavation and Aboveground Treatment

If this is considered the most appropriate remediation for a site, a UST permit addendum must be filed with the DEQ and approved before the remediation program can begin. If laboratory analysis of verification samples shows that soil hydrocarbon concentrations are below the Matrix cleanup levels for the site, a Matrix closure report, as described above, is submitted to the DEQ. When the Matrix criteria are met, the site is closed and an NFAR letter is issued by the DEQ. A parallel procedure is followed if the soil is hauled and treated offsite.

In Situ

If soil will be treated in situ, a corrective action plan (CAP) or a DEQ-approved alternative must be prepared. The procedure for preparing, reporting, and implementing a CAP are contained in Oregon Cleanup Rules for Leaking Petroleum Underground Storage Tank Systems (OAR 340-122-205 through 340-122-260) and are not discussed in this chapter.

Copies of Oregon UST rules and regulations may be obtained from the DEQ UST cleanup staff in the following regional offices:

Northwest Region
811 SW Sixth Avenue
Portland, OR 97204
503-229-5263
503-378-8240

Willamette Valley Region
750 Front Street NE
Suite 120
Salem, OR 97310

Southwest Region
201 W Main Street
Suite 2-D
Medford, OR 97501
503-776-6010

Central Region
2146 NE 4th
Bend, OR 97701
503-388-6146

Eastern Region
700 SE Emmigrant
Suite 330
Pendleton, OR 97801
503-276-4063

CONCLUSIONS

The Matrix expedites the assessment, remediation, and closure of simple (no groundwater) petroleum hydrocarbon contaminated sites.

The expeditious process facilitated by the Matrix enables resources to be focused directly on the assessment, remediation, and closure of sites rather than on the generation of voluminous reports or litigation.

Minimal involvement is required of DEQ regulators, which reduces their workload and makes responsive turnaround possible. Minimizing DEQ involvement also reduces the taxpayer's burden.

When the analytical methods were revised in March 1991, the cleanup concentrations were not raised. Because these standards were originally set low to compensate for the assumed inaccuracy and imprecision of U.S. EPA Method 418.1, it seems reasonable that they should have been correspondingly raised with the implementation of improved analytical methods. Consequently, the cleanup levels originally established in the Matrix in July 1989 should be reexamined for a possible increase because of the improved analytical methods implemented in March 1991.

The development of a ''Ground Water Matrix'' counterpart to the Numerical Soil Cleanup Levels Matrix would be a worthwhile future endeavor for the DEQ and a new citizen advisory committee.

REFERENCES

1. Oregon Administrative Rules 430-122-305 through 340-122-360 (July 21, 1989 Version).
2. Oregon UST Citizen Advisory Committee meeting notes, 1989–1990.
3. Oregon Administrative Rules 340-122-305 through 340-122-360 (Revised March 1, 1991).

PART V

Remediation Assessment and Design

CHAPTER 14

Bioremediation of PAH Compounds in Contaminated Soil

Raymond C. Loehr, Environmental and Water Resources Engineering Program, The University of Texas, Austin, Texas

INTRODUCTION

Bioremediation is a managed treatment process that uses microorganisms to degrade and transform organic chemicals in contaminated soil, aquifer material, sludges, and residues. Bioremediation also reduces the toxicity of the organics and the migration potential of hazardous constituents in the material being treated. Thus, bioremediation can be considered a source control, pollution prevention, and risk reduction process that can reduce or eliminate sources of groundwater contamination and thereby reduce the need for more costly and long-term ground-water treatment processes.

The advantages of contaminated soil bioremediation processes are that such processes:

- are used where the problem is located
- do not require transporting large quantities of contaminated material offsite
- minimize long-term liability
- are ecologically sound and an extension of natural processes
- generally are cost-effective and competitive with other decontamination technologies for organics.

Bioremediation processes have been recommended at many abandoned sites requiring cleanup. An evaluation of decisions for technical remedies at such sites indicates that bioremediation has been recommended at over 20 major sites. The types of bioremediation processes proposed for these sites are noted in Table 14.1. The chemicals to be treated primarily were polycyclic aromatic hydrocarbons (PAH), volatile organics (benzene, toluene, ethylbenzene and xylene-BTEX), pentachlorophenol, phenols, and other chlorinated and nonchlorinated organics.

Table 14.1. **Bioremediation Processes Recommended for Use at Sites Requiring Cleanup[a]**

Type of Process	Number of Times Recommended
Excavation followed by land treatment	8
Excavation with onsite treatment	6
In situ treatment	5
Lagoon aeration	2
To be determined and other	3

[a]From review of recommendations through 1989; some decisions specified multiple remedies; adapted from Reference #1.

If bioremediation can occur and is effective, why do some organics, such as those in contaminated soils, persist in the environment? This is a reasonable question that is answered by identifying the factors that affect microorganisms and therefore bioremediation processes. Even biodegradable organics may persist if adverse factors and nonoptimum conditions exist. The factors that may prevent microbial degradation and bioremediation include:

- chemical concentrations that are toxic to microorganisms
- inadequate type or numbers of microorganisms, such as due to toxic conditions
- conditions too acid or alkaline
- lack of nutrients such as nitrogen, phosphorus, potassium, sulfur, or trace elements (most organic chemicals, for example, are not nutritionally balanced)
- unfavorable moisture conditions (too wet or too dry)
- lack of oxygen or other electron acceptors.

Bioremediation processes are biological treatment processes that improve or stimulate the metabolic capabilities of microbial populations to degrade organic residues. Therefore, it is important to understand those conditions and reactions so that bioremediation processes can be successful. With such knowledge, it is possible to modify the nonoptimum conditions so that microbial degradation can occur.

In this study, laboratory treatability studies evaluated key parameters that affect contaminated soil bioremediation. The objective of the research was to determine the feasibility of bioremediating soils contaminated with wastes from a petrochemical facility.

MATERIALS AND METHODS

Contaminated Soil

The soil came from the bottom of a pond used to store primary clarifier sludge that resulted from the treatment of industrial wastewater. Approximately 25 kg were received from a petrochemical plant and were stored at 4°C until used. Subsamples were taken as required for the laboratory tests. The contaminated soil was analyzed for organic carbon, total carbon, nitrogen, phosphorus, and metals as well as for organic chemicals suspected of being present—BTEX, phenol and chlorophenols, and PAH compounds.

In the research, the contaminated soil was mixed with varying quantities of uncontaminated soil to evaluate different loading rates. The uncontaminated soil was obtained from a rural area approximately four miles southwest of Austin, Texas. The soil was air dried, sieved, and stored at 4°C in the dark until used. The uncontaminated soil had not had previous exposure to industrial chemicals or wastes, and did not receive any pretreatment such as soil amendments or specially acclimated biological cultures prior to these experiments. Thus, the naturally occurring soil microbial consortium was responsible for the bioremediated removal of the chemicals.

Toxicity screening analyses estimated the potential impact of the material on soil biota. These analyses estimated the toxicity of the contaminated soil, the initial application rates used in the degradation studies, and the detoxification that occurred. The Microtox System® (Microbics Corporation) was the toxicity screening method.

Degradation Test Protocol

Many preliminary and six detailed degradation tests were performed, each evaluating a different combination of environmental parameters that may affect the degradation rates of the organic compounds. The loss rate of specific compounds was determined using small soil microcosms and a consistent protocol. Each test consisted of sets of triplicate microcosms for each sampling interval. Controls containing only uncontaminated soil also were included in the sets.

Quantitative assessment of the loss rate of an individual chemical in a bioremediation process must involve measurement of the changes in the chemical concentration with time. This was done in this study; however, no evaluation of specific chemical loss mechanisms was made. Thus, the noted chemical loss rates could have been due to one or more of the following: biodegradation, chemical degradation, hydrolysis, and volatilization.

The rate of chemical loss was determined by measuring the difference between the amount of chemical initially added to the soil and that which was recovered after specified time intervals. The general protocol used to determine these differences is presented in Table 14.2. This protocol approximated a batch contaminated soil bioremediation process. Loss rates using this protocol have been found to be comparable to loss rates identified in the literature and in field units.

Table 14.2.   Experimental Procedures Used in the Chemical Loss Microcosm Studies

Each experiment consisted of eight sample sets. Each sample set contained four beakers (triplicates for a sample and one blank). The experimental procedure used was as follows:

(a) Place 10 g of the contaminated soil mixture into each 150 mL beaker and adjust the soil moisture content to 80% of field capacity with distilled deionized water. Place a glass stirrer in each beaker. Record all weights, including beaker, soil, water, and glass stirrer.

(b) Mix soil and water thoroughly, cover the beakers with aluminum foil, and place the beakers in the dark at 20°C. The cover minimizes water loss and the possible addition of contaminated dust.

(c) Incubate the beakers in the dark to prevent photodegradation of the added chemicals.

(d) Adjust the moisture content in each beaker weekly and maintain between 60% and 80% of field capacity.

(e) Sacrifice sample sets at selected time intervals for chemical analyses.

Based on the protocol noted in Table 14.2, the soil was maintained at a moisture content of about 80% of field capacity. This provided adequate moisture for the biodegradative reactions but avoided saturated conditions. For the soils used, a moisture content of about 16% was the equivalent of 80% field capacity.

The microcosms were arranged in sample sets for sacrifice at days 0, 4, 9, 15, 21, 28, and 35. The intervals provided at least seven data points to establish the loss rate of the compounds under the experimental conditions. The reactors were incubated at 20°C in the dark.

At the selected time intervals, the concentrations of the remaining chemicals in a set of microcosms were determined. At these sampling periods, a sample set (four beakers—one blank and three with chemical) was taken from the constant temperature room and extracted for 16 hours with methylene chloride in a Soxhlet extraction apparatus (Method 3540).[2] The concentration step was conducted in a water bath maintained at 60°C. The concentrated extracts were dried by passing them through disposable sodium sulfate columns and then refrigerated at 4°C until analysis by gas chromatography (GC).

Mass Spectrometry Analyses

The contaminated soil contained a complex mixture of organic compounds. A sample of the concentrated extract from one of the experiments was characterized by a GC/MS analysis. The contaminated soil contained primarily two-, three-, and four-ring PAHs. Several were selected for evaluation of loss in the degradation studies.

Gas Chromatography

The concentrated methylene chloride extracts were analyzed by capillary column gas chromatography using Method 8040.[2] The method consisted of injection of a 1 μL sample into the gas chromatograph which was equipped with an electronic integrator, a methyl silicone capillary column, and a flame ionization detector.

All compounds were monitored for changes in concentration during each test. All the tracked PAHs were confirmed by spiking a sample with known standards of each PAH or by cross-referencing with the GC/MS chromatogram. A reference mixture of PAH compounds, each at a known concentration of known retention time, was a standard against which each analyzed degradation test sample was compared. The reference mixture was analyzed in the GC each day.

Soil Characterization

The contaminated soil had an organic carbon content of 2.0%, a pH of 8.1, TKN of 640 mg/kg, and total phosphate of 550 mg/kg dry weight. The dominant elements were calcium, iron, aluminum, and magnesium, typical elements in soil. Present in lower concentrations were potassium, silicon, phosphorus, and manganese, also common soil elements. Potentially toxic elements were present at concentrations within reported background levels for soils.

pH values for biodegradation should be between 6 and 8. All degradation experiments in this study were conducted with the pH "as received," since that pH did not require modification.

DEGRADATION STUDIES

These studies evaluated the potential of bioremediation to reduce the concentration of organics in the contaminated soil. Preliminary experiments explored the general feasibility of the test protocol and identified the most important environmental factors that affected the loss of the PAHs in the contaminated soil. Six detailed studies evaluated the effect of: (a) loading rates; (b) moisture level; and (c) need for amendments.

The parameters evaluated in each of the six detailed tests are defined in Table 14.3. The results of the testing, described in terms of compound half-life—based upon first-order reaction kinetics, as well as published half-life data, are compiled in Table 14.4. Half-life is the time to reduce the compound mass by 50%. Statistical analysis of the chemical loss data indicated that first-order loss rates represented the data satisfactorily.

Table 14.3. Degradation Test Parameters

Test[a] Designation	Mixture[b]	Loading Rate %EHC[c] Ratio	Moisture Content % of Field Cap.	% by Weight	Amendments
15	1:1	1.3	80	16	None
16	5:1	2.3	80	16	None
17	5:1	2.3	40	8	None
18	5:1	2.3	40	8	1% Activated sludge[d]
20	1:0	2.7	80	16	None
21	1:0	2.7	80	16	1% Activated sludge

[a] Tests 1 to 14 were preliminary evaluations that included characterization, nutrient and micro-tox testing; test 19 was not completed due to laboratory difficulties.
[b] Mixture ratio defined as parts contaminated soil by weight to parts uncontaminated soil by weight. In tests 20 and 21 no uncontaminated soil was used and only the contaminated soil was evaluated.
[c] Methylene Chloride extractable hydrocarbons (EHC) expressed in terms of percent of EHC in mixture or in contaminated soil.
[d] Municipal dry activated sludge as percent of air-dry mixture or contaminated soil.

Table 14.4. Summary of Loss Rate Results—Half-Life (Days) (Based on First-Order Reaction Rates)

Compounds	15	16	17	18	20	21	Literature Data[a]
Naphthalene	9	9	14	10	35	6	15–30
Acenaphthylene	63	25	33	c	c	d	14–78
Acenaphthene	c	8	c	17	c	99	45–96
Fluorene	18	10	120	11	c	13	39–64
Phenanthrene	19	13	170	11	c	11	7–69
Anthracene	c	50	53	43	c	50	11–180
Phenylnaphthalene	41	120	170	17	c	140	—
Fluoranthene	58	69	140	77	c	170	13–350
Pyrene	c	87	120	c	c	230	13–350
Unknown 5[b]	43	c	87	c	c	d	—
4-Methylpyrene	41	c	120	c	c	69	—
Benz(a)anthracene	58	c	c	87	c	d	18–250
Chrysene	c	c	c	c	c	d	16–70

Experiment Number spans columns 15–21.

[a] Data derived from laboratory tests on refinery, creosote, or PCP wastes or contaminated sediments as reported in the literature; range resulted from a detailed literature review.
[b] Mass spectrometry analysis indicated Unknown 5 was either 1-Methylpyrene or Benzo(b) fluorene.
[c] No compound loss from soil microcosms during the 37-day test duration.
[d] No loss of compound from soil microcosm during the 118 day tests.

DISCUSSION

In this study, the primary emphasis was on determining degradation (loss) rates of representative polyaromatic hydrocarbons (PAH). Characterization of the contaminated soil had indicated that these were the predominant organic chemicals present.

Moisture Content

The aerobic degradation of organics in soil can be a function of the available soil moisture. Commonly, this is expressed in terms of percent of field capacity, i.e., the amount of water retained by the soil micropores. Several authors[3-5] have cited ranges in which biodegradation is optimum. Some indicate that 30% to 90% of field capacity is needed. Others indicate that 50% to 80% is a better range. At moisture levels below 30% to 40%, activity is restricted due to low water concentrations, while at high moisture levels, aerobic microbial metabolism is reduced because the soil pores are near saturation and transfer of oxygen is inhibited.

Several experiments evaluated the hypothesis that the moisture content needed to be controlled. This was examined in test 20 in which the moisture content was maintained at 80% of field capacity. Tests 16 and 17 were conducted with contaminated soil diluted by uncontaminated soil and had moisture levels maintained at 80% and 40% of field capacity, respectively.

The results of test 20 (Table 14.5) indicate that moisture was not the only factor affecting degradation of the PAH. Had moisture been the only factor, the PAH, after adjusting the moisture content in the test to optimum conditions, would have been degraded considerably in the 37-day test. The data in test 20 show that only naphthalene exhibited any significant loss. Loss of two-, three-, and four-ring PAH occurred in tests 16 and 17 but not in test 20.

Comparison of results (Table 14.5) when moisture content was 80% of field capacity (test 16) to those obtained in test 17, when the moisture content was 40% of field capacity, demonstrated the need for adequate moisture. The removals of the compounds were considerably greater in test 16 when the moisture content was 80% of field capacity. The conclusions from tests 20, 16, and 17 were:

- The moisture level was a critical parameter for the degradation of two-, three-, and four-ring PAH. If moisture content is maintained at 80% of field capacity, degradation rates will be considerably greater than at 40% of field capacity.
- For contaminated soil alone, maintaining the moisture content at 80% of field capacity did not result in PAH degradation. Other parameters such as acclimated microorganisms and nutrients also appear to play an important role in achieving degradation of PAH compounds in this contaminated soil.

Amendments

The total carbon content of the contaminated soil was 4.3% and the organic carbon content was 2.0%, both expressed on a dry basis. An organic amendment, dry waste-activated sludge, was added to tests 18 and 21 at 1% by weight of contaminated soil. Tests 17 and 18 were performed under identical conditions except that test 18 contained the 1% dried activated sludge.

Generally, the removals (Table 14.5) that occurred with the added activated sludge (test 18) were greater than those without the amendment (test 17). There were some exceptions. For all three-ring PAH, the removals were greater when the activated sludge was added. It must be remembered that both tests 17 and 18 had moisture contents of 40% of field capacity, i.e., less than the 80% of field capacity identified earlier as better for removal in these studies.

Table 14.5. Percent Compound Removal in Specific Tests[a]

| | % Removal | | | | | | No. |
| | Experiment Number | | | | | | |
Compound	15	16	17	18	20	21	Rings
Naphthalene	81	94	51	63	50	80	2
Phenylnaphthalene	37	20	13	44	0	42	2
Acenaphthylene	33	42	0	35	0	0	3
Acenaphthene	0	80	0	43	0	54	3
Fluorene	64	83	15	59	0	68	3
Phenanthrene	62	76	13	60	0	52	3
Anthracene	0	39	37	41	0	23	3
Fluoranthene	28	30	16	23	0	38	4
Pyrene	0	25	19	0	0	28	4
Unknown 5[b]	35	0	18	0	0	0	4
4-Methylpyrene	46	0	19	0	0	21	4
Benzo(a)anthracene	35	0	0	25	0	0	4
Chrysene	0	0	0	0	0	0	4
Conditions of Tests[c]							
Loading rate	1:1	5:1	5:1	5:1	1:0	1:0	
Moisture content	16	16	8	8	16	16	
Amendments	None	None	None	1% A.S.[d]	None	1% A.S.	

[a] Percent is based on average mass of compound in soil at beginning and end of tests. Duration of tests 15–20 ranged from 35.5 to 37 days. Duration of test 21 was 118 days.
[b] UNK 5 was either 1-Methylpyrene or Benzo(b)fluorene.
[c] From Table 14.3.
[d] Dry waste activated sludge.

Tests 20 and 21 had identical conditions except that test 21 was amended with 1% dry activated sludge. Results in Table 14.5 showed improvement in two- and three-ring PAH removal in test 21 when compared to test 20. Removal of four-ring PAHs was somewhat improved in test 21. Based on the results of tests 17, 18, 20, and 21, it was apparent that amendments that add nutrients, organic matter and active organisms can be of benefit in removing PAHs in these contaminated soils.

Loading Rates

Initial toxicity screening demonstrated that the contaminated soil was not toxic to indicator bacteria. Two contaminated soil: uncontaminated soil loading rates, 1.3 and 2.3% methylene chloride extractable hydrocarbons, were evaluated in tests 15 and 16, respectively. Dry uncontaminated soil was mixed with the contaminated soil to yield the desired loading rates. This loading rate evaluation identified whether the dilution and the indigenous microorganisms in the uncontaminted soil affected the degradation rates.

Test 16 demonstrated greater percent removals of the two- and three-ring PAHs than did test 15 (Table 14.5). The percent removal of some, but not all, of the four-ring compounds also was improved in test 16. Based on test 15 and 16 results, an increase in the waste-loading rates, within the limits of tests, increased the overall degradation rates of the PAHs in the contaminated soils.

Acclimated Microorganisms

Acclimated microorganisms are important to successful bioremediation. Test 20 (contaminated soil only) evaluated the potential for PAH degradation, without amendments. In that test, the moisture content was maintained at 80% field capacity, a level that yielded PAH degradation in tests 15 and 16. Of all PAHs monitored, only naphthalene exhibited any loss. This loss of naphthalene may have been due to volatilization.

Test 21 on contaminated soil amended with an organic supplement (dry waste activated sludge) was performed under conditions identical to test 20. The results (Table 14.5) indicated that removal of the two- and three-ring PAH in test 21 was similar to the removals in test 18. Somewhat different degradation of the four-ring PAH occurred in test 21. These results demonstrate that, within limits of the research and under a range of environmental conditions, the uncontaminated soil contained an indigenous microorganism population capable of quickly initiating PAH degradation. This result is significant in that the uncontaminated soil was gathered from a rural location and had not been exposed previously to the PAH compounds in the contaminated soil. Specially developed microoganisms were not required to degrade the hydrocarbon contaminants. Combining the contaminated soil with uncontaminated soil resulted in significant hydrocarbon degradation.

CONCLUSIONS

These treatability studies indicated that bioremediation could result in the loss of PAH compounds in these contaminated soils when certain environmental factors were controlled. The factors included: (a) aerobic conditions; (b) adequate moisture; (c) nontoxic loading rates; (d) indigenous, acclimated organisms; and (e) amendments that add nutrients and degradable organic matter.

ACKNOWLEDGMENTS

Support for this research was provided by the Union Carbide Corporation and the Texas Water Development Board. The assistance of Frank Hulsey and other staff and students at The University of Texas at Austin was greatly appreciated. The data presented resulted from studies conducted by Robert B. Jacques as part of his Master of Science degree. His hard work and dedication also was appreciated. Details of the research and the results and their implications are available.[6] A portion of the material reported in this paper has been presented previously.[7]

REFERENCES

1. "Selected Data on Innovative Treatment Technologies for Superfund Source Control and Ground Water Remediation," Technology Innovation Office, Office of Solid Waste, U.S. Environmental Protection Agency, Washington, DC, August 1990.

2. *Test Methods for Evaluating Solid Waste,* SW-846, 1–4, Third Edition, U.S. Environmental Protection Agency, National Technical Information Service, Springfield, Virginia, 1986.

3. Ryan, J. R., M. L. Hanson, and R. C. Loehr, "Land Treatment Practices in the Petroleum Industry," *Land Treatment, A Hazardous Waste Management Alternative,* R. C. Loehr and J. F. Malina, Jr., Eds. Center for Research in Water Resources, The University of Texas at Austin, 1986, pp. 319–346.

4. Bossert, I., W. M. Kachel, and R. Bartha, "Fate of Hydrocarbons During Oily Sludge Disposal in Soil," *Applied Environ. Microbiol.,* 47: 763–767 (1984).

5. Huddleston, R. L., C. A. Bleckman, and J. R. Wolfe, "Land Treatment—Biological Degradation Processes," in *Land Treatment, A Hazardous Waste Management Alternative,* R. C. Loehr and J. F. Malina, Jr., Eds., Center for Research in Water Resources, The University of Texas at Austin, 1986, pp. 41–62.

6. Jacques, R. B., "Land Treatment of Soils Contaminated with Petroleum Hydrocarbons," Master of Science thesis, Civil Engineering Department, The University of Texas at Austin, 1990.

7. Jacques, R. B., and R. C. Loehr, "Biological Detoxification of Contaminated Soil," Proceedings, Joint CSCE-ASCE National Conference on Environmental Engineering, Vancouver, B.C., American Society of Civil Engineers, 1988, pp. 377–384.

CHAPTER 15

A State-of-the-Art Review of Remedial Technologies for Petroleum Contaminated Soils and Groundwater: Data Requirements and Efficacy Information

Melitta Rorty, Roy F. Weston, Inc., Walnut Creek, California
Lynne M. Preslo, ICF Kaiser Engineers, Inc., San Francisco, California
Raymond A. Scheinfeld, Roy F. Weston, Inc., West Chester, Pennsylvania
Mary E. McLearn, Electric Power Research Institute, Palo Alto, California

INTRODUCTION

This chapter summarizes the results of a study funded by the Electric Power Research Institute (EPRI) to update EPRI CS-5261, *Remedial Technologies for Leaking Underground Storage Tanks.* [1] That document was published in book form by Lewis Publishers. [2] The present study describes, evaluates, and updates available technologies for remediating soils and groundwater containing petroleum products released from underground storage tanks (USTs) or from other discharges, leaks, or spills. In particular, this study provides an update of the technologies included in the first study, presents a number of new case studies, and focuses on efficacy data as well as data needed to properly assess and design the remedial alternatives for a particular site. Regulatory acceptance of specific technologies also is included.

The genesis of federal regulations governing USTs began when the 1984 Hazardous and Solid Waste Amendments (Public Law 98-616) to the Resource Conservation and Recovery Act (RCRA) mandated that the United States

Environmental Protection Agency (U.S. EPA) establish a national program for regulating USTs and their associated piping. The legislation addressed tank systems containing either petroleum products or substances designated as hazardous (other than hazardous wastes regulated under Subtitle C of RCRA) under the Comprehensive Environmental Response Cleanup and Liability Act of 1980 (CERCLA). The final technical regulations concerning USTs were promulgated by the U.S. EPA on September 23, 1988, and became effective on December 22, 1988. The regulations are presented in Title 40, Part 280 of the Code of Federal Regulations (40 CFR 280), and the requirements for state UST programs, which must be no less stringent than the federal program, are presented in 40 CFR 281. 40 CFR 280 Subpart F addresses release response and corrective action for UST systems that must be implemented in the event of a release. The regulations allow considerable flexibility in the manner in which cleanup of soil and groundwater and/or institutional controls are performed. Cleanup and/or control must be performed adequately to protect human health and safety and the environment.

The report focuses on long-term remediation and site restoration that is performed after initial measures, such as free product removal, have been implemented. The study was not intended to be a remediation design and implementation manual, but rather was intended to provide sufficient information to enable users to become familiar with the methods that may be most applicable to their specific problems, and eliminate from further consideration those methods that are not appropriate.

APPLICABLE REMEDIAL TECHNOLOGIES

The 13 remedial technologies discussed here are divided into two general categories: in situ treatment and non-in situ treatment. In situ treatment refers to treatment of soil or groundwater in place without excavation or pumping. Non-in situ technologies are those that are applied to contaminated soil or groundwater, either onsite or offsite, after excavation or pumping. The in situ technologies include:

- soil vapor extraction
- biodegradation for soils and groundwater
- leaching and chemical reaction (soil flushing)
- vitrification
- passive remediation
- isolation/containment.

The non-in situ technologies include:

- land treatment
- thermal treatment
- asphalt incorporation
- solidification/stabilization
- groundwater extraction and treatment

- chemical extraction (soil washing)
- excavation and landfilling.

SELECTION OF REMEDIAL TECHNOLOGIES

All soil and groundwater remediation problems require site-specific considerations in choosing, designing, and implementing the most appropriate remedial measures. A summary description and evaluation of remedial methods for petroleum-contaminated soils and groundwater is presented in the report. There are many technical, economic, institutional, and related factors to be considered in designing or evaluating remedial techniques. Many of these factors are highly site-specific; therefore, no single method is best for all or even most circumstances. Further, it should be noted that EPRI neither endorses nor recommends any particular cleanup method.

Selecting and implementing a remedial method depends first of all on a determination of cleanup goals for the site. These are the contaminant levels or concentration limits to which the site must be cleaned. These, in turn, are usually based on either an assessment of potential site-specific risk or general regulatory standards as determined by the regulatory agency having jurisdiction at the site. Remedial options are then chosen by assessing the feasibility of each option, as it applies to the particular site conditions, to achieve the desired cleanup goal and evaluating the relative cost and acceptability of the method.

Numerous site-specific factors influence the selection of remedial measures. These factors may include the quantity of hydrocarbons released, the type of petroleum product released, the type of geologic materials present, and the depth to first groundwater. With knowledge of the site hydrogeologic characteristics, the distribution of hydrocarbons, the physical-chemical properties of the hydrocarbons, and the volume released, it is possible to assess the predominant transport mechanisms operating at a site. Depending on the nature of both the site and the hydrocarbons, transport commonly may take place in the dissolved phase, vapor phase, or as nonaqueous phase liquid (NAPL). Hydrocarbons that have low solubility, low volatility, and strong adsorption characteristics will be most persistent within site soils. The compounds with high solubility will be most persistent in the groundwater, and the compounds with high volatility will be found with the highest incidence in soil gases. An understanding of these mechanisms at a particular site is essential to defining the contaminant source, predicting contaminant fate and transport, and designing effective and appropriate remedial measures.

DATA REQUIREMENTS

Collection of data for screening remedial action technologies can be divided into (a) characterization, which is similar for all technologies, and (b) treatability investigations, which differ for each technology.

Site characterization includes definition of the following:[3]

- site physical characteristics
- sources of contamination
- nature and extent of contamination
- contaminant fate and transport.

Site physical characteristics include surface features, geology, soil types and properties of the vadose zone, surface water, hydrogeology, meteorology, receptor populations, land uses, and ecology. Source characterization describes the facility characteristics and the waste characteristics, including type, quantity, chemical and physical properties, and concentrations. Data requirements for treatability investigations vary depending on the technology selected. Certain technologies have been demonstrated sufficiently so that site-specific information collected during the site characterization is adequate to evaluate those technologies without conducting treatability testing. An example is extraction of groundwater containing volatile organic compounds and subsequent treatment with air stripping and/or carbon adsorption. Frequently, however, technologies have not been sufficiently demonstrated or characterization of the waste is insufficient to predict treatment performance or estimate the size and cost of appropriate treatment units. In these cases, bench- and/or pilot-scale treatability studies are required. An example of such a technology is bioremediation, which requires at least bench-scale testing to examine the compatibility and effectiveness of the native, augmented, or induced microbial population with the waste type and site soils. Various nutrient mixes also would be tested. Data requirements for the 13 technologies included in this study are summarized in Table 15.1.

EFFICACY INFORMATION

Numerous factors affect the ability of a remedial technology to reduce contaminants to the desired cleanup level or to nondetectable levels. These factors include site characteristics such as moisture content of soils and grain size, waste characteristics, available time, and weather, among others. Case studies and efficacy information for selected remedial technologies are presented below.

Case Study: Pilot Study of Groundwater Extraction and Treatment Using Ultrox International Ultraviolet Radiation/Oxidation Technology

The U.S. EPA[4] funded a demonstration of the Ultrox International UV/oxidation technology through the Superfund Innovative Technology Evaluation (SITE)

Table 15.1.   Typical Data Requirements for Treatability Investigations for Remedial Technologies

Technology	Waste Matrix	Data Required
Soil vapor extraction	Soil	Soil type Particle size distribution Depth to groundwater Concentration of VOCs Presence of SVOCs Soil air permeability Radius of influence
Biodegradation	Soil	Climate Nutrient availability Waste volume Waste characteristics Soil pH Gross organic components (BOD, TOC) Soil moisture Waste concentration Indigenous bacteria Population
Biodegradation	Groundwater	pH Nutrient availability Gross organic components Total suspended solids Dissolved oxygen Waste concentration Waste characterization
Leaching and chemical reaction	Soil	Grain size analysis Soil permeability Soil organic content Waste analysis (VOCs and SVOCs) Waste concentration Hydrogeology
Vitrification	Soil	Soil moisture Grain size analysis Presence of buried metals Presence of void volumes
Passive remediation	Soil and groundwater	Risk analysis Topography Soil conditions Waste concentrations Waste characteristics Depth to groundwater

continued

Table 15.1. *continued*

Technology	Waste Matrix	Data Required
Isolation/containment	Soil and groundwater	Topography
		Depth to impermeable strata
		Soil conditions
		Groundwater depth
		Groundwater flow velocity
		Groundwater flow direction
		Soil chemistry
		Chemistry of waste and groundwater (compatibility)
		Depth to impermeable strata
		Grain size distribution
Land treatment	Soil	Climate
		Waste volume
		Waste characteristics
		Waste concentration
		Soil pH
		Nutrient availability
		Gross organic components (BOD, TOC)
		Depth to groundwater
		Soil moisture
		Indigenous bacteria
		Population
Thermal treatment	Soil	Moisture content
		Heat value
		Chlorine content
		Destruction efficiency
		Particle size
		Ash content (high temperature)
Asphalt incorporation	Soil	Hydrocarbon characteristics
		Grain size analysis
Solidification/stabilization	Soil	Soil properties
		Waste characteristics
		Waste constituents
		Climate
Groundwater extraction and treatment	Groundwater	Climate
		Waste concentration
		Waste characteristics
		pH
		Total suspended solids
		Water temperature
		Geology
		Grain size analysis

continued

Table 15.1. *continued*

Technology	Waste Matrix	Data Required
Chemical extraction	Soil	Waste characteristics Waste concentrations Grain size analysis
Excavation and landfilling	Soil	Waste characteristics Waste volume

Program. The *Ultrox International Ultraviolet Radiation/Oxidation Technology: Applications Analysis Report* included the following case study describing treatment of groundwater contaminated with BTEX. Sponsorship of this particular case study was provided by the California Department of Health Services Hazardous Waste Reduction Grant Program.

The pilot study was performed at the Hewlett Packard facility in Palo Alto, California, where groundwater contamination was discovered in 1982 and was traced to leaking underground chemical storage tanks. In 1988, a GAC filter system was installed to treat groundwater extracted from three wells.

To accommodate the existing GAC treatment system, the Ultrox unit was installed upstream of the carbon filters so that the usual treatment system could operate when the Ultrox unit was not being tested.

The main objective of this pilot demonstration was to meet the discharge limitations specified by regulatory agencies. These requirements were to reduce the level of total toxic organics (TTO) to less than 1,000 μg/L with no specific constituent greater than 750 μg/L and to reduce the BTX, 1,1,1-TCA, and 1,2-DCA contaminants to less than 5 μg/L.

Twelve tests were conducted over a 2.5 week period with two or three test runs each day. The samples were analyzed the day after the test runs. The initial test was conducted with the Ultrox unit operating in a batch mode to determine the retention time and oxidant dosage required. Based on these results, subsequent runs were performed in a continuous, flow-through operation. The tests were conducted to examine removal efficiencies with the Ultrox reactor operating with various oxidant dosages. Removal efficiencies varied from 96.7% to 99.9%.

Case Studies: Thermal Treatment

Rotary kiln incineration has been used for many years as a remedial technology. Use of the fluidized bed and low temperature technologies has been increasing. Selected case studies are presented here. Although they are presented for specific chemical compounds, the techniques can be used for most volatile compounds and petroleum hydrocarbons.

Case Study: Rotary Kilns

Stationary or fixed rotary kilns have been in use for many years and qualify as a proven technology. Coupled with an afterburner, rotary kilns can meet destruction and removal efficiencies (DRE) of 99.99% ("four nines") or greater for petroleum products.

The use of transportable or mobile rotary kilns for the removal and destruction of hydrocarbons from soils is a more recent development. The U.S. EPA, Chemical Waste Management, Ensco, IT Corporation, Kimmins, Vertac, Vesta, and Weston Services, Inc. (WSI) have developed and operate mobile or transportable rotary kiln incinerators.

The U.S. EPA mobile incineration system (MIS) is capable of handling up to 1.8 metric tons (2 tons) of soil per hour at 10 million Btu/hour, and reaches temperatures as high as 2,500°C (4,532°F) in the secondary combustion chamber.

The system can be transported on four semitrailers and consists of the rotary kiln as well as a secondary combustion chamber and air pollution control equipment. Replacement parts are transported with the incinerator on six additional trailers. In a 1982–1983 test of this unit, the incinerator destroyed 99.9999% of the influent polychlorinated biphenyls (PCBs) and 99.999999% of the carbon tetrachloride.

The WSI transportable incineration system (TIS) is capable of handling up to 9.1 metric tons (10 tons) of soil per hour at 30 million Btu/hour. Maximum temperatures reach about 1,316°C (2,400°F) in the secondary combustion chamber. The TIS demonstrated a DRE of 99.9999% for PCBs in a remediation conducted at an Illinois State Superfund site. More than 8,500 pounds of contaminated soil were successfully treated. The TIS is currently being used on a second State Superfund site in Illinois. Trial burn testing demonstrated that the unit is capable of exceeding a DRE of 99.99% on the waste matrix which consists of an oily, heavy chlorinated sludge. The cleanup criteria is based on the level of total polynuclear aromatic hydrocarbon compounds.

Based on the successful results of the many systems available, the rotary kiln incinerator is considered appropriate for petroleum products that are easily combusted.

Case Study: Circulating Bed Incinerator

Circulating bed incinerators or combustors (CBC) are employed internationally for a variety of users including waste incineration, steam, and electric generation. Commercial plants currently in operation are incinerating peat, wood, coal, oil, and sewage sludge. Ogden Environmental Services, Inc. has four transportable units available for remediation. Two sites have been addressed for remediation using the transportable CBC system. At a site in California, 16,300 metric tons (18,000 tons) of an oil-contaminated waste material were successfully treated. The CBC system is currently being used for treatment of PCB-contaminated soil at a site in Alaska. To date, more than 40,800 metric tons (45,000 tons) have been treated; over 31,700 metric tons (35,000 tons) remain to be treated.

Case Study: Oxygen Enrichment

The Linde System has been proven and commercially applied in a number of steel reheating furnaces, copper smelters, and glass furnaces. The Linde System was fitted into the EPA Mobile Incineration System (MIS) at the Denney Farm in Missouri. The study evaluated the feasibility of treating dioxin-contaminated soils.[5] The MIS has been demonstrated for various waste streams, but its relatively low throughput and dust carryover limited its use. Field testing of the Linde System showed an increase in contaminated soil throughput from 2,000 lb/hr with a conventional air burner to more than 4,000 lb/hr with the Linde System. The DRE was improved and auxiliary fuel consumption was reduced by 63%.

Case Study: X*TRAX System

The first commercial X*TRAX system has been fabricated by Chemical Waste Management and is currently being used at a Superfund site in New England. The site is contaminated with PCBs, with maximum concentrations as high as 50,000 ppm. The unit is designed to handle 136 metric tons/day (150 tons/day). During the one year remediation, 31,700 metric tons (35,000 tons) of soil will be treated.

Case Study: Enhancing Biodegradation Through Soil Venting

Miller et al.[6] have documented a field evaluation and demonstration of enhanced biodegradation through soil venting that was conducted at the site of an abandoned tank farm located on Tyndall Air Force Base (AFB), Florida. The site was contaminated with fuel, primarily JP-4 jet fuel, and free product was observed to be floating on the shallow groundwater table. The objective of the study was to investigate the potential for enhanced biodegradation of JP-4 in the vadose zone by providing oxygen through soil venting combined with moisture and nutrient addition. Volatilization of contaminants through soil venting alone is not effective for removing nonvolatile or low volatility components of JP-4. A further objective of this study was to research the possibility of reducing or eliminating expensive off-gas treatment while remediating low volatility jet fuel contamination of vadose zone soils through enhancing in situ biodegradation.

The seven-month-long field investigation was conducted at Tyndall AFB, Florida, where past jet fuel storage had resulted in contamination of a sandy soil. The contaminated area was dewatered to maintain an approximately 1.6 m thick vadose zone. Initial soil hydrocarbon concentrations ranged from 30 to 23,000 mg/kg (ppm). Contaminated and uncontaminated test plots were vented for 188 days, with five interruptions during operation to allow for measurement of biological activity (CO_2 production and O_2 consumption) under varying moisture and nutrient conditions.

Moisture addition had no significant effect on soil moisture content or biodegradation rate. Soil moisture content ranged from 6.5% to 9.8%, by weight, throughout the field investigation. Nutrient addition also was shown to have no statistically significant effect on biodegradation rate. Initial soil sampling results indicated that naturally occurring nutrients were adequate for the amount of biodegradation observed. Acetylene reduction studies, conducted in the laboratory, indicated a biological nitrogen fixation potential capable of fixing the organic nitrogen, observed in initial soil samples, in five to eight years under anaerobic conditions. Biodegradation rate constants were shown to be affected by soil temperature and followed predicted values based on the van't Hoff-Arrhenius equation.

In one treatment cell, approximately 26 kg of hydrocarbons volatilized and 32 kg biodegraded over the seven-month-long field test. Although this equates to 55% removal attributed to biodegradation, a series of flow rate tests indicated that biodegradation could be increased to 85% by managing the air flow rate. Off-gas from one treatment cell was injected into clean soil to assess the potential for complete biological remediation. Based on biodegradation rate data collected at this field site, a soil volume ratio of approximately four parts uncontaminated to one part contaminated soil would have been required to completely biodegrade the off-gas from the soil venting.

Case Study: Asphalt Incorporation

The use of fuel oil contaminated soil in a cold mix asphalt has been described by Sciarrotta,[7] a Senior Research Scientist at Southern California Edison (SCE). SCE's comprehensive feasibility study (which included construction of a two-lane, 3/4-mile-long demonstration road using 1,100 cubic yards of fuel oil contaminated soil) showed that paving which uses contaminated sandy soil represented no more environmental risk than pavement containing clean sandy soil.

Samples from batches of both a clean and a contaminated soil cold mix pavement were analyzed for polynuclear aromatic hydrocarbons (PAH) and toxic metals, and were subjected to aquatic bioassay toxicity testing by SCE. Laboratory results indicated that neither sample contained detectable concentrations of PAH compounds nor regulated concentrations of toxic metals. A zero mortality rate for fathead minnows was recorded during bioassay testing of both samples.

The petroleum contaminated sandy soil used in the cold mix asphalt on the SCE project had been excavated during the repair of fuel oil delivery pipelines. After excavation, the soils were held in several stockpiles on plastic tarpaulins. Before beginning the paving process, these stockpiles were blended to obtain a homogeneous mixture, and aerated to reduce moisture content. Prior to blending, testing showed that the oil content of the stockpiles ranged from between 1.3% and 4.7%. Moisture content of the soil ranged from 3.1% to 10.7%. Hydrocarbon emissions normally associated with the blending and aerating used in the hot mix asphalt process were reduced because the mixing occurred at ambient temperatures, avoiding potential emissions from fuel combustion and volatilization of the lighter asphalt fractions.

Before the paving project began, grain size analysis was performed on the soil. This analysis indicated that the soil should be supplemented with imported aggregate in a 70:30 ratio (virgin aggregate to sandy soil) to achieve an ideal gradation. Since the addition would more than triple the project cost, virgin aggregate was eliminated from the mix formula, and the designated pavement use was downgraded to "light duty." Light duty use includes all weather surfacing of light duty access roads, storage yards, and parking lots.

To compensate for the gradation shortfall, a more viscous liquid asphalt was required to improve stability. Calculations showed that this required a complex mixture of rapid curing (RC) and slow curing (SC) asphalts. With cost in mind, ultimately a mix formula consisting simply of 4% to 5% SC-3000 liquid asphalt combined with the contaminated sandy soil was established. Stability was compromised somewhat in the interest of greater economy, durability, flexibility, and workability.

During the actual paving process, the soil was further mixed by blading and dragging over the graded road bed. After the soil was graded to a uniform thickness, it was sprayed with the SC-3000 asphalt and then bladed into a windrow. Finally, the material was picked up from the windrow with a traveling pugmill for thorough mixing of soil with the recently applied SC-3000. The mix was then spread across the surface of the road at a uniform three-inch depth and compacted. Complete curing was expected to occur within 6 to 12 months.

Based on these experiences, SCE made the following conclusions:

- This technology cannot be universally applied to all petroleum contaminated soils, and is not yet applicable to contaminated clay soils.
- The applicability of this technology to soils contaminated with other petroleum hydrocarbons such as gasoline was not determined by this study.
- Bench-scale testing for unique site conditions is critical.
- Blending of less suitable materials with more suitable feedstock could create paving that meets design criteria for pavement with a designation higher than "light duty."

The cost associated with the remediation of the 1,100 cubic yards of contaminated soils was approximately $54,000. This cost did not include environmental testing costs associated with the project.

REGULATORY ACCEPTANCE

Regulatory acceptance of a particular technology depends on many factors; often nontechnical issues, such as community perception, and which agency (federal, state, or local) has jurisdiction over the particular site, will greatly affect the ultimate technology choice and regulatory acceptance. Additionally, recent legislation, including RCRA land disposal restrictions (''Land Ban'') and the National Oil and Hazardous Substances Pollution Contingency Plan (NCP), influence regulatory acceptance of a remedial technology. In general, regulatory agencies prefer remedial technologies that have the following characteristics:

- are proven and reliable
- provide long-term protection of human health and the environment
- permanently and significantly reduce the volume, toxicity, or mobility of contaminants
- are performed onsite rather than offsite, and in some jurisdictions, in situ rather than non-in situ
- result in the destruction of contaminants rather than the transfer of contaminants from one medium to another.

SUMMARY AND CONCLUSIONS

A variety of options are available to remediate petroleum-contaminated soils and groundwater. The selection of a particular technology depends on numerous factors, including site-specific conditions, regulatory acceptance of the technology, the type of petroleum hydrocarbon released, and the volume released, among others.

REFERENCES

1. *Remedial Technologies for Leaking Underground Storage Tanks,* EPRI CS-5261, Electric Power Research Institute (EPRI), Palo Alto, California, 1987.
2. Preslo, L. M., M. Miller, W. Suyama, M. McLearn, P. T. Kostecki, and E. J. Fleischer. ''Available Remedial Technologies for Petroleum Contaminated Soils,'' in *Petroleum Contaminated Soils,* Volume I. P. T. Kostecki and E. J. Calabrese, Eds. (Chelsea, MI: Lewis Publishers, Inc., 1989).
3. ''Guidance for Conducting Remedial Investigations and Feasibility Studies Under CERCLA: Interim Final,'' EPA/540/G-89/004. U.S. Environmental Protection Agency, 1988.
4. ''Ultrox International Ultraviolet Radiation/Oxidation Technology: Applications Analysis Report,'' EPA/540/A5-89/012. U.S. Environmental Protection Agency, September 1990.
5. Ho, M., and M. G. Ding. ''Proposed Innovative Oxygen Combustion System for the Incineration of Hazardous Waste,'' Union Carbide Corporation, 1986.
6. Miller, R. N., R. E. Hinchee, C. M. Vogel, R. R. Dupont, and D. C. Downey. ''A Field Scale Investigation of Enhanced Petroleum Hydrocarbon Biodegradation in the

Vadose Zone at Tyndall AFB, Florida,'' in Proceedings of the NWWA/API Conference on Petroleum Hydrocarbons and Organic Chemicals in Ground Water—Prevention, Detection, and Restoration, Houston, Texas, 1990.

7. Sciarrotta, T. "Contaminated Soil Recycled in Road Mix," in W. E. Neeley, Ed. *Soils,* Vol. 1, No. 1, 1990.

CHAPTER 16

The Need for a Laboratory Feasibility Study in Bioremediation of Petroleum Hydrocarbons

W. T. Frankenberger, Jr., Department of Soil and Environmental Sciences, University of California, Riverside

INTRODUCTION

Bioremediation is becoming a popular approach in the cleanup of petroleum hydrocarbons because it is simple to maintain, applicable to a large area, cost-effective, and most importantly, it leads to the complete destruction of the contaminant. Native microflora can use petroleum hydrocarbons as a carbon (C) and energy source for growth and proliferation in soil. Microbial degradation of hydrocarbons results in the release of innocuous products such as carbon dioxide (CO_2), water (H_2O) and cellular biomass as the final products.

Biodegradation of oils can be dramatically stimulated by manipulation of the environment such as adding nutrients, promoting aeration, and supplying moisture. However, before such a program can be implemented, it is important to obtain background information on the contaminated site to optimize conditions for biodegradation. For this program to be successful, basic information is needed such as the inventory residual oil concentration, population density of the oil-degrading microorganisms, various chemical and physical properties of the contaminated soil (e.g., pH, inorganic nitrogen [N] and phosphorus [P], salinity, and particle size analysis) and the biodegradation potential with optimum rates of chemical fertilizers applied. All contaminated sites vary in their potential to carry out biodegradation depending on the soil type, the biological diversity, prevailing environmental conditions, type of contamination, and amount of spillage. A bench-scale feasibility study is needed to enumerate the hydrocarbonoclastic population and to monitor the natural unamended biodegradation rates vs accelerated rates upon the addition of biostimulating agents (e.g., fertilizers and alternate terminal electron acceptors). Once the optimum conditions have been determined through these laboratory tests, biostimulating agents can be applied directly onsite to initiate a bioremediation approach.

FEASIBILITY OF BIOREMEDIATION

Leading popular technologies in the remediation of hazardous waste include chemical fixation, incineration, soil washing, excavation/and transport to land-fills, and bioremediation. Chemical fixation or solidification involves the conversion of wastes into a solid stable form. This technology should not be applied to hazardous organics, but is useful for the fixation of heavy metals. Incineration involves the exposure of waste to high temperatures to ensure destruction. However, the ash resulting from incineration must be disposed of by some other means. The extraction efficiency of soil washing may be somewhat low depending on the soil type and may require the use of surfactants involving the addition of hydrocarbons to extract petroleum hydrocarbons. The transport of polluted soil to a landfill involves the risk of spill during transportation, the long-term liability associated with Class II landfills, and prohibitive costs. The Comprehensive Environmental Response, Compensation, and Liability Act (CERCLA) has created long-term liability for disposal of all substances in landfills. The newly enacted ''land ban'' prevents disposal of polluted soil into landfills requiring onsite treatment. These remedial practices listed above do not completely destroy the contaminant. Pollution may be either transferred from one medium to another or remain in place through fixation. The Superfund Amendments and Reauthorization Act (SARA) of 1986 imposes cleanup standards emphasizing permanent remedies rather than simply moving hazardous waste and burying it somewhere else.

Bioremediation is a technology which is very cost-effective and decomposes hazardous materials into nontoxic components. Bioremediation involves the use of natural (indigenous) or inoculated microorganisms to mineralize or metabolize toxic substances. The use of microbes in the cleanup of toxic waste is not a new concept. Bioremediation has been practiced for several decades in treating municipal wastewater, municipal sewage, refinery waste, and other specific chemical waste streams. The United States Environmental Protection Agency (U.S. EPA) Administrator, William Reilly, recently emphasized the importance of bioremediation in the cleanup of harmful chemicals at hazardous waste sites. He recognized the potential of bioremediation strategies during the massive oil spill in Alaska's Prince William Sound caused by the damaged *Exxon Valdez*. Reilly estimated that it could cost up to $20 million for each of the 1,200 Superfund sites by traditional cleanup means. He called for a biotechnology breakthrough in the cleanup of hazardous wastes to reduce this cost.[1]

The Superfund Innovative Technology Evaluation (or SITE) allows environmental companies to demonstrate the viability of their remedial methods in cleanup of hazardous waste without having to go through EPA's long approval procedure. This allows vendors of environmental technology to test their new methods through pilot studies. There are currently 39 technologies being tested, of which 6 to 8 are biorelated.[2] EPA's Biosystems Technology Development Program is a $4 million multilaboratory effort including several EPA research and development laboratories which conduct research through bench-scale and field demonstrations. EPA is emphasizing the development of laboratory bench-scale "flask" protocol that can be used to show the feasibility of a bioremediation strategy working on a specific site. However, many of the biotechnology developers have shown frustrations with the lack of familiarity among government regulators on bioremediation, leading to long regulatory delays.

Bioremediation is an effective remedial measure, particularly for petroleum hydrocarbons, because soil microorganisms can utilize hydrocarbons as a carbon and energy source. The breakdown is most active in the presence of oxygen (aerobic degradation); however, it may also occur under anaerobic conditions but at a much less efficient rate. Water and nutrients are often added to increase the rate of biodegradation.

The guidance and limits of soil cleanup varies from state to state and is highly dependent upon site-specific factors. Often a risk assessment is performed to evaluate the exposure and toxicity potential in characterizing the cumulative risk to both the public and the environment.[3] However, a risk assessment involves a fragile mix of science and art.[4] This assessment deals with masses of scientific data often unstructured, and contains substantial uncertainties. The risk assessment should include the identity of the pollutant, toxicology, fate, transport, and exposure (inhalation, dermal contact, and injection). Common assumptions are frequently made and applied for each situation. Appropriate models are constructed with extrapolation from observed tests known to occur in the environment.

The major advantages of bioremediation, and particularly landfarming operations include: (a) low operating costs; (b) relatively safe environmentally; (c) utilizes natural processes which recycle carbon; (d) relatively simple, with no dependence on high maintenance; (e) a 24-hour workday; and (f) improvement in soil fertility and soil physical conditions.[5] Landfarming has been practiced for many years by the petroleum industry in the disposal of oily solid waste without any reported environmental problems or operational difficulties. Landfarming has also been referred to as land spreading, land application, sludge farming, land disposal, and land treatment. It is estimated that approximately half of all petroleum oily wastes are being landfarmed today.[5]

One predominant source of petroleum hydrocarbon contamination is caused by underground storage tanks (USTs). Leaky tanks can contaminate the surrounding soil and groundwater. The EPA estimates that 35% of the existing USTs in the United States are not liquid tight and are leaking. However, this figure has been contested by the American Petroleum Institute, which reports that a more realistic number is 2%. There are currently between 1 to 4 million USTs in the

United States. The Resource Conservation and Recovery Act (RCRA) required all tank owners to report the location, age, size, and use of their USTs beginning in May 1986. Many of these tanks have been removed in recent years and replaced with internal double-walled or fiber-glass-lined tanks which protect against corrosion. However, much of the soil surrounding these USTs is contaminated with petroleum hydrocarbons and must be treated through some type of remedial measure. Bioremediation would be an ideal technique to treat the small volume of soil onsite by a landspreading operation. Field demonstrations which show the success of landfarm and landspreading operations have been cited in Table 16.1.

BENCH-SCALE TREATABILITY STUDIES

A bench-scale treatability study should involve the characterization of the contaminated soil in terms of its physical and chemical properties, toxicity of oils, heavy metals, microbial enumeration, and determination of the optimum environmental factors enhancing degradation of petroleum hydrocarbons.

Collection of Soil Samples

Soil samples should be collected in the field and placed in containers without headspace to prevent the loss of volatile organic compounds (VOCs). The California LUFT Field Manual[19] recommends that soil samples be collected in a thin wall, stainless steel or brass cylinder at least 3 " long by 1 " in diameter. No headspace should be present in the cylinder once the sample is collected. The soil samples should be protected from light. Containers should be used which minimize adsorption of the contaminant to the container itself and the cap. Sometimes glass jars with Teflon liners are used in the collection of hazardous waste materials. These samples should be maintained on blue ice and transported at $<5°C$. Soil samples should be maintained at field moist conditions and under no circumstances be allowed to air dry. No chemical preservatives should be added to the soils. The treatability study should be conducted immediately after collection of the samples with no more than 7 days of storage. Chain-of-custody procedures should be followed in collection and delivery of all samples in the event of a legal challenge. Once received by the laboratory, the soils should be sieved to remove large debris such as plant roots or rocks. Sieving thoroughly mixes the soil, giving it a homogeneous consistency in which subsamples can be divided and set up into microcosm studies with the same initial hydrocarbon concentration.

Physical and Chemical Properties of the Contaminated Soil

An ideal soil subject to bioremediation would be sandy in texture with high porosity, allowing diffusion of oxygen. This high porosity allows adequate aeration for the oxidation of hydrocarbons, yet there should be enough silt and clay in the soil matrix to hold water and keep the soil moist. The pH should be in

Table 16.1.   Field Demonstrations of Successful Landfarm Operations

Type of Oil	Degree of Contamination	Removal Rate	Comments	References
Crankcase oil, crude oil, home heating oil (No. 2) and residual fuel oil (No. 6)	11.9 m^3/4 × 10^3 m^2 soil	48.5–90% over one year	Aeration and added fertilizers led to high population counts of oil-degrading microorganisms.	6
Oil wastes and machine coolants	330 gal of vacuum pump oil/60 ft^3 soil and 600 gal of machine coolant/60 ft^3 soil	1.1 lb of waste oil/ft^3/mo and 0.34 lb of machine coolant/ft^3/mo	Applied lime to adjust pH, and fertilizer (726–772 kg/ac of N and 159–204 kg/ac of P and K) for 4 months. Subject to continuous cultivation and irrigation.	7
Industrial oil waste	Applied 3 times at 0.17–0.5 kg oil/m^2 soil	Half-life ranged from 260–400 days	Application of oil increased pH and organic matter content. Aromatics were subject to volatilization ($T_{1/2}$ = 30 days). The refractory residue ranged from 20–50% of the applied oil.	8
API separator sludge	3–13% wt with loading rate frequencies of 1 to 2 per year over a 1.5 year period	Oil loss was proportional to amount applied averaging 54%	n-Alkanes degraded rapidly followed by aromatics, and asphaltenes. Biodegration followed first-order kinetics. Heavy metals accumulated in the surface soil. Rototilling and proper moisture were maintained.	9
Diesel fuel	1084 mg total petroleum hydrocarbons (TPH)/kg soil	After 88 days TPH content was 2 mg/kg	Plot received nutrient supplements, proprietary inoculum and vigorous aeration. However, the control which did not receive nutrients, inoculum or aeration was as effective in reducing TPH.	10
Kerosene	0.87%	Innocuous levels after 21 months	Treated with lime, fertilizers (N, P, K at 10:1:0.85) and frequent tillage to a depth of 46 cm. Kerosene degradation was greatest in the summer months. n-Alkanes were most susceptible to degradation.	11

continued

Table 16.1. *continued*

Type of Oil	Degree of Contamination	Removal Rate	Comments	References
Heavy oil, kerosene, mineral oil	1000–20,000 mg/kg	After one growing season the oil content ranged from 3–1000 mg/kg	Recommend frequent tillage and addition of fertilizers. Optimum C:N:P ratio for degradation of oil was roughly 250:10:3. Leaching of NO_3–N was considerable (150–250 mg/L) with little to no P in leachate.	12
Cutting oils	2300–2500 mg/kg	120–420 mg/kg remained after 51 weeks (95% removed)	The most successful treatment was with the addition of bark and fertilizer. Microflora had adapted to degradation of cutting oil.	13
Oily sludge	100 m^3 was spread over 1000 m^2 soil (8–10 kg/m^2)	After 1 year the oil content was 4–5 kg/m^2	Tillage was repeated every 3 months. Number of bacteria increased from 72 million/g soil to 2,100 million/g soil. Oil was decomposed rapidly at first and more slowly later on. Fertilizer was not applied.	14
Oily sludge from gravity oil-water separators	Spread over soil for 16 years. Sludge thickness was 4–5″ (equivalent to 400–500 tons/ac)	5–60 lbs/ft^3/month	Two tons of limestone, 50 lbs N/ac and 60 lbs P/ac were added. The average growing season was 301 days (Houston, Texas).	15
Oily waste from an oil-refining operation	1,700,000 gallons as oily residues	After 1 year, oil degradation on fertilized plots approached 80% removal	High respiration rates and increases in microbial populations indicated that the oil was being rapidly degraded.	16

continued

Table 16.1. *continued*

Type of Oil	Degree of Contamination	Removal Rate	Comments	References
Diesel fuel	2,800 ppm TPH	After the first several weeks, TPH dropped to 1,100 ppm. After forced aeration and nutrient delivery, TPH dropped to 800 ppm	The landspreading operation consisted of a 6' high soil pile spread out to 200' × 60' with forced aeration duct work and water/nutrient delivery system.	17
Oil sludge	10–30% by weight	0.5 lb/ft³ soil/mo without fertilizer and 1.0 lb/ft³ soil/mo when fertilized during an 18-month experiment	Active microorganisms included *Arthrobacter, Corynebacterium, Flavobacterium, Nocardia* and *Pseudomonas.*	18

the neutral range (pH 7–8), with a low soluble salt (ECe) concentration (<4 dS m^{-1}). Often soils require fertilization with nitrogen (N) and phosphorus (P) to promote degradation of petroleum hydrocarbons. In our experience, soils which contain >50 mg/kg of inorganic N (NH$_4$-N plus NO$_3$-N) usually require no additional N supplements unless severely contaminated. The quantities of P needed to mineralize petroleum hydrocarbons are much less than N.

Toxicity of Hydrocarbons

Although the final product of hydrocarbon mineralization is CO_2, H_2O and cellular biomass, biodegradation of the higher molecular weight hydrocarbons may give rise to toxic intermediates which can accumulate at inhibitory levels. Degradation of aromatic hydrocarbons can yield phenolics and benzoic acid intermediates. The growth of marine *Pseudomonas* strains utilizing naphthalene and methyl-naphthalene as a C source was limited by the accumulation of salicylic derivatives.[20] Lauric acid was reported as the principal agent of a growth-inhibitory factor produced by yeast cultures growing on n-hexadecane.[21] The production of long-chain n-alkanes (waxes) as intermediates during biodegradation of some crude oils could also affect microbial degradation.

Other intermediates derived from biodegradation can have profound effects on microbial populations. Liu[22] found that C$_5$ through C$_9$ alkanes were not toxic to a bacterial population but the alcohols derived from these hydrocarbons were inhibitory. Various oxidation products of aromatics have been found to be more toxic than the original hydrocarbon.[23] Petroleum hydrocarbons may affect populations of algae and protozoa. Atlas et al.[24] found that coccoid green algae and amoeboid protozoa populations decreased in the presence of Prudhoe crude oil in coastal ponds along the Prudhoe Bay. Microbial metabolic activities such as nitrogen fixation and algae photosynthesis also decrease in the presence of crude oil. Bacterial chemotaxis is affected by the addition of petroleum hydrocarbons.[25]

The inhibitory effects of various petroleum components are dependent upon their solubility and concentration. Various microbial populations exposed to petroleum hydrocarbons may be inhibited by such compounds as toluene and phenol, particularly at high concentrations. These two chemicals are often used as disinfectants in the laboratory. Toluene- and phenol-degrading microorganisms have been isolated from soil when exposed to low concentrations but both compounds may be biocidal when found in high concentrations.

The effect of petroleum pollution on plants can be very severe, particularly at high concentrations. Hydrocarbons have a solvent effect on the lipid membranes of plant cells. The volatile fraction can have high penetrating power, entering plants with a narcotic effect. However, upon decomposition, the oil contaminated soil is often improved in terms of its organic matter content, physical structure (granulation), aggregate stability, water-holding capacity, and microbial diversity. One to several years following contamination often promotes greater growth of plants as a result of increased soil fertility. Baker[26] reported plant growth stimulation following oil pollution with increases in shoot length and dry weights of the salt marsh grasses, *Puccinellia maritima* and *Festuca rubra*. This growth stimulation was attributed to the nutrients released by oil-killed organisms, growth regulatory compounds present in the oil, and an increase in nitrogen fixation. Organic acids from various petroleum wastes have been shown to stimulate plant growth. A weak solution of napthenic acid significantly increased root length of cotton, cucumbers, onions, and winter wheat.[27] Carr[28] reported that crude oil added to a sandy soil at a rate of 0.75% stimulated the growth of soybeans. Plice[29] found that petroleum addition of 0.5% to 1% by weight produced more luxuriant plants than those growing in check plots where no oil was applied. Damage to certain plants can occur upon contamination of 1 kg of oil per m^{-2} of soil.[30]

The toxicity of petroleum hydrocarbons in soil can be determined in terms of EC_{50} using the Microtox® toxicity analyzer. This system makes use of a bioassay in which bioluminescent bacteria produce light as a result of a complex set of energy-producing reactions. Inhibition in any one of a multiple number of enzymes involved in this process causes a change in the rate of light emission. However, this system is only applicable for the water-soluble fraction of a soil extract.

Frankenberger and Johanson[31] utilized the soil dehydrogenase assay to monitor the influence of crude oil and refined products on the biological activity of soil. Dehydrogenase activity in the contaminated soil was dependent upon the amount and type of oil added. The highest level of dehydrogenase activity in soil was observed with crude oil, 30 days after incubation. Dehydrogenase activity in soils contaminated with refined oils (leaded gasoline, kerosene, diesel fuel, and motor oil) was extremely low when compared with the crude-oil treated soils. The lowest level of dehydrogenase activity was observed in the kerosene-treated soil. The additives present in the refined oils most likely had a major influence on the dehydrogenase reaction. This method is a very effective technique to assess the microbial toxicity of oil-contaminated soil.

Influence of Heavy Metals on Oil Biodegradation

The inhibitory effects of heavy metals can influence biodegradation of organic materials. Haanstra and Doelman[32] used glutamic acid decomposition as a sensitive measure of heavy metal pollution in soil. They found that 400 ppm of Cd, Cr, Cu, Ni, Pb, and Zn had adverse effects on decomposition of glutamic acid. Bhuiya and Cornfield[33] and Tyler[34] reported inhibitory effects of Pb on the decomposition of natural plant materials. The presence of heavy metals in oil sludge, motor oil, and used crankcase oil may also have deleterious effects on the hydrocarbon oxidizers in decomposing petroleum hydrocarbons. Jensen[35] studied the effects of Pb on biodegradation of oily waste in soil and found that Pb caused certain changes in the population of the soil microbiota. Reduction in the bacterial population was evident, particularly at the highest Pb concentration of 5,000 ppm. Measurements of oxygen consumption revealed increased microbial activity after the addition of oil to soils but the presence of Pb markedly reduced this activity with a prolonged lag phase in the biodegradation of oil sludge.

The availability of Pb for microbial uptake decreases with increasing pH and humus content. Tornabene and Edwards[36] found that Pb can be immobilized and is largely associated with microbial cell membranes but not with the cytoplasmic fraction. Soils generally contain about 15 ppm of Pb and rarely above 100 ppm. By maintaining a soil pH of 7 to 8, the solubility of Pb in soil solution will decrease.

Other elements of concern include Zn, Cu, Cr, Ni, and Cd. In gasoline and diesel fuel, Vazquez-Duhalt[37] reported Pb concentrations of 7,500 and 75 ppm, respectively. Heavy metals in used motor oil can amount to as high as (ppm): Pb, >13,000; Zn, 2,500; Cu, 50; Cr, 20; Ni, 5; and Cd, 0.1. Trace elements often associated with crude oil include Pb, Cr, Hg, Ba, V, and Ni. With repeated applications of oily sludge to a landfarm operation, heavy metals may accumulate at levels in which biodegradation may be reduced. However, many of the elements are strongly immobilized in soils as a result of complexed formation with organic matter and adsorption to clay minerals.

It is highly recommended that the metal and metalloid content in contaminated soils be assessed before a bioremediation program is implemented. The laboratory should conduct a CAM metal analysis which includes Sb, As, Ba, Be, Cd, Cr, Co, Cu, Pb, Hg, Mo, Ni, Se, Ag, Ti, V, and Zn, or RCRA metals including As, Ba, Cd, Cr, Pb, Hg, Se, and Ag. The metals and metalloids can be determined by test methods for evaluating solid wastes as outlined by EPA SW-846 by atomic absorption spectrophotometry or inductively coupled argon plasma emission spectrometry.

In our work, we found that the addition of Pb at 10, 100, 500, and 1,000 ppm all inhibited the mineralization of diesel fuel added to soil compared with the control (untreated) (Fig. 16.1). However, at the highest concentration (1,000 ppm Pb), the rate of CO_2 evolution was considerably enhanced after 17 days of incubation. At the end of incubation (32 days), mineralization of diesel fuel in the presence of 1,000 ppm of Pb almost approached the control in terms of cumulative amount of CO_2 produced. Figure 16.2 shows the effect of Cd on the mineralization of diesel fuel in soil. There was no significant difference in CO_2 evolution upon the addition of 0, 10, and 100 ppm of Cd. However, at 500 ppm Cd, there was a prolonged lag and considerably less CO_2 evolved during incubation. At 1,000 ppm Cd, the hydrocarbon mineralization activity was completely inhibited. Cadmium ions are generally more toxic to microorganisms than Pb. The bioavailability of Cd in soil is often greater than Pb at equal concentrations.[34]

Microbial Population Counts

In fertile soils, the bacteria biomass can comprise approximately 0.015% to 0.5% of the soil mass.[38] The first study involving the isolation of hydrocarbon-oxidizing organisms was carried out by Sohngen[39] and Kaserer.[40] Sohngen[39] isolated a methane bacilli which was named, *Methanomonas methanica*. He later reported that gasoline, kerosene, paraffin oil, and paraffin wax could be

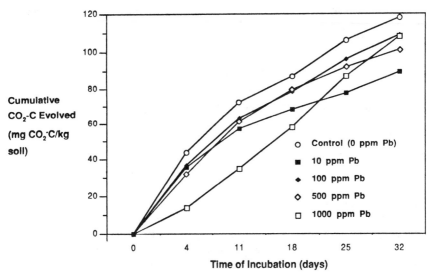

Figure 16.1.   Influence of lead on mineralization of diesel fuel in soil.

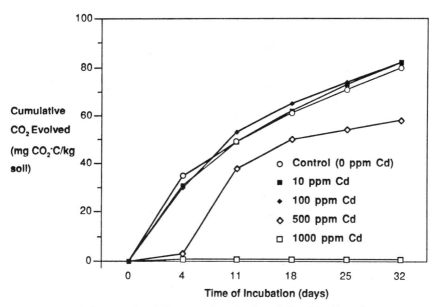

Figure 16.2.   Influence of cadmium on mineralization of diesel fuel in soil.

metabolized to CO_2, H_2O, and traces of organic acids by bacteria isolated from garden soil, ditch water, and compost.[41] The organisms isolated from these materials were primarily *Mycobacterium* and *Pseudomonas* spp.

The classical medium used to enumerate hydrocarbon-oxidizing organisms is the Bushnell-Haas medium[42] which consists of $MgSO_4$ (0.2 g), $CaCl_2$ (0.2 g), KH_2PO_4 (1.0 g), K_2HPO_4 (1.0 g), NH_4NO_3 or $(NH_4)_2SO_4$ (1.0 g), $FeCl_3$ (2 drops conc. solution) and distilled water (1 L). The medium is adjusted to pH 7.0 to 7.2 with dilute NaOH. Two percent of washed agar is added for a solid medium. Bushnell and Haas[42] added volatile hydrocarbons such as gasoline to the inverted lids of Petri dishes to enrich, cultivate, and enumerate hydrocarbon-oxidizing microorganisms. The volatile hydrocarbon vapors were sufficient to support microbial growth. For nonvolatile hydrocarbons, such as the heavier oils, the material was poured over the surface of the inoculated agar. For liquid cultures, hydrocarbons were layered on the surface of the medium. When hydrocarbons were added to the liquid medium, there was no evidence of retarded diffusion of oxygen to prevent growth of aerobic cultures.

Another method that may be used for the enumeration of hydrocarbon-oxidizing microbes in soil is the sprinkled soil plating method.[43] Dry soil samples are weighed and applied as evenly as possible on the surface of mineral salt washed agar plates. The plates may be rotated on a turntable when the soil is slowly applied from the edge of the rotating plate to its center. Soil plates were incubated at 33°C under 40% ethane in air in large desiccators. The principal organisms isolated with this method were the ethane oxidizers such as *Mycobacterium paraffinicum* having a high degree of specificity for ethane and higher paraffinic hydrocarbons.

More recent methods for enumeration of petroleum-degrading microorganisms include plating on oil agar,[44-45] plating on silica gel-oil media,[46-47] and inoculating liquid media by the most probable number (MPN) method.[48-50] Most petroleum degraders appear to be lipolytic in nature. Walker and Colwell[51] reported that high counts were obtained on silica gel-oil medium compared with oil agar or MPN techniques. Their results indicated that not all organisms appearing on an oil-containing solid medium were capable of petroleum degradation. The MPN method using a liquid salt medium with oil added was concluded to be the most reliable method for enumeration of petroleum-degrading microorganisms in estuarine and marine environments.[52] Some of the advantages of the MPN medium over other enumeration techniques include its simplicity compared with the silica-gel-based media, and the MPN media is placed in test tubes which will not dehydrate during incubation as solid preparations frequently do.[52] Microorganisms cultivated on agar-solidified petroleum-amended media are not always capable of degrading oil, thereby giving overestimations of the number of oil utilizers.

In a laboratory feasibility study, it is often a prerequisite to conduct a population count of the hydrocarbon-degrading microorganisms in soil before establishing a bioremediation program. Total bacterial counts give no indication of the bioremediation potential of a contaminated soil. Many laboratories frequently use a minimal basal salt medium consisting of macro- and micro-nutrients with agar to enumerate the population of hydrocarbon oxidizers. Hydrocarbon vapors are used as the sole carbon and energy source. A small sterile square of cheesecloth (1 ″ × 1 ″) or filter paper discs are dipped into the oil and placed in the lid of inverted glass petri dishes. The microorganisms inoculated onto the minimal medium through dilution blanks utilize vapors from gasoline, diesel fuel, or kerosene emitted from the cheesecloth lids of these dishes.

Another approach is to place petri dishes containing the minimal basal salt medium into a desiccator containing a beaker of volatile hydrocarbons derived from gasoline, kerosene, or diesel fuel. The lighter fractions of VOCs are available as a carbon and energy source within the enclosed desiccator. The problem with these two methods in estimating the population of hydrocarbon oxidizers is that they only account for those organisms which can utilize hydrocarbon vapors and not microorganisms active on residual nonvolatile hydrocarbons. Furthermore, when the medium is treated with soil dilutions and incubated in the absence of hydrocarbon vapors, often greater microbial numbers are encountered. Apparently there are many soil microorganisms that can utilize agar as a carbon and energy source. There are numerous reports on agar-digesting bacteria present in seawater.[46,51,53-54] We recently conducted a study to determine if impurities from the agar were supporting oligotrophic organisms in soil. A high-grade purified agar along with phosphate buffer made with HPLC water (dilution blanks) were used to enumerate the hydrocarbonoclastic organisms in soil. For a highly contaminated soil without a carbon source added to the medium, 26×10^3 cfu's g^{-1} of soil were recorded, while the addition of diesel fuel vapors supported only 20×10^3 cfu's g^{-1} of soil. The results clearly indicate that soil microorganisms are capable of using agar as a carbon and energy source. There is also the possibility that oligotrophs not active in hydrocarbon oxidation are being isolated and counted.

We have conducted extensive studies to investigate the use of different solidifying agents, including several seaweed extracts, to avoid this problem. Our results indicate that the hydrocarbon-oxidizing population in soil cannot be quantified on a solid matrix. Thus, the only way to determine the hydrocarbon-oxidizing population is to use the MPN technique (liquid medium).

Walker and Colwell[51] suggested that the number of petroleum degraders should be expressed as a percentage of the total bacterial population. Nutrient agar is often used for total counts and MPN for the hydrocarbon oxidizers. Total heterotrophic bacteria may also be enumerated on Trypticase soy agar (BBL). Frequently, anywhere from < 1% to 10% of the total bacteria consist of hydrocarbon oxidizers. Generally, if the hydrocarbon-oxidizing population is greater than 3% of the total bacterial count, the soil is considered to have a high potential in hydrocarbon degradation. The indigenous microbial population is often acclimated to the presence of organic contaminants and degradation will naturally occur but may be limited by certain environmental factors.

For enumeration of mesophilic heterotrophic microorganisms, incubation should be conducted at room temperature (23 ± 2°C) for 3 days to one week. For psychrophilic microorganisms, incubation should be at 5°C for at least two weeks.

Other techniques used for the determination of potential heterotrophic activity in soil includes assimilation and respiration of glutamate, enumeration by epifluorescence direct counts, and MPN determinations in the presence of hexadecane or naphthalene.[55] The metabolic activity of microorganisms in soil may be monitored by respiratory activity measured by the reduction of 2-(p-iodophenyl)-3-p-phenyl)-5-phenyl tetrazolium by cytochromes[56] and hydrolysis of fluoresceine diacetate.[57]

Design of a Microcosm Reactor

Treatability tests should provide an evaluation of the rate of degradation of petroleum hydrocarbons removed through bioremediation. The experimental design of the microcosm study should account for mass balances in the removal or loss of hazardous waste materials as a result of degradation, but also through abiotic reactions including volatilization and sorption to soil particles. Such a design is illustrated in Figure 16.3, where a continuous air stream flushes the VOCs released from the contaminated soil into a water trap or activated carbon. The soil is kept well aerated and the VOC-enriched water can be reapplied to the soil

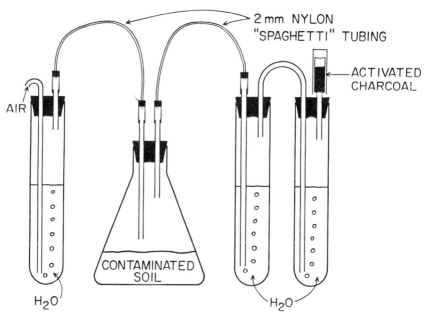

Figure 16.3. Microcosm reactor designed to account for volatilization losses of hydrocarbons and biodegradation of residual compounds.

to maintain moist conditions. Once incubation is terminated, the soil, water and activated carbon can be analyzed for hydrocarbons.

Various environmental parameters can be tested to determine the optimum conditions for biodegradation such as moisture levels, microbial inocula, and different treatments, including biostimulating agents such as fertilizers and alternate electron acceptors. The ideal fertilizer application rates can be determined by varying the loading rates of nutrients added as well as the ratio of nutrients. Biodegradation rates should be compared under treated vs. untreated conditions. The microcosm study should closely mimic the proposed bioremediation strategy in the field to determine if a microbial inoculum or nutrient amendments are required to accelerate biodegradation over time. However, some caution should be used in extrapolating the data collected from a bench-scale treatability study to the actual rates which may occur in the field.

To determine the rates of decomposition, a minimum of five sampling points should be collected in order to graph the trend in decomposition. This graphical representation should include a time zero analysis. The soil should be maintained at a moisture tension near -0.3 bars (-33 kPa) unless soil moisture is a variable to be evaluated. The water content of the soil can be determined by oven drying at $105°C$ for 24 h. If the moisture tension of the sample is greater than -0.3 bars, water should be added to the microcosm. The temperature of incubation should simulate that encountered under field conditions. If the soils collected are not typical of the contaminant concentration found in the field, a composite soil may be spiked. One treatment should consist of sterilization to account for loss of hydrocarbons through volatilization.

Depletion of Total Petroleum Hydrocarbons

Frequently, shallow pan experiments are conducted to monitor the depletion of total petroleum hydrocarbons (TPH) over a short period of time (e.g., 6–8 weeks). This experiment accounts for volatilization of the lighter fraction of hydrocarbons and the disappearance of the residual carbons through biodegradation. The experiment design may consist of shallow pans ($7'' \times 11''$) in which soil (ca. 750 g) is added. The soil is then treated with (a) water alone as one treatment, (b) the application of nutrients (N and P) as another treatment, and (c) soil sterilization by gamma irridation or treatment with a biocide such as NaN_3 or $HgCl_2$ (1% to 2% solution) as the control (Figure 16.4). Irradiation is employed to kill the soil microorganisms and to account for the loss of TPH through volatilization. Eastcott et al.[58] refers to a $HgCl_2$ treatment as 'hindered biodegradation' because the biocide does not completely inhibit all microbial populations in soil. However, the combination of irradiation and treatment with a biocide to kill the indigenous microflora and hinder recolonization

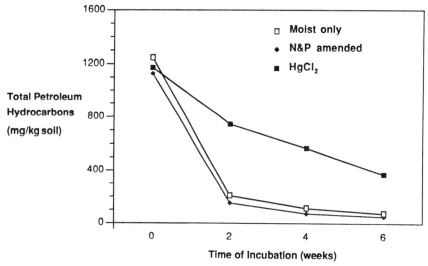

Figure 16.4.   Influence of water, nutrients and HgCl$_2$ on volatilization and biodegradation of diesel fuel.

permits an assessment of the loss of TPH through abiotic reactions. The soils should be maintained under moist conditions. The pans are incubated at specific temperatures to simulate field conditions. The soil within the pan should be thoroughly mixed at least twice a week to simulate mechanical disking in a field operation.

Within each pan, a grid should be made to sample the soils at certain time intervals. Subsamples are composited to account for the heterogeneity of TPH within the soil. TPH is determined by the EPA Method 418.1 with infrared spectrophotometric analysis. Samples are extracted with 1,1,2,-trichloro-1,2,2-trifluoroethane. Interferences are removed from extraction with silica gel to account for biogenic components. The detection limit for measurements of TPH in soil is typically at 5 mg kg^{-1} with a 20% error margin.

There are many inherited limitations in the practical use of the EPA Method 418.1 analysis to establish cleanup goals. First, there is at least 20% inherited variability with the analysis; second, there is an unknown amount of sampling error in which a subsample may not be representative of the whole soil fraction; third, the measurement of EPA Method 418.1 is highly dependent upon the moisture content. Moist soils (18% g/g) tend to show lower TPH concentrations than drier and saturated soils (Figure 16.5). Fourth, Method 418.1 is standardized with only 25% aromatics.[59] Thus if a petroleum mixture has more or less than 25% aromatics, the 418.1 determination is an inaccurate measurement. In crude oil, the aromatic fraction can vary depending on the origin and conditions during oil generation; however, an average of about 25% is often estimated.[60] With refinery products such as kerosene, diesel fuel, and gasoline, the aromatic

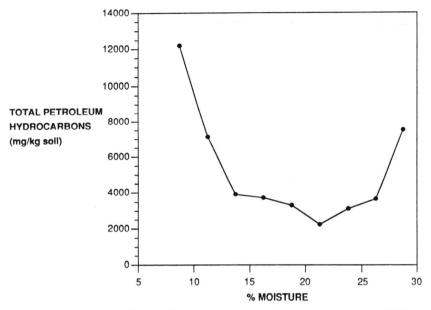

Figure 16.5. Total petroleum hydrocarbon content under various moisture regimes.

hydrocarbons are present in amounts of 5% to 20%.[61] Another inherited problem with the 418.1 analysis is that it detects all aliphatic hydrocarbons in soil, including the long chain hydrocarbons known to be constituents of plant waxes. For this reason, the n-alkanes detected by EPA Method 418.1 should by no means be regarded as hazardous waste compounds attributed to only petroleum hydrocarbons unless a clean background sample is analyzed.[59] Another disadvantage in using EPA Method 418.1 is the use of Freon (1,1,2,-trichloro-1,2,2-trifluoroethane) as the extracting solvent. Freon evaporates in the atmosphere damaging the ozone layer. An attractive method to aid in the extraction of TPH is the use of sonication or Soxhlet extraction. The Soxhlet method has high extraction recoveries, fast sample turnaround times, high sample throughput, and use of a minimum CFC-113 solvent.[62] Supercritical fluid extraction with high pressure CO_2 should also be evaluated in its extraction efficiency of petroleum hydrocarbons.

The purpose of the shallow pan experiment is to assess the actual amount of hydrocarbons lost through volatilization vs. that degraded by microorganisms. The emission of volatile hydrocarbons is highly dependent upon the air and soil temperature, humidity, wind speed, soil type, and moisture content.[63] Jury et al.[64] developed a model to classify and screen organic chemicals for their relative susceptibility to different loss pathways, including volatilization, leaching and degradation in soil, which requires knowledge of the organic carbon partition coefficient (K_{oc}), Henry's constant (K_H) and first-order degradation rate coefficients. Short[65] developed a mathematical model on land treatment processes to account for volatilization of contaminants such as oily waste. The model assumed to no dispersion, local equilibrium, linear partitioning, volatile losses by a modified Thibodeaux-Hwang approach, first-order degradation kinetics, no migration of oil after application, uniform soil profile and constant water flux throughout the treatment zone.

McGill et al.[66] estimated that up to 20% to 40% of crude oils may volatilize from soil. High summer temperatures will enhance volatilization, particularly when soils begin to dry out. Figure 16.6 illustrates the volatilization capacity of BTEX from a gasoline contaminated soil under different moisture regimes. Volatilization of BTEX tends to increase with decreasing moisture content. The rate of volatilization is a function of temperature, oil composition, solar radiation, and thickness of the oil layer.[67] Bossert and Bartha[38] mentioned that n-alkanes greater than C_{18} exhibit no substantial volatilization at ambient temperatures. However, the lighter fractions ($<C_{18}$) are subject to evaporation. Bioventing may have applications to treat fuel spills from USTs upon soil vacuum extraction.

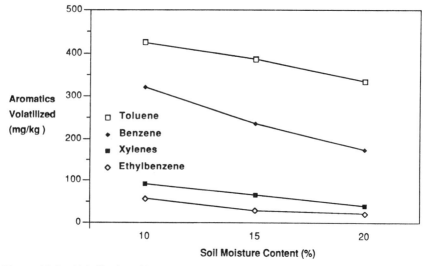

Figure 16.6. Volatilization of benzene, toluene, ethylbenzene and xylenes from soil under various moisture regimes.

Volatilization and biodegradation both tend to selectively remove the lighter hydrocarbon fractions. This fraction includes the short chain aliphatics and aromatic hydrocarbons such as benzene, toluene, and xylenes. Each of these aromatic compounds have a relatively high vapor pressure: benzene, 12.7 kPa; toluene, 3.8 kPa; and xylene, 0.9 to 1.2 kPa indicating that they are readily volatilized into the atmosphere.

Other Methods to Measure Petroleum Biodegradation in Soil

Biodegradation of oil has also been assessed by determining the characteristic change in oil composition as a decrease in the ratio of n-alkanes to pristane or phytane. Often this ratio tends to decrease with the addition of inorganic fertilizers, increased temperature, aeration, and sometimes microbial inoculants.[38] Gas chromatographic (GC) analysis of a solvent extract of hydrocarbons is frequently conducted to evaluate the progress in biodegradation of oils. However, hydrocarbons above C_{35} have a very low volatility and are difficult to analyze by GC analysis. For long-term incubation, CO_2 production should be monitored either through a static or continuous air flow system. Carbon dioxide can be measured by taking a headspace sample and injecting it into a GC equipped with a thermal conductivity detector.

Shaikh et al.[68] used wax pregnated graphite electrodes and platinum wire electrodes to monitor the redox potential (E_h) and oxygen diffusion rates during biodegradation of glucose and phenol in soil. This method may have applications in landfarm operations where several probes could be employed over a large area. The remote sensing electrodes could be connected to a microcomputer for data collection while monitoring E_h and oxygen diffusion upon biodegradation of hydrocarbons. In the laboratory, biosensors with attached bacteria or yeast cells at the surface of electrodes consisting of a monolayer of immobilized cells could be used to measure oxygen uptake. Multi-species sensors could be used for the change in oxygen concentration.

Some laboratory feasibility studies have been carried out with slurry phase reactors to reduce concentrations of petroleum hydrocarbons. In these systems, soil particles are kept in suspension in a soil-water-air system with liquid motion. The turbulent action enhances the aeration capacity of the reactor. Kleintjens et al.[69] studied the biokinetics of hexadecane which is a diesel component during aerobic growth of a *Pseudomonas* sp. Hexadecane degradation during exponential growth of the bacterium was calculated at 3 g hexadecane per kg of soil per hr. In similar studies, Bachman and Zehnder[70] studied the mineralization of hexachlorocyclohexane isomers in soil-slurried reactors with mixed aerobic microbial populations. The soil slurry consisted of 100 g of soil per L of water and was continuously mixed. Aerobic conditions led to complete mineralization in about 25 days. With this type of reactor system, pH, redox, aeration, homogeneity, and temperature can be closely monitored and controlled to enhance decomposition.

Mineralization of Petroleum Hydrocarbons

Petroleum hydrocarbons mineralized into CO_2 as an end product results in the complete destruction of the contaminant. Monitoring mineralization of hydrocarbons should be a standard test in all treatability studies to determine the optimum biostimulating agents added to enhance decomposition of the oil. Detection of

CO_2 is much more sensitive than monitoring the disappearance of TPH upon the addition of nutrients. Also, depletion of specific substrates is not a good means of monitoring biodegradation since toxic intermediates could possibly arise. Often at the initiation of incubation, degradation begins after a 2 to 4 day lag period, reaching a maximum rate within one to two weeks. Atlas and Bartha[71] indicated that n-alkanes were preferentially degraded when compared with branch-chain hydrocarbons based upon CO_2 evolution.

The consumption of oxygen has also been used to assess biodegradation of hydrocarbons. In many cases, the volume of oxygen consumed by hydrocarbon oxidizers exceeds the volume of CO_2 produced.[72] Stone et al.[73] reported that a respiratory quotient (RQ) for light oils to be approximately 0.63, indicating that a large percentage of the hydrocarbons attacked were completely oxidized to CO_2. An estimated figure for theoretical oxygen demand in the breakdown of hydrocarbons is approximately 3.5 mg of oxygen per mg of oil.[74] The problem with respiratometric measurements of oxygen uptake is that this is a measure of biotransformations and not necessarily mineralization.

The release of CO_2 by hydrocarbon oxidizers tends to emulsify the oil. Other microbial emulsifying agents include organic acids and long chain fatty acids which increase the interface for microbial utilization of the water soluble mineral fraction and the water insoluble components of the oil.

Besides production of CO_2, some of the hydrocarbons may be incorporated into microbial cellular biomass. It was estimated that approximately half of the methane exposed to *Methanomonos methanica* was oxidized to CO_2 and the other half was converted into cellular substances, or possibly other products of metabolism.[75] Buttner[76] reported that 10% to 20% of the C of n-alkanes decomposed by soil organisms in 21 to 31 days was not accounted for as CO_2.

Dobson and Wilson[77] found that soils amended with hydrocarbons had greater respiration rates than untreated soil. When a bench-scale treatability study is conducted, a background sample which is not contaminated with oil should be collected to compare the rates of respiration to that of a contaminated sample. This will reflect the degree of contamination and the biodegradation rate in the presence of hydrocarbons. It has been proposed that enumeration of microbial populations in soil contaminated with oils can be utilized to determine the limitations in nutrients to enhance decomposition. However, the inaccuracy of microbial counts does not allow a high degree of response to different application rates of biostimulating agents. The evolution of CO_2 or oxygen uptake would be a much better method to characterize the limitations in environmental parameters.

The presence of free carbonates in calcareous soils may give rise to CO_2, thus the proper controls should be run on noncontaminated samples. Nitrification of ammonium fertilizers results in the release of H^+ ions which may affect the soil reaction with the subsequent release of CO_2 derived from free CO_3^{2-}. However, most soils tend to have a high buffering capacity to resist any change in pH. To ensure that this reaction is not occurring, it is suggested that the soil pH, CO_3^{2-},

NH_4^+ and NO_3^- content be measured before and after the treatability study. For acidic and neutral soils, the abiotic release of CO_2 should not be a concern. It is highly recommended that calcareous soils be neutralized to dissipate the free CO_3^{2-} before initiation of the treatability study.

Often previous exposure to petroleum hydrocarbons results in greater degradation rates. Rowell[78] found that the rate of soil respiration increased progressively with increasing loading rates of oil and failed to reach a maximum value. His results indicate that decomposition of oil will occur even at concentrations which appear to saturate the soil. Stone et al.[79] found that after 2 to 3 transfers of bacterial cultures exposed to oil, the breakdown of the hydrocarbons proceeded faster and the period of incubation was shortened appreciably. After several culture transfers, the medium hydrocarbon chain fractions in crude oils were quite rapidly broken down and completely emulsified in 3 to 5 days at room temperature. Interestingly enough, continuous cultivation on nutrient agar caused a marked decrease in the ability of the cultures to attack hydrocarbons, indicating that many of the biotransformations may have been plasmid-mediated. *Pseudomonas* spp. were the most active in the breakdown of n-alkanes. However, a mixed culture was required for the breakdown of the aromatics. It was apparent from their study that the lighter fraction of hydrocarbons were more readily attacked than the heavier fraction.

The presence of heavy oils often gives a much lower CO_2 to oxygen ratio. The heavy oils become more difficult to attack as the viscosity and their molecular weight increases. The viscous oils are hard to disperse in liquid medium and have less surface area exposed for growth of microorganisms. It is apparent that compounds in the range of C_{10} to C_{16} are more readily attacked than those of larger molecular weight.[80] Bushnell and Haas[42] suggested that kerosene was more susceptible to biodegradation than gasoline based upon bacterial counts. Stone et al.[79] used manometric studies as well as bacterial counts to propose that paraffinic oils are more easily broken down than the aromatic fraction.

Radioassays have been used in monitoring mineralization of oils in which radioethane has been employed in measuring radio-labeled CO_2 evolution by ethane oxidizers with subsequent radioautographic measurements of radioactive colony formation. The labeled CO_2 can be trapped and adsorbed in alkali, followed by precipitation with $BaCl_2$ as $BaCO_3$ and measured for radioactivity. The radioactive colonies can be detected by use of agar pads from petri plates placed on x-ray film and incubated for one week.[43]

Biodegradation Kinetics

Biodegradation of petroleum hydrocarbons often follows an asymptotic function; that is, as the amount of contaminant decreases, the rate of degradation tends to decrease. At some point, the residual hydrocarbons provide insufficient carbon and energy for the microorganisms to degrade the remaining hydrocarbons. However, McCarty et al.[81] indicate that trace organic compounds may be utilized by microorganisms if supported by other organic materials present in sufficiently high concentrations. The phenomena is termed as 'secondary utilization' in which the trace organic material is not the primary source of energy for maintenance. 'Secondary utilization' is a form of cometabolism in which the secondary substrate is not used for energy but is degraded or transformed by enzymes by primary substrate metabolism.

The microbial breakdown of hydrocarbons in soils is often characterized by first-order kinetics.[82] However, based upon respiration rates, Verstraete et al.[83] indicate that biodegradation best fits asymmetric sigmoidal kinetics. Soil microorganisms adapt to the presence of pollution, and once this adaptation takes place, biodegradation occurs much more rapidly, often following a linear rate with time. Therefore, the rate of degradation may fall in between a first-order (exponential) and zero order (linear) kinetics. Song et al.[84] report that the kinetics of hydrocarbon degradation are complicated by the fact that there are numerous hydrocarbons within these oils which may be utilized at different rates. The lower rate of utilization may not only be caused by substrate depletion but also by the fact that the remaining hydrocarbons may be less degradable than the ones that disappeared.

Environmental Parameters

Bioremediation is dependent upon several environmental factors including aeration, pH, moisture, temperature, nutrient concentrations, and the microbiota.

Aeration

The influence of aeration on the degradation of petroleum hydrocarbons is reported in Table 16.2. In almost all cases, the presence of oxygen is essential for effective biodegradation of the oil. Anaerobic decomposition of petroleum hydrocarbons leads to extremely low rates of degradation (Figure 16.7). Oxidation of the n-alkanes proceeds through terminal oxidation and subsequent β-oxidation to yield C_2 units. Formation of a primary alcohol is the first stable intermediate. This is followed by the formation of an aldehyde and fatty acid which is subsequently oxidized by β-oxidation. The net result is the formation of a new monocarboxylic acid which is two carbon units less than the parent hydrocarbon and an acetyl coenzyme A that is eventually converted to CO_2. Subterminal oxidation of n-alkanes also occurs, leading to the formation of secondary alcohols and ketones.

The degradation of aromatic hydrocarbons requires ring fission through a dihydroxylation reaction. This reaction involves the introduction of two atoms of oxygen into the ring and the formation of a cis-cis diol.[97] Oxidation of aromatics such as benzene, toluene, ethylbenzene, anthracene, and naphthalene have been extensively studied in microbial systems. Oxidation of benzene by *Pseudomonas putida* and *P. aeruginosa* gives rise to catechol and cis, cis-muconic acid.[92] Microbial oxidation of toluene is initiated at either the methyl group or aromatic ring, depending on the organisms. Oxidation of m- and p-xylenes at the methyl

Table 16.2.   Influence of Aeration on Biodegradation of Petroleum Hydrocarbons

Aeration	Remarks	References
The presence of free O_2 is essential for biodegradation of oil.	Soil plow-layer is aerobic under normal conditions. Adequate drainage is needed to prevent waterlogging.	5
Aeration improved microbial degradation of oil in soil.	Plowing accelerated oil decomposition.	85
Bio-oxidation is increased with the use of H_2O_2.	Problems are encountered with gaseous oxygen transfer or ozone to liquid. It took 3 hours for decomposition of H_2O_2 to occur.	86
Toluene-grown soil bacteria oxidized toluene, benzene, catechol, 3-methylcatechol, and benzyl alcohol.	*Pseudomonas* and *Achromobacter* grew on toluene and benzene as a sole C source. 3-Methylcatechol is an intermediate in toluene oxidation.	87
Less O_2 uptake occurred upon decomposition of heavy fractions of hydrocarbons compared with the lighter fractions. Less O_2 occurred upon decomposition of the aromatic fraction compared with the n-alkanes.	The theoretical respiratory quotient (RQ) for complete oxidation of long chain paraffin hydrocarbons [$CH_3(CH_2)_nCH_3$] is 0.67. The RQ for light oils is approximately 0.65.	73
Aerobic conditions promote greater degradation than anoxia.	Aerobic degradation leads to the release of CO_2, H_2O and biomass while anaerobic decomposition leads to mainly CH_4, CO_2, organic acids, alcohols and biomass. In groundwater approx. 8–12 mg/L of oxygen are dissolved in water. To degrade 1 litre of diesel fuel, about 3,800 g of oxygen is required. The use of H_2O_2 and ozone (O_3) can increase the oxygen content in water to approx. 40 mg/L.	88
Soil venting systems can be implemented to supply aeration.	For partly oxidized hydrocarbons, NO_3^- can be used as an alternate electron acceptor.	89

continued

Table 16.2. *continued*

Aeration	Remarks	References
Microbial cooxidation of nongrowth gaseous hydrocarbons can lead to water-soluble oxidation products.	Cooxidation is important in the microbial utilization of pesticides and various hydrocarbons (e.g., utilization of cyclohexane by propane-grown microbial cells).	90
n-Alkane oxidation gives rise to a primary alcohol, aldehyde and acid. Subterminal oxidation of n-alkanes also occurs leading to the formation of secondary alcohols and ketones.	Cell-free extracts of *Corynebacterium* sp. oxidized *n*-octane to octan-1-ol. *n*-Tetradecane was oxidized to tetradecan-1-ol and tetradecanoic acid by a thermophilic *Bacillus* sp.	91
Biological degradation of aromatics involves oxidation to form catechol intermediates.	Oxidation of benzene by *Pseudomonas putida* and *P. aeruginosa* give rise to catechol and *cis, cis*-muconic acid. Microbial oxidation of toluene is initiated through either the methyl group or aromatic ring depending on the organism. Oxidation of m- and p-xylenes at the methyl group yields 3- and 4-methyl catechol, respectively.	92
Anaerobic petroleum degradation is negligible.	Use of NO_3^- as an alternate electron acceptor is not energetically favorable. Intermediates of aerobic hydrocarbon metabolism may be further broken down under anaerobic conditions using NO_3^- or SO_4^{2-} as an electron sink.	38
O_2 consumption is a reliable method to monitor oil degradation.	Oil contaminated soils display greater oil degradation when aeration is maximized.	38
Bacterial degradation of aromatics involves a dihydroxylation reaction which 2 atoms of oxygen are introduced into the ring forming a *cis*-diol.	*Pseudomonas putida* oxidizes benzene, toluene and ethylbenzene as sole sources of carbon and energy. *P. putida* oxidizes p- or m-xylene but these compounds cannot serve as a C source.	110

continued

Table 16.2. *continued*

Aeration	Remarks	References
Oxidation of *n*-alkanes proceeds through terminal oxidation and subsequent β-oxidation to yield C_2 units.	Branched chain hydrocarbons are also metabolized into dicarboxylic acids and subsequent β-oxidation.	110
Cooxidation results in microbial degradation of cycloparaffins.	Growth of *Nocardia* sp. on hexadecane allowed partial oxidation of ethylbenzene to phenylacetic acid.	110
Alternative sources of aeration include pure oxygen, H_2O_2 and ozone to stimulate hydrocarbon degradation.	Depending on the temperature of groundwater, 8–12 mg/L of dissolved O_2 may be achieved by air sparging.	93
Biodegradation of petroleum hydrocarbons requires molecular oxygen.	Marine oil spills rely on wave action for aeration. Algae may also provide O_2. Frequent tillage of soil increases the diffusion of oxygen.	94
A series of two-week interval injections of H_2O_2 (100 ppm) and nutrients were made in an *in situ* treatment of an underground diesel fuel spill for 4 months.	Substantial diesel fuel reduction was evident after initiation of the *in situ* bioremediation program.	95
Microorganisms in groundwater have the ability to degrade benzene, toluene, xylenes under anoxic (denitrifying) conditions.	The addition of NO_3^- to groundwater is not be allowed by the California Water Quality Control Board.	96

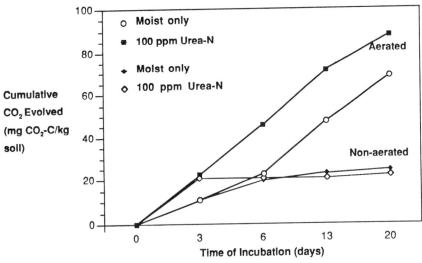

Figure 16.7.   Mineralization of diesel fuel in the presence and absence of air.

group yields 3- and 4-methyl catechol, respectively.[92] Co-oxidation has been shown important in degradation of cycloparaffins by soil microorganisms.[98]

In a few cases reported in the literature, alternative electron acceptors (e.g., NO_3^- and SO_4^{2-}) have been suggested to promote biodegradation of hydrocarbons, particularly with intermediates released through aerobic metabolism. However, in most cases, anaerobic degradation has negligible effects in reclamation of oil polluted soils. When soils are contaminated with oil, biodegradation is often enhanced by disc harrowing or rototilling. Loynachan[99] found that disturbance of the soil surface increased hydrocarbon utilization because of increased surface area for diffusion of molecular oxygen. In a bench-scale treatability study, soils should be frequently mixed, at least twice a week, to simulate field tillage operations and to ensure aeration is not a limiting factor.

Aeration is obviously going to be a limiting factor in biodegradation of oils in the subsurface environment. A bench-scale treatability study should simulate these conditions by incubating the soil samples under anoxic conditions. The oxygen content in the subsoil can be increased by the addition of hydrogen peroxide (H_2O_2). In the presence of catalase, H_2O_2 breaks down into O_2 and water. Hydrogen peroxide was used as a biostimulating agent for an in situ bioremediation program to accelerate the decomposition of underground diesel fuel in California.[95] The catalytic decomposition of H_2O_2 is not only dependent upon catalase activity, but also the presence of heavy metals or minerals in soil.[100] The presence of Fe, and especially Mn compounds can serve as nonbiological catalysts. Other heavy metals which are known to promote decomposition of H_2O_2 include Pb, Co, Ni, Cu, Zn, Mo, W, and Cr. The presence of insoluble phosphate compounds are known to decrease the rate of decomposition. The breakdown rate of H_2O_2 tends to vary from site to site. A batch test conducted in the laboratory may help determine the spacing for H_2O_2 injection wells as well as the concentration to apply in the field. The application of excessive H_2O_2 can result in destruction of soil structure, leading to poor soil permeability. Generally, the optimum concentration of H_2O_2 to add to soil falls in the range of 100 to 300 ppm.

Soil pH

The ideal pH range to promote biodegradation of oils in soil is within the neutral to slightly alkaline range. Most studies indicate that pH 7 to 8 is optimum for degradation of petroleum hydrocarbons (Table 16.3). Dibble and Bartha[101] found that increasing the soil pH with the addition of $CaCO_3$ to pH 7.8 promoted the rate of CO_2 evolution from soils receiving oil sludge. Previous work in this laboratory has shown that at pH values at or above 9.5, hydrocarbon degradation is inhibited.

Table 16.3. Influence of Soil pH on Biodegradation of Petroleum Hydrocarbons

Optimum pH	Remarks	References
pH 7.0–7.8	Biodegradation of *n*-alkanes was found to be minimal in acidic soils (pH 3.7). Liming with $CaCO_3$ favored biodegradation of oil.	101
pH 7–9	Most heavy metals are insoluble at pH values above 7. The leachability of metals increase at pH values below 7.	5
Neutral pH	Biodegradation activity was stimulated at neutral pH.	102
pH 6–9	The effect of pH on degradation of asphalt corresponded to the pH effect on growth of *Pseudomonas* spp.	103
pH 7.5–7.6	Lime ($CaCO_3$) was added to soil to enhance hydrocarbon degradation.	84
pH 7.4	The rate of breakdown of *n*-alkanes and aromatics increased at pH 7.4. The number of bacteria increased sharply with increasing pH.	83
pH 8.0	Biodegradation of octadecane and naphthalene was greater at pH 8.0 than at pH 5.0 and 6.5.	104
pH 6.5–9.5	Hydrocarbon oxidizers are not extremely sensitive to changes in hydrogen-ion concentration.	82
pH 6–10	Luxuriant growth of hydrocarbon oxidizers at this pH range.	105
pH 5–8	Hydrocarbon oxidizers were active at pH 5–8 but grew best at pH 7.6–8.0	106

Contaminated soils that have a relatively high pH can be treated with a dilute solution of H_3PO_4 to lower the pH within the neutral range. By the addition of H_3PO_4, the nutrient, P is also supplied to enhance decomposition. Soil pH can also be lowered by applying elemental sulfur, ferrous sulfate or aluminum sulfate (alum). The application of nitrogen fertilizers also tend to lower the soil pH. Ammonium sulfate gives twice as much acidification per pound of N than ammonium nitrate or urea. The amount of amendment required to acidify a soil is calculated based upon the cation exchange capacity of the soil, the existing and desired pH, and percent base saturation.

Acidic soils below pH 7 can be treated with lime ($CaCO_3$) to enhance the biological activity. The amount of lime required to raise the pH can be determined by the pH-base saturation method, which requires knowledge of the cation exchange capacity and soil pH. An alternative method is the buffer solution technique, which is based upon the ability of a soil to lower the pH of a buffered solution. If too much lime is applied to the contaminated soil, the availability of P, Fe, Mn, B, and Zn will be reduced as macro- and micro-nutrients for hydrocarbon oxidation. Sandy soils are most likely to be deficient in minor elements whose availability will be reduced by overliming. Liming materials which may be considered in amending the soil to a neutral pH include ground limestone, consisting mainly of $CaCO_3$ and dolomitic varieties containing $MgCO_3$. The use of Na_2CO_3 should be avoided because it has too strong an alkalizing effect, supplying a harmful cation (sodium) and being very expensive. Upon biodegradation of oil very little change in soil pH occurs with time.[16]

Moisture

The presence of oil will reduce soil wettability but biological decomposition of the hydrocarbons will return the soil back to its normal wettability. The moisture content of the contaminated soil will affect biodegradation of oils due to dissolution of the residual compounds, dispersive action, and the need for microbial metabolism to sustain high activity. The moisture content in soil affects microbial locomotion, solute diffusion, substrate supply, and the removal of metabolic byproducts. Excessive moisture will limit the gaseous supply of oxygen for enhanced decomposition of petroleum hydrocarbons. Most studies indicate that the optimum moisture content is within 50% to 70% of the water-holding capacity (Table 16.4) or 15% to 20% dry weight (Figure 16.8). Precautions should be made not to saturate the soil in promoting leaching of the residual hydrocarbons or water-soluble nitrogen fertilizers. Both extremes, waterlogging and desiccation, will reduce the effectiveness of a bioremediation approach. The unavailability of water could have a limiting effect on biodegradation, particularly in summer months. If water does not infiltrate uniformly in a landspreading operation, sand may be added to increase the percolation rate of the soil. In a feasibility study, the soils should be maintained under moist conditions by reapplying water at daily intervals.

The solubility of oils generally decreases with increasing molecular weight of the hydrocarbons, but branching of hydrocarbon isomers and condensed ring

Table 16.4.   Influence of Soil Moisture on Biodegradation of Petroleum Hydrocarbons

Moisture level	Comments	References
Moist near field capacity without saturation	Irrigated almost daily during dry summer months. Irrigation was determined from soil tensiometer readings, soil moisture analyses and precipitation and evaporation records. Typical irrigation ranges from ¼ to ⅜".	107
50–80% water-holding capacity	Anaerobic petroleum biodegradation is negligible.	38
20% (dry weight)	Established loose, friable crumb structure. No activity occurred in a water-saturated soil.	14
60% water-holding capacity	Added water regularly for 8 months.	85
50–70% water-holding capacity	Less water results in inadequate bioactivity while too much affects soil aeration. The presence of hydrocarbons reduces the water-holding capacity of soil.	101
25–85% of the water-holding capacity	Attempts should be made to avoid migration of inorganic nutrients, organics and heavy metals.	93
15% moisture content	Peak of microbial activity in mineralization of oil to CO_2.	108

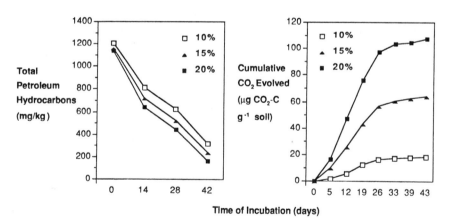

Figure 16.8.   Influence of moisture content on biodegradation of diesel fuel.

formation tends to increase their solubility in water.[109] The alkenes are usually more soluble than the alkanes, and aromatics are often more soluble than the cyclic-alkanes or -alkenes. Gutnick and Rosenberg[110] reported that approximately 0.02% of crude oil or refined oil can be extracted in the aqueous phase. Compounds extracted in this phase include phenols, anilines, and alkylated derivatives of benzene and naphthalene, which are the most toxic components of petroleum.

The application of nutrients may enhance the bio-emulsifying activity of micro-organisms, releasing water-soluble intermediates capable of migrating at great depths within the soil profile.[83] However, most of the hydrocarbons are hydrophobic in nature which absorb to organic matter and are not readily transported through soil (Table 16.5). The partitioning of these hydrophobic organics between the water phase and the organic particulates can be estimated from the octanol/water partition coefficient.[81] Migration of the hydrocarbons is highly dependent upon soil texture (permeability, diffusivity, and hydrodynamic dispersitivity), uniformity and configuration of soil layers, moisture content of the soil, and the fluid viscosity of the oil itself.[112] Attempts should be made to keep the soil at field capacity to promote diffusion of oxygen and high biodegradation rates.

Some of the low molecular weight aromatic hydrocarbons such as benzene and cyclohexane in water may be toxic to microorganisms, particularly when found in high concentrations. Their high solubility in water can result in extensive damage to the lipoid membranes of microbial cells.[115] However, low concentrations of these soluble toxic components can favor microbial degradation. van der Linden[115] reported that hydrophobic forces explain the remarkable ability of microbes to utilize hydrocarbons in the low ppm concentration range. Some of the high molecular weight hydrocarbons which are insoluble in water are attacked by microbes growing at the oil-water interface. Emulsification plays an extremely important role in increasing this interface.

Temperature

All biological transformations are affected by temperature. Generally, as the temperature increases, biological activity tends to increase up to a temperature where enzyme denaturation occurs. Hydrocarbon oxidizers have been isolated at temperatures as low as $-1\,°C$ to as high as $70\,°C$.[20] The presence of oil should increase soil temperature, particularly at the surface. The darker color increases the heat capacity to adsorb more radiation. The optimum temperature for biodegradation of hydrocarbons ranges from $18°$ to $30\,°C$ (Table 16.6). Minimum rates would be expected at $5\,°C$ or lower. However, microorganisms active in degradation of organic materials often adapt to these extreme temperatures. Lower temperatures decrease the volatilization and increase the water solubility of volatile hydrocarbons, which may have adverse effects on microbial populations involved in degradation of oils.

Mineralization of crude oil within the $5°$ to $20\,°C$ range usually shows an initial lag, but afterward, the rate tends to double with each $5°$ to $10\,°C$ increase in temperature.[116] Generally, seasonal conditions tend to show the highest counts of petroleum oxidizers in the summer months compared with the winter months when all other environmental factors are equal. When oily wastes are to be deposited on soil for disposal, such as a landfarm operation, direct sunlight should be considered as an important parameter for site location. Photooxidation may render some of the high molecular weight hydrocarbons to become more susceptible to biodegradation.

Table 16.5. Transport of Hydrocarbons Upon Added Moisture

Hydrocarbons	Leaching capacity	Remarks	References
Gasoline	Relatively small and not promoted by application of detergents.	Soil column was subject to clogging.	102
Oil	Displacement of contaminants.	Infiltration of water in unsaturated zone stimulates biological activity.	89
Halogenated hydrocarbons and substituted benzenes	Most organic pollutants were transported readily through soil.	Retardation of chlorinated benzenes increased with decreasing water solubility of chemicals. Exhibited linear sorption isotherms. Sorption increased with increasing organic matter.	111
Hydrophobic organic materials	Sorbed to organic materials.	Estimated from the octanol/water partition coefficient for each organic compound. Charged species may be subject to ion exchange by clays.	81
Oily sludge	No leaching of oil or Pb was observed.	Addition of oil sludge to soil prevented water from infiltrating into the soil. Soil treated with fertilizers (600 kg N/ha/yr) had infiltration rates 8 times greater than without fertilizer addition.	79
Gasoline	Soil columns treated with N and P mobilized the hydrocarbons due to the bio-emulgating activity of soil microbiota.	The movement of oil increased with increasing soil tension. Soil columns fumigated with methylbromide did not show any evidence of hydrocarbon leaching.	83

continued

Table 16.5. *continued*

Hydrocarbons	Leaching capacity	Remarks	References
Hydrophobic hydrocarbons	Affected by soil texture (related to permeability, diffusivity and hydrodynamic dispersivity), vertical uniformity or nonuniformity and configuration of soil layers.	More migration occurs in sandy and gravelly soils. Layered soils retard migration. Horizontal layers within concave depressions often retain perched bodies of contaminated fluid.	112
Motor fuel	Dependent on soil particle size, moisture content, fluid viscosity.	Solubility of gasoline is in the range of 50–150 mg/L. Solubility of diesel fuel is in the range of 0.4–8.0 mg/L.	113
Oil wastes	Rainfall causes deep percolation.	A mathematical model was developed based on linear partitioning of the contaminant between phases, volatile losses by a modified Thibodeaux-Hwang approach, first-order degradation kinetics, uniform soil profile and constant water flux throughout the treatment zone.	65
Benzene, toluene and xylenes	Adsorption was proportional with organic matter content. Adsorption isotherms were linear up to 20 mg/L in the water phase.	Active microorganisms lead to greater adsorption.	114
Gasoline, fuel and cutting oil	Little oil was leached. Oil content of drainage water varied from 20–400 μg/L.	Leaching of nitrate fertilizer was considerable (150–250 mg/L) while no P was leached.	12

Table 16.6. Influence of Temperature on the Biodegradation of Petroleum Hydrocarbons

Temperature	Remarks	References
Mineralization of Sweden crude oil was greater at 20°C and 15°C and progressively lower at 10° and 5°C. Rates roughly doubled with each 5°C rise in temperature in the 5–20°C range. Lower temperatures caused increasing lag periods in biodegradation.	Volatile components of oils may have been inhibitory to oil-degrading microorganisms. Evaporation occurred slowly at low temperatures retarding biodegradation.	116
Oxidation of hydrocarbons occurs as low at −1°C, but maximum degradation occurs between 25–37°C.	Below 10°C the rate of oxidation is very slow.	117, 118
Oxidation of oil at 4°C was not measurable.	At 10, 20 and 25°C, 20–30%, 30–50% and 50–80% per week of commercial oil was biologically degraded when inoculated with sewage sludge, respectively.	119
At 2, 18 and 30°C, 30, 40 and 50% of added oil was degraded in 8 weeks, respectively.	Biodegradation of fuel oil was temperature dependent.	120
Approximately 20 mg/L of Cook Inlet crude oil was almost completely degraded after 2 months at 10°C.	Biodegradation was active at low temperatures.	121
Bacterial growth on n-alkanes can occur up to 70°C.	Thermophilic conditions result in high growth rates but cell yields are low because of thermal damage to cells.	122, 123
Biodegradation of a 'model petroleum' consisting of n-paraffins (86%) and some branched, aromatic and alicylcic hydrocarbons was greater at 0°C than at higher temperatures (5° and 10°C) when added to winter-collected enriched seawater.	Decreased toxicity of some of the volatile hydrocarbons at lower temperatures was proposed as a possible explanation.	124
Bacterial growth and cellular uptake of hexadecane increased with temperature reaching a maximum at 25°C in estuarine and marine environments.	The percentage of hexadecane mineralized to CO_2 decreased as the temperature increased from 5 to 35°C.	51
Microbial populations metabolize similar quantities of crude oils under psychrophilic and mesophilic conditions.	Populations at 4°C failed to degrade isoprenoids, phytane and pristane, whereas the 30°C enrichments metabolized these compounds.	125

continued

Table 16.6. *continued*

Temperature	Remarks	References
Bacterial growth was reduced at low temperatures (6 and 12°C).	Emulsification was almost unaffected by a decrease in temperature.	126
Hydrocarbon decomposition occurs rapidly at 37°C.	Microbial attack of *n*-alkanes occurred within 2 days at 28° or 37°C while at 20°C a week was required for oxidation.	41
Heated water enhances biodegradation.	Warm temperatures promote biodegradation and leaching processes.	89
Heterocyclic aromatics undergo photolytic transformations.	Photooxidation may render some hydrocarbons more susceptible to biodegradation.	127
Optimal temperature for fungal degradation of hydrocarbons was between 15–37°C.	At <15°C, fungal degradation occurred more slowly.	128
Optimum temperature for microbial growth on asphalt was 30°C.	More degradation of asphalt occurred at 20–25°C.	103
Optimum temperature for degradation of oily sludge in Norwegian soils was about 18°C.	About 2/3 of the optimum activity (at 18°C) was achieved at 12°C. Minimum degradation was found at 6°C.	79
Oil removal was at a minimum during the winter months.	Degradation of oil was very slow at <20°C.	18
Psychrophilic mixed cultures (5°C) readily degraded gas oil in fertilized seawater.	A generation time of 13.5 h was reported for an adapted culture during the first few months of degradation.	115
Thermophilic aerobic spore-forming bacteria with optimum growth at 45–65°C assimilated hydrocarbons and fixed atmospheric nitrogen.	97 strains of thermophilic *Bacillus* were capable of utilizing hydrocarbons as a sole source of carbon isolated from petroleum-impregnated soil in the Soviet Union.	129
Oil contaminated soils should be exposed to warm areas and direct sunlight.	Promotes rapid metabolism by mesophilic microorganisms.	94
Oil-utilizing microorganisms in Alaskan soils were active at 10°C.	Time required for doubling of the indigenous aerobic bacterial population at 10°C was 2 days.	99

continued

Table 16.6. continued

Temperature	Remarks	References
Black plastic film increases soil temperature during the winter months promoting biodegradation of waste oil. In the summer months, transparent polythene would reduce water evaporation.	Use of film would interfere with tillage of soil and decrease aeration.	130
Bacterial growth, emulsification and utilization of crude oil occurred readily at 4° and 30°C.	Under psychrophilic conditions very little of the aromatic fraction was degraded. The isoprenoid compounds, phytane and pristane were resistant to degradation.	131
Carbon dioxide production from soils with and without the addition of oils could be measured at 5°C.	A long lag period occurred at low temperatures.	14
Optimum temperature for most hydrocarbon-oxidizing organisms is 30°C.	Thermophilic temperatures affect microbial membranes.	132
Degradation of hydrocarbons was maximum at 27°C.	Half-lives of fuels were longer at 17°C than 27°C. A further increase in temperature to 37°C increased the half-lives.	84
UV energy from sunlight can promote photooxidation of hydrocarbons near the soil surface.	Photooxidation renders some compounds more susceptible to biodegradation.	65
Biodegradation of petroleum hydrocarbons occurred as low as −1.1°C as long as soil solution remains liquid.	High degradation rates generally occurred between 30–40°C.	133
High biodegradation rates occurred between 20–37°C.	Elevated temperatures (>37°C) in the presence of certain hydrocarbons may increase membrane toxicity.	101
No biodegradation occurs in frozen soils.	Hydrocarbon-degrading microorganisms isolated in Alaska were able to grow at 5°C.	134, 135, 136
Q_{10} values for decomposition of oils range from 1.7 to 2.7 at 5 to 26°C.	Oil sludge utilization in soil had a Q_{10} of 1.7 at 5–20°C.	38

Composting may also be an effective means to promote biodegradation of oils. By adding readily degradable C sources (e.g., animal manure, alfalfa, animal feed), along with bulky materials (wood chips, tree bark, sawdust) to enhance aeration with strict control over the moisture content and inorganic nutrients, the contaminated soil subject to elevated temperatures often leads to high hydrocarbon oxidizing activity. Perforated pipes within the compost pile may be used to promote aeration. The leachate resulting from added moisture can be re-applied to the compost. With this bioremediation approach, it is expected that many of the hydrocarbons will be incorporated into complex organic matter as a result of the rich carbon sources added to the compost.

Bioremediation should not be ruled out as a possible remedial practice for removal of petroleum hydrocarbons in the colder regions of the world. Figure 16.9 shows substantial rates of mineralization of arctic diesel spiked (8,000 ppm TPH) to an Alaska soil at 5° to 20°C, particularly with the addition of nutrients [250 ppm N (NH_4NO_3) and 25 ppm P (K_2HPO_4)]. Atlas and Bartha[137] found a high number of hydrocarbon-utilizing microorganisms capable of growth at 5°C in Raritan Bay, New Jersey during the winter months. The rate of hydrocarbon mineralization at 5°C was considerably greater in water samples collected in the winter months than those collected in the summer. Walker and Colwell,[124] using a "model petroleum mixture" incubated with estuarine water collected during the winter, found slower but more extensive biodegradation occurring at 0°C

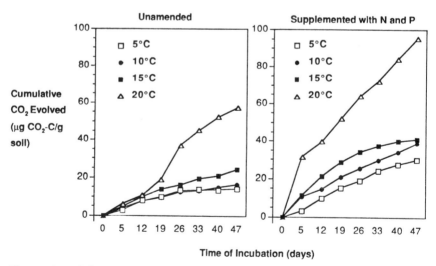

Figure 16.9.   Influence of temperature on the mineralization of arctic diesel fuel added to soil.

than at higher temperatures. Colwell et al.[138] found greater degradation of Metula crude oil at 3°C than at 22°C with mixed microflora cultures in beach sand samples. When 0.1% oil was added, 48% of the hydrocarbons were degraded at 3°C compared with only 21% degraded at 22°C. They indicated that temperature was not a limiting factor for petroleum degradation in the Antarctic marine ecosystem affected by the Metula spill.

A laboratory feasibility study should be incubated at a temperature to simulate the median temperature in the field. There will obviously be some daily and seasonal variations in temperature affecting the subsequent oxidation of petroleum hydrocarbons. Attempts should be made to calculate Q_{10} values over a psychrophilic to mesophilic temperature range to represent daily and nightly fluctuations in biodegradation of hydrocarbons.

Nutrients

Nitrogen is usually the main limiting nutrient governing the rate of decomposition of petroleum hydrocarbons. However, small amounts of phosphorus fertilizers may also stimulate biodegradation. The accepted C:N and C:P ratio to convert 100% of petroleum hydrocarbons into microbial biomass are 10:1[139] and 100:1,[140] respectively. Nitrogen is a key building block of proteins and nucleic acids (components of amino acids, purines, amino sugars, and vitamins) while P is needed to produce enough ATP to carry out metabolic functions. Phosphorus is also an important constituent of nucleic acids, phospholipids, and techoic acids. Fertilizer addition will alleviate the nutritional imbalances in a petroleum contaminated soil. Application of N and P often enhances microbial biomass, stimulating biodegradation of the residual hydrocarbons as a luxurious carbon source. Biodegradation rates have frequently been reported to double upon application of fertilizers (Table 16.7).

Common N fertilizers applied to oily soils include ammonium nitrate, ammonium sulfate, and urea. Urea contains a considerably high percentage of N (46%). When urea is applied to soil, it is rapidly hydrolyzed to ammonia and CO_2. However, petroleum hydrocarbons, particularly refined oils, can inhibit the hydrolysis of urea requiring more time for the N to become available to the hydrocarbon oxidizers.[152] It is recommended that an inorganic N source be used in landfarm and landspreading operations rather than urea. We have found that when the inorganic N content (NH_4-N and NO_3-N) is maintained at >50 ppm, an adequate supply of N is usually available for biodegradation of oil. There are some recent indications that NH_4^+ is preferred over NO_3^- as an N source in the microbial breakdown of hydrocarbons. Further research is needed to evaluate the effectiveness of slow-release fertilizers in enhanced hydrocarbon degradation. As the hydrocarbons are broken down and mineralized to CO_2, N is recycled under aerobic conditions.

Table 16.7.   Influence of Nutrients on Biodegradation of Petroleum Hydrocarbons

Nutrients	Remarks	References
Nitrate (10^{-2}M) and phosphate (3.5×10^{-4}M) added in combination promoted biodegradation of petroleum in seawater.	Only 1% of the hydrocarbons were mineralized in an unamended sample without nutrients while 42% was mineralized upon nutrient addition in 18 days.	71
Nitrogen fixation will not overcome N limitations in oil metabolism.	Azotobacter strains utilized tetradecane, toluene and naphthenic acids while fixing atmospheric nitrogen.	20
'Oleophilic fertilizers' are commercially available to promote oil-degrading organisms and minimize algal bloom formation.	Paraffinized urea (26.8% N) and octylphosphate (14.8% P) supplied as N and P accelerated decomposition of oil 30- to 50-fold compared with untreated seawater.	141
Paraffinized MgNH$_4$PO$_4$-stimulated oil degradation.	Approximately 63% of the oil disappeared in 3 weeks upon treatment with MgNH$_4$PO$_4$ vs. 40% in the untreated control.	142
Nitrogen, phosphorus and iron supplements stimulated biodegradation of crude oil in seawater.	Without supplementation, biodegradation was negligible. Addition of N and P allowed rapid biodegradation (72% in 3 days). Ferric octoate also stimulated degradation in less polluted and less iron-rich water.	143
Advantage in seeding oil-polluted areas with organisms having the genetic capacity to fix molecular nitrogen (Klebsiella, Azotobacter, and Rhizobium).	Nitrogen fixation alone will not overcome N limitations.	110
Bacteria utilize hydrocarbons as a C and energy source and require N and P with minor quantities of K, Ca, S, Mg, Fe and Mn.	Certain organic growth factors may also stimulate hydrocarbon oxidizers.	29
Maximum efficiency of degradation requires N and P at a ratio of 2–3:1.	Solutions of (NH$_4$)$_2$HPO$_4$ and NH$_4$NO$_3$ were added.	29

continued

Table 16.7. *continued*

Nutrients	Remarks	References
Addition of phosphate stimulated microbial degradation of hydrocarbons.	Seawater enriched with 0.1% $FeNH_4PO_4$ stimulated the population of hydrocarbon oxidizers. The presence of carbonate or phosphate in the medium is desirable to buffer a change in pH since microbial degradation of hydrocarbons leads to acid production.	72
Application rates of 1600–1700 lb/ac of N and 350–450 lb/ac of P and K were added to a landfarm operation for a four month period.	Approximately 70% of the oil was biodegraded in 3 weeks. Appreciable amounts of NO_3-N were collected in the leachate.	7
Gas oil was degraded rapidly upon the addition of 6 kg of P and 60 kg of N per ton of oil.	Little differences in degradation were observed when N was supplied as $(NH_4)_2SO_4$, NH_4NO_3 or urea. Phosphorus was supplied as disodium phosphate.	115
0.5 mg of N is required for the decomposition of 1 mg of n-hexadecane.	Hexadecane is readily degraded.	49
4 μmoles of N were required/mg crude oil degraded.	Approximately 44% was degraded.	144
Field experiments with the application of N, P, K and S were effective in reclaiming oil contaminated soils.	Ammonium nitrate, ammonium sulfate, monoammonium phosphate and potassium chloride were added at a ratio of 10:3.1:2:1 N:P:K:S. Soils receiving fertilizers improved structurally and began to wet normally. Unfertilized soils were dispersed and resisted wetting. The added nutrients accelerated oil decomposition and shorten the time when new plants could colonize the polluted soil.	78
Stimulating indigenous microbiota with nutrients is more effective than inoculation with commercial biological products such as dried bacterial cultures.	Inoculated cultures were limited in growth without added nutrients.	145

continued

Table 16.7. *continued*

Nutrients	Remarks	References
Nutrient amendments of $(NH_4)_2SO_4$, disodium phosphate and monosodium phosphate were injected into an aquifer contaminated with gasoline.	A laboratory study indicated that air, inorganic N and phosphate salts increased the population of hydrocarbon-oxidizing organisms by 1000-fold. Biodegradation of 1 liter of gasoline was estimated to require 44 g of N, 22 g of P and 730 g of oxygen.	146
Nutrient requirements can be determined in laboratory studies.	An increase in number of total and hydrocarbon-degrading bacteria can be used to determine nutrient limitations.	6, 147
Nitrogen and phosphorus are required in high concentrations.	Sulfur, iron, magnesium, and calcium are required in lesser amounts.	94
Oil sludge biodegradation was optimum at C:N and C:P ratios of 60:1 and 800:1, respectively.	The quantity of N and P required to convert 100% of petroleum carbon into biomass corresponds to a C:N ratio of 10:1 and C:P of 100:1.	101
Application of urea-phosphate (N:P:K, 27:27:0) stimulated bacterial numbers and utilization of *n*-alkanes.	Nutrient-deficient conditions may be responsible in part for the persistence of oil in soil.	148
N requirement for decomposition of oily waste is approximately 0.001–0.1 part N/25 parts of waste carbon to maintain maximum biodegradation rates.	Excessive fertilizer application may result in high salt content and NO_3^- in the leachate.	5
Significant growth on crude oil did not occur unless seawater was supplemented with N and P.	Fungi utilized crude oil more effectively than did bacteria when supplemented with nutrients.	128
Rehabilitation of oil spills requires the addition of N and seeding with leguminous plant species to establish vegetation.	The addition of NH_4NO_3 increased microbial respiration by 2.3-fold with 50% disappearance of an oily waxy cake within 50 days upon vegetative cover. Without vegetative cover, degradation was slow even in the presence of N.	130

continued

Table 16.7. *continued*

Nutrients	Remarks	References
Addition of N-P-K fertilizer stimulated oil degradation.	Oil content in soil was reduced by 4% in control plots (no fertilizer added) and to 9–26% in fertilized plots. In plots receiving the highest N addition (600 kg N/ha/yr) about 80% of the oil was degraded in 33 months. Nitrogen is the major limiting nutrient in decomposition of oil spills.	79
Addition of N fertilizer stimulated decomposition of oil markedly, whereas K and P were not effective.	Decomposition rates of oil could be increased 3- to 4-fold with the application of N fertilizers.	14
Soil columns treated with inorganic N and P fertilizers mobilized entrapped hydrocarbons in the soil matrix.	Hydrocarbons were leached out as a result of bio-emulgating agents.	83
Fertilizer applications should be applied in small frequent doses.	Manure was added as an N source and to absorb hazardous waste. Recommended to maintain C:N ratio between 50:1 to 25:1.	107
N and P fertilizers stimulated biodegradation of *n*-alkanes of Prudhoe Bay crude oil.	The addition of N and P fertilizers were less effective on decomposition of aromatics.	149
Manure plus NO_3^- stimulated CO_2 production in oil-polluted soils more than NO_3^- alone.	The greater CO_2 evolved was most likely derived from the manure.	150
Biodegradation of Prudhoe Bay crude in an arctic northern soil was stimulated by low N applications.	Nitrogen levels (NH_4NO_3) above 100 ppm depressed soil respiration; however, $NaNO_3$ or NH_4Cl stimulated CO_2 evolution. There was a 4-fold increase in respiration with samples receiving 300 ppm of N applied with $NaNO_3$ or NH_4Cl.	151
Diesel fuel, kerosene, motor oil and leaded gasoline inhibited urea hydrolysis in soil.	The greater the oil contamination, the less effective urea became as an N fertilizer. It was recommended that an inorganic N source be used in landfarm operations rather than urea because the availability of NH_4^+ derived from urea would be greatly delayed by providing it as a source of N.	152

continued

Table 16.7. *continued*

Nutrients	Remarks	References
Nitrogen-phosphorus oleophilic fertilizer *Inipol* EAP-22 and granular *Customblen* was applied after the *Exxon Valdez* oil spill at Prince William Sound, Alaska.	The number of hydrocarbon-oxidizing microorganisms increased upon fertilizer application. Periodic application of fertilizers promoted biodegradation with the upper limit being 2–5 times greater than the natural rate. Monitoring degradation rates indicated that decomposition was occurring at 5 to 10 g hydrocarbon/kg of beach sand/yr. Beaches treated with *Inipol* lost 8 to 20% more oil over the winter than untreated beaches.	153, 154
Nitrogen was added in the vapor phase as anhydrous ammonia.	100 ppm of anhydrous ammonia in the air stream increased microbial counts by a factor of 100.	155
Synergistic effects were noted when inorganic nutrients were supplemented with molasses and proteins.	Various organic proteins were tested for their enhancement in oil degradation.	108
Urea and phosphate were added at a rate of 735 to 3540 lb/ac and 250–750 lb/ac, respectively.	Oil in fertilized plots degraded by approximately 80% while the nonfertilized control had degraded by only 55% after a 1-year period.	16
NH_4NO_3 added to gasoline-contaminated groundwater enhanced biodegradation of benzene and toluene.	For toluene, 17 h was required to reach <100 ppb upon amendment with N compared with 23 h for the unamended sample. For benzene, the unamended water decreased from 480 to 218 ppb in 48 h while that treated with N dropped to 25 ppb.	156

Phosphorus fertilizers should be added at approximately one-fifth to one-tenth the rate of nitrogen fertilizers. Phosphate fertilizers do not tend to move in soil at pH 7 to 8, thus mixing is very important with the application of P. Mechanisms which restrict the movement of P include sorption by soil constituents and precipitation by multivalent cations. Organic P compounds which are water-soluble and may have applications for in situ bioreclamation include potassium diphenylphosphate, potassium diphenyl-pyrophosphate, calcium diethyl-phosphate, and glycerol phosphate. Ammonium phosphate is a common combined N and P fertilizer used in landfarm operations. The addition of P fertilizers can improve the wetting characteristics of oily soils.

Analysis of total N, P and S gives no indication of the nutrients available for immediate uptake by the hydrocarbon-oxidizing microbiota. Thus, these particular tests are not recommended for a feasibility study.

Bioaugmentation

Several bacterial cultures are sold as liquid preparations, air-dried or freeze-dried products. They are being marketed specifically for the cleanup of certain waste materials. The effectiveness of specialized bacterial cultures has been proven with the operation of activated sludge, trickling filters, and lagoon treatment plants providing a greater response to such problems as startup, shock loads, and cold weather operations.[157] However, the success in bioaugmentation in degradation of petroleum hydrocarbons is less promising as indicated in Table 16.8. Often many of these studies show enhanced decomposition of hydrocarbons but then fail to account for the nutrients and growth factors added along with the seeded organisms. Under most circumstances, the indigenous microflora are in adequate numbers to decompose petroleum hydrocarbons, particularly upon the addition of biostimulating agents.

The main group of organisms considered to be active hydrocarbon oxidizers in soil include *Pseudomonas, Flavobacterium, Nocardia, Corynebacterium,* and *Arthrobacter.*[18] Often there is a limiting factor which prevents the indigenous heterotrophic microbiota from utilizing hydrocarbons as a source of carbon and energy such as nutrients, aeration, and moisture. Cultivation of hydrocarbonoclastic organisms in soil follows Liebig's "Law of the Minimum" in that a limitation in any one of these environmental factors will affect the overall degradation rate of petroleum hydrocarbons.

Once the environmental conditions have been optimized, the cell density of the indigenous microbiota will increase. If bioaugmentation is practiced by adding a microbial inoculum, any one of the limiting environmental factors could prevent the success of the seeded organisms.

Although bioremediation of petroleum hydrocarbons is not a new approach in the cleanup of hazardous waste, the *Exxon Valdez* oil spill in Prince William Sound, Alaska has created a boom in the bioaugmentation business of hazardous waste cleanup. Many companies are attempting to sell 'superbugs.' Despite the fact that many of these products are not effective, some companies have marketed

Table 16.8.   Success of Bioaugmentation in Decomposition of Petroleum Hydrocarbons

Nutrients	Remarks	References
Ekolo-Gest and *DBC-Bacteria*.	Ineffective in promoting oil biodegradation in seawater.	158
No single microorganism possesses the enzymatic capacity to metabolize all compounds found in oil.	Problems in method of application (viz., wet slurry, dry powder, pelletization or aerosol).	159
Inoculation of specialized microbial species may not be effective in degradative activity.	Concentration of target compound required to support activity may be be limiting. High density inocula may be grazed by predators. Adequate mixing of the organism with the pollutant is difficult in the subsurface environment.	93
Intensity of hydrocarbon degradation by inoculated *Candida* spp. decline gradually with time.	Effectiveness on hydrocarbons was highly dependent on the yeast species and nature of the hydrocarbon itself. Introduction of certain yeast species stimulated degradation of aromatic hydrocarbons.	160
First genetically modified oil-degrading bacterium to be patented by A. Chakrabarty has not been used in the field.	It has no real value in complete cleanup of oil.	161
Yeast addition accelerated hydrocarbon degradation by supplying additional nutrients.	Yeast autolysate promoted the growth of hydrocarbon-decomposing bacteria.	85
Degradation activity was enhanced by the addition of seeded microbes to a landfarm.	In addition to inoculation, moisture and the addition of N and P promoted degradative activity.	102

continued

Table 16.8. *continued*

Nutrients	Remarks	References
Addition of microorganisms originating from a microcosm reactor might be a good seed inoculum.	Biorestoration should also involve stimulation of the indigenous microbes.	89
A 'superblend' of over 150 types of hydrocarbon-oxidizing microorganisms with a fertilizer-based catalyst (*Alpha Environment Inc.*) was sprayed in the Gulf of Mexico (off the coast of Texas) caused by the leaking supertanker *Mega Bora* releasing 4 million gallons of light Angolan crude oil.	Laboratory experiments revealed that the added microbes degraded 94% of the hydrocarbons in 24 h. Preliminary observations of the post-treatment oil slick indicated no adverse reactions such as oxygen depleting-algae blooms which could be lethal to marine life. Post-microbe treated oil had concentrations of 67% after 30 min, 30% after 2.5 h, 62% after 4 h and 1% after 7 h. The non-treated control had oil concentrations of 130% after 2.5 h, 120% after 4 h and 22% after 7 h.	162, 163, 164
Combination of nutrients and added microbes enhanced biodegradation of oil at a locomotive maintenance yard.	Isolates demonstrated heavy growth in oil supplemented media.	155
32-acre abandoned refinery site contaminated with oil (1,500 to 30,000 ppm TPH) was treated with a consortia of microbes supplied by Solmar Corp., Orange, CA.	Over 40 years there was little to no sign of decontamination by the indigenous microbiota. However, nutrients, moisture and aeration were most likely limited. Treatment was conducted over 6 months. The area was certified as clean within one year.	165

promising products based upon scientific research. The EPA has sponsored a program that evaluates some of these new promising products, including research conducted by the National Environmental Technology Application Center (NETAC) at the University of Pittsburgh. Some promising products have been applied to sections of the contaminated beach in Alaska as a result of the *Exxon Valdez* oil spill. However, criticism has been made of this EPA study on the basis that some products were applied in amounts that exceeded those recommended for field applications. Companies selling many of these products have created a tremendous credibility gap in the biotechnology industry. However, bioremediation is now recovering from this lack of credibility by supporting scientific research.

CONCLUSIONS

An ideal bench-scale treatability study should include thorough investigations on the physical and chemical properties of the contaminated soil, microbial population dynamics, heavy metal content, and a kinetic study monitoring the depletion of TPH as well as mineralization of petroleum hydrocarbons yielding CO_2 (Table 16.9). The physicochemical properties of the contaminated soil should be characterized in terms of its pH, inorganic nitrogen (NH_4-N and NO_3-N), inorganic phosphorus (orthophosphate), salinity, and particle size analysis. An assessment of each of these parameters is necessary to optimize the biodegradation rates to initiate a bioremediation approach. Information collected on total N, P, and S will not provide useful information simply because the total amount of these elements gives no information on the immediate availability of the nutrients for microbial uptake in decomposition of oils.

Microorganisms should be enumerated by two different methods, including the total heterotrophic bacterial count, using a broad spectrum medium, such as nutrient agar, and hydrocarbon-oxidizing population with a narrow spectrum medium. The population of hydrocarbon-oxidizing organisms cannot be quantified on a solid matrix with the use of agar as a solidifying agent. A liquid medium is required with basal salts and the specific hydrocarbon contaminant added directly to the medium. The ratio of the population of hydrocarbon-utilizing organisms to the total bacterial count provides useful information on the potential of a bioremediation operation. If this percentage falls above 3%, the soil has a relatively high population of organisms capable of attacking petroleum hydrocarbons. The heavy metals within the soil should be assessed since the presence of trace elements are known to inhibit biodegradation of petroleum hydrocarbons. Trace elements that should be of concern include Pb associated with leaded gasoline, as well as Ba, V, Ni, Hg, and Cr associated with crude oil. The kinetic study should consist of monitoring the depletion of TPH for heavy oils by using shallow pans. This will account for the loss of the lighter volatile hydrocarbons as a result of volatilization. Frequently $HgCl_2$ is added

Table 16.9.   Ideal Test Parameters for a Bench-Scale Feasibility Study in Bioremediation of Petroleum Hydrocarbons

I. Physical and Chemical Properties of the Contaminated Soil
   pH
   Inorganic nitrogen
      NH_4-N
      NO_3-N
   Inorganic phosphorus (orthophosphate)
   Salinity (EC_e)
   Particle size analysis

II. Enumeration of Micororganisms
   Total heterotrophic bacterial count
   Total hydrocarbon-oxidizing population

III. Heavy Metal Content
   CAM metals (Sb, As, Ba, Be, Cd, Cr, Co, Cu, Pb, Hg, Mo, Ni, Se, Ag, Tl, V and Zn)
      or RCRA metals (As, Ba, Cd, Cr, Pb, Hg, Se and Ag)

IV. Kinetic Study
   A. Decomposition of total petroleum hydrocarbons (EPA 418.1 or modified 8015)

| | Time (weeks) | | | |
Treatment	0	2	4	6
Sterile (gamma irridation)	xx	xx	xx	xx
Unamended (moist only)	xx	xx	xx	xx
Nutrient-amended	xx	xx	xx	xx

xx = duplicates

   B. Degradation of petroleum hydrocarbons yielding CO_2 as the final product of decomposition to determine the optimum load of nutrients to add to soil

| | Time (weeks) | | | | | | |
Treatment	0	1	2	3	4	5	6
Sterile (gamma irridation)	xxx	xxx	xxx	xxx	xxx	xxx	xxx
Unamended (moist only)	xxx	xxx	xxx	xxx	xxx	xxx	xxx
Nitrogen (250 ppm)	xxx	xxx	xxx	xxx	xxx	xxx	xxx
Phosphorus (25 ppm)	xxx	xxx	xxx	xxx	xxx	xxx	xxx
Nitrogen/phosphorus (100/10 ppm)	xxx	xxx	xxx	xxx	xxx	xxx	xxx
Nitrogen/phosphorus (250/25 ppm)	xxx	xxx	xxx	xxx	xxx	xxx	xxx
Nitrogen/phosphorus (500/50 ppm)	xxx	xxx	xxx	xxx	xxx	xxx	xxx

xxx = triplicates

as a biocide to the control to hinder biodegradation. Other treatments should consist of the application of water alone and nutrient amendments. Mineralization of the oil can be monitored in a closed microcosm reactor. One should account for mass balances with the loss of VOCs as well as sorption directly to soil particles. The amount of CO_2 released gives a reliable index on the amount of petroleum hydrocarbons degraded. Treatments which should be considered for the microcosm study include a sterile control, an unamended soil treated with moisture only, the application of N alone and P alone, and the combined application of N and P at various rates (e.g., 100 to 10, 250 to 25, and 500 to 50 ppm N to ppm P).

Once the optimum environmental conditions have been characterized, biostimulating amendments can be applied directly onsite in the field. The laboratory study will provide the necessary information to make bioremediation the most cost-effective remedial approach with complete destruction of the contaminant. The time saved upon bioremediation with the recommendations based on this laboratory study will be well worth the expenditures for the up-front costs. The credibility of a bioremediation approach can be greater with a bench-scale treatability study in showing that the mechanism of disappearance of TPH is through biodegradation. Often regulatory agencies require some type of preliminary assessment in remediation before a full-scale operation. A well-planned bench-scale treatability study would satisfy this requirement.

REFERENCES

1. Ezzell, C. M. "Promising Biotechnology Solutions to Environmental Cleanup Face Resistance," *BioWorld.* 2:1–3 (1990).
2. Berstein, K. "Bioremediation's Use of Microbes May Hold Key to Toxic Waste Clean Up," *BioWorld.* 2(52):1–4 (1990).
3. Bell, C. E., P. T. Kostecki, and E. J. Calabrese. "An Update on a National Survey of State Regulatory Policy: Cleanup Standards," in P. T. Kostecki and E. J. Calabrese, Eds., *Petroleum Contaminated Soils,* Vol. 3, (Chelsea, MI:Lewis Publishers, Inc., 1990), pp. 49–72.
4. Goldstein, B. D. "Risk Assessment and Risk Management," *Environ. Toxicol. Chem.* 4:1–2 (1985).
5. Huddleston, R. L. "Solid-waste Disposal: Landfarming," *Chem. Eng.* 86:119–124 (1979).
6. Raymond, R. L., J. O. Hudson, and V. W. Jamison. "Oil Degradation in Soil," *Appl. Environ. Microbiol.* 31:522–535 (1976).
7. Francke, H. C., and F. E. Clark. "Disposal of Oil Wastes by Microbial Assimilation," Report Y-1934. U.S. Atomic Energy Commission, Washington, D.C (1974).
8. Loehr, R. C., J. H. Martin, Jr., E. F. Neuhauser, R. A. Norton, and M. R. Malecki. "Land Treatment of an Oily Waste: Degradation, Immobilization and Bioaccumulation," EPA/600/2-85/009. U.S. Environmental Protection Agency, Ada, Oklahoma, 1985.
9. Streebin, L. E., J. M. Robertson, H. M. Schornick, P. T. Bowen, K. M. Bagawandoss, A. Habibafshar, T. G. Sprehe, A. B. Callender, C. J. Carpenter, and V. G. McFarland. "Land Treatment of Petroleum Refinery Sludges," EPA-600/2-84-193. U.S. Environmental Protection Agency, Ada, Oklahoma, 1984.

10. California Department of Health Services. "Biological Remediation of a Fuel Contaminated Soil Site in Carson, California," in Remedial Technology Demonstration Project Report. Sacramento, CA, 1990.

11. Dibble, J. T., and R. Bartha. "Effect of Environmental Parameters on the Biodegradation of Oil Sludge," *Appl. Environ. Microbiol.* 37:729–739 (1979).

12. Scozo, E. R., and J. J. M. Staps. "Review of Biological Soil Treatment Techniques in the Netherlands," in K. Wolf, W. J. van den Brink, and F. J. Colon, Eds., *Contaminated Soil,* Vol. 1 (Dordrecht, Germany: Kluwer Academic Publishers, 1988), pp. 63–67.

13. de Kreuk, J. F., and G. J. Annokkee. "Applied Biotechnology for Decontamination of Polluted Soils," in K. Wolf, W. J. van den Brink, and F. J. Colon, Eds., *Contaminated Soil,* Vol. 1 (Dordrecht, Germany: Kluwer Academic Publishers, 1988), pp. 679–686.

14. Jensen, V. "Decomposition of Oily Wastes in Soil," in *Biodegradation et Humification: Rapport du ler Colloque International* (France, 1975).

15. Dotson, G. K., R. B. Dean, W. B. Cooke, and B. A. Kennar. "Land Spreading, A Conserving and Non-polluting Method of Disposing of Oily Wastes," in *International Association on Water Pollution Resources,* Vol. 1, (New York:Pergamon Press, 1970), pp. 11–36/1- 11–36/15.

16. Snyder, H. J., Jr., G. B. Rice, and J. J. Skujins. "Disposal of Waste Oil Re-Refining Residues by Land Farming," EPA/600/9-76-015. U.S. Environmental Protection Agency, 1976.

17. Fogel, S., M. Findlay, and A. Moore. "Enhanced Bioremediation Techniques for In Situ and Onsite Treatment of Petroleum Contaminated Soils and Groundwater," in P. T. Kostecki and E. J. Calabrese, Eds., *Petroleum Contaminated Soils,* Vol. 2, (Chelsea, MI:Lewis Publishers, Inc., 1989), pp. 201–209.

18. Kincannon, C. B. "Oily Waste Disposal by Soil Cultivation Process," EPA-R2-72-110, U.S. Environmental Protection Agency, Washington, D.C., 1972.

19. California Department of Health Services. *Leaking Underground Fuel Tank (LUFT) Field Manual: Guidelines for Site Assessment, Cleanup, and Underground Storage Tank Closure,* Sacramento, CA , 1989.

20. Bartha, R., and R. M. Atlas. "The Microbiology of Aquatic Oil Spills," *Adv. Appl. Microbiol.* 22:226–266 (1977).

21. Aida, T., and K. Yamaguchi, K. "Studies on the Utilization of Hydrocarbons by Yeasts," *Agri. Biol. Chem.* 33:1244–1250 (1969).

22. Liu, D. L. S. Microbial Degradation of Oil Pollution, Workshop, 1972 Louisiana State University Publication Number LSU-SG-73-01, 1973, pp. 95–104.

23. Calder, J. A., and J. Lader. "Effect of Dissolved Aromatic Hydrocarbons on the Growth of Marine Bacteria in Batch Culture," *Appl. Environ. Microbiol.* 32:95–101 (1976).

24. Atlas, R. M., E. A. Schofield, F. A. Morelli, and R. E. Cameron. "Effects of Petroleum Pollutants on Arctic Microbial Populations," *Environ. Pollut.* 10:35–43 (1976).

25. Young, L. Y., and R. Mitchell. "Negative Chemotaxis of Marine Bacteria to Toxic Chemicals," *Appl. Microbiol.* 25:972–975 (1973).

26. Baker, J. M. "Growth Stimulation Following Oil Pollution," in *The Ecological Effects of Oil Pollution on Littoral Communities: Proceedings of a Symposium Organized by the Institute of Petroleum and held at the Zoological Society of London.* (Institute of Petroleum, London, 1971), pp. 72–77.

27. Huseinov, D. M. "The Influence of Organic Compounds of Petroleum Origin Upon the Growth of Roots and Crop Capacity of Agricultural Plants," in *7th International Congress of Soil Science,* Madison, WI, 1960, pp. 253–259.

28. Carr, R. H. "Vegetative Growth in Soils Containing Crude Petroleum," *Soil Sci.* 8:67–68 (1919).

29. Plice, M. J. "Some Effects of Crude Petroleum on Soil Fertility," *Soil Sci. Soc. Am. Proc.* 13:413–416 (1948).

30. Schwendinger, R. B. "Reclamation of Soil Contaminated by Oil," *J. Inst. Petrol.* 54:182–197 (1963).

31. Frankenberger, W. T., Jr., and J. B. Johanson. "Influence of Crude Oil and Refined Petroleum Products on Soil Dehydrogenase Activity," *J. Environ. Qual.* 11:602–607 (1982).

32. Haanstra, L., and P. Doelman. "Glutamic Acid Decomposition as a Sensitive Measure of Heavy Metal Pollution in Soil," *Soil Biol. Biochem.* 16:595–600 (1984).

33. Bhuiya, M. R. H., and A. H. Cornfield. "Effects of Addition of 1000 ppm Cu, Ni, Pb and Zn on Carbon Dioxide Release During Incubation of Soil Alone and After Treatment with Straw," *Environ. Pollut.* 3:173–177 (1972).

34. Tyler, G. "Heavy Metals Pollute Nature, May Reduce Productivity," *Ambio* 1:52–59 (1972).

35. Jensen, V. "Effects of Lead on Biodegradation of Hydrocarbons in Soil," *Oikos* 28:220–224 (1977).

36. Tornabene, T. G., and H. W. Edwards. "Microbial Uptake of Lead," *Science* 176:1334–1335 (1972).

37. Vazquez-Duhalt, R. "Environmental Impact of Used Motor Oil," *Sci. Total Environ.* 79:1–23 (1989).

38. Bossert, I., and R. Bartha. "The Fate of Petroleum in Soil Ecosystems," in R. M. Atlas, Ed., *Petroleum Microbiology* (New York: Macmillan Publishing Company, 1984), pp. 546–574.

39. Sohngen, N. L. "Uber Bakterien, welche Methan als Kohlenstoffnahrung Energiequelle genrauchen," *Zentr. Bakt. Parasitenk,* Abt. II, 15:513–517 (1906).

40. Kaserer, H. "Uber die Oxydation des Wassertstoffes und des Methans durch Mikroorganismen. *Zentr. Bakt. Parasitenk,* Abt. II, 15:573–576 (1906).

41. Sohngen, N. L. "Benzin, Petroleum Paraffinol und Paraffin als Kohlenstoff und Energiequelle fur Mikroben," *Zentr. Bakt. Parasitenk,* Abt. II, 37:595–609(1913).

42. Bushnell, L. D., and H. F. Haas. "The Utilization of Certain Hydrocarbons by Microorganisms," *J. Bact.* 41:653–672 (1941).

43. Davis, J. B., R. L. Raymond, and J. P. Stanley. "Areal Contrasts in the Abundance of Hydrocarbon Oxidizing Microbes in Soils," *Appl. Microbiol.* 7:156–165 (1959).

44. Ehlers, S. E., and H. G. Hendrick. "Detection and Identification of Bacteria Utilizing Petroleum Waste in Dilution Waters," *Dev. Ind. Microbiol.* 13:283–289 (1972).

45. Walker, J. D., and R. R. Colwell. "Microbial Ecology of Petroleum Utilization in Chesapeake Bay," in API/EPA/USCG Conference on Prevention and Control of Oil Spills: American Petroleum Institute, Washington, D.C. (1973).

46. Seki, H. "Silica Gel Medium for Enumeration of Petroleumlytic Microorganisms in the Marine Environment," *Appl. Environ. Microbiol.* 26:318–320 (1973).

47. Walker, J. D., and R. R. Colwell. "Factors Affecting Enumeration and Isolation of Actinomycetes from Chesapeake Bay and Southeastern Atlantic Ocean Sediments," *Mar. Biol.* 30:193–201 (1975).

48. Gunkel, W., and H. H. Trekel. "On the Method of Quantitative Determination of Oil-Decomposing Bacteria in Oil-Polluted Sediments and Soils, Oil-Water Mixtures, Oils and Tarry Substances," *Helgol. Wiss. Meeresunters.* 16:336–348 (1967).

49. Mulkins-Philips, G. J., and J. E. Stewart. "Distribution of Hydrocarbon-Utilizing Bacteria in Northwestern Atlantic Waters and Coastal Sediments," *Can. J. Microbiol.* 20:955–962 (1974).

50. ZoBell, C. E., and J. F. Prkop. "Microbial Oxidation of Mineral Oils in Bacteria Bay Bottom Deposits," *Z. Allg. Mikrobiol.* 6:143–162 (1966).

51. Walker, J. D., and R. R. Colwell. "Measuring the Potential Activity of Hydrocarbon-Degrading Bacteria," *Appl. Environ. Microbiol.* 31:189–197 (1976).

52. Mills, A. L., C. Breuil, and R. R. Colwell. "Enumeration of Petroleum-Degrading Marine and Estuarine Microorganisms by the Most Probable Number Method," *Can. J. Microbiol.* 24:552–557 (1978).

53. Stanier, R. Y. "Studies on Marine Agar-Digesting Bacteria," *J. Bacteriol.* 42:527–558 (1941).

54. Veldkamp, H. "A Study of Two Marine Agar-Decomposing, Facultatively Anaerobic Myxobacteria," *J. Gen. Microbiol.* 26:331–342 (1961).

55. Wyndham, R. C., and J. W. Costerton. "Heterotrophic Potentials and Hydrocarbon Biodegradation Potentials of Sediment Microorganisms Within the Athabasca Oil Sands Deposit," *Appl. Environ. Microbiol.* 1:783–790 (1981).

56. Balkwill, D. L., and W. C. Ghiorse. "Characterization of Subsurface Bacteria Associated with Two Shallow Aquifers in Oklahoma," *Appl. Environ. Microbiol.* 50:580 (1985).

57. Federle, T. W., D. C. Dobbins, J. R. Thornton-Manning, and D. D. Jones. Microbial Biomass, Activity, and Community Structure in Subsurface Soils," *Ground Water* 24:365 (1986).

58. Eastcott, L., W. Y. Shiu, and D. Mackay. "Modeling Petroleum Products in Soils," in P. T. Kostecki and E. J. Calabrese, Eds., *Petroleum Contaminated Soils* Vol. 1 (Chelsea, MI: Lewis Publishers 1978), pp. 63–80.

59. Puttman, W. "Microbial Degradation of Petroleum in Contaminated Soil-Analytical Aspects," in K. Wolf, W. J. van den Brink, and F. J. Colon, Eds., *Contaminated Soil,* (Dordrecht, Germany: Kluwer Academic Publishers, 1988), pp. 189–199.

60. Tissor, B. P., and D. Welte. "Petroleum Formation and Occurrence" 2nd Ed., (Springer Verlag: Berlin, 1984).

61. Howard, H. E. "Composition of Petroleum Middle Distillates and Methods of Analysis," Prep. Div. Pet. Chem. *American Chemical Society Aug–Sept. Meeting* 17:54–65 (1972).

62. Hayes, P. D., Jr., M. L. Bruce, and M. W. Stephens. "Test Method to Extract TPHs from Soil," *Amer. Environ. Lab.* (1990).

63. Wetherold, R. G., J. L. Randall, and K. R. Williams. "Laboratory Assessment of Potential Hydrocarbon Emissions from Land Treatment of Refinery Oily Sludges," EPA-600/2-84-108. U.S. Environmental Protection Agency, Ada, Oklahoma, 1984.

64. Jury, W. A., W. F. Spencer, and W. J. Farmer. "Behavior Assessment Model for Trace Organics in Soil: I. Model Description," *J. Environ. Qual.* 12:558–564 (1983).

65. Short, T. E. "Movement of Contaminants from Oily Wastes During Land Treatment," EPA/600/D-86/005. U.S. Environmental Protection Agency, Ada, Oklahoma, 1985.

66. McGill, W. B., M. J. Rowell, and D. W. S. Westlake. "Biochemistry, Ecology and Microbiology of Petroleum Components in Soil," in E. A. Paul and J. N. Ladd, Eds., *Soil Biochemistry* 5:229–296 (New York: Marcel Dekker, 1981).

67. Regnier, Z. R., and B. F. Scott. "Evaporation Rates of Oil Components," *Environ. Sci. Technol.* 9:469–472 (1975).

68. Shaikh, A. U., R. M. Hawk, R. A. Sims, and H. D. Scott. "Redox Potential and Oxygen Diffusion Rate as Parameters for Monitoring Biodegradation of Some Organic Wastes in Soil," *Nucl. Chem. Waste Management* 5:337–343 (1985).

69. Kleintjens, R. H., K. Ch. A. M. Luyben, M. A. Bosse, and L. P. Velthuisen. "Process Development for Biological Soil Decontamination in a Slurry Reactor," in O. M. Neijssel, R. R. van der Meer and K. E. Luyben, Eds., *Proc. 4th European Congress on Biotechnology,* Vol. 1 (Amsterdam: Elsevier Sci. Publ., 1987), pp. 252–255.

70. Bachman, A., and A. J. B. Zehnder. "Engineering Significance of Fundamental Concepts in Xenobiotics Biodegradation in Soil," in K. Wolf, W. J. van den Brink, and F. J. Colon, Eds., *Contaminated Soil,* Vol. 1 (Dordrecht, Germany: Kluwer Academic Publishers, 1988).

71. Atlas, R. M., and R. Bartha. "Degradation and Mineralization of Petroleum by Two Bacteria Isolated from Coastal Waters," *Biotechnol. Bioeng.* 14:297–308 (1972).

72. ZoBell, C. E. "Action of Microorganisms on Hydrocarbons," *Bacteriol. Rev.* 10:1–49 (1946).

73. Stone, R. W., M. R. Fenske, and A. G. C. White. "Bacteria Attacking Petroleum and Oil Fractions," *J. Bacteriol.* 44:169–178 (1942).

74. Gibbs, C. F., and S. J. Davis. "The Rate of Microbial Degradation of Oil in a Beach Gravel Column," *Microbial Ecol.* 3:55–64 (1976).

75. Soghngen, N. L. "Sur le role du methanc dans la vie organique," *Rec. Trav. Chim.* 29:238–274 (1910).

76. Buttner, H. "Zur Kenntnis der Mykobakterien, Inabesondere ihres Quantitativen Stoffwecchsels auf Paraffinnahrobed," *Arch. Hyg.* 97:12–27 (1926).

77. Dobson, A. L., and H. A. Wilson. "Respiration Studies on Soil Treated with Some Hydrocarbons," *Soil Sci. Soc. Proc.* 28:536–539 (1964).

78. Rowell, M. J. "Restoration of Oil Spills on Agricultural Soils," Proceedings on the Environmental Effects of Oil and Salt Water Spills on Land. Banff, Alberta, 1975, pp. 250–276.

79. Sandvik, S., A. Lode, and T. A. Pederson. "Biodegradation of Oily Sludge in Norwegian Soils," *Appl. Microbial. Biotech.* 23:297–301 (1986).

80. Stone, R. W., A. G. C. White, and M. R. Fenske. "Microorganisms Attacking Petroleum and Petroleum Fractions," *J. Bact.* 39:91 (1940).

81. McCarty, P. L., B. E. Rittmann, and M. Reinhard. "Processes Affecting the Movement and Fate of Trace Organics in the Subsurface Environment," in T. Asano, Ed., *Artificial Recharge of Groundwater* (Boston: Butterworth Inc., 1985), pp. 627–646.

82. Antoniewski, J., and R. Schaeffer. "Recherches sur les Reactions de Coenoses Microbiennes des Sols Impregnes par des Hydrocarbures: Modification de l'activite Respiratoire," *Ann. Inst. Pasteur* 123:805–819 (1972).

83. Verstraete, W., R. Vanloocke, R. de Borger, A. Verlinder. "Modelling of the Breakdown and the Mobilization of Hydrocarbons in Unsaturated Soil Layers," Proceedings of the International Biodegradation Symposium, 1975, Kingston, Rhode Island (London: Applied Science Publ., 1976), pp. 99–112.

84. Song, H.-G., X. Wang, and R. Bartha. "Bioremediation Potential of Terrestrial Fuel Spills," *Appl. Environ. Microbiol.* 56:652–656 (1990).

85. Lehtomaki, M., and S. Niemela. "Improving Microbial Degradation of Oil in Soil," *Ambio.* 4:126–129 (1975).

86. Chin, C., G. Hicks, and C. A. Geisler. "Bio-Oxygen Stabilization Using Hydrogen Peroxide," *Water Pollut. Control Fed. J.* 45:283–291 (1973).

87. Claus, D., and N. Walker. "The Decomposition of Toluene by Soil Bacteria," *J. Gen. Microbiol.* 36:107–122 (1964).

88. Schwefer, H. J. "Latest Development of Biological In Situ Remedial Action Techniques, Portrayed by Examples from Europe and USA," in K. Wolf, W. J. van den Brink, and F. J. Colon, Eds., *Contaminated Soil* (Dordrecht: Kluwer Academic Publishers, 1988), pp. 687–694.

89. Satijn, H. M. C., and P. A. de Boks. "Biorestoration, A Technique for Remedial Action on Industrial Sites," in K. Wolf, W. J. van den Brink, and F. J. Colon, Eds., *Contaminated Soil* (Dordrecht: Kluwer Academic Publishers, 1988), pp. 745–753.

90. Vestal, J. R. "The Metabolism of Gaseous Hydrocarbons by Micro-Organisms," in R. M. Atlas, Ed., *Petroleum Microbiology* (London: Macmillan Publ., 1984), pp. 129–152.

91. Klug, M. J., and A. J. Markovetz. "Utilization of Aliphatic Hydrocarbons by Microorganisms," *Adv. Microb. Phys.* 5:1–43 (1971).

92. Hou, C. T. "Microbial Transformation of Important Industrial Hydrocarbons," in J. P. Rosazzo, Ed., *Microbial Transformations of Bioactive Compounds* (Boca Raton, FL: CRC Press, 1982).

93. Thomas, J. M., M. D. Lee, P. B. Bedient, R. C. Borden, L. W. Canter, and C. H. Ward, "Leaking Underground Storage Tanks: Remediation with Emphasis on in situ Biorestoration," EPA/600/2-87/008. U.S. Environmental Protection Agency, Ada, Oklahoma, 1987.

94. Atlas, R. M., "Stimulated Petroleum Biodegradation," *Crit. Rev. Microbiol.* 5:371–386 (1977).

95. Frankenberger, W. T., Jr., K. D. Emerson, and D. W. Turner. "In Situ Bioremediation of an Underground Diesel Fuel Spill: A Case History," *Environ. Management* 13:325–332 (1989).

96. Gersberg, R. M., W. J. Dawsey, and H. F. Ridgway, "Biodegradation of Dissolved Aromatic Hydrocarbons in Gasoline-Contaminated Groundwaters Using Denitrification," in E. J. Calabrese and P. T. Kostecki, Eds., *Petroleum Contaminated Soils,* Vol. 2 (Chelsea, MI:Lewis Publishers, Inc., 1989), pp. 211–217.

97. Gibson, D. T. "The Microbial Oxidation of Aromatic Hydrocarbons," *Crit. Rev. Microbiol.* 1:199–223 (1971).

98. Beam, H. W., and J. J. Perry, "Microbial Degradation of Cycloparaffinic Hydrocarbons via Co-Metabolism and Commensalism," *J. Gen. Microbiol.* 82:163–169 (1974).

99. Loynachan, T. E. "Low-Temperature Mineralization of Crude Oil in Soil," *J. Environ. Qual.* 7:494–500 (1978).

100. Lawes, B. C. "Soil-Induced Decomposition of Hydrogen Peroxide: Preliminary Findings," in P. T. Kostecki and E. J. Calabrese, Eds., *Petroleum Contaminated Soils,* Vol. 3 (Chelsea, MI: Lewis Publishers, Inc., 1989), pp.239–249.

101. Dibble, J. T., and R. Bartha. "Effect of Environmental Parameters on the Biodegradation of Oil Sludge," *Appl. Environ. Microbiol.* 37:729–739 (1979).

102. Verheul, J. H. A. M., R. van den Berg, and D. H. Eikelboom. "In Situ Biorestoration of a Subsoil, Contaminated with Gasoline," in K. Wolf, W. J. van den Brink, and F. J. Colon, Eds., *Contaminated Soil* (Dordrecht: Kluwer Academic Publishers, 1988), pp. 705–715.

103. Traxler, R. W. "Microbial Degradation of Asphalt," *Biotechnol. Bioeng.* 4:369–376 (1962).

104. Hambrick, G. A., III, R. D. DeLaune, and W. H. Patrick, Jr. "Effect of Estuarine Sediment pH and Oxidation-reduction Potential on Microbial Hydrocarbon Degradation," *Appl. Environ. Microbiol.* 40:365–369 (1980).

105. ZoBell, C. E., C. W. Grant, and H. F. Haas. "Marine Microorganisms Which Oxidize Petroleum Hydrocarbons," *Bull. Am. Assoc. Petroleum Geol.* 27:1175–1193 (1943).

106. Strawinski, R. J. "The Dissimilation of Pure Hydrocarbons by Members of the Genus, Pseudomonas," Dissertation, Pennsylvania State College, State College, PA, 1943.

107. Lynch, J., and B. R. Genes. "Land Treatment of Hydrocarbon Contaminated Soils," in P. T. Kostecki and E. J. Calabrese, Eds., *Petroleum Contaminated Soils,* Vol. 1 (Chelsea, MI: Lewis Publishers, Inc., 1989), pp. 163–174.

108. Ying, A., J. Duffy, G. Shepherd, and D. Wright. "Bioremediation of Heavy Petroleum Oil in Soil at a Railroad Maintenance Yard," in P. T. Kostecki and E. J. Calabrese, Eds., *Petroleum Contaminated Soils,* Vol. 3 (Chelsea, MI: Lewis Publishers, Inc., 1989), pp. 227–238.

109. Curl, H., Jr., and K. O'Connell. "Chemical and Physical Properties of Refined Petroleum Products," NOAA Technical Memorandum ERL MESA-17, 1977, pp. 135.

110. Gutnick, D. L., and E. Rosenberg. "Oil Tankers and Pollution: A Microbiological Approach," *Ann. Rev. Microbiol.* 31:379–396 (1977).

111. Wilson, J. T., C. G. Enfield, W. J. Dunlap, R. L. Cosby, D. A. Foster, and L. B. Baskin. "Transport and Fate of Selected Organic Pollutants in a Sandy Soil," *J. Environ. Qual.* 10:501–506 (1981).

112. Hillel, D. "Movement and Retention of Organics in Soil: A Review and a Critique of Modeling," in P. T. Kostecki and E. J. Calabrese, Eds., *Petroleum Contaminated Soils,* Vol. 1 (Chelsea, MI: Lewis Publishers, Inc., 1989), pp. 81–86.

113. Bauman, B. J. "Soils Contaminated by Motor Fuels: Research Activities and Perspectives of the American Petroleum Institute," in P. T. Kostecki and E. J. Calabrese, Eds., *Petroleum Contaminated Soils,* Vol. 1 (Chelsea, MI: Lewis Publishers, Inc., 1989), pp. 3–17.

114. Harmsen, J., P. Bouter, F. Bransen, and G. J. Versteeg. "Adsorption of Benzene, Toluene and Xylenes in Soil Material," in K. Wolf, W. J. van den Brink, and F. J. Colon, Eds., *Contaminated Soil* (Dordrecht: Kluwer Academic Publishers, 1988), pp. 109–111.

115. van der Linden, A. C. "Degradation of Oil in the Marine Environment," in R. J. Watkinson, Ed., *Developments in Biodegradation of Hydrocarbons* (London: Applied Science Publ. Ltd., 1978), pp. 165–200.

116. Atlas, R. M., and R. Bartha. "Biodegradation of Petroleum in Seawater at Low Temperatures," *Can. J. Microbiol.* 18:1851–1855 (1972).

117. ZoBell, C. E. "Microbial Modification of Crude Oil in the Sea," in *Proceedings, Joint Conference on Prevention and Control of Oil Spills,* A.P.I.F.W.P.C.A. (New York: American Petroleum Institute Publication, 1969), pp. 317–326.

118. ZoBell, C. E., and J. Agosti. "Bacterial Oxidation of Mineral Oils at Sub-zero Celsius," *Bacteriol. Proc.* E-11 (1972).

119. Ludzack, F. L., and D. Kinkead. "Persistence of Oil Wastes in Polluted Water Under Aerobic Conditions," *Ind. Eng. Chem.* 48:263–267.

120. Gunkel, W. "Experimentell-okologische Untersuchungen Uber Die Limitierenden Faktoren des Mikrobiellen Olabbaues in Marinen Milieu," *Helgolaender wiss. Meeresunters.* 15:210–225 (1967).

121. Kinney, P. J., D. K. Button, and D. M. Schell. "Kinetics of Dissipation and Biodegradation of Crude Oil in Alaska's Cook Inlet," in *Proceedings, Joint Conference on Prevention and Control of Oil Spills,* A.P.I.F.W.P.C.A. (New York: American Petroleum Institute Publication, 1969), pp. 333–340.

122. Klug, M. J., and A. J. Markowetz. "Thermophilic Bacterium Isolated on n-Tetradecane," *Nature* 215:1082–1083 (1967).

123. Mateles, R. I., J. N. Baruah, and S. R. Tannenbaum. "Growth of a Thermophilic Bacterium on Hydrocarbons: A New Source of Single-Cell Protein," *Science* 157:1322–1323 (1967).

124. Walker, J. D., and R. R. Colwell. "Microbial Degradation of Model Petroleum of Low Temperatures," *Microbial Ecol.* 1:63–95 (1974).

125. Westlake, D. W. S., A. Jobson, R. Phillippe, and F. D. Cook. "Biodegradability and Crude Oil Composition," *Can. J. Microbiol.* 20:915–928 (1974).

126. Zajic, J. E., B. Supplisson, and B. Volesky. "Bacterial Degradation and Emulsification of No. 6 Fuel Oil," *Environ. Sci. Technol.* 8:664–668 (1974).

127. Andreesen, J. R., M. Nagel, I. Sigmund, K. Konig, and W. Freudenberg. "Prerequisites for Bacterial Degradation of N-,S, and O- heterocyclic, Aromatic Compounds," in K. Wolf, W. J. van den Brink, and F. J. Colon, Eds., *Contaminated Soil* (Dordrecht: Kluwer Academic Publishers, 1988), pp. 755–757.

128. Cerniglia, C. E., and J. J. Perry. "Crude Oil Degradation by Microorganisms Isolated from the Marine Environment," *Zeitschrift fur Allg. Mikrobiologie* 13:199–306 (1973).

129. Kvasnikov, E. I., A. M. Zhuravel, and T. M. Klyushnikova. "Thermophilic Aerobic Spore-Forming, Hydrocarbon-Assimilating Oligonitrophilic Bacteria and Their Distinctive Physiological Features," *Microbiology* 40:443–449 (1974).

130. Gudin, C., and W. J. Syratt. "Biological Aspects of Land Rehabilitation Following Hydrocarbon Contamination," *Environ. Pollut.* 8:107–112 (1975).

131. Jobson, A., F. D. Cook, and D. W. S. Westlake. "Microbial Utilization of Crude Oil," *Appl. Microbiol.* 23:1082–1089 (1972).

132. Beerstecher, E., Jr. *Petroleum Microbiology* (Houston, TX: Elsevier Press, 1954).

133. Huddleston, R. L., and L. W. Cresswell. "Environmental and Nutritional Constraints of Microbial Hydrocarbon Utilization in the Soil," in *Proceedings of 1975 Engineering Foundation Conference: The Role of Microorganisms in the Recovery of Oil* (NSF/RANN, Washington, 1976), pp. 71–72.

134. Atlas, R. M., A. Sexstone, P. Gustin, O. Miller, P. Linkins, and K. Everett. "Biodegradation of Crude Oil by Tundra Soil Microorganisms," in T. S. Oxley, D. Allsop, and G. Becker, Eds., *Biodeterioration: Proceedings of the 4th International Symposium, Berlin* (London, 1978), pp. 21–28.

135. Sexstone, A., K. Everett, T. Jenkins, and R. M. Atlas. "Fate of Crude and Refined Oils in North Slope Soils," *Arctic* 31:339–347 (1978).

136. Sexstone, A., P. Gustin, and R. M. Atlas. "Long-term Interactions of Microorganisms and Prudhoe Bay Crude Oil in Tundra Soils at Barrow, Alaska," *Arctic* 31:348–354 (1978).

137. Atlas, R. M., and R. Bartha. "Abundance, Distribution and Oil Biodegradation Potential of Microorganisms in Raritan Bay," *Environ. Pollut.* 4:291–300 (1973).

138. Colwell, R. R., A. L. Mills, J. D. Walker, P. Garcia-Tello, and V. Campos. "Microbial Ecology Studies of the Metula Spill in the Straits of Magellan," *J. Fish. Res. Board Can.* 35:573–580 (1978).

139. Waksman, S. A. "Influence of Microorganisms Upon the Carbon: Nitrogen Ratio in the Soil," *J. Agric. Sci.* 14:555–562 (1924).

140. Thompson, L. M., C. A. Black, and J. A. Zoellner. "Occurrence and Mineralization of Organic Phosphorus in Soils with Particular Reference to Associations with Nitrogen, Carbon, and pH," *Soil Sci.* 77:185–196 (1954).

141. Atlas, R. M., and R. Bartha. "Stimulated Biodegradation of Oil Slicks Using Oleophilic Fertilizers," *Environ. Sci. Technol.* 7:538–541 (1973).

142. Olivieri, R., P. Bacchin, A. Robertiello, N. Oddo, L. Degen, and A. Tonolo. "Microbial Degradation of Oil Spills Enhanced by a Slow-Release Fertilizer," *Appl. Environ. Microbiol.* 31:629–634 (1976).

143. Dibble, J. T., and R. Bartha. "Effect of Iron on the Biodegradation of Petroleum in Seawater," *Appl. Environ. Microbiol.* 31:544–550 (1976).

144. Gibbs, C. F. "Quantitative Studies on Marine Biodegradation of Oil. I. Nutrient Limitation at 14°C," *Proc. Royal Soc.* (London) Ser. B 188:61 (1975).

145. Davis, J. B., V. E. Farmer, R. E. Kreider, A. E. Straub, and K. M. Reese. "The Migration of Petroleum Products in Soil and Ground Water: Principles and Countermeasures," API Publication 4149, American Petroleum Institute, Washington, D.C., 1972.

146. Raymond, R. L., V. W. Jamison, and J. O. Hudson. "Final Report on Beneficial Stimulation of Bacterial Activity in Ground Water Containing Petroleum Products," Committee on Environmental Affairs, American Petroleum Institute, Washington, D.C., 1975.

147. Raymond, R. L., V. W. Jamison, J. O. Hudson, R. E. Mitchell, and V. E. Farmer. "Final Report on Field Application of Subsurface Biodegradation of Gasoline in a Sand Formation," American Petroleum Institute Project No. 307-77, Washington, D.C., 1978.

148. Jobson, A., M. McLaughlin, F. D. Cook, and D. W. S. Westlake. "Effect of Amendments on the Microbial Utilization of Oil Applied to Soil," *Appl. Microbiol.* 27:166–171 (1974).

149. Fedorak, P. M., and D. W. S. Westlake. "Degradation of Aromatics and Saturates in Crude Oil by Soil Enrichments," *Water Air Soil Pollut.* 16:367–375 (1981).

150. McGill, W. B., and M. J. Rowell. "The Reclamation of Agricultural Soils After Oil Spills," in J. A. Toogood, Ed., Alberta Institute of Pedology, Publication No. M-77-11. University of Alberta, Edmonton, 1977, pp. 69–132.

151. Hunt, P. G., W. E. Rickard, F. J. Deneke, F. R. Koutz, and R. P. Murman. "Terrestrial Oil Spills in Alaska: Environmental Effects and Recovery," in Prevention and Control of Oil Spills API/EPA-USCG. American Petroleum Institute, Washington, D.C., 1973.

152. Frankenberger, W. T., Jr. "Use of Urea as a Nitrogen Fertilizer in Bioreclamation of Petroleum Hydrocarbons in Soil," *Bull. Environ. Contam. Toxicol.* 40:66–68 (1988).

153. Frost, G. "Exxon Reinstates Its Controversial Bioremediation Effort as Year Two of the Exxon Valdez Oil Spill Cleanup Begins," *BioWorld* 102:1–5 (1990).

154. Kelso, D. D., and M. Kendziorek. "Alaska's Response to the Exxon Valdez Oil Spill," *Environ. Sci. Technol.* 25:16–23 (1991).

155. Dineen, D., J. P. Slater, P. Hicks, J. Holland, and L. D. Clendening. "In Situ Biological Remediation of Petroleum Hydrocarbons in Unsaturated Soils," in P. T. Kostecki and E. J. Calabrese, Eds., *Petroleum Contaminated Soils,* Vol. 3 (Chelsea, MI: Lewis Publishers, Inc., 1989), pp. 177–187.

156. Karlson, U., and W. T. Frankenberger, Jr. "Microbial Degradation of Benzene and Toluene in Groundwater," *Bull. Environ. Contam. Toxicol.* 43:505–510 (1989).

157. Deutsch, D. J. "Waste Treatment Boosted by Bacterial Additions," *Chem. Engineering* 86:100–102 (1979).

158. Atlas, R. M., and R. Bartha. Microbial Degradation of Oil Pollution Workshop. Louisiana State University Publication Number LSU-SG-73-01 (1973).

159. Colwell, R. R., and J. D. Walker. "Ecological Aspects of Microbial Degradation of Petroleum in the Marine Environment," *CRC Crit. Rev. Microbiol.* 5:423–445 (1977).

160. Ismailov, N. M. "Biodegradation of Oil Hydrocarbons in Soil Inoculated with Yeasts," *Microbiology* 54:670–675 (1985).

161. Roberts, L. "Discovering Microbes with a Taste for PCBs: Microbial Ecologists and Microbiologists are Finding New Organisms in the Environment with Unexpected Abilities to Degrade Toxic Chemicals," *Science* 237:975–977 (1987).

162. Gaffney, J. "Microbes Chow Down on Gulf Oil Slick," *BioWorld* 119:1–2 (1990).

163. Gaffney, J. "Study Finds That Slick-Eating Microbes Proved Their Worth in Gulf Oil Cleanup," *BioWorld* 133:1–3 (1990).

164. Gaffney, J. "Scientists Try Bioremediation on Norwegian Tanker Oil Spill," *BioWorld* 116:1–3 (1990).

165. Molnaa, B. A., and R. B. Grubbs. "Bioremediation of Petroleum Contaminated Soils Using a Microbial Consortia as Inoculum," in E. J. Calabrese and P. T. Kostecki, Eds., *Petroleum Contaminated Soils,* Vol. 2 (Chelsea MI: Lewis Publishers, Inc., 1989), pp. 219–232.

CHAPTER 17

Technological Limits of Groundwater Remediation: A Statistical Evaluation Method

William A. Tucker and **Erin F. Parker,** Environmental Science & Engineering, Inc., Gainesville, Florida

INTRODUCTION

The most commonly used groundwater restoration technology at sites where petroleum hydrocarbons have been released is extraction and aboveground treatment of the groundwater (pump-and-treat). At many sites where this technology has been used, hydrocarbon removal rates follow a relatively consistent pattern: after a period of initially steady reductions, groundwater contaminant concentrations tend to level off and remain fairly constant, with random fluctuations around a ''leveling off'' concentration, or asymptotic limit.

The asymptotic concentration levels may be higher than specified cleanup target concentrations. Figure 17.1 illustrates the phenomenon at a site undergoing pump-and-treat groundwater restoration. The data are from one of several representative data sets utilized in this study.

Based upon the observed pattern, the site remediation manager might reevaluate whether the continued operation of the existing remedial system would be cost-effective. Decisions to discontinue or modify remedial activities at a site should be made in a broad context considering regulatory standards, technological feasibility and cost-effectiveness, and site-specific exposure and risk assessment.

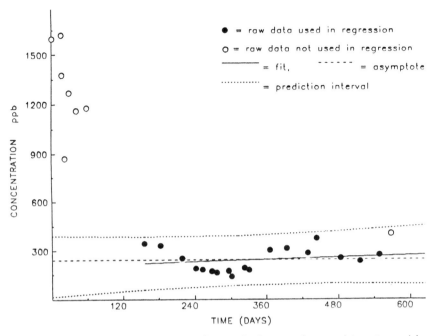

Figure 17.1.   Linear asymptotic regression results for one of seven data sets used for method evaluation.

The apparent leveling off of groundwater quality is indicative of a technological limitation on the effectiveness of the in-place remedial technology. It may be infeasible to further improve groundwater quality via the technology in use at the site.

In 1988, the American Petroleum Institute (API) funded Environmental Science & Engineering, Inc., to develop and evaluate alternative statistical procedures that can be used to document and quantify the asymptotic condition. Preliminary results of this study were published in 1989, and an API report documenting the method evaluation was published in 1991.[1,2] The second, ongoing, phase of that study is to develop personal computer (PC) software to support wider dissemination of the statistical procedures. This chapter summarizes the previous work and presents additional results and hypothetical applications based on the to-be-released PC software.

METHODS: PHASE 1

Groundwater quality data from seven sites undergoing active remediation were selected to test alternative statistical procedures. These data sets were selected from approximately 20 site data sets provided by API. The data sets represent observed concentrations of benzene in a recovery well. If the remedial objectives were based on some other contaminant, this method still could be applied.

Data from a recovery well were utilized principally because it is the most frequently sampled well at many sites undergoing remediation. Extrapolation of trends in recovery wells may not correctly predict the time to achieve acceptable quality in all monitoring wells, considering such factors as well placement, screened intervals, and the horizontal and vertical distribution of contamination in the aquifer. Nonetheless, the recovery well is the focal point of the pump-and-treat technology, and its potential limitations were evaluated statistically in this study.

The seven characteristic data sets used for method evaluation are presented in the API report.[2] An eighth data set was submitted subsequent to that work and has also been used for subsequent evaluation of the PC software. The data records range in duration from 9 to 22 months. All data sets exhibit an asymptotic region, with average concentrations of benzene at the asymptote ranging from 2 to 300 parts per billion (ppb).

Two alternative statistical techniques were evaluated in Phase 1. The first technique was based on regressing the concentration, C, versus time, t. The second technique was based on regressing the natural logarithm of concentration, lnC, versus t. For both techniques, an asymptotic condition is indicated if the slope (dC/dt) is not significantly different from zero and if the 95% upper and lower confidence bounds on the slope are both near zero. The logarithmic regression is based on the full data set, whereas the linear technique includes an objective identification of the asymptotic portion of the data set. This step for the linear case was necessary because, with a data set of the type illustrated in Figure 17.1, a flat, straight line can only be fit through data after the initial period of concentration decline.

These two alternatives were selected because it was unclear at the initiation of the study whether the apparent asymptote was actually greater than zero. Several mechanisms that have been documented to affect the fate and persistence of hydrocarbons in soil and groundwater are expected to exhibit first order kinetics.[3] If the overall restoration process during remedial action were a first-order removal process (i.e., contaminant is being eliminated from the system in direct proportion to its concentration), then the logarithmic regression would closely fit the data. A first-order process is asymptotic, but the asymptotic concentration limit should be at zero. At study initiation it was unclear whether the appearance of a nonzero asymptote simply resulted from a small trend coupled with large data variability. The results of the method evaluation indicate that this is not the case, and that a nonzero asymptote exists.

The alternative linear regression technique is expected to fit the data better if a valid nonzero asymptote exists. This would also tend to indicate that fundamentally distinct processes control the initial rapid reduction as contrasted with the subsequent asymptotic interval. For example, the initial reduction may result from the rapid removal of contamination from the aquifer by the pump-and-treat technology, whereas the asymptotic period could be attributable to the slow release of contaminant from soils, a process that is not accelerated by the pump-and-treat system.

Each technique was applied to the seven characteristic data sets. The two techniques were compared by a single criterion: which technique could demonstrate an existing asymptotic condition with the least data.

Method 1—Linear Asymptotic

Method 1 focuses on the most recent portion of the time series data set, because that is the portion of the record that is anticipated to be approaching an asymptote. This latter portion of the time series (e.g., t > 150 days in Figure 17.1) is linear with a slope near zero (dC/dt ≈ 0) and is referred to as the asymptotic region of the time series record. Method 1 selects a subset from the latter portion of the time series record and performs linear regression analysis of that subset to determine its asymptoticity. In performing this linear regression of the asymptotic region, it is assumed that the data fit an equation of the following form:

$$C = a + bt \qquad (1)$$

where: a = concentration intercept at t = 0 (ppb),
      b = slope or trend of regression line [parts per billion per day (ppb/day)], and
      t = time (days).

However, it is postulated that b is not significantly different from zero. The identification of the asymptotic region must be objective if the results of the statistical analysis are to be reproducible. An objective selection procedure was developed as follows. Sequential linear regression analyses were performed on the most recent, consecutive data records (e.g., subsets of the last five data points, then the last six data points, then the last seven data points, etc.) until the final data set regressed included all the data. Upper and lower 95% confidence intervals were calculated for the slope for each regression in the sequence. For many regressions, the upper bound was greater than zero and the lower bound was less than zero, signifying that the slope was not significantly different from zero. For each regression with a slope not different from zero, the absolute values of the upper and lower bound slopes were compared, and the larger was defined as a critical value (CV) characterizing the trendlessness of the data subset for that regression. The most trendless subset was the subset that had the smallest CV.

The uncertainty in a regression line slope decreases as more and more data points are used in a regression, assuming the additional data points approximately conform to a straight line. When data from the nonasymptotic, early region were included in regressions in the sequence, the absolute value and uncertainty in slopes increased. By finding the regression with the minimum upper bound on the slope (minimum CV), this procedure objectively selected the longest trendless subset of the data as defining the asymptotic region.

From the data set shown in Figure 17.1, the asymptotic region selected using the sequential regression procedure was the period from day 157 to day 568. The regression results involved in selecting the asymptotic region for this data set are summarized in Table 17.1. The asymptotic region is identified as the period following day 157. The regression of the data subset following day 157 has the following characteristics: (1) the apparent slope was not significantly different from zero; (2) both the best estimate of the slope and the standard error in the slope were among the smallest, indicating trendlessness; and, as a consequence, (3) regression had the smallest CV. With 95% confidence, the slope is not greater than 0.50 ppb/day.

Method 2—Exponential Regression

In Method 2, the data are assumed to be samples from an aquifer in which the concentration is declining according to the following equation:

$$C = C_o e^{-kt} \tag{2}$$

where: C_o = concentration (ppb) when t equals 0,
     e = 2.718,
     k = first-order rate constant (day^{-1}), and
     t = time in days.

In such a system, the natural logarithm of the concentration, ln C, would be expected to be linear in time:

$$\ln C = \ln C_o - kt \tag{3}$$

It was hypothesized that the complete data record could be fit to this function. With this hypothesis, the apparent asymptotic region was simply a region in which concentrations were declining slowly, but the rate of decline could not be distinguished amid the noise associated with natural fluctuations, heterogeneity, or sampling and analysis errors. Benefits of this approach would be utilization of all the available data and the potential ability to anticipate a reduction in slope as time increased.

If the underlying aquifer contamination is declining according to Equation 2, then the rate of decline decreases with time. This function is asymptotic, but the asymptote is zero. If variability in the observed concentrations is large, the trend may be obscured as the rate of decline decreases.

By regressing the natural logarithm of concentration versus time and applying classical rules of statistics and calculus,[4,5] upper and lower confidence bounds on the slope were estimated based on Equation 2.

The criterion used to compare the two methods was identification of trendlessness at the earliest time (with the least data). Given a data set like that of Site 1, which to most observers clearly exhibits an asymptotic tendency, which of these methods recognizes that the slope is insignificantly small at the earliest time (i.e., with the least data)? To answer this question, each method was applied to each

Table 17.1.   Method 1—Asymptotic Region Identification Using Linear Regression for Site 1 (Selected results; every other concluding subset).

All Data for Time Greater Than t (t = days)	Best Estimate of Slope (ppb/day)	Standard Error of Estimate of Slope (ppb/day)	95% Confidence		Critical Value (ppb/day)	Slope Significantly Different from Zero
			Lower Bound (ppb/day)	Upper Bound (ppb/day)		
445	0.67	0.89	-1.67	1.81	1.81	No
395	0.06	0.44	-0.80	0.91	0.91	No
331	0.31	0.29	-0.25	0.87	0.87	No
301	0.52	0.22	0.08	0.96	—	Yes
277	0.58	0.17	0.25	0.91	—	Yes
253	0.56	0.14	0.29	0.8	—	Yes
221	0.46	0.13	0.21	0.71	—	Yes
157	0.22	0.14	-0.06	0.50	0.50	No
42	-1.01	0.38	-1.76	-0.26	—	Yes
23	-1.41	0.35	-2.10	-0.72	—	Yes
17	-1.90	0.37	-2.63	-1.17	—	Yes

of the seven characteristic data sets using only its first five data points, then its first six data points, then its first seven data points, etc. This comparison is equivalent to evaluating the data sets for trendlessness as the data are being collected.

RESULTS: PHASE 1

Summarizing results presented in detail elsewhere,[1,2] the linear asymptotic method was superior to the exponential method for most of the seven characteristic data sets evaluated in Phase 1, i.e., it identified an asymptotic condition with less data than the exponential regression method. The exponential method assumes a zero asymptote, while most of the data sets clearly exhibited a nonzero asymptote. Presence of a nonzero asymptote has been interpreted to indicate that two distinct processes combine to produce the characteristic concentration:time patterns.[1,2] The initial decline represents the active removal of contamination from the contaminated groundwater, while the asymptotic region represents slow release from residual hydrocarbon contaminated soils or a continuing slow leak.

It was also shown[1,2] that statistical variability in data in the asymptotic region could be explained, in large part, by sampling and analytical variability, and that these sources of variability obscure precise determination of trends in the asymptotic region. The effect of analytical variations on definition of the trend were analyzed and are predictable.[1,2]

PHASE 2: ADDITION OF NONLINEAR METHOD

Results of Phase 1 suggested that an alternative nonlinear method representing an exponential decay with a nonzero asymptote may be useful. Regression analysis is used to fit a function of the form:

$$C = C_o e^{-kt} + C_f$$

The nonlinear regression procedure uses an iterative Gauss Newton method.[6]

The initial estimates of C_o and k are determined by the exponential regression procedure while the initial estimate of C_f is zero. By an iterative process, more realistic values of C_o, k, and C_f are determined that minimize the sum of squared residuals.

Development of PC Software

The nonlinear, linear asymptotic, and exponential methods are supported by a user-friendly PC software package to be released by API in 1991. The program is written in "C" with graphics displays and plotting created with the Microcompatibles Grafmatic and Printmatic function libraries. Fitted equations and data points are displayed when the regression is complete, with goodness of fit statistics. Graphic displays show plots of the data, the regression line, the asymptote, residuals, and regression prediction intervals, where applicable.

Hypothetical applications of the software are illustrated in the following section.

APPLICATIONS

Certain states have adopted or proposed rules that recognize achievement of an asymptotic limit as a basis for termination of groundwater pump-and-treat remediation systems. Specifically, the State of Florida rule 17-770, Florida Administrative Code (F.A.C.) specifies that after asymptotic conditions have been observed for one year, remediation may be halted based on a statistical evaluation that includes both the nonlinear and linear regressions described in this chapter. These procedures are applied to a hypothetical data set presented in the State of Florida guidance pursuant to that regulation as well as actual site data provided by API.

Florida Example

After 17 months of remediation (see Figure 17.2) concentrations appeared to be leveling off; for the previous 12 months the concentration of benzene in the

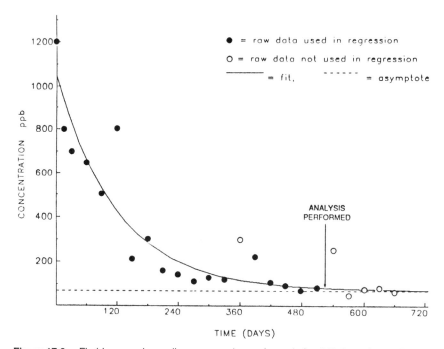

Figure 17.2.   Florida example: nonlinear regression performed after 510 days of remediation.

recovery well ranged from 70 to 300 ppb. The nonlinear regression was fit by $C = 980\ e^{-0.0085\ t} + 70$ where t is in days ($r^2 = 0.87$). Thus, the asymptotic concentration was estimated to be 70 ppb. The linear asymptotic procedure (see Figure 17.3) identified the asymptotic period as Days 210 through 510. The data from day 360 (300 ppb) was removed as an outlier following Florida guidance insofar as the concentration exceeded the mean of the asymptotic region by more than two standard deviations. Data from the asymptotic period exhibit a slope of −0.20 ppb/day which was not significantly different from zero. The average concentration in the asymptotic region was 120 ppb. The linear asymptotic regression predicted that concentrations would not exceed 200 ppb during the next quarter.

The following quarter's data failed to confirm the linear asymptotic regression: on day 540, the observed concentration of 250 is outside the prediction interval. Repeating the analysis after day 600, however, shows that the nonlinear regression yielded stable results: $C = 980\ e^{-0.0083\ t} + 65$. The linear asymptotic procedure (see Figure 17.4) identified the asymptotic period as days 420 through 600 with an asymptotic concentration of 78 ppb, a slope of −0.20, and predicted that concentrations would not exceed 100 ppb during the next quarter.

The final analysis, after day 660, confirmed the predictions made after day 600. New data fell within the prediction interval estimated after day 600. The nonlinear regression (see Figure 17.5) yielded $C = 980\ e^{-0.0083\ t} + 66$ ($r^2 = 0.88$). The linear asymptotic procedure, however, provided much stronger

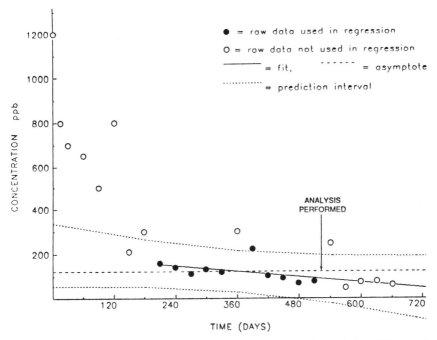

Figure 17.3. Florida example: linear asymptotic regression performed after 510 days of remediation.

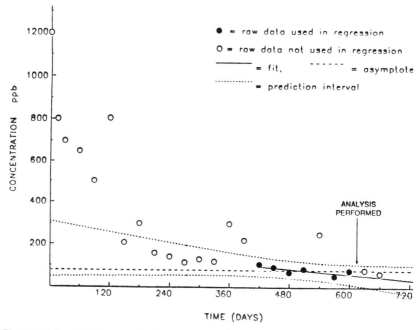

Figure 17.4.   Florida example: linear asymptotic regression after day 600.

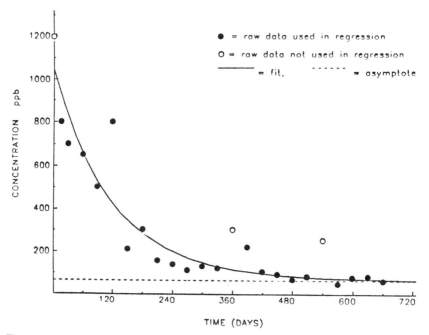

Figure 17.5.   Florida example: nonlinear regression performed after 660 days of remediation.

evidence for leveling off, attributable to the longer data record in the asymptotic region. After day 600 it yielded a slope of -0.068 ppb/day, not significantly different from zero and an asymptotic concentration of 73 ppb (see Figure 17.6). The upper bound on the difference between the 90% confidence interval and the best fit was 15 ppb. Based on this analysis, the criteria specified by Florida guidance are met. These results are similar to the example analysis provided in the Florida guidance.

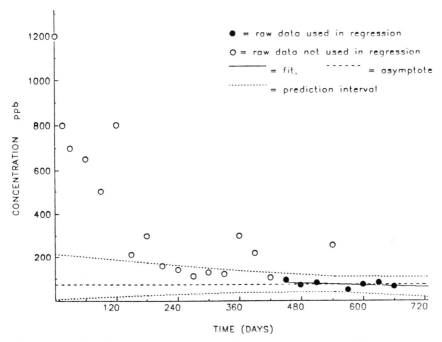

Figure 17.6.   Florida example: linear asymptotic regression after day 660.

Actual Site Data Set

The second example represents actual site data provided by API. After 570 days of pump-and-treat remediation, concentrations appeared to have leveled off. The nonlinear regression yielded C = 2300 $e^{-0.010\,t}$ + 300 (r^2 = 0.91, see Figure 17.7). The linear asymptotic procedure identified the period from 420–570 days as the asymptotic region (see Figure 17.8). The result from day 540 (900 ppb) was rejected as an outlier insofar as 900 ppb exceeded the asymptotic region mean by more than two standard deviations. The best estimate slope was -0.29 ppb/day, which was not significantly different from zero. The asymptotic concentration was estimated to be 270 ppb. At that time the linear regression predicted concentrations to remain below 300 ppb for the next quarter.

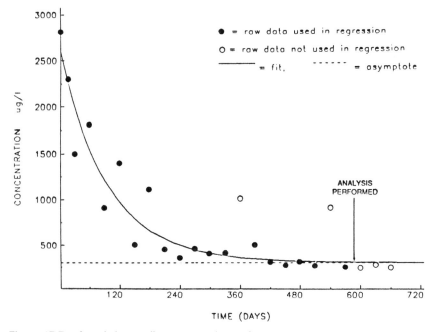

Figure 17.7. Actual site: nonlinear regression performed after 570 days of remediation.

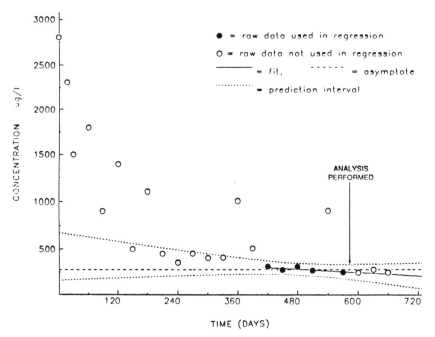

Figure 17.8. Actual site: linear asymptotic regression performed after 570 days of remediation.

The following quarter confirmed the predictions. All data fell within the prediction interval of the linear regression. Nonlinear regression of data to day 660 yielded $C = 2300 \, e^{-0.0010 \, t} + 280$. The linear asymptotic procedure confirmed that the asymptotic period began, statistically, at day 420, and estimated the slope at -0.17 ppb/day, and the asymptotic concentration at 270 ppb (see Figure 17.9). The maximum difference between the 90% confidence interval and the best fit was 19 ppb. At that time (660 days), the asymptotic analysis met the criteria established by the State of Florida and could be used to petition the state to terminate the pump-and-treat system under Rule 17-770, F.A.C.

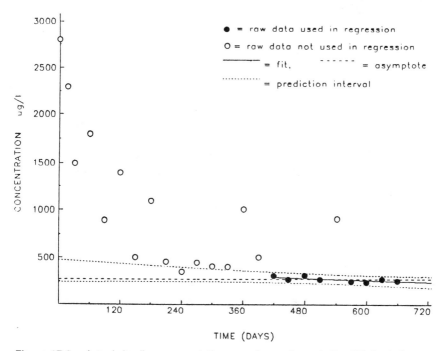

Figure 17.9.   Actual site: linear asymptotic regression performed after 660 days of remediation.

SUMMARY OF FINDINGS

Statistical regression procedures to test for an asymptotic condition in groundwater quality data records have been developed. The linear asymptotic technique defines an asymptotic condition as one in which some concluding time period in a time series data set exhibits a negligible trend or slope that is not significantly different from zero. Additionally, both the upper and lower bounds (95% confidence) on the slope should be small, so that the data is virtually trendless. As previously reported,[1,2] this method was compared with an alternative exponential regression technique over seven characteristic data sets and was found to be more efficient in identifying an asymptotic condition with less data. The weakness of the exponential regression technique appears to be related to its assumption of a zero asymptote, while most of the characteristic data sets clearly exhibit a nonzero asymptote.

In Phase 2 of this study, a nonlinear regression procedure has been developed to address the weakness of the exponential procedure. PC software has been developed to facilitate broader application of the procedures. This software, to be released by API during 1991, has been applied to two additional data sets: (1) a hypothetical data set included in a guidance document by the State of Florida, and (2) an additional data set provided by API.

Florida Rule 17-770, F.A.C. provides for conclusion of groundwater pump-and-treat remediation based on a finding that an asymptotic limit has been reached. The PC software is designed to meet the requirements of the Florida rule while also providing options for alternative analyses. A typical application of the software would include:

1. subjective determination that concentrations in a recovery well have leveled off and remained steady with random fluctuations for more than five sampling events,
2. application of the nonlinear regression procedure to verify that the data generally fit an equation of the form $C = C_o e^{-kt} + C_f$; (this conclusion can be based on r^2 or on subjective interpretation),
3. application of the linear asymptotic procedure to objectively identify the asymptotic region,
4. remove outliers and repeat steps 2 and 3 (optional),
5. use the linear asymptotic results to predict the range expected in upcoming sample results, and
6. verify the model by checking that a minimum of one calendar quarter's sampling results (not less than two subsequent data points) confirm the model's prediction.

A variety of additional applications are supported by the PC software. The linear regression can be used to estimate the time to achieve specific concentration targets. The user may subjectively identify the asymptotic region and perform linear regression analysis (distinct from the objective procedure referred to as the linear asymptotic analysis). The exponential regression analysis is still supported, but is of limited utility (authors' opinion). The first-order removal rate constant, k, calculated by the nonlinear regression has broader implications regarding the operating conditions of the remedial system and contaminant fate and transport.[1,2]

The development of the asymptotic limit statistical approach was prompted by the observation that groundwater pump-and-treat systems at petroleum contaminated sites are effective initially at reducing groundwater contaminant levels but eventually reach a technological limit of reduced effectiveness. Once an asymptotic region of a groundwater data record is defined, the average asymptotic concentration can be compared to target cleanup levels to evaluate the restoration program. When the asymptotic average is above a target cleanup concentration, a technologically limiting condition may have been reached that would signal a reevaluation of the ongoing remedial response.

Decisions about the further course of action should be based on a variety of factors: cost-effectiveness of continuing inplace technology, technical feasibility and cost-effectiveness of alternative remedial technologies, and the health and environmental risks posed by the site. In addressing the latter issue, it would be important to determine whether the primary objective in maintaining the system was to contain further migration rather than restore the aquifer.

ACKNOWLEDGMENTS

Financial support for this research was provided by the American Petroleum Institute (API).

REFERENCES

1. Tucker, W. A., M. W. Kemblowski, and D. E. Draney. "Technological Limits of Groundwater Remediation: A Statistical Evaluation Method," in Proceedings of Petroleum Hydrocarbons and Organic Chemicals in Groundwater (National Water Well Association, Dublin, OH, 1989).
2. Environmental Science & Engineering, Inc. "Technological Limits of Groundwater Remediation: A Statistical Evaluation Method" American Petroleum Institute, Publication No. 4510, 1991.
3. Tucker, W. A., C. T. Huang, J. M. Bral, and R. E. Dickinson. "Development and Validation of the Underground Leak Transport Assessment Model (ULTRA)," in Proceedings of Petroleum Hydrocarbons and Organic Chemicals in Groundwater (National Water Well Association, Dublin, OH, 1986).
4. Beers, Y. Introduction to the Theory of Error (Reading, MA: Addison-Wesley Publishing Company, Inc., 1957).
5. Gumbel, E. J. Statistics of Extremes (New York, NY: Columbia University Press, 1958).
6. Meyers, R. H. Classical and Modern Regression with Applications (Duxbury, MA: Duxbury Press, 1986).

CHAPTER 18

Development of Post-Treatment Cleanup Criteria for In Situ Soils Remediation

Phil La Mori and **Jon La Mori,** NOVATERRA [formerly Toxic Treatments (USA)], San Pedro, California

INTRODUCTION

The verification of meeting the agreed-upon cleanup criteria is an important final step in any soils remediation project. It is particularly difficult for soils remediation technologies which treat in situ. Heterogeneities in the contaminant distribution, soil type, and treatment efficiencies, as well as the critical need in obtaining representative samples, conspire to make verification of cleanup difficult. This chapter describes the approach used to develop post-treatment cleanup criteria at a site containing a mixture of 49 volatile organic compounds (VOCs) and semi-volatile organic compounds (SVCs) which was subjected to in situ hot air steam remediation by the Detoxifier Process.

TECHNOLOGY

The Detoxifier technology used to remediate the site mixes steam and hot air into the soil through two 5′ diameter counter-rotating overlapping blades. The process consists of four main elements; they are:

1. The two 5′ diameter drills which are capable of penetrating the soil to a depth of 30 feet.

2. A metal shroud which thoroughly encloses the area above the blades and captures the vaporized hydrocarbons.
3. A closed loop process train which condenses and removes the liquids and returns the air to the ground via a compressor.
4. Process control instrumentation.

The condensed liquid containing hydrocarbons and water is further distilled so that the captured hydrocarbons can be recycled or destroyed. The process operates in a stepwise manner to completely remediate the site. The process has advantages over passive Vapor Extraction Systems in that it actively mixes the soil with the vapor and steam. The process is designed to operate at large sites free of above- and below-ground obstructions. The site must also be fairly level.

RESULTS TO DATE

The Detoxifier technology has been used at a former industrial chemical storage facility in San Pedro, California, to remove chlorinated and nonchlorinated volatile and semivolatile hydrocarbons from soils, both in the vadose and saturated zones. A total of 49 VOCs and SVCs have been identified as contaminants of concern at the site. The VOC contamination ranges from 50 ppm to over 20,000 ppm, with a similar range for SVCs. The five major areas of contamination are shown in Figure 18.1.

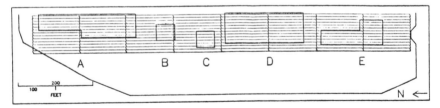

Figure 18.1. The five major areas of contamination and sampling grid configuration.

Additional examination of the site characterization shows that the contamination of the southern half of the site is quite different from the northern half. The southern half of the site is almost entirely contaminated with tetrachloroethylene (PERC) and chlorobenzene in very high concentrations. The total VOC can be as high as 20,000 ppm, and contains very low concentrations of SVCs. The soil matrix here is 80% plus clay which, since gaseous diffusion is directly related to the air filled porosity of the soil, makes removal of the PERC difficult. The northern half is much more heterogeneous in contamination both in SVCs and VOCs. The total hydrocarbon concentrations are similar in the northern half but the more sandy/silty soil makes remediation easier. These considerations may be important in the final site cleanup evaluation because they control the compounds and concentrations remaining after treatment.

Depending on the season, the top of the saturated zone is at 5 to 6 feet. Site characterization and treatment studies indicated that for the majority of the site the contaminant concentration is at a maximum between 6 and 8 feet and falls to background levels by 10 feet. The decision was therefore made to actively treat the site to 10 feet.

Because the Detoxifier technology was a new technology, the California Department of Health Services (DHS), the regulatory body responsible for the site, required an extensive test program to determine the capabilities of the technology before approving its use to remediate the site. This test program had three major phases: (1) experiments conducted by NOVATERRA to develop operational and control parameters; (2) a DHS-sponsored Remediation Test, QA/QC by Harding Lawson Associates; and (3) The Site Program Test managed by SAIC. The summary results for the latter two test programs are presented in Tables 18.1 and 18.2. Analysis for the VOC was by EPA 8240 and SVC by EPA 8270, thus detailed speciation was available for all the chemicals of concern. The results of the test programs became the input to a posttreatment risk assessment.

Table 18.1. Summary of Chemical Analysis

Treatment Block	Concentration (ppm)		% Reduction
	Pre-Treatment	Post-Treatment	
Volatile Organic Compounds:			
A-8-g	1,149	18	93
A-9-g	824	7	99
A-10-g	1,368	11	99
B-50-n	1,123	23	98
B-51-n	1,500	13	99
B-51-m	1,872	55	97
B-52-m	917	29	97
D-92-b	2,305	53	98
D-93-b	3,720	163	96
D-94-b	5,383	203	97
Semivolatile Hydrocarbons:			
A-8-g	1,794	637	64
A-9-g	2,510	653	74
A-10-g	7,020	592	92
B-50-n	22,829	1,670	93
B-51-n	14,924	2,304	85
B-51-m	10,040	2,495	75
B-75-m	669	594	11
D-92-b	707	55	92
D-93-b	1,465	90	94
D-94-b	869	111	87

Table 18.2.   EPA Demonstration (SITE Program)

Block Number	Concentration (ppm)		% Removal
	Pre-Treatment	Post-Treatment	
Volatile Organic Compounds			
Test One:			
A-25-e	54	14	73
A-26-e	28	12	56
A-27-e	642	29	96
A-28-e	444	34	92
A-29-e	850	82	90
A-30-e	421	145	65
A-31-e	788	61	92
A-32-e	479	64	87
A-33-e	1,133	104	91
A-34-e	431	196	54
A-35-e	283	60	79
A-36-e	153	56	64
Avg	475	71	78
Test Two:			
A-26-n	NA	16	NA
A-27-n	NA	22	NA
A-28-n	NA	36	NA
A-29-n	NA	80	NA
A-30-n	NA	119	NA
A-31-n	NA	45	NA
Avg	NA	53	NA
Semivolatile Organic Compounds			
A-25-e	595	82	86
A-26-e	1,117	172	86
A-27-e	1,403	439	69
A-28-e	1,040	576	45
A-29-e	1,310	726	45
A-30-e	1,073	818	24
A-31-e	781	610	22
A-32-e	994	49	95
A-33-e	896	763	15
A-34-e	698	163	77
A-35-e	577	192	67
A-36-e	336	314	7
Avg	902	409	53

RISK ASSESSMENT

At the request of DHS, a risk assessment was made for the site using all the chemicals found in the post-treatment soil sampling results. The risk assessment considered various scenarios. The most conservative scenario resulted from construction workers at the site. The major risks were caused by the ingestion and dermal pathways. While residential uses give a higher risk, residential use is not permitted because the site is located in the Port of Los Angeles. As a result of further analysis, DHS agreed to the breakdown of the chemicals of concern into those over 1% risk as major (Table 18.3) and the others as minor. This resulted in the setting of different cleanup criteria for the two chemical sets.

Table 18.3. Noncarcinogenic and Carcinogenic Risk Chemicals MEI[a]

Chemical	Hazard Index	% of Total Risk	ppm Site Avg	Grid Max
		Noncarcinogenic Risk MEI[a]		
DEHP	.024	10	681	6,807
Phth Matrix	.018	8	510	5,103
UniGlycol Ethers	.081	36	63	633
Butyl Cellosolve	.004	2	61	606
2-2MoEt-2EtOH	.008	4	125	1,248
C5-C11	.025	11	123	1,233
1,1-DCE	.004	2	2.0	20
Cis-1,2-DCE	.009	4	3.2	32
Trans-1,2-DCE	.004	2	2.1	21
C9-C10 Arom.	.008	4	90	900
Phenanthrene	.006	3	4.9	49
Total	.23			

Chemical	Cancer Risk/10^{-6}	% of Total Risk	ppm Site Avg	Grid Max
		Carcinogenic Risk Chemicals MEI[a]		
DEHP	.13	43	681	6,807
Phth Matrix	.10	33	510	5,103
1,1-DCA	.01	2	2.3	23.3
CH_2Cl_2	.01	2	16.1	161
1,1,2-TCA	.02	5	1.9	19.4
1,1-DCE	.02	7	2.0	20.0
1,2-DCA	.01	3	3.2	32.4
Perc.	.02	5	19.5	195
Site Total	.31	100		

[a]MEI = Maximally Exposed Individual.

DISCUSSION

At this point in the project the technology had been demonstrated and the risk assessment showed that treatment resulted in an acceptable risk. The problem was to agree with DHS, the lead agency, on an acceptable cleanup criteria. Because the risk assessment was done in a forward direction; i.e., it calculated the risk resulting from the chemistry of remediated soil, it could not be used directly to set the cleanup criteria. Statistical methods require that for high confidence in the site meeting the cleanup levels, set by previously remediated soil, a level of approximately 0.8 times that of the previous level must be attained.[1] Also, the large number of chemicals at the site complicated any risk-based criteria. The agency was also concerned by the heterogeneity of the site as related to chemical contamination, i.e., concentration (spill points) and species location (storage points). This was particularly true for two SVC chemicals, the glycol ethers and the phthalate esters, which constitute the major risk.

The agency recommended that the cleanup criteria for the site be the arithmetic mean of the post-treatment chemical analysis compared to the input values of the risk assessment, and also indicated they wanted to evaluate potential hotspots. This presented two serious problems. The first was that the input data to the risk assessment was the geometric mean of the post-treatment soil chemistry since the distribution of all the post-treatment data is log normal. Sample results in environmental fields often show concentrations that are distributed log normally.[2] Because the log normal mean, i.e., mean of the log of the values, is less than the arithmetic mean, even duplication of the previous results would result in failure to meet the cleanup criteria. The second reason was that even if the log mean were used (i.e., redo the analysis in log space) failure would occur on a statistical basis again even if the results of the previous remediation were duplicated, e.g., achieving a reasonable 95% upper confidence level and a sufficient power curve for the mean. A third, not so serious problem, occurred from the additional concern of DHS for hotspots by making the post-treatment sampling expensive if the area of concern was too small. It would typically require 200 sampling locations over the San Pedro site to find an elliptical hotspot, with 90% confidence, of area 500 square feet (17.70 feet by 35.40 feet). To decrease the hotspot area by half, the number of samples would need to be doubled.[3]

The attempt to reach agreement resulted in a series of meetings in which both sides worked to find a solution which met all the concerns. One alternative was to remediate the site and then do an additional risk assessment to evaluate those portions which had been underremediated. Although the feeling was that this would result in acceptable results, the agency had staff limitations and did not want to review another risk assessment.

The final agreement was reached when the agency proposed the following:

1. The chemicals of concern (majors) would be the chemicals which constituted over 1% of the risk. For those chemicals, a three-tiered cleanup criteria was developed:
   — The site arithmetic average cleanup level would be three times the risk assessment value (approximately 10^{-6} risk).
   — The site would be divided in 104 grids, each approximately 1000 square feet and the concentration of major chemicals in each grid would be less than ten times the site cleanup level (approximately 10^{-5} risk) This resulted in 208 samples which, while costly, was acceptable.

— An intermediate level was established for three contiguous length-wise grids or 8 grids wide as five times the cleanup level (approximately 5×10^{-6} risk).

2. The cleanup level for the minors can be twice that of the majors and there is no intermediate level.

3. If grid failure, i.e., hotspot, occurred additional soil samples were to be taken and all the samples averaged to determine passage or failure.

4. Samples and analysis will be done for the 3′ and 8′ levels, i.e., one analysis for the vadose zone and one for the saturated zone.

This was agreed to by Toxic Treatments (USA) (TTUSA) and the responsible party (RP), even though they knew there would be some small percentage of failures. This fact was determined from an analysis of the original input data set, tested against the final cleanup criteria. This analysis is discussed in the next section, but the reason for it is directly related to using arithmetic means in the post-treatment criteria as opposed to geometric means in the risk assessment. The analysis did show that the failure rate would be low and could be removed by additional sampling, additional remediation, or negotiation if the failure was only a few percent or additional post-treatment risk assessment. In no case where there was failure did the revised risk of all the chemicals exceed the risk approximations given above, i.e., if one chemical exceeded the criteria, others were sufficiently less than their criteria to give an acceptable overall risk. These results should be expected for a site as complicated as this one. Analysis using geometric means showed no samples that failed the site cleanup criteria.

RESULTS

Because of DHS concerns for hotspots, patterned sampling units of 1,056 square feet were created over the site (such concerns were related to the area a typical construction worker might encounter in a day, or the area a groundskeeper might be in close contact with). The resulting areas, called grids, measure 2 by 18 treatment blocks (8′ × 132′). This oblong shape was chosen over a square grid configuration for two reasons. First, should a second treatment pass be required, the Detoxifier could cover the entire grid with only slight shifting. Second, baseline testing treatment data showed that regions of high (and low) contamination have similar oblong shapes. Thus the grids were designed to be as homogenous as possible, under the constraints of the site as a whole.

The resulting grids for post-treatment sampling are shown in Figure 18.1. The grids are numbered 1 to 104, left to right and top to bottom. Locations of post-treatment samples are determined by a random number selection of 1 of the 36 blocks in a grid. Evaluation of existing post-treatment data, which could not be used to officially determine the compliance with the cleanup criteria, consisted of the input into the risk assessment, plus some more recent samples, and was done to pre-evaluate the level of the remediation effort against the proposed cleanup standards. Due to the complex nature of the site and the three-tiered cleanup levels proposed by DHS, this evaluation was an important interim measure of compliance and to the decision to go forward with the treatment.

The existing data was analyzed relative to the cleanup standards. For those grids where multiple samples existed, the arithmetic average of all samples was used. For chemicals which had no reported value, half the detection value was used as a conservative estimate. The results are shown in Figures 18.2 and 18.3. In these figures a distinction of above or below the intermediate level criteria has been suppressed, since no standards are exceeded. Data taken from ground level to six feet and below six feet were considered separately, since there is a similar stratification in DHS criteria. A detailed look at chemicals and grids which did not meet the hotspot standard is shown in Figure 18.4. In each case the failures to meet the criteria are the result of samples from specific treatment blocks. It was felt that these failures could be dealt with by averaging three additional samples or further remediation.

Results of chemicals with respect to the site average are shown in Figures 18.5 and 18.6. Site average failures above six feet occurred for tetrachloroethylene, 2-phenoxyethanol, and unidentified glycol ethers. The failure of 2-phenoxyethanol and glycol ethers can be traced to one analysis. Site average failures below six feet occurred for tetrachloroethylene and phthalate ester matrix. The phthalate ester matrix failure was in the same boring as the 2-phenoxyethanol and glycol ethers. This strongly suggests additional remediation will be of value to remove the excess material in this limited area. Additionally, the site average is examined using only north (south) data. Failures are clearly shown to be driven by a north (south) break. The one VOC failure is in the southern area, while the SV failures are in the north.

A special case is tetrachloroethylene (PERC). A detailed look at grid averages shows there is no one grid analysis which causes failure of the site average relative to the cleanup level. Also, the PERC problem is limited to the southern half of the site and is great enough to negate the overtreatment in the north half. The remediation to date has concentrated on the highly contaminated portions of the site. Complete remediation of the site may reduce the PERC site averages to an acceptable level. DHS knew this when they established the cleanup criteria and suggested that the post-treatment averaging of the cleaner areas of the site with the dirtier would result in meeting the cleanup criteria. However, it may turn out that given the complexity of the site, the performance of the technology, and nature of soils at the site, levels of tetrachloroethylene could still fail the overall site criteria. Failure of tetrachloroethylene to meet average site cleanup values would present a problem in how to resolve this failure.

Resampling is an valid option and worth consideration. As mentioned above, the failure of a grid often results from a limited sampling population. The heterogenic nature of the site encourages resampling.

Retreatment of several highly contaminated grids is another valid option. The grids were created to allow ease of retreatment. After retreatment a new set of samples would be taken and a new analysis performed.

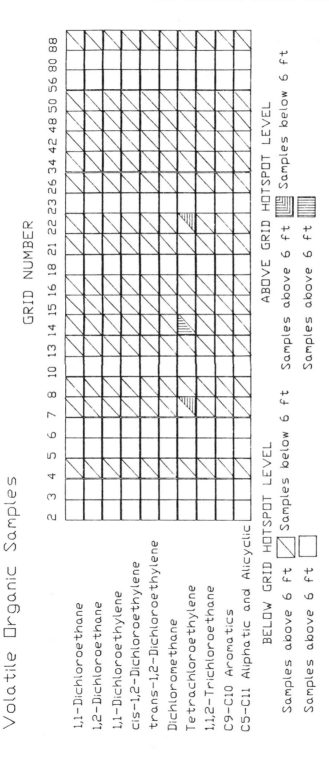

Figure 18.2. Results from volatile organic samples.

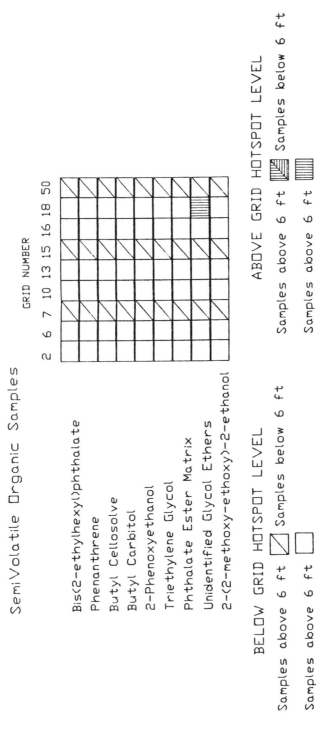

Figure 18.3.   Results from semivolatile organic samples.

Grid Hotspots

Chemical	Grid #	Reason
Tetrachloroethylene (0-6 Ft Samples)	14	1 Sample: 200 PPM (D97E)
Unidentified Glycol Ethers (0-6 Ft Samples)	18	6 Sample AVG: ◆ 567 PPM 2 Samples AVG: 4000 PPM (A36G, A35G)
Tetrachloroethylene (Below 6 Ft Samples)	7	4 Sample AVG: ◆ 170 PPM 1 Sample: 310 PPM (E118B)
	22	1 Sample: 220 PPM (D94F)

◆ = HOTSPOT

Figure 18.4.   Chemicals and grids which did not meet the hotspot standard.

Another option might be a combination of the above with an additional risk assessment for the remaining PERC in the southern half. Table 18.3 shows the risk of PERC at 5% of the total. Thus, if PERC was the only contaminant of significance remaining in the southern half, a value of two or three times the cleanup criteria would still result in insignificant risk.

SUMMARY

Verification of meeting the cleanup criteria is one of the most important final steps in any remediation project. At a very heterogenetic site even agreeing upon the cleanup levels can be a major step. A three-tiered risk-based approach was presented as a practical method to satisfy the concerns of both regulators and the responsible parties, as well as taking into account the performance of an

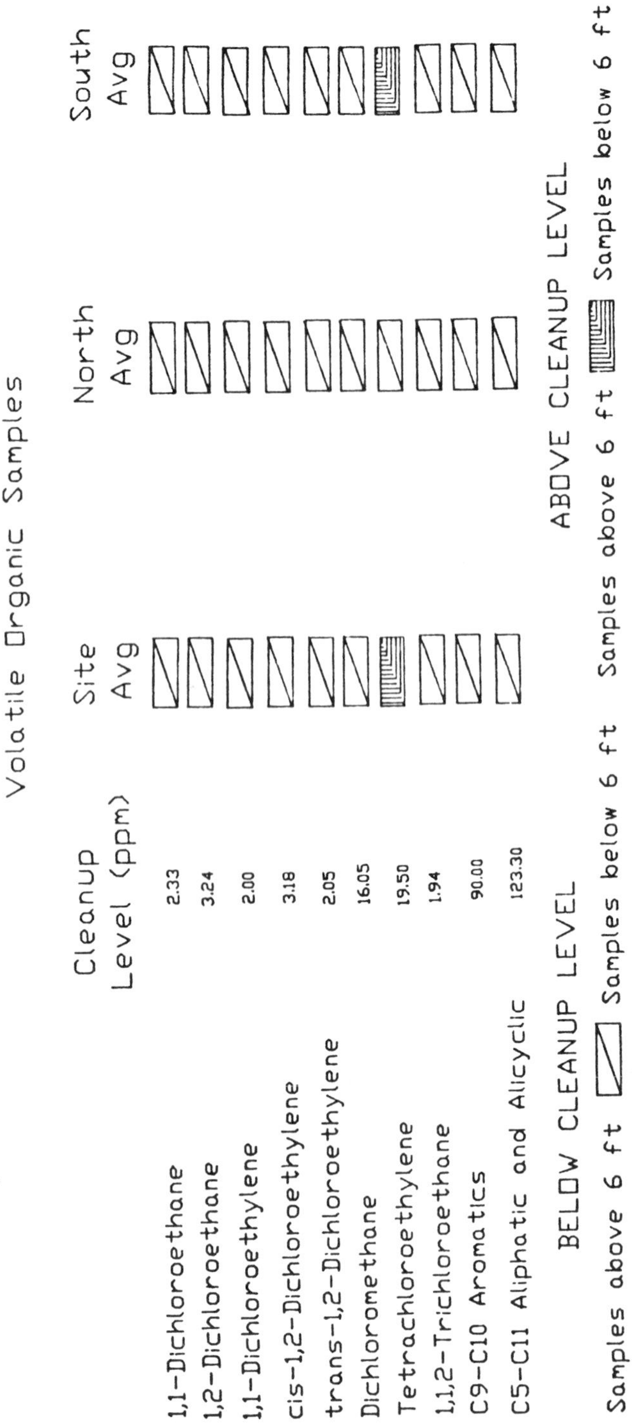

Figure 18.5.   Results of VOC samples with respect to the site average.

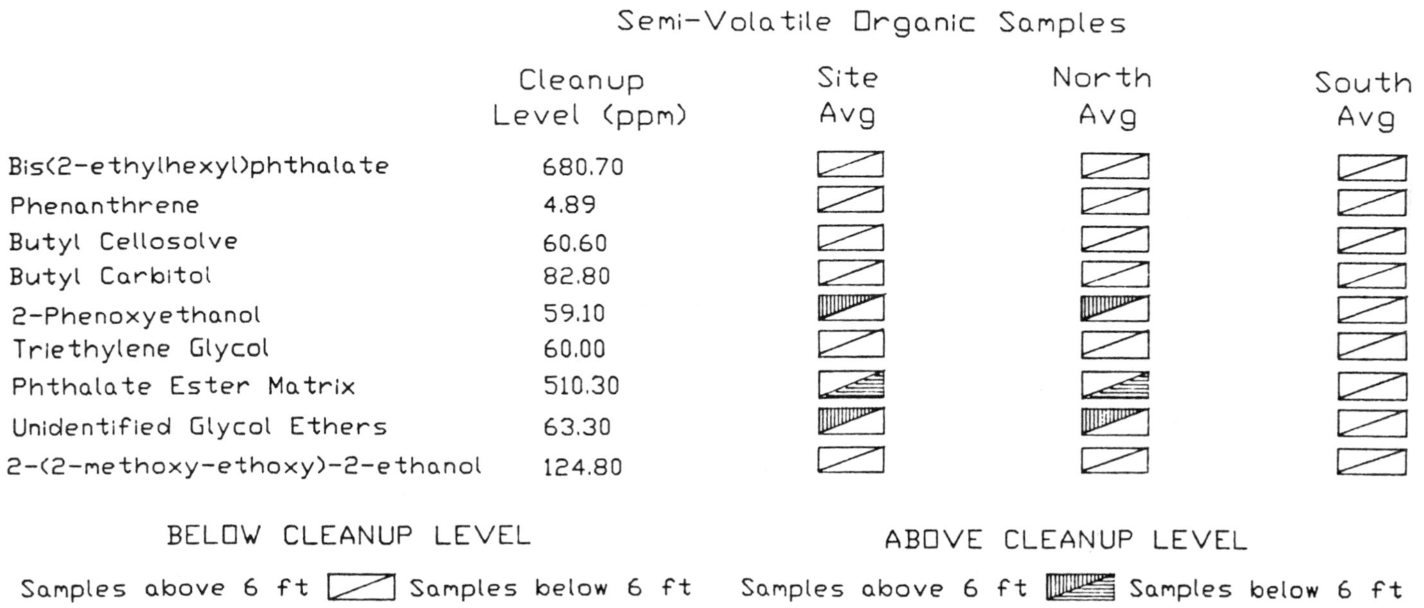

Figure 18.6. Results of SVC samples with respect to the site average.

innovative technology at a chemically diverse site. The result is a verifiable and cost-effective post-treatment sampling plan where cleanup standards were met or exceeded in most cases.

REFERENCES

1. "Methods for Evaluating the Attainment of Cleanup Standards, Volume 1: Soils and Solid Media," U.S. Environmental Protection Agency, 1989.
2. Ott, W. R. "A Physical Explanation of the Lognormality of Pollutant Concentrations," *J. Air Waste Manage. Assoc.* 40:10 (Oct 1990).
3. Gilbert, R. O. *Statistical Methods for Environmental Pollution Monitoring* (New York, NY: Van Nostrand Reinhold Company, 1987).

PART VI

Remediation Case Studies

CHAPTER **19**

Microencapsulation of Hydrocarbons in Soil Using Reactive Silicate Technology

Tom McDowell, Siallon Corporation, Los Angeles, California

INTRODUCTION

There are many methods being proposed or actually used for stabilizing, solidifying or fixing organic contaminants in a soil matrix.[1] Some of these processes show good short-term results; very few show good long-term stability.[2] With the exception of asphalt blending and other thermoplastic encapsulation techniques, the majority of stabilization methods utilize pozzolanic materials (portland cement, fly ash, kiln dust) as the major ingredient. The result is a cementitious reaction and entombment of the organics within a solid matrix. It is this very reaction that defeats most attempts to solidify moderate to high levels of hydrocarbons and produce a material that shows long-term stability.

It is well known in the cement industry that various hydrocarbons (i.e., diesel) can be used in minor amounts to retard the set of concrete. This occurs due to absorption of the hydrophobe onto the various oxide crystal faces which effectively block the infusion of water necessary for crystal hydration. This dichotomy of trying to form cement around a hydrocarbon that is trying to prevent the formation is in part responsible for the less than enthusiastic regulatory acceptance[3] of solidification or stabilization of organics.

The Siallon process for microencapsulation of hydrocarbons takes a completely different approach. It uses two water-based products, an Emulsifier and a Reactive

Silicate. This eliminates the use of pozzolanic materials and also the resultant volume increase (usually 15% to 50%) associated with conventional solidification.

Because the Siallon process uses water-based materials it can be applied in situ as well as used in an ex situ mode. Treatment efficiency has been shown either full-scale or pilot-scale for gasoline, diesel, waste motor oil, crude oil, coal tars and PCBs.

SIALLON PROCESS OVERVIEW

The Siallon process is a simple two step procedure. The first stage desorbs the hydrocarbon and emulsifies it. The second stage is the application of the Reactive Silicate. This reacts with the Emulsifier to form a nonsoluble silica cell that measures less than 10 microns across. One of the advantages to this form of microencapsulation is that there is no change in the soil's physical characteristics; permeability, porosity, grain size, etc. are the same before and after treatment.

It is the first step, the desorption and emulsification, that is the key to both the efficiency and completeness of the process. A range of emulsifiers has been developed for different hydrocarbons, soil types, and application techniques, all of which are predicated on the same chemistry.

The emulsifier is based on the use of surfactants with acidic moieties, with a resulting pH of 3–4 as a concentrate and pH 4–5 when diluted with water for use.

Conventional emulsion theory shows the emulsifier orienting itself with the hydrophobic tail toward the oil and the hydrophilic head and acidic moiety toward the water phase. This orientation has the effect of producing an emulsified hydrocarbon micelle with an outer shell or surface of reactive acidic sites. The average size of the micelles as determined by optical microscopy was in the 2–4 micron range. With micelles of this size, a very stable emulsion is formed which allows the application of the second stage Reactive Silicate over a protracted period of time, and hence the ability to work on an in situ basis.

The Reactive Silicate is a water-based solution having a pH of 10.5 as a concentrate and a pH of 9.0 to 9.5 when diluted to use concentration. When the slightly alkaline silicate is added to the slightly acidic emulsifier, a simple acid-base reaction takes place. The neutralization of the silicate produces silica, water, and a trace amount of by-product salt. The formation of the silica starts at the active sites on the emulsifier. The emulsifier is already surrounding the hydrocarbon as a micelle so the silicate/emulsifier reaction produces a silica cell around the hydrocarbon. A great deal of research effort was directed at characterization of both the reaction and the resultant silica cell.

SILICA CELL CHARACTERIZATION

The micron-sized silica cell, containing hydrocarbon, was characterized by both chemical and morphological analysis. The objectives of this analytical sequence were:

- to determine chemical composition both surface and interior
- to evaluate the hydrocarbon encapsulation mechanism
- to examine particle morphology with respect to durability and lifespan.

To provide sufficient sample for this work, 1 liter of used motor oil was encapsulated using the Siallon process. The encapsulated material was homogenized, and sequentially quartered to obtain a representative sample.

OPTICAL MICROSCOPY

All of the optical microscopic work was performed on a Zeiss Axioplan Universal Research Microscope equipped with a Contax RTS 11 35mm camera. Samples were mounted in Cargille immersion oil type Ai 5150.

The particles were found to be irregular in shape and mostly of a granular nature. Particle size was measured and found to be less than 10 microns, with most of the unaglommerated particles in the 2–6 micron range.

More than 99% of the particles were isotropic, showing either an amorphous or cubic crystal nature. Under polarized light, fewer than 1% of the particles appeared bright, indicating crystallinity (other than cubic) or an elevated degree of polymerization with preferred molecular orientation.

X-RAY DIFFRACTION (XRD)

To further define the crystallography, samples were analyzed by XRD using a Phillips X-Ray diffractometer with a cobalt source. The diffractogram showed the material to be amorphous with only a slight amount (less than 1%) of crystalline material. The crystalline material was examined by Selected Area Electron Diffraction and identified as sylvite (KCL).

SCANNING ELECTRON MICROSCOPY (SEM)

Samples were prepared for SEM by mounting on carbon stubs and then carbon coated with an Edward carbon evaporator. Samples were examined at a range of magnifications using a JEOL JS35 scanning electron microscope operating at 25 kV and equipped with a Tracor Northern TN5500 energy dispersive X-ray analyzer (EDXA). Shown in Figure 19.1 is a secondary electron image of a representative particle at a magnification of 565X. It shows a substantially solid material with a nonporous surface. EDXA was used to analyze several areas across these particles (Figure 19.2); in all cases the silica peak was greater than 95%. The peak heights showing X-ray intensity were used for semiquantitative analysis. The Cu and Zn peaks originate from within the SEM and should be ignored.

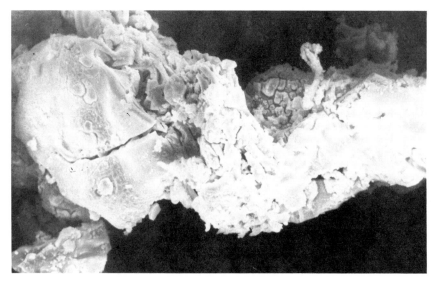

Figure 19.1.   SEM of silica cell.

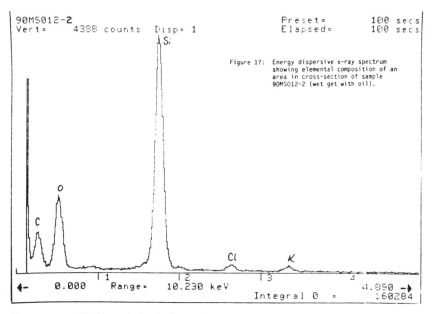

Figure 19.2.   EDXA analysis of silica cell.

TRANSMISSION ELECTRON MICROSCOPY (TEM)

In an effort to delineate the interior structure, samples were analyzed by TEM using a JEOL CX-100 transmission electron microscope operating at 60kV, equipped with a Tracor Northern TN5500 energy dispersive X-ray analyzer. As shown in Figure 19.3, no detail in internal structure could be discerned.

Figure 19.3.   TEM of silica cell.

In order to determine the cell interior morphology and chemistry with a high degree of reliability, the micron-sized silica cells were sectioned. Individual silica cells were infiltrated and embedded in Epon resin in BEEM capsules, dried, and then trimmed and sectioned using a diamond knife on an LKB ultramicrotome.

Suitable sections were lifted on copper/rhodium grids and evaluated by TEM and EDXA. The sections were analyzed by EDXA and once again showed only silica as a major component. (Figure 19.4). The micrographs from TEM analysis (Figure 19.5 at 15,000 × magnification and Figure 19.6 at 60,000 × magnification) show an almost equal ratio of void space and solid. There is no discernible pattern to either the void or solid space of the section, partly due to the two-dimensional limitation of TEM. To further define the interior structure, an image of the negative for Figure 19.6 was transferred to a LeMont OASYS Optical Analysis System. This image was processed through three passes of the fine structure enhancement program. After inverting the data, the resultant image (Figure 19.7) has a definite maze-like pattern.

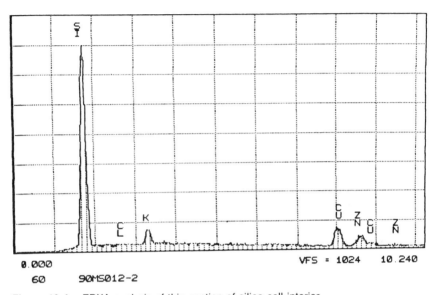

Figure 19.4. EDXA analysis of thin section of silica cell interior.

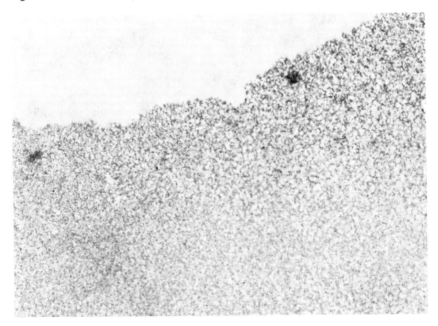

Figure 19.5. Photomicrograph of section at 15,000× magnification.

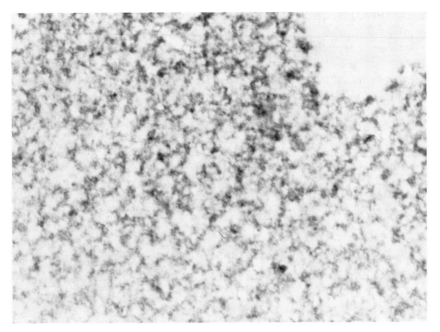

Figure 19.6. Photomicrograph of section at 60,000× magnification.

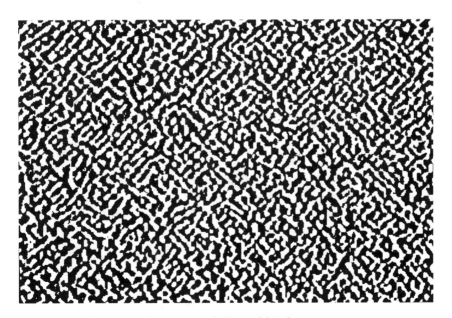

Figure 19.7. Computer enhancement of silica cell interior.

X-RAY PHOTOELECTRON SPECTROSCOPY (XPS)

Sectioned samples were mounted on indium substrates and analyzed by XPS using a Surface Science Labs S-100 XPS with monochromatized Al Ka X-rays. The results of this, expressed in atomic percent, show an average of 30% as carbon (Figure 19.8). This indicates clearly that the interior of the silica cell was infused with hydrocarbon.

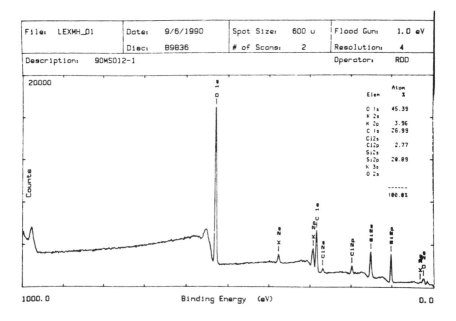

Figure 19.8.   XPS analysis of section of silica cell.

CHARACTERIZATION RESULTS

Chemical analysis by both XPS and EDXA show the Siallon cell to be essentially pure (SiO_2) silica both on the surface and interior. The trace amounts of potassium and chloride (reaction by-products) were found only on the surface and as such would not affect the nonsolubility of the silica cell. Crystallography by plane polarized light, XRD, and SAED all show the final reaction product to be amorphous, with the only crystalline material being the trace of sylvite (potassium chloride).

Surface morphology shows the cell to have a relatively monolithic nature and a nonporous skin or surface.

The interior is a honeycomb-like maze of channels with the hydrocarbon trapped within the channels or voids. The honeycomb effect is the result of using a water-based silica. After the first or emulsion stage, a drop of hydrocarbon is held within a micelle that has a silicate reactive surface.

As the silicate reacts with the emulsifier and converts to silica, the growth is three-dimensional and the silica grows toward the hydrocarbon. Since the silica/silicate materials are water-based and hydrophilic, the growth is around the hydrocarbon and into it as two distinct phases; the result being a total microencapsulation of hydrocarbon in a particle of almost infinite tortuosity.

SITE REMEDIATION

The true test of any remediation process is how well it works in real life. The Siallon process has applicability to a wide variety of hydrocarbon contaminants. A brief description of the remediation of three sites: gasoline, diesel, and PCBs, will show the applicability of the process to various sites.

REMEDIATION OF GASOLINE IMPACTED SOIL

Site Description

A southern California service station site was found to have gasoline contamination in the soil as a result of a leaking underground storage tank. The tanks and associated piping were removed and approximately 400 tons of contaminated soil was excavated. The site provided minimal work space and the client required remediation to be completed in less than a week.

Site Analysis

The soil was analyzed for TPH and BETX both prior to excavation and throughout the excavated pile. Results of the analysis were as shown in Table 19.1.

Table 19.1. Results of Soil Analysis for TPH and BETX

Sample ID	Benzene	Toluene	Ethyl Benzene	Xylene	TPH
SP-2	1.0	21	12	87	1900
SP-3	1.1	16	8.7	63	750
T1-W-14	120	420	350	2300	57,000
T1-M-14	12	68	45	190	5200
T1-W-14	7.6	39	15	83	1400
T2-M-14	7.0	41	23	110	2990

NOTE: 1. All concentrations are in mg/kg.
2. BTEX were analyzed by EPA method 8020.
3. TPH was analyzed by CA-DHS modified method 8015.

Remediation Process

Since the contaminated soil was already excavated, a mobile Pug Mill was used for remediation. Production rate was approximately 20 tons per hour. As the soil was conveyed to the pug mill hopper, it was immediately mixed with the Siallon Emulsifier solution. This step desorbed the hydrocarbon from the soil and emulsified the hydrocarbon. This step also suppressed the volatilization of any VOC, as was evidenced by OVA analysis.

As the soil moved down the pug mill, at the midpoint, the Siallon Reactive Silicate was added. The soil exiting the pug mill was sampled on a routine basis for analysis.

Post-Treatment Analyses

Since the Siallon process is a "real time" procedure, a mobile, onsite lab was used to provide analysis both for closure and process monitoring. Results of these analyses are shown in Table 19.2.

Table 19.2. Results of Analysis for Closure and Process Monitoring

Sample #	Benzene	Toluene	Ethyl Benzene	Xylene	TPH
1	ND	0.061	ND	0.21	ND
2	ND	0.047	ND	0.14	10
3	ND	0.020	ND	0.088	ND
4	ND	0.031	ND	0.094	4.3
5	ND	ND	ND	ND	ND

NOTES: 1. ND = Not Detected.
2. BTEX analyzed by EPA method 8020 direct purge-and-trap.
3. TPH analyzed by CA-DHS modified method.
4. All concentrations in mg/kg.

To arrive at a percent treatment efficiency, we averaged the before and after results and expressed these as a percentage shown in Table 19.3.

Table 19.3. Percent Treatment Efficiency Averages, Before and After

	Benzene	Toluene	Ethyl Benzene	Xylene	TPH
Before	24.8	100.8	75.06	472.2	11,540
After	ND	0.031	ND	0.10	2.86
% Treatment efficiency	100	99.97	100	99.98	99.98

The Siallon process was extremely effective in remediating the gasoline contaminated soil. Upon completion, the treated soil was disposed of at a Class 3 sanitary landfill.

REMEDIATION OF DIESEL-IMPACTED SOIL

A train derailment resulted in spillage of diesel fuel. The spill area encompassed the rail bed bordered by a creek on one side and marsh area on the other. This site presented several key challenges:

- concern for maintenance of track bed soil characteristics
- multiple-contaminated mediums; soil, sand, gravel
- high water table
- environmentally sensitive creek and marsh lands
- high traffic rail line that could not be shut down

Site Analysis

Soil samples were taken from several areas and analyzed for total petroleum hydrocarbons. The TPH in soil were as shown in Table 19.4.

Table 19.4.   Results of TPH Soil Sample Analysis

Sample	Concentration (ppm)
SW	1360
MW	1645
NW	1510
NE	590
ME	1012
SE	1415
S	1260

These samples were taken mainly from the edges of the plume area. The plume was fairly easy to delineate due to odor and discoloration. It was accepted that the middle areas, or source areas, would have much higher concentrations.

Treatability Study

Two composite samples were prepared and subsequently split in half to provide two untreated control samples and two treated samples.

The four samples (treated and untreated) were then analyzed by TCLP, measuring the total hydrocarbon in the leachate by GC/MS as per EPA 624. The results were as shown in Table 19.5.

From the treatability study, we proceeded to full-scale, in situ remediation.

Table 19.5. Results of TCLP Analysis of the Leachate

Sample	TPH in Leachate
MW Control	149.0 mg/L
SW Control	22.5 mg/L
MW Treated	<1.0 mg/L
SW Treated	<1.0 mg/L
MDL (method detection limit)	1.0 mg/L

Remediation

Attention was given to the concern for possible mobilization of the hydrocarbon material during treatment (ASTM Freshwater Dispersants Revision 6, March 3, 1989). To eliminate this concern, the plume area was first "zippered" on both sides by the injection of Siallon Reactive Silicate. The Siallon Emulsifier was then injected into the soil at 10-foot centers at 8 psi. Any mobilization toward the water areas of the emulsified hydrocarbon material would then be reacted with the previously injected silicate. This procedure proved to be 100% effective as no migration into the fresh water was observed. Following the emulsifier injection process, the Reactive Silicate was injected, using the same 10-foot centers and pressures.

Results

The site was segmented into four quadrants and two samples taken from each quadrant. Samples were taken:

- just prior to remediation
- just after remediation
- two weeks after remediation

The samples were analyzed for Total Petroleum Hydrocarbon (TPH), Benzene, Toluene and Xylene, with average results shown in Table 19.6.

Soil tests were conducted in order to verify that the Siallon treatment process did not change any of the soil characteristics. No difference before and after were found in porosity, falling-head permeability, or grain size analysis.

Table 19.6. Results of Sample Analysis for TPH, Benzene, Toluene, and Xylene

Average	Before Treatment	After Treatment
TPH	26,500	132
Benzene	4	<0.1
Toluene	6	0.2
Xylene	17	0.4

NOTES: 1. All results in mg/kg.
2. Test methods—EPA 8015/8020.

REMEDIATION OF PCB CONTAMINATED SOIL

Background

The surface soil of a former electric transformer manufacturing plant was contaminated with PCB-containing oil. The plume of contamination was an area of 25 ft × 100 ft where transformers had been loaded for shipment. Depth of contamination ranged from 6″ to 18″ below surface.

Soil analysis (Table 19.7) showed mostly a silty soil with some sand present. An analysis of the contaminated soil in three areas labeled south, middle, and north are as shown in Table 19.8.

Table 19.7. Results of Soil Analysis

Soil Properties	
Bulk Density	1.52
Total Porosity	0.35
Permeability (Darcy)	0.1

Table 19.8. Analysis of Contaminated Soil for Three Areas

	South	Middle	North
Total PCBs (Parts per billion)	115,000	583,000	235,000

Remediation

Since the site was now overgrown with brush, a considerable amount of harvesting of vegetation was required prior to remediation.

As the contamination only impacted the top two feet of soil, an augering system was used and the Siallon solutions were applied by spray during the auger process.

The Siallon Emulsifier was diluted 1:1 with water and the diluted solution sprayed on at 0.1% by weight of soil. Following the application of the Emulsifier solution, the Siallon Reactive Silicate was sprayed on as a concentrate at a rate of 0.05% by weight of soil.

Results

An independent laboratory was contracted to:

1. obtain proper composite samples
2. provide chain-of-custody
3. analyze samples
4. provide final report

The results of the TCLP and resultant PCB analysis of the leachate are shown in Table 19.9. Based on these results, the material treated is no longer considered a Registerable Solid Waste, a Leachate Toxic Waste, nor a Hazardous Waste.

Table 19.9.   Results of TCLP and PCB Analysis of Leachate

	South	Middle	North
Total PCBs (Parts per billion)	1.01	0.692	6.80

CONCLUSIONS

We have shown that the reaction product of the Siallon process, the silica cell, has the following characteristics:

- essentially pure silica
- a nonporous surface
- a relatively solid particle
- less than 10 microns in size
- a honeycomb or maze-like interior
- reduces the mobility and toxicity of hydrocarbons

These characteristics make the Siallon process an excellent tool for remediation of hydrocarbons both in situ and ex situ.

REFERENCES

1. "Stabilization Solidification of CERCLA and RECRA Wastes," U.S. Environmental Protection Agency, EPA/625/6-89/022, May 1989.
2. "Technology Evaluation Report: SITE Program Demonstration Test," U.S. Environmental Protection Agency, EPA/540/5-89/005a, February 1990.
3. Funderbruk, R. *Hazmat World* February 1991, pp. 57–59.

CHAPTER 20

The CROW© Process and Bioremediation for In Situ Treatment of Hazardous Waste Sites

Lyle A. Johnson, Jr., Western Research Institute, Laramie, Wyoming
Alfred P. Leuschner, Remediation Technologies, Inc., Concord, Massachusetts

INTRODUCTION

The Western Research Institute (WRI) and Remediation Technologies Inc. (ReTeC) are actively pursuing development of a process for in situ treatment of contaminated soil. The result of these efforts has been a process which combines physical and chemical extraction of contaminants from soil followed by in situ biological treatment of any remaining residuals. To date, the effectiveness of this technology has been demonstrated in one-dimensional laboratory flow tests and in three-dimensional flow tests at a small (approximately 2 ton) pilot scale. Currently, field-scale demonstration programs are in the planning and design or permitting stages.

WRI has been developing technologies for recovery of hydrocarbons from underground formations for the past 50 years under funding from various entities of the United States government. In the 1980s, WRI began to adapt technologies for energy production to the recovery of dense, nonaqueous-phase liquids from contaminated industrial sites. One such process, the Contained Recovery of Oily Wastes (CROW©) process uses hot water displacement to reduce concentrations of oily wastes in subsurface soils and underlying bedrock. In this process, the contaminated area is hydraulically isolated so that contamination is not spread into uncontaminated areas. The downward penetration of dense organic liquids is reversed by controlled heating of the subsurface to float the oily wastes in water. The buoyant wastes are displaced to production wells by sweeping the subsurface with hot water. Waste flotation and vapor emissions are controlled by maintaining both temperature and concentration gradients near the ground surface. Reducing waste concentrations to residual saturation immobilizes the oily wastes. At this point, the only remaining transport mechanism for these oily wastes is through dissolution into the groundwater.

In situ bioremediation processes remediate the soils and groundwater contaminated with biodegradable contaminants by enhancing the growth and activity of aerobic bacteria. These microorganisms use the contaminants as a source of carbon and energy while converting them to carbon dioxide and water. The process involves installing a groundwater injection and extraction system that allows for the controlled transport of an oxygen source (typically hydrogen peroxide) and water-soluble nutrients (typically nitrogen and phosphorus) between injection and extraction points. Oxygen and nutrients are required to promote and sustain the microorganisms needed to effect the degradative process.

The combination of the CROW© and in situ bioremediation processes offers the potential for complete site restoration. The CROW© process is effective at removing free product from soils. It is not effective at substantially reducing contaminant levels in soils where free product is not found. Additionally, residual levels of contaminants remaining after the CROW© process has been completed, in many instances, can be above action levels, thus requiring additional treatment. In situ biological treatment is effective at treating soils that are not heavily contaminated with free product, but the treatment is ineffective in dealing with hotspots, essentially due to the accessibility of the contaminant to the microbes. Thus, operating these two processes in sequence should provide complete site treatment. There also exists compatibility between these two technologies in the capital expenditures required for installation. Injection and recovery wells, the aboveground piping network, tanks, and pumps are all equipment common to both processes. Finally, the physical site characteristics such as contaminant source(s), soil type, permeability, and aquifer depth and thickness are compatible for the effectiveness of each technology.

TESTED MATERIALS

Material from two contaminated field sites has been used in process development and testing. Site 1 is a former manufacturer gas plant site in Pennsylvania with soils contaminated with coal tars. Site 2 is a wood treatment plant in Minnesota with soils contaminated with creosote, PCP, and petroleum products. The total quantity of material taken from Site 1 was 11 55-gallon drums. This material was used throughout the entire program for laboratory and small-scale pilot testing. An additional 15 gallons of free organic product was also obtained from the site. The samples taken from Site 2 were 2 5-gallon pails of contaminated soil and 2 gallons of free organic product. Soil samples were taken by means of a clam bucket on a rotary drilling rig, and the free liquid was pumped from existing wells located near the sampling locations.

CROW© PROCESS RESULTS

One-Dimensional Reactor

The reactor system used for the one-dimensional displacement tests is a standard tube reactor (Figure 20.1). The disposable CPVC reactor tube (3.75-inch i.d. × 36-inch long) is uniformly packed with approximately 30 lb of contaminated soil and is vertically oriented within a series of insulated shield heaters. Auxiliary equipment includes inlet water injection and metering devices, a water heater, product collection equipment, and a gas chromatograph. The entire system is instrumented and interfaced to a data acquisition computer that records temperatures, pressures, and flow rates every five minutes.

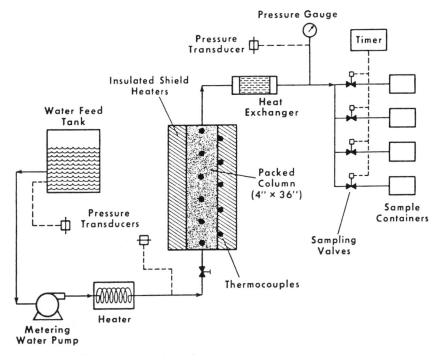

Figure 20.1. Tubes reactor schematic.

Water is metered into the bottom of the reactor by a positive displacement pump. The injected water passes through a heater to generate steam or hot water. Product water samples are collected from the automatic sampling system. The reactor backpressure is maintained at atmospheric pressure by venting the product collection vessels to the atmosphere through the gas collection system. Product gas is collected from the sample vessels and the gas composition is analyzed as needed by an online gas chromatograph.

Site 1 Simulations

Eleven ambient temperature or straight hot water tests were conducted on samples with 0.13 wt % organics or on samples with the organic content increased to ~ 3 wt %, which was the approximate saturation determined for one of the original grab samples from the site.

An ambient temperature displacement test was conducted at approximately 64°F (18°C) using both original and resaturated soil. The water injection rate was approximately twice the natural groundwater flowrate. With the high porosities of the site and the packed tubes, the flow will always be in the laminar regime. The reduction in oily waste saturation was 15 wt % for the original soil and 21 wt % for the higher saturated soil. These two tests show that pump-and-treat processes or cold water flushing will leave up to 85% of the contaminant in place.

Initial hot water displacement tests were conducted at nominal temperatures of 100°, 120°, and 140°F (38°, 49°, and 60°C). As the temperature of the flushing water increased, the percentage decrease in the saturation of the oily material increased from 23 wt % at 100°F (38°C) to 42 wt % at 140°F (60°C). The increase in the removal of the organic material with increasing temperature was also reflected by an increase in the amount of organic material dissolved in the flushing water as the temperature increased. The trend produced by the initial tests suggested that a temperature higher than the 140°F (60°C) may further enhance the reduction in oily waste content.

To test the higher temperature, five tests were conducted at temperatures between 155° and 165°F (68° and 74°C). Four of the tests used soil with 1.9 to 2.8 wt % contaminant saturation with the fifth test conducted on original material. The percentage reductions in the oily waste saturation ranged from 55 to 63 wt % with the highest reduction at approximately 156°F (68°C). The test with the original material had a 61 wt % reduction compared with the 15 wt % for the ambient temperature flushing test (Figure 20.2). These results show the desirable effect of hot water flushing.

Eight one-dimensional, chemical-enhanced flushing tests were conducted at approximately the optimum flushing temperature. The testing consisted of preliminary screening tests to determine which of three chemicals to use in the final flushing tests. The final four tests used Igepal CA-720 and contaminated soil with and without added contaminant. For chemical concentrations in the flush water of 0.45 to 0.95 vol %, the oily waste saturation was reduced by 64 to 83 wt %, respectively. These results indicate that increased reduction in oily waste saturations can be realized if chemical enhancement can be incorporated (Figure 20.2).

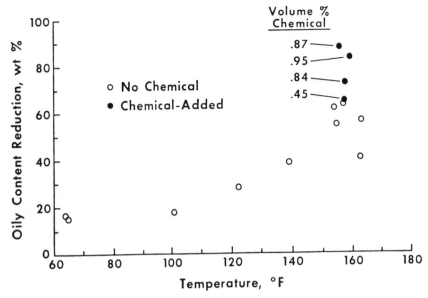

Figure 20.2.   One-dimensional CROW© tests.

Site 2 Simulations

Two elevated-temperature flushing tests were conducted on the original material from the site. The organic content of the test samples varied from 7.4 to 2.9 wt % with PCP content of 3,200 and 1,500 ppm, respectively. Both tests were conducted without chemical added to the hot water and at a flow rate approximately twice the natural groundwater rate. The higher flow rate can be used because of the high porosity of the site.

The two tests were operated at 140° and 120°F (60° and 49°C) and both produced residual organic saturations of 0.5 wt %. Therefore, the organic removal for the tests was 94% for the 7.4 wt % sample and 84% for the 2.9 wt % sample. The residual PCP concentration for both samples was below 2.5 ppm or a 99.9% reduction in the saturation. These results show that the lighter, more mobile creosote mixture can be removed more efficiently and at a lower temperature than the MGP contaminant.

Three-Dimensional Site 1 Simulations

The experimental apparatus used for the three-dimensional tests is a large pressure vessel into which an encapsulated sample is placed (Figure 20.3). The pressure vessel is a thick-walled, horizontal, cylindrical unit that is 9 ft long and 6 ft in diameter. Test samples are encased in a steel reaction box with up to 80

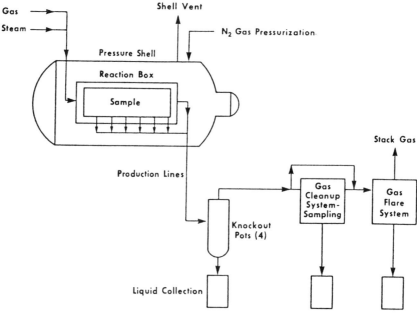

Figure 20.3. Three-dimensional reactor schematic.

thermocouples installed in and around the sample. The fluid-handling system consists of a water-metering, heating, and pumping system, and a product collection and sampling system.

For these simulations, the reactor sample was reconstructed from actual site material so that it resembles the field site. The sample consisted of four layers, an impervious 1-ft thick base with a 6-inch layer of highly saturated oily sand over it. The next layer from the bottom was 6 inches of a lower saturated oily sand that was capped with 1 ft of clean dry sand. The reconstructed sample was fully instrumented with a systematic array of 80 thermocouples so that the progression of the displacement front could be monitored. The well pattern for the sample consisted of two injection wells and a single central production well to simulate the entire CROW© concept.

Following the tests, the samples were partitioned both horizontally and vertically into individual samples, and each sample was analyzed for organic content. These samples permit the displacement efficiency of the process to be determined.

Two three-dimensional tests were conducted. The first test used only hot water injection, where as the second had chemical addition to the hot water. The target temperature for both tests was the optimum temperature determined during the one-dimensional testing. The first test was conducted for approximately 100 hours. The test demonstrated that a cooler water cap could be maintained above the hot water zone and that the hot water could sweep an area consistent with those predicted by streamline analysis. The percentage reduction in oily waste saturation changed with position but was fairly consistent with the developed isotherms.

The 60% oily waste saturation reduction gradient corresponds reasonably well with the 140°F (60°C) isotherm (Figure 20.4). This reduction is higher than for the 140°F (60°C) one-dimensional test, but it is not unexpected in that the larger sample may permit the formation of a larger oil bank, which produces a better displacement of the oily material.

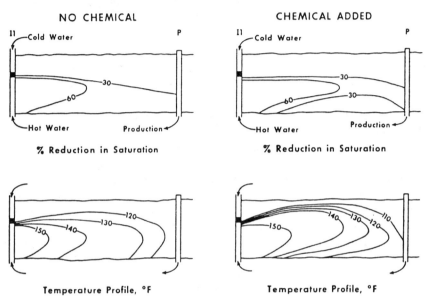

Figure 20.4. Cross-sectional saturation reduction gradients and isotherms for three-dimensional tests.

The second three-dimensional flushing test, with chemical (Igepal CA-720) added at a concentration of 1 vol % to the initial hot water flush was operated at the same conditions as the first three-dimensional test. The test was conducted for approximately 50 hours because of plugging problems caused by migration of fine solid material, which was disturbed when the sample was taken and homogenized. This problem should not be serious in the field because fine material will not be disturbed from its depositional site. Even with the shorter flushing period than was used in the first test, the saturation reduction gradients and the isotherms are more advanced than the test without chemical addition (Figure 20.4). This tends to verify the results of the one-dimensional tests in that the addition of a chemical will increase the effectiveness of the flushing process.

BIOTREATMENT RESULTS

Once the soil was treated by the CROW© process, the next step was to evaluate in situ biological degradation of the residual organics. This included biological treatment of the residual soils in bench-scale slurry reactors and inocula development using the flush waters.

The primary chemicals of interest at MGP sites are polynuclear aromatic hydrocarbons (PAHs). The 16 PAH compounds included on the United States Environmental Protection Agency's (U.S. EPA's) Priority Pollutant List were the focus of this study. The following section discusses the laboratory results from biologically treating the flush water with subsequent inocula development and the treatment of residual PAHs in the soil in bench-scale slurry reactors.

Inocula Development/Flushed Water Treatment

The resultant liquid from the CROW© process was initially subjected to oil/water separation. The water was still significantly contaminated with PAHs after separation. This water was subsequently evaluated to determine if it could be treated biologically as well as serve as substrate source to grow microbial inocula, which could be used to treat the residual soil. Figure 20.5 presents a schematic of the apparatus used. Both aerobic and anaerobic (denitrifying) systems were evaluated. Continuous culture reactors were operated at a 5-day hydraulic retention time (HRT) in the aerobic system and 10.4-day HRT in the denitrifying system. Table 20.1 shows the results of treatment. Although aerobic treatment was somewhat more efficient in degrading PAHs than anaerobic (denitrifying) treatment, greater than 95% degradation of total PAHs was achieved in both cases, and greater than 90% degradation of individual PAHs was achieved for most of the PAHs present.

As such, it can be concluded that PAHs in the flush water can be reduced via aerobic or anaerobic biological treatment. Subsequently, this flush water is also capable of acclimating and sustaining a robust microbial population.

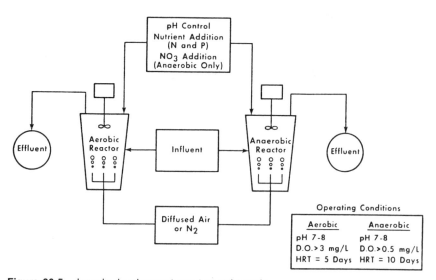

Figure 20.5. Inocula development reactors schematic.

Table 20.1.   Aerobic and Anaerobic Inocula Development Reactors—PAH Treatment

Parameters	Influent	Aerobic Effluent	Anaerobic Effluent	Aerobic (Percent Removed)	Anaerobic (Percent Removed)
TOC (mg/L)	23.8	10.4	16.4	56	31
PAHs (ug/L)					
Carbazole	<800	<80	23	90	97
Naphthalene	8180	86.9	227	99	97
Acenaphthylene	4450	19.6	108	100	98
Acenaphthene	7170	<80	<20	99	100
Fluorene	3260	11.6	34.5	100	99
Phenanthrene	6710	16.1	63.1	100	99
Anthracene	1610	6.55	49.9	100	97
Fluoranthene	2550	10.9	87.5	100	97
Pyrene	5960	<54.4	193	99	97
Benzo(A)anthracene	1280	<0.8	58.8	100	95
Chrysene	3110	<8	163	100	95
Benzo(B)fluoranthene	382	15.6	37.4	96	90
Benzo(K)fluoranthene	184	5.08	12.8	97	92
Benzo(A)pyrene	481	18.8	44.5	96	91
Dibenzo(A,H)anthracene	1490	74.8	174	95	88
Benzo(G,H,I)perylene	959	50.5	120	95	87
Indeno(1,2,3-C-D)pyrene	178	10.1	42.6	94	76
Dibenzofuran	777	<80	69.4	90	91
TOTAL PAHs	47934	467.71	1436.1	99	97

NOTE: All results are on a dry weight basis.
    The total PAHs were calculated without carbazole, 2-Methylnaphthalene and Dibenzofuran.

Slurry Reactor Treatment of Residual PAHs

The second part of these laboratory tests was to evaluate the microbial degradation of the residuals remaining after the CROW© process has removed the bulk of the contaminants from the soil. This test consisted of evaluating treatment of the residual soils in a slurry reactor both aerobically and anaerobically. The soils were mixed to a 15 to 20% slurry by weight, and microbial inocula were added. Inocula developed by biologically treating the flush water was used as a bacterial seed material for these reactors. Treatment of PAHs were monitored over a six-week period. The discussion below details the rationale for using bench-scale slurry reactors for this study, the configuration and operation of the reactors used, and the results of the treatability tests.

Results from previous treatability studies (and full-scale work) suggested that the biological treatment of an organic-contaminated soil decreases the initial organic concentration to a residual plateau concentration. The magnitude of the plateau concentration is not dictated by the quantity or robustness of the microbial population but rather by mass transfer limitations, which are a strong function of the specific soil-waste matrix. The dominance of mass transfer effects has been observed in comparisons of traditional pan microcosm studies and slurry reactor studies. In a study performed using the site soil as is, the two laboratory configurations achieve the same residual plateau concentration, but the slurry reactor achieves it in one-fourth the time or less. This is due to the design of the slurry reactor, which maximizes the rate of transfer of organics from the solids to the aqueous phase. As such, bench-scale slurry reactors have become a valuable laboratory tool. Where a pan study or an in situ study may require seven to eight months to complete, a slurry reactor study only requires one and one-half to two months to complete. This includes time for laboratory work, chemical analyses, and data interpretation. In this way, potential treatment endpoints can be determined much more rapidly than with traditional soil-pan or in situ treatability studies.

Soils were subjected to treatability testing in laboratory-scale slurry reactor vessel. Soils were mixed with buffered water in ratios of 1:9 to 1:4 (weight basis), depending on the ability of the reactor unit to suspend the soil. Figure 20.6 illustrates the laboratory slurry reactor used. The primary reactor consists of a 3-gallon stainless steel vessel with side ports from which samples of the slurry can be obtained. The slurry is mixed with a variable-speed mixer. Oxygen is provided with the introduction of air through a submerged diffuser (dissolved oxygen levels are typically 7 mg/L). An additional port is provided on top of the reactor for the addition of acids or bases (for pH control) and nutrients (nitrogen and phosphorus). Mixing speed, temperature, air or nitrogen inflow, and pH are monitored daily. Additions of nutrients are made accordingly. Typically, an aerobic reactor is maintained at a mixing speed of 1500 rpm, pH between 7 and 7.5, air inflow at 1 to 2 L/min, nutrient levels equivalent to 15 to 25 mg/L of nitrogen and phosphorus, and dissolved oxygen greater than 0.5 mg/L.

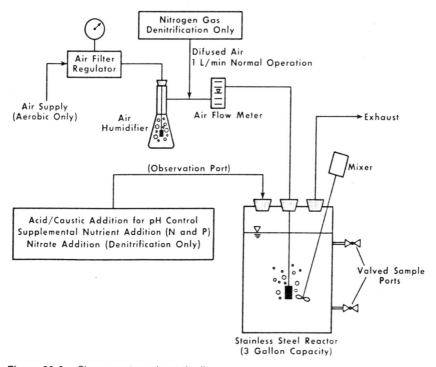

Figure 20.6. Slurry reactor schematic diagram.

Both aerobic and denitrification slurry reactor treatments were evaluated for PAHs remaining in the hot water-flushed soil. Figure 20.7 shows aerobic degradation of PAHs (total and by ring) for these soils. Ninety-five percent confidence intervals are shown for the initial and final samples. Aerobic biological treatment for six weeks further reduced the total PAHs from an average concentration of 17 mg/kg to 4 mg/kg. Figure 20.7B illustrates that biological treatment was successful in further degrading 2-, 3-, and 4-ring PAHs.

Anaerobic denitrification treatment was also evaluated for the hot water flushed soils (Figure 20.8). Ninety-five percent confidence intervals are shown for the initial and final concentrations. The degradation pattern for anaerobic treatment similar to aerobic treatment with a final PAH concentration of approximately 4 mg/kg achieved after six weeks. Figure 20.8B illustrates that biological treatment (under denitrifying conditions) was successful in further reducing the 2-, 3-, and 4-ring PAHs.

These results can also be compared with slurry reactor treatment of the site soil as is (i.e., soil that was not subjected to the CROW© process). Figure 20.9 shows aerobic and anaerobic (denitrifying) degradation for as is soil of total PAHs. Initial PAH concentrations of 160 mg/kg were reduced to approximately 6 and 30 mg/kg after six weeks by aerobic and anaerobic slurry reactor treatment,

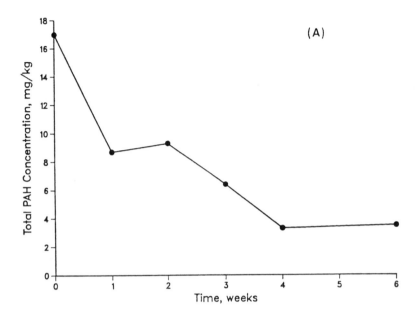

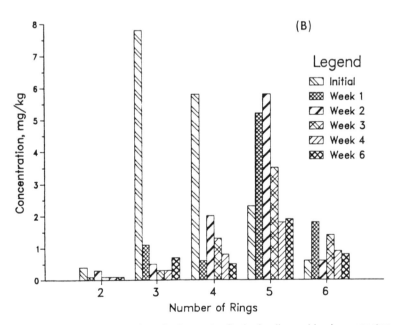

Figure 20.7.   PAH concentrations for hot water-flushed soil, aerobic slurry reactor.

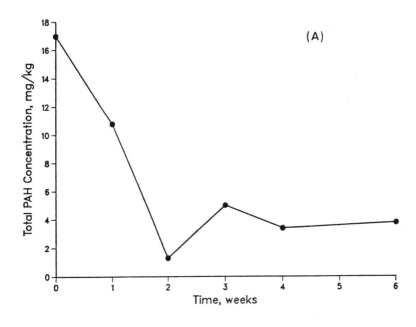

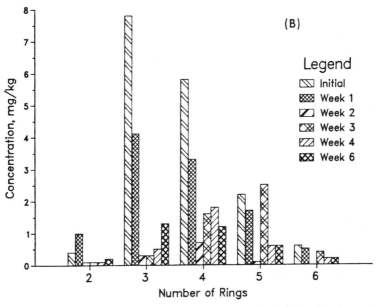

Figure 20.8.   PAH concentrations for hot water-flushed soil, denitrification slurry reactor.

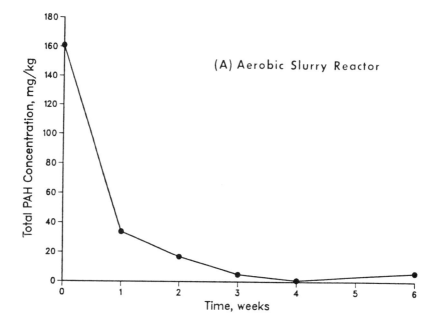

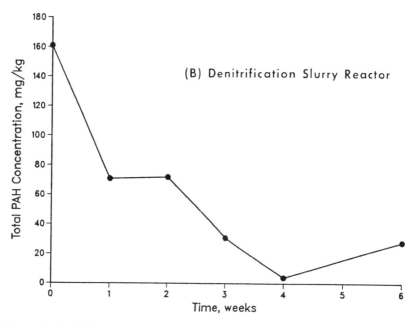

Figure 20.9.   PAH concentrations for site soil as is.

respectively. The PAHs in the soil were predominantly three and four rings. All PAHs were shown to be degraded.

CONCLUSIONS

Several preliminary conclusions can be drawn from the results of the these tests:

- The use of the CROW© process achieved over 60 wt % removal of MGP coal tars at a temperature of 156°F (69°C) and over 80 wt % removal of creosote-wood treatment waste at a temperature of 120°F (49°C) from contaminated soils.
- PCP concentrations in a soil contaminated with a wood-treatment waste can be reduced by over 99%.
- Removal efficiency of MGP coal tars was increased to over 80 wt % with surfactant addition at approximately 1% by volume at 156°F (69°C).
- CROW© process waters are treatable by both aerobic and anaerobic microorganisms.
- Both aerobic and anaerobic (denitrification) treatment was successful in degrading CROW© process residuals to levels below 5 mg/kg total PAHs.
- The CROW© process alone was capable of reducing PAHs to a level similar to the endpoints achieved by slurry reactor treatment of the site soil (total PAH concentrations of 17 mg/kg compared to 6 mg/kg). By subjecting the CROW© process-treated soil to slurry reactor treatment, the residual PAHs were reduced to nearly identical levels (4 mg/kg).

CHAPTER 21

Onsite Thermal Treatments: A Case Study in Multisource Petroleum Contamination

Robert W. Blanton and **Jeffrey Powell,** City of Richmond, Virginia

BACKGROUND

The City of Richmond operates a number of maintenance, supply, and construction service facilities throughout the city. Many of these facilities have underground and aboveground storage tanks for petroleum products and gasoline. The Bureau of Equipment Management has already embarked upon an aggressive program to precision-test all underground storage tanks (USTs) at these facilities. As a result of this annual testing, and in an effort to ensure compliance with federal, state, and local regulations governing USTs, a number of leaking tanks, fuel lines, and overspill areas have been identified by the Bureau. During calendar year 1990, over 20 USTs were removed. In some cases, evidence of past petroleum hydrocarbon contamination was found. Currently, the Bureau has identified 8,100 tons of soil that was contaminated by petroleum hydrocarbons in concentrations greater than the State Water Control Board action levels for these constituents. These contaminated soils, as well as any additional soil found to be contaminated during subsequent removal activities, must be treated, processed, and disposed of in accordance with federal, state, and local regulations. Based on past experience in tank removal, it is estimated that as many as 15,000–20,000 tons of contaminated soil may ultimately have to be treated and disposed of by the City of Richmond as it continues its compliance program. In May of 1990, the Bureau executed a contract with TPS Technologies, Inc. of

359

Apopka, Florida to thermally remediate approximately 15,000 tons of petroleum contaminated soils. A thorough examination of the project, from Request for Proposal phase to Site Closure will be presented in layman's terms.

Current Regulations

The principal regulations governing USTs within the State of Virginia are those promulgated by the State Water Control Board, VR 680-13-02. These regulations cover the design, construction, installation, operation, and closure requirements for USTs located within the state. Of particular significance for this project are Part 4 (Release Detection); Part 5 (Release Reporting, Investigation, and Confirmation); Part 6 (Release Response and Corrective Action for UST Systems Containing Petroleum or Hazardous Substances); and Part 7 (Out-of-Service UST Systems and Closure). The State Water Control Board published new guidance on contamination levels and disposal options for petroleum hydrocarbon contaminated soils associated with USTs. This guidance became effective on April 1, 1990. As a result of this guidance, petroleum hydrocarbon contaminated soils can no longer be disposed of in sanitary landfills within the state if total petroleum hydrocarbon (TPH) concentrations are greater than 500 ppm. In addition, soils with TPH concentrations greater than 50 ppm must be treated or disposed of off-site. The effect of these guideline changes has been to increase the quantity of soil that must be treated or disposed of, and has reduced disposal options available to owners and operators of such tanks and systems. The latter effect has significantly increased the cost of transportation and disposal of soils. The overall effect of these recent guideline changes has been to make it extremely difficult to dispose of such contaminated soils within the Commonwealth of Virginia unless some form of treatment is performed prior to disposal. Unless treated, most soils will have to be transported, at great expense, outside the state. The State Water Control Board has established the following action levels for the treatment of petroleum hydrocarbon contaminated soils:

- TPH concentration—< 50 ppm
- Benzene, toluene, ethylbenzene, xylenes (BTEX)—< 1 ppm
- Lead—EP toxicity lead (Pass)—< 5 ppm

Any treatment option considered for petroleum hydrocarbon contaminated soils must be able to meet these action limits. Additionally, the Virginia Department of Waste Management promulgated solid waste management regulations (VR 672-20-10) that establish siting, design, and construction requirements that must be met by any soil reclamation facility operated within the Commonwealth. These standards are covered in Sections 6.4.B and 6.4.C, and operational requirements are contained in Section 6.4.D. Finally, if any soil treatment and reclamation facility involves discharges to air, land, or water, additional permits will be required. The most directly relevant permit requirement associated with a thermal treatment facility is an air permit from the Virginia Air Pollution Control Board.

Such permit will specify limitations that must be met to operate a thermal treatment unit within the Commonwealth. Presently, these permits are granted by the local area control region, and once granted, can be used for state-wide operation.

Since regulations pertaining to the treatment and disposal of contaminated soils are evolving, we selected TPS Technologies because their Soil Remediation Unit (SRU) is flexible, has built-in safety margins when compared to existing regulatory remediation endpoints, and was prepared to respond to changing requirements. For example, it is entirely possible that air emissions permits will, at some time, be issued on a region-specific basis. In addition, local noise ordinances need to be considered in selecting treatment options and equipment. Finally, offsite disposal and transportation costs are escalating almost daily. As a result, treatment options that may not be cost-efficient or effective today could well become so in the very near future. This would be particularly true if states place severe restrictions on out-of-state disposal. In such a circumstance, soils from the State of Virginia that do not meet Commonwealth requirements for in-state disposal may not be acceptable in other locales. In short, regulatory programs are changing daily. Practices and solutions that are acceptable and economically sensible today may not be so tomorrow.

Faced with a large identified quantity of petroleum hydrocarbon contaminated soils and an evolving regulatory and cost environment, the City of Richmond astutely decided to seek a more stable and permanent solution to its problem.

PROJECT APPROACH

In this section of the chapter, the technical plan is described, demonstrating how the task at hand was accomplished and including the schedule followed, the specific equipment and specialized services that were provided by our consultants, and the resources provided the Department of Public Works. A Gantt chart is provided in Figure 21.1.

Technical Plan Overview

During the first five months of 1990, the Project Manager conducted a detailed review of soil thermal treatment technologies. This review included not only the detailed design specifications and unit processes of the equipment, but also major operation and permitting considerations that have an important impact on overall project planning. As a result, we selected TPS Technologies, whose SRU would give us the capacity, flexibility, and margins of safety to prepare a plan and schedule which we were confident met the stringent requirements of the Request for Proposal (RFP).

As part of this review, site visits were made to ongoing thermal soil treatment projects. We observed and discussed with the operators the problems that their projects experienced with materials handling, fugitive dust control, site selection, and site layout. We also discussed the actual time required for these units

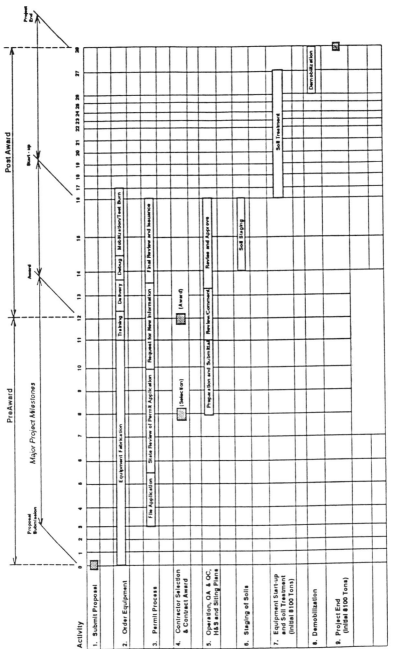

Figure 21.1. Soil Remediation Plan for city of Richmond (GANTT chart).

to obtain air and special use permits. We factored this information into our equipment selection and project approach.

Under the direction and control of the Project Manager, we also had our environmental consulting firm, Froehling & Robertson, Inc., conduct their own review of the permitting requirements associated with the operation of thermal soil treatment equipment. Mario Kuhar, Environmental Engineer, met with State Air Pollution Control Board staff members to discuss the time frames required for permitting, and to incorporate a realistic schedule for obtaining the required permits. Mr. Kuhar also met with the Department of Solid Waste Management staff members to explore in detail the requirements for system siting and operation in compliance with Sections 6.4.B and 6.4.C of Virginia's Department of Solid Waste Management Regulations, December 1988 and Virginia Regulation 672-20-10. As a result of these meetings and research, we possessed the technical expertise and confidence to comply with the required regulations.

Finally, we carefully analyzed potential problems which may be encountered during projects of this nature and selected equipment to reduce the possibility of problems to an acceptable level, or we developed simple equipment modifications to eliminate these problems. This was accomplished with Froehling & Robertson's magnitude of knowledge of local soil and geological conditions, current regulations, and our combined knowledge of the regulatory community. These contacts and the knowledge obtained facilitated anticipation and rapid resolution of potential problems.

In summary, we conducted a thorough review of all aspects of setup and operation of a thermal soil treatment unit. We possessed the knowledge and confidence to meet the RFP standards and the actual machine to accomplish our goal.

Treatment Options

A wide range of emerging technologies and options are available for the treatment of petroleum hydrocarbon contaminated soils. These options include:

- soil washing
- soil venting
- vapor extraction
- thermal treatment
- bioremediation
- steam cleaning

Most of these options can be implemented on the various sites identified to have TPH contaminated soils. Some can be done in situ. Each of these options has advantages and disadvantages. A complete, comprehensive evaluation of the relative costs and benefits of each option is beyond the scope of this chapter. With the exception of bioremediation and thermal treatment, all of the above techniques remain to be reduced to commercial state-of-the-art practice. In other words, they have promise, but have not been employed in large-scale practice across the country. As such, we did not consider them to be appropriate solutions for the City of Richmond's immediate problem of as much as 20,000 tons of contaminated soils. Moreover, the actual effectiveness and cost of these other techniques have not been well documented nor understood across a wide range of varying soil types, conditions, and locations. Accordingly, we decided that none is sufficiently developed to be a serious alternative to solve the City of Richmond's current problem.

Although bioremediation of petroleum hydrocarbon products and wastes is a developed technique, and one that appears to work well, we believe that certain features of bioremediation make it a less attractive option when compared to thermal treatment. Bioremediation remains an art, not a science. The costs, time frames for removal, and remediation endpoints are difficult to predict, and vary dramatically with soil types and local conditions. Moreover, full-scale bioremediation on site requires a large amount of space that often precludes its effective use. Finally, permit requirements and controls are difficult to predict as the process is still evolving. In short, bioremediation may be attractive, but a bioremediation plan and program would take a significant amount of time to develop and obtain approval for from state regulatory agencies. Accordingly, we did not consider it to be a feasible solution for the City.

On the other hand, thermal treatment systems have been used effectively in a number of locations to remediate petroleum hydrocarbon contaminated soils. The basic technology has been applied within the cement, asphalt, and related building products industries for nearly 100 years. Operating regimes, cost, and efficiency of the basic unit processes are well known. The principal advantages of a soil volatilization treatment techniques are:

- The basic drying and combustion technology is proven.
- A significant body of experience in using such systems has developed over the past five years.
- Soil volatilization is cost-competitive with offsite disposal, particularly now that in-State disposal has been severely restricted.
- Pollution control requirements for such technology have been in place and proven, over the past five years.
- Onsite volatilization reduces the long-term liability associated with offsite disposal options.
- The accident and other risks associated with long distance transportation are eliminated.

While thermal treatment appeared to be an effective, economical way to approach the City's problem, there were some remaining issues that will need to be resolved. These issues are discussed in more detail in the following section.

Problems Associated with Thermal Treatment Techniques

Even though the basic unit processes applied in thermal treatment techniques have been in existence for well over 100 years and some of these techniques have been applied to soil remediation, particularly during the past 10 years, not all of the problems associated with the thermal treatment of petroleum hydrocarbon contaminated waste have been resolved. Nevertheless, these problems were manageable because of the design of the TPS SRU.

The economics and efficiency of thermal treatment depends, to a large extent, on the type of the soil being processed. Moisture content, in particular, drives the economics of the system. Throughput capacity is directly related to moisture content in the soils being processed. Reductions in rated capacity can often be as high as 50% when moisture content goes above 15%. As a result, operating costs change dramatically as a function of soil moisture percentage. Many smaller thermal treatment units being developed and operated today have rated capacity in the 15 ton per hour range. In my opinion, such units will not be price-competitive with other disposal options for soil moisture levels typical of the Richmond area. Accordingly, we have selected the TPS SRU rated at 25 tons per hour to ensure that we will be price-competitive with alternate treatment techniques even at moisture contents of 18% and higher. In addition to moisture content, clay and lumpy feed material creates both materials handling problems as well as dust and fugitive emission control problems. Water quenching is a straightforward, effective solution to the dust and fugitive problem. The TPS SRU has a water quenching system incorporated onto the discharge auger. To overcome the material feed and operating problems associated with clays and lumpy material, TPS invested additional capital in conveyers and super shakers to ensure that their feed rates are optimal.

Meeting the remediation endpoints specified in the RFP should not be difficult for any properly operated thermal treatment unit. With the TPS SRU one would expect a degree of remediation for TPH and BTEX a factor of five below the RFP specifications. We chose this higher quality, more capable system because we wanted to ensure the project could continue when, and if, regulatory requirements are adjusted downward. Moreover, the types of soils to be burned for the City vary greatly. This is true because the City is located on the fall line. As a result, the thermal treatment unit will have to process a wide range of soils that are more characteristic of the Piedmont geographic region. The unit we have selected will ensure that all types of soils can be remediated properly while maintaining good throughput capacity.

Meeting air emission requirements will be essential for any treatment unit. The size, design, and operation of the baghouse and afterburner unit are the critical elements necessary to meet these requirements. The SRU we selected has a baghouse capable of handling 12,000 cubic feet per minute of air flow. This will be more than adequate to handle the most rigorous operating conditions without jeopardizing compliance with particulate standards. Here again, we believe we have a margin of safety of approximately five in comparison to present permit requirements. Afterburner temperatures in the 1600°F range will ensure complete destruction of all semivolatile organic emissions. The one area of emission concern that will remain for any thermal treatment unit is lead. None of the systems today will completely remove lead from the air stream. While remediating the soils to meet the EP toxicity lead test will not be difficult, lead removed from the soils will not be destroyed by the system. In fact, the system will volatilize the lead in the dryer into the air stream that flows through the baghouse. Not all the lead will be removed in this system, and once the air stream reaches the afterburner, the lead will be dispersed into the atmosphere. Although inorganic lead, naturally occurring, is not of concern, tetraethyl lead is a factor that must be considered in the operation of the unit. Based on preliminary dispersion calculations, we believe that any tetraethyl lead present in the soil will be dispersed in a pattern dictated by prevailing wind conditions over an area potentially as large as a quarter mile in radius. Even though the public health and environmental risk associated with tetraethyl lead dispersion is considered to be low, we believe this issue was addressed in the permitting process.

Services Provided by TPS Technologies

- Obtain permit from Department of Air Pollution Control (DAPC) to operate a portable 25 ton per hour rotary thermal kiln at a site to be designated by the Project Manager.
- Provide a certification from a professional engineer and principal corporate officer that the thermal treatment unit has been sited, and will be operated to meet the standards of 6.4.B and 6.4.C of the Commonwealth of Virginia's Department of Waste Management Regulations, December 1988 and Virginia Regulation 672-20-10.
- Certify that the soil reclamation facility will be operated in accordance with all requirements of Section 6.4.D of DWMSWMR, December 1988 and VR 672-20-10. This certification was provided by Robert W. Blanton, Project Manager, City of Richmond.
- Certify that the operating plan and contingency plans required under state regulations will be implemented and adhered to during operation of the thermal treatment unit. Once again, Robert Blanton provided this certification.
- Provide a TPS SRU rotary kiln dryer equipped with a baghouse, after-

burner, and materials handling equipment to thermally treat petroleum hydrocarbon contaminated soils to the levels specified within the RFP. The equipment provided has a rated throughput capacity of 25 tons per hour. The unit will remediate petroleum hydrocarbon contaminated soils to levels five times lower than those specified in the RFP; is equipped with state-of-the-art air pollution control devices, including a baghouse and an afterburner; has noise suppression devices to ensure below 100 decibels during operation; and is equipped with a water quench system to control fugitive dust and emissions.

- Provide all fuel to operate the treatment unit and allow provision for purchase of surplus fuel from the City, as it may become available.
- Provide all materials handling equipment and operators to process and load the contaminated soil into the treatment unit and to remove or place the treated soils at locations as directed by the Project Manager.
- Provide a Field Service Manager, site supervisor, and a crew of four technicians and equipment operators for thermal treatment unit operation and materials handling.
- Provide front-end loaders for handling contaminated feed and clean soil placement.
- Collect and analyze one sample per hour for each eight-hour period of operation. Composite these samples and analyze them using EPA methods 418.1 (TPH), EP Toxicity Lead, and BTEX (EPA Method 8020). Perform sampling and analysis necessary for feed materials control (e.g., moisture percent, particulate/grain size, and TPH, BTEX, and lead concentrations). A minimum of one soil sample will be taken from each independent soil stream.
- Conduct all sampling and analysis necessary to conform to and comply with DAPC requirements.
- Maintain all operating records and reports, including but not limited to operating temperatures, air flow, and other design parameters of the thermal treatment unit. Provide monthly progress reports to the Project Manager covering production parameters, analytical results, and any problems encountered.
- Provide site access control, visit logs, and documentation, as well as training certificates and staff documentation necessary to implement and comply with the Health & Safety Plan.
- Conform to the City's QA/QC program that specifies the exact number of trip, field, and method blanks to ensure quality analytical data.
- Stockpile all treated soils at the treatment site specified by the City.
- Provide ancillary support services such as toilet and refuse disposal/collection.

Services Provided by the City of Richmond

- A site consisting of 1.5 acres with concrete or asphalt pad.
- Potable water, approximately 25 gallons per minute at 50 pounds per square inch.
- Transportation of contaminated soil from remote sites to the SRU site.

Project Summary

Soils were trucked in from the following sites by the Department of Public Works, Simons Hauling Company, Inc., Mason Trucking Company, and various other contractors:

Department of Public Safety
  Engine Company Number 8
  Public Safety Building
Department of Public Works
  Sauers Site, Future Site of Engine Company Number 10
  Public Works Equipment Yard
Department of Recreation and Parks
  Facilities Maintenance, Forest Lawn Road
Port Commission
  Port of Richmond
Richmond Public Schools
  Bellemeade School
  Blackwell School
  Mary Munford School
  Robert E. Lee School
  Whitcomb Elementary School
  Whitcomb Court School Bus Compound
Richmond Metropolitan Authority
  Colonial Avenue Parking Deck (Under construction)

PROJECT STATISTICS

Total Tons of Soil Remediated	15,000
Average number of tons per day	150
Total Days SRU onsite	116

Figure 21.2 exhibits a typical site where soils were identified and remediated. Figure 21.3 may be used to identify potential soils for thermal remediation.

CITY OF RICHMOND
DEPARTMENT OF GENERAL SERVICES
BUREAU OF EQUIPMENT MANAGEMENT
SOIL REMEDIATION PLAN
LOCATION WORKSHEET

LOCATION 506, 810 FOREST LAWN ROAD, RECREATION & PARKS FACILITIES MAINTENANCE

RESULTS OF CHEMICAL ANALYSIS OF SOIL SAMPLES
DURING TANK REMOVAL

Sample Number	Depth (Ft.)	TPH - G. C. (mg/kg)
S-1	9.0	2430 mg/kg
S-2	9.0	1617 mg/kg
C-1	N. A.	1463 mg/kg

Note: * Sample C-1 was a composite sample obtained from the stockpiled soils.

RESULTS OF CHEMICAL ANALYSIS OF SOIL SAMPLES
AFTER ADDITIONAL EXCAVATION

Sample Number	Depth (Ft.)	TPH - G. C. (mg/kg)
S-1	18.0	< 10.0 mg/kg
S-2	18.0	< 10.0 mg/kg
S-3	18.0	< 10.0 mg/kg
S-4	18.0	< 10.0 mg/kg
S-5	18.0	< 10.0 mg/kg

RESULTS OF CHEMICAL ANALYSIS OF SOIL SAMPLES
FOR THERMAL REMEDIATION

TCLP:	
Arsenic (As)	0.01 mg/l
Cadmium (Cd)	<0.02 mg/l
Chromium (Cr)	<0.2 mg/l
Lead (Pb)	0.48 mg/l
EOX	< 30 mg/kg

SOIL REMEDIATION CHEMICAL RESULTS

DATE (SAMPLE PICKED UP)	DATE (SAMPLE RESULTS)	SAMPLE ID NUMBER	TPH - IR, mg/kg	EP TOXICITY LEAD (Pb), mg/l
11-7-90	11-12-90	C45	< 25	< 0.2
"	"	C46	< 25	< 0.2
"	"	C47 *	< 25	< 0.2

* NOTE - RESULT FOR TPH - GC IS <10 mg/kg.

Figure 21.2.   Soil Remediation Plan and worksheet for a typical site where soils were identified and remediated.

SOIL CONTAMINATED WITH PETROLEUM FROM UNDERGROUND STORAGE TANKS
THAT MAY BE THERMALLY TREATED

TESTING REQUIREMENTS

SOIL CONTAMINATED WITH	TPH	BTEX	EOX	PAINT FILTER LIQUID	EP-TOXICITY LEAD	EP-TOXICITY (FOR ALL METALS)	PCB'S	TCLP
DIESEL FUELS								
GASOLINE								
FUEL OIL								
HYDRAULIC FLUIDS								
JP-4								
KEROSENE								
MOTOR OIL								
WASTE OIL								

SOIL CONTAMINATED WITH PETROLEUM THAT MAY BE THERMALLY TREATED

TESTING REQUIREMENTS

SOIL CONTAMINATED WITH	TPH	BTEX	EOX	PAINT FILTER LIQUID	EP-TOXICITY LEAD	EP-TOXICITY (FOR ALL METALS)	PCB'S	TCLP
DIESEL FUELS								
GASOLINE								
FUEL OIL								
HYDRAULIC FLUIDS								
JP-4								
KEROSENE								
MOTOR OIL								
WASTE OIL								

NOTE, ALL PETROLEUM CONTAMINATED SOIL SHALL BE ANALYZED BY APPROVED EPA TEST METHODS.

ARSENIC, BARIUM, CADMIUM, CHROMIUM, LEAD, MERCURY, SELENIUM, AND SILVER SHALL BE ANALYZED FOR TCLP.

Figure 21.3. Worksheet for identifying potential soils for thermal remediation.

CHAPTER 22

Full-Scale, In Situ Bioremediation at a Superfund Site: A Progress Report

Michael R. Piotrowski, Woodward-Clyde Consultants, San Diego, California

INTRODUCTION

Soil and groundwater at a site in northwestern Montana were contaminated by uncontrolled releases of wood preservative wastes (primarily creosote and pentachlorophenol) between 1946 and 1969. A large groundwater plume (in excess of one mile in length) has formed in the shallowest (upper) aquifer beneath and downgradient from the site and the site was placed on the U.S. Environmental Protection Agency's (EPA's) National Priorities List (NPL) in 1983. Remedial investigations (RIs) were then initiated and conducted through 1986.[1-3]

In 1985, the feasibility study (FS) process to evaluate remedial options for treatment of the site was initiated. During the development of the FS report, a series of onsite, pilot-scale studies were performed which indicated that biological treatment could produce appreciable reductions in contaminant concentrations in the upper aquifer and in soil. As a result, the 1988 feasibility study, which contained the results of the pilot-scale studies, concluded that land (biological) treatment appeared to be the most cost-effective solution for soil remediation and that in situ biological treatment was likely the most effective option for treatment of the upper aquifer.[4] The FS report also contained a conceptual plan for full-scale remediation of the site which included the use of the two bioremediation approaches as well as a pump-and-treat system for the heavily contaminated groundwater.

In December 1988, the EPA handed down the Record of Decision (ROD) for the site which stipulated that land treatment would be used for soil remediation and that an in situ enhanced biorestoration program would be used for treatment of the upper aquifer.[5] The latter selection was a first in the Superfund program.

The remedial action plan (RAP) for treatment of the site was prepared[6] and approved by the U.S. Department of Justice and the EPA in October 1989 by consent decree (Civil Action No. CV 89-127-M-CCL). In this chapter, the RAP is described and details of remedial activities conducted to date are summarized. The chapter begins with brief descriptions of the site, the extent of contamination, the pilot-scale studies, the ROD, and the remedial action goals established for the site by the EPA.

SITE DESCRIPTION

The site is an active lumber mill located just southeast of the town of Libby in northwest Montana (Figure 22.1). It is located in an alluvial valley adjacent to the Kootenai River, and surface soils vary from clays to gravels. As a result of alluvial processes and past glaciation, the stratigraphy beneath the site is complex and consists of intermingled deposits of clays, silts, sands, and gravels.

Two groundwater-bearing units lie beneath the site. The shallowest unit (the upper aquifer) extends from approximately 12 to 70 feet below ground surface (bgs). Regional groundwater flow in the upper aquifer is primarily to the northwest, although the complex stratigraphy in the aquifer produces flow paths of variable directions on a microscale. Similarly, the complex geology of the aquifer produces broad ranges of aquifer characteristics such as transmissivity and conductivity.

Beneath the upper aquifer is a discontinuous, semi-impermeable, fine-grained layer that restricts, but does not preclude, vertical migration of groundwater. The layer is roughly 30 feet in thickness and occurs from approximately 70 to 100 feet bgs. The second water-bearing unit (the lower aquifer) lies below the fine-grained layer and extends from 100 to ~180 feet bgs. Regional groundwater flow in this unit is to the north. This aquifer exhibits indications that it is a more consolidated water-bearing unit than the upper aquifer.

EXTENT OF CONTAMINATION

Wood preservative wastes (creosote, pentachlorophenol, a diesel-like carrier, and waste sludges) were released in three primary areas of the site: an unlined waste pit, a former tank farm area, and an unlined butt dip area (Figure 22.2). The waste pit received waste sludges and other organic residues derived from the woodtreating operations and wood preservative liquids that were out of specification. Surface soils in the former tank farm area were contaminated during chemical transfer operations. In the former butt dip area, wood preservatives were

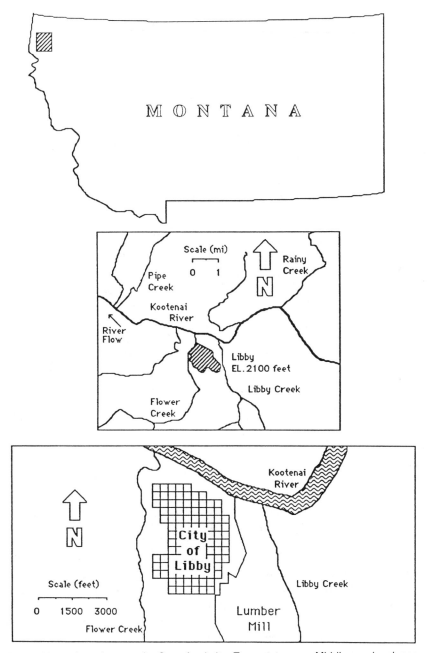

Figure 22.1. Location map for Superfund site: Top—state map; Middle—regional map; Bottom—town map.

released directly to the ground in an unlined excavation and used to treat the ends of telephone poles.

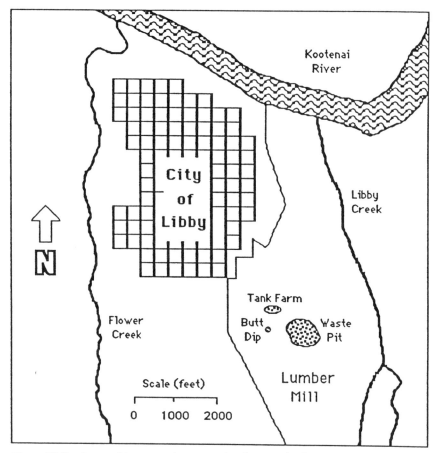

Figure 22.2.   Areas of known and suspected soil contamination.

The quantities of wood preservatives released at each area were not recorded, and therefore are not accurately quantifiable. The primary contaminants of concern are the 16 polycyclic aromatic hydrocarbon (PAH) compounds found in creosote and pentachlorophenol (PCP). Total contaminant concentrations in the soils of the three source areas range to over 5000 mg/kg.[3]

The released wood-treating fluids migrated downward and encountered the upper aquifer. Lighter-than-water organic fluids then spread out over the surface of the upper aquifer and were dispersed by groundwater flow. Light nonaqueous phase liquids (LNAPLs) occur in some areas of the site. The heavier-than-water organic fluids continued to descend through the upper aquifer, passed through the more permeable regions of the fine-grained intervening layer, and entered the lower aquifer. During downward passage of the dense, nonaqueous phase liquids (DNAPLs), sediments in the upper aquifer, the fine-grained layer, and the lower aquifer became contaminated by sorption of organic components of the liquids.

These sorbed contaminants are long-term sources of groundwater contamination. In addition, pools of DNAPL occur in the lower aquifer.

Groundwater flow distributed the wood-treating fluids downgradient from the locations of release, and by 1986, the plume of contaminated groundwater in the upper aquifer was over 1 mile in length (Figure 22.3[3]). Contaminant concentrations in the upper aquifer plume range from percent levels at the primary source area (the waste pit) to parts-per-billion (ppb) levels at the downgradient ends of the plume.

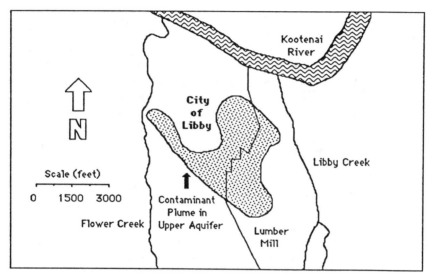

Figure 22.3.   Groundwater contaminant plume in upper aquifer. Plume in lower aquifer not shown, but it is oblong and trends to the northwest.

A number of offsite, private drinking and irrigation wells were contaminated by plume migration, and complaints by the residents of Libby with affected wells led to an EPA investigation and eventual placement of the site on the NPL. Among the first actions taken was implementation of a "Buy Water" plan in which Libby residents with contaminated wells were reimbursed for using municipally supplied water in lieu of the contaminated water from their private wells. This plan was initially implemented by the owner of the lumber mill and the city of Libby in the spring of 1986 and subsequently mandated by the EPA as the first operable unit in the fall of 1986.[7] The "Buy Water" plan has essentially resulted in the elimination of human exposure to the contaminated groundwater.

Due to concerns over the potential risks posed by the contaminant plume, and because additional private, offsite wells may become contaminated, the upper aquifer has become the immediate focus of remedial activities. In addition, because the surface soils were the three primary contaminant source areas, remediation of the soils was also deemed a priority. After consideration of various remedial technologies for treatment of contaminated groundwater and soil, two biological approaches (in situ aquifer treatment for groundwater and land treatment for soil) were selected for additional evaluations. These evaluations took the forms of onsite, pilot-scale studies.

THE PILOT-SCALE STUDIES

Aquifer Study

The pilot-scale study of in situ bioremediation of the upper aquifer was initiated in July 1987 and conducted for over one year.[4,8] A preliminary feasibility assessment had indicated that dissolved oxygen was a primary limiting factor for contaminant biodegradation within the aquifer.[9,10] Therefore, the pilot-scale study involved operation of an injection system to supply dissolved oxygen to an onsite, aquifer plume region downgradient from the waste pit area. A detailed monitoring program evaluated the effect of oxygen injection on dissolved contaminant concentrations in the treatment area.

Two injection sites were constructed within the plume approximately 750 feet downgradient from the waste pit area (Figure 22.4). Each site consisted of paired wells such that water was injected at depths approximately 15 and 30 feet bgs. Injection occurred under atmospheric pressure and gravity. The injection sites were connected by subsurface piping to an injection system that delivered a total flow rate of 40 gallons-per-minute (gpm) of filtered pond water containing 100 mg/L hydrogen peroxide to the four injection wells. Hydrogen peroxide decomposes to dissolved oxygen in the environment at a ratio of two peroxide molecules to one oxygen molecule. Therefore, the approximate oxygen concentration in the injected water was 50 mg/L.

A pair of 4-inch diameter monitoring wells was installed approximately 200 feet downgradient from the two injection sites. Each well had an extended vertical screen (15 or 20 feet in length) that was sited at a depth to sample the aquifer region under the influence of the injection system (the treatment zone). These wells served as the primary downgradient monitoring wells during the pilot-scale study. A monitoring well located just upgradient of the injection system was selected for monitoring during the study to provide information on water quality characteristics of the groundwater entering the treatment zone.

Before injection of hydrogen peroxide began, a bromide tracer study was performed to evaluate groundwater flow conditions during operation of the injection system and to verify that monitoring wells located downgradient from the injection sites were hydraulically connected to the injection system. The results of the tracer study indicated that the injection system produced a groundwater flow rate of 100 feet per day in the upper aquifer region immediately downgradient from the injection system. As such, the injected water was reaching the primary downgradient monitoring wells in less than 2 days. The tracer study also verified

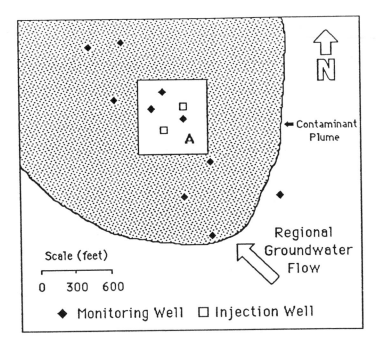

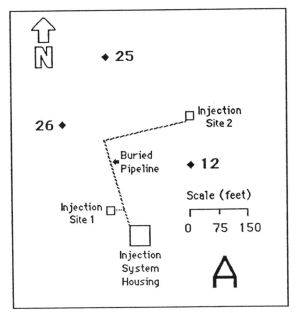

Figure 22.4.   Location of injection sites and primary monitoring wells in contaminant plume for pilot-scale study of in situ bioremediation. Inset A in top figure is expanded in lower diagram to show positions of injection sites 1 and 2; the primary monitoring wells 25, 26, and 12; and the injection system housing.

that the onsite monitoring wells downgradient from the injection sites were hydraulically connected to the injection system.

Injection of hydrogen peroxide began in July 1987, and the monitoring program began immediately thereafter. The three primary monitoring wells were purged and sampled biweekly and analyzed for dissolved oxygen (YSI Dissolved Oxygen Meter and Probe System, Yellow Springs Instrument Company, Yellow Springs, Ohio), PAH (EPA Method 8310), and PCP (EPA Method 8040) concentrations. In addition, a number of other groundwater quality parameters were periodically assessed during the study, including nutrient concentrations and total microbial densities. Finally, samples were periodically collected from a number of other monitoring wells located downgradient from the injection system and analyzed for various constituents. In all cases, strict EPA-mandated analytical and quality assurance/quality control (QA/QC) procedures were followed. Because the results of this study have been reported elsewhere,[4,8,11] only the results and implications of the dissolved oxygen and contaminant data collected at the three primary monitoring wells will be summarized.

For the first four months of the study, dissolved oxygen concentrations in the three primary monitoring wells generally remained below 1 mg/L. Such low oxygen concentrations are typically observed in organically contaminated aquifers (Piotrowski, personal observations). The low oxygen concentrations persisted in the two primary downgradient monitoring wells (Wells 25 and 26, Figure 22.5)

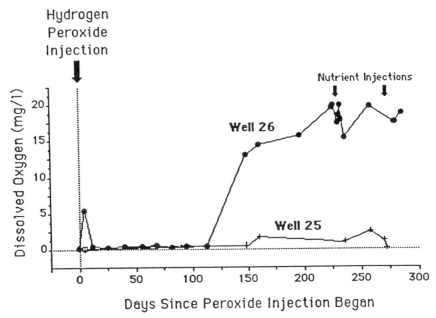

Figure 22.5.   Dissolved oxygen concentrations in groundwater over time in monitoring wells 25 and 26. Nutrient injections may have produced transient decreases in dissolved oxygen concentrations in well 26. Dissolved oxygen was measured in situ using a polarographic electrode system.

even though water containing approximately 50 mg/L of dissolved oxygen was being injected approximately 200 feet upgradient from the wells and this water was reaching the wells within two days.

Approximately 150 days after the injection program began, the dissolved oxygen concentration in one of the primary downgradient monitoring wells (Well 26) was approximately 12 mg/L (Figure 22.5). Moreover, dissolved oxygen concentrations in this well continued to increase and remained between 15 and 20 mg/L for over one year. These observations indicated that dissolved oxygen breakthrough had occurred in the aquifer region sampled by the well and that the source of oxygen was the injection system. The results suggested that a large-scale, persistent zone of elevated oxygen (an oxic zone) had been created in the aquifer that extended at least 200 feet downgradient from one of the injection sites (Figure 22.6).

The concentrations of dissolved organic contaminants in groundwater samples collected from this well showed the inverse pattern. During the first four months

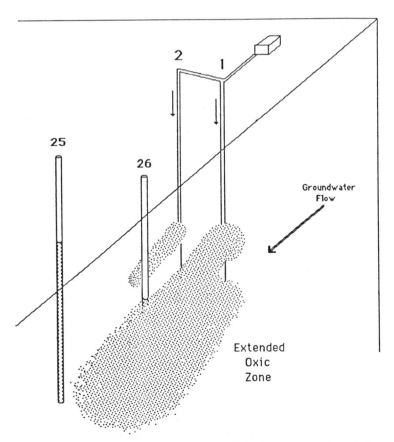

Figure 22.6. Hypothetical three-dimensional view of injection system and monitoring wells 25 and 26: view after 148 days of injection.

of the study, PAH and PCP concentrations were similar to levels observed in the upgradient control well (generally > 1000 μg/L total contaminant concentration[8]). However, once the dissolved oxygen concentration increased in the well, dissolved contaminant concentrations declined to < 20 μg/L (Figure 22.7). Furthermore, contaminant concentrations continued to decline so that after 454 days of injection, PAH compounds were not detectable in a groundwater sample collected from the well and the PCP concentration was 1 μg/L. This was the only well that exhibited oxygen breakthrough and produced groundwater samples containing reduced contaminant concentrations, indicating that the subsurface influence of the injection system varied across the treatment zone of the aquifer.

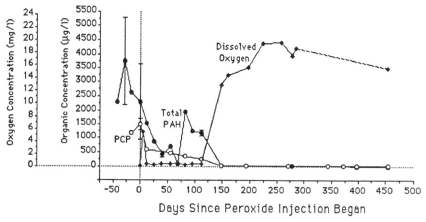

Figure 22.7.   Total PAH, PCP, and dissolved oxygen concentrations in groundwater samples collected from monitoring well 26 over time. Only detected compounds were summed for total PAH calculation. Concentrations below detection limits were set to zero. Error bars represent ± 1 standard error for occasional collections of replicate samples from a well. Dashed line for dissolved oxygen represents time period when measurements were suspended. PAH analysis by high performance liquid chromatography, PCP analysis by gas chromatography. Dissolved oxygen was measured in situ using a polarographic electrode system. EPA-approved sampling techniques and quality control/quality assurance procedures were used during all sample collections.

The results of this study demonstrated that continuous application of elevated levels of dissolved oxygen (via injection of a dilute solution of hydrogen peroxide) to an organically contaminated aquifer can produce a persistent large-scale oxic zone within the contaminant plume and that the formation of the oxic zone coincided with significant reductions in dissolved contaminant concentrations, even when multiple contaminants are present. Furthermore, under appropriate conditions, a large-scale (> 200 feet in length) oxic zone can be created in an aquifer within a matter of months.

Soil Study

In 1988, the pilot-scale study of biological treatment of the contaminated soils was conducted.[4] In this study, an 80 feet × 40 feet area adjacent to the waste pit was converted to a bermed, land treatment demonstration unit (LTDU). Baseline sampling of PAH and PCP concentrations in the soil layers below the LTDU was conducted. Approximately 60 cubic yards of screened (<3 inch) contaminated soil was placed on the unit and spread to a uniform lift depth of 6 inches. Initial PAH and PCP concentrations were measured in the soil lift.

The soil was periodically tilled and irrigated with a nutrient-microbial solution. The nutrients consisted of a dilute solution of inorganic nitrogen and phosphorus. The microorganisms were derived directly from the site's soils and grown as a mixed culture in 7000-gallon batches with molasses as the organic substrate (individual microbial species were not identified). Soil pH and moisture content were monitored and moisture adjustments were made when necessary (no pH adjustments were necessary). Triplicate composited soil samples were collected monthly for three months to quantify changes in contaminant levels in the soil lift during treatment. Soil layers below the LTDU were sampled at the end of the study, analyzed for PAH and PCP, and the results were compared to the baseline data collected before lift treatment began to evaluate contaminant migration into the underlying soil during lift treatment.

The mean (± 1 standard error) initial concentrations of PCP, total PAH, and carcinogenic PAH were 750 ± 292, 786 ± 134, and 238 ± 48 mg/kg, respectively (Table 22.1). After 100 days of treatment, the mean PCP concentration was 22 ± 0 mg/kg (97% reduction), the mean total PAH concentration was 74 ± 2 mg/kg (91% reduction), and the mean concentration of total carcinogenic PAH compounds was 63 ± 2 mg/kg (74% reduction) (Table 22.1). Little vertical migration was observed below the LTDU, indicating that contaminant reductions had been primarily the result of enhanced biodegradation. Inspection of the monthly data indicated that most of the biodegradation of the contaminants had taken place during the first 48 days of treatment.

The results of this study demonstrated that the contaminated soil could be effectively treated biologically and that large reductions in contaminant concentrations could be produced within 50 days.

1988 RECORD OF DECISION

The combined results of the FS report were submitted to the EPA by the owner of the lumber mill and Woodward-Clyde Consultants in mid-1988 with the recommendations that land (biological) treatment be used to decontaminate the source area soils and that in situ bioremediation be used to treat the contaminated regions of the upper aquifer. The EPA approved these recommendations and handed down a Record of Decision (ROD) for the site in December 1988.[5] This was the

Table 22.1. Summary of Land Treatment Demonstration Unit Results for Contaminant Concentrations in Treated Soil Over Time: 1988[a]

Date	PCP[b]	Total PAH[c]	Carcinogenic PAH[d]
July 1	750 ± 292^d	785.6 ± 133.9	237.6 ± 47.8
August 17	90 ± 13	171.2 ± 15.4	137.1 ± 15.7
September 12	24 ± 8	96.0 ± 5.2	68.6 ± 6.9
October 10	22 ± 0	73.6 ± 1.9	62.8 ± 1.8
Percent Reduction (Overall)	97	91	74

[a] Adapted from Woodward-Clyde Consultants, Reference 4.
[b] PCP = Pentachlorophenol; EPA Method 604.
[c] Total PAH = Total Concentrations of 16 Priority Pollutant Polycyclic Aromatic Hydrocarbon (PAH) Compounds. Values below detection limits set at one-half respective limit. EPA Method 610.
[d] Carcinogenic PAH = Combined concentration of 12 of the 16 Priority Pollutant Polycyclic Aromatic Hydrocarbon (PAH) Compounds considered to be or potentially carcinogenic. Values below detection limits set at one-half respective limit. EPA Method 610.
[e] Mean ± 1 Standard Error of analytical results from three composite replicates collected on each date.

first ROD in the Superfund program to support in situ bioremediation for aquifer treatment.

Summary of Remedial Action Goals

Based on risk and remedial technology performance assessments provided in the FS, the EPA's 1988 ROD for the site also established the remedial action goals that were to be achieved for the site's soil and groundwater (Table 22.2).

Table 22.2. Summary of Remedial Action Goals Established by EPA for Soil and Groundwater at the Montana Site[a]

Medium	Contaminant	Remedial Goal
Soil	Carcinogenic PAH[b]	88 mg/kg
	Naphthalene[c]	8.0 mg/kg
	Phenanthrene[c]	8.0 mg/kg
	Pyrene[c]	7.3 mg/kg
	Pentachlorophenol[c]	37.00 mg/kg
	2,3,7,8-TCDD Equivalency[d]	0.001 mg/kg
Groundwater	Total PAH[e]	400 ng/L
	Carcinogenic PAH	40 ng/L
	Pentachlorophenol	1.05 mg/L
	Benzene	5 μg/L
	Arsenic	50 μg/L
	Other Organics/Inorganics	Risk $< 10^{-5}$

[a] *Source:* EPA 1988, Reference 5.
[b] Carcinogenic PAH = (suspected) Carcinogenic Priority Pollutant Polycyclic Aromatic Hydrocarbon (PAH) compounds.
[c] Goal based on proposed treatment requirement for land disposal of K001 RCRA waste.
[d] 2,3,7,8-TCDD Equivalency = Combined concentrations of chlorinated dibenzo-p-dioxins and dibenzofurans assessed in terms of toxicity equivalency to 2,3,7,8-tetrachloro-dibenzo-p-dioxin.
[e] Total PAH = Total concentrations of 16 Priority Pollutant PAH compounds.

With respect to the soil goals, the concentrations of 2,3,7,8-TCDD ("dioxin") and other chlorinated dibenzo-p-dioxins and dibenzofurans in the site's soil have not proved to be of concern at the site. For groundwater, two remedial goals deserve mention. The concentration goals for total PAH and total carcinogenic PAH compounds are in the ng/L (parts-per-trillion) range. Current analytical capabilities are not available to consistently attain the detection limits required to observe such low concentrations of individual PAH compounds. Therefore, the final assessment of the performance of the in situ bioremediation system relative to these goals will not be possible until analytical technologies improve or the goals are altered based on realistically attainable cleanup levels which are also deemed to be protective of human health.

SUMMARY OF THE REMEDIAL ACTION PLAN

The Remedial Action Plan is presented in a conceptual diagram in Figure 22.8 and summarized below:

Contaminant Source Area Soils

Excavate contaminated soils in the three primary source areas down to the groundwater table, derock the excavated soils, stockpile and begin biological pretreatment of the screened soils in the former waste pit area. Construct a one-acre, double-lined, Land Treatment Unit (LTU) and begin treatment of contaminated soil lifts in the LTU. Use screened rocks to construct subgrade infiltration galleries upgradient from former waste pit area.

Extraction and Treatment of Heavily Contaminated Groundwater

Install groundwater recovery wells in the region downgradient from the waste pit area. Construct an aboveground, biologically based treatment system to separate and store organic phase liquids recovered by the groundwater recovery wells, and treat the aqueous phase recovered concurrently. Use the effluent from the treatment system to irrigate the LTU and the screened soils in the former waste pit area, to initiate biological treatment of the contaminated rocks, and as the influent to the rock infiltration galleries.

Phased In Situ Bioremediation Program

Over a sequence of phases, install and operate three in situ bioremediation systems for treatment of the upper aquifer. The first system installed will be sited at an intermediate location in the contaminant plume and connected to an upgraded version of the original pilot-scale system (termed the intermediate injection system). The second system installed will be sited along the site boundary and its operation is intended to treat offsite plume areas (the boundary injection system).

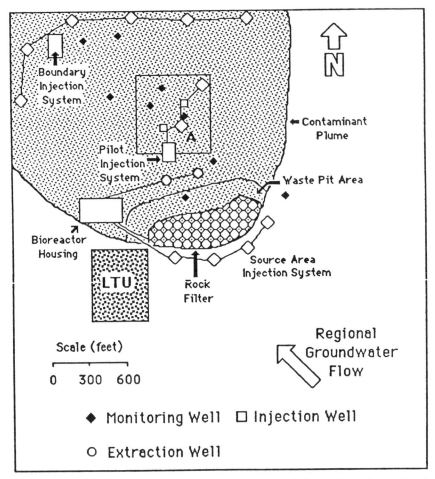

Figure 22.8. Conceptual diagram of integrated bioremediation activites for treatment of contaminated soils and groundwater at the Superfund site. See text for discussion.

The third system installed will be sited just upgradient from the waste pit area and its operation is intended to treat the most heavily contaminated region of the upper aquifer (the source area injection system).

SUMMARY OF THE REMEDIAL ACTIONS COMPLETED TO DATE

Contaminant Source Area Actions, 1989–1990

The contaminated soils in the three primary source areas have been excavated down to the water table at each area, passed through screens to remove rock debris greater than 1 inch in diameter, and the screened soils have been stockpiled in the former waste pit area. The rock debris has been stockpiled near the former waste pit area, and a portion of the rocks has been used to construct the subgrade rock infiltration galleries. Approximately 75,000 cubic yards of materials were excavated, with approximately 45,000 cubic yards of this total consisting of screened soil. Excavation activities were completed in the spring of 1990.

Land Treatment Demonstration, 1989

In 1989, construction of the one-acre LTU was completed and a land treatment demonstration was conducted. The purpose of the demonstration was to gather information on contaminant degradation rates, further evaluate the potential for the contaminants to migrate downward during treatment, and collect data to support a "No Migration Petition" (see next section).

The LTU is double-lined (with a compacted clay layer, geotextiles, and geomembranes), bermed, and contains a leachate collection system. Shallow groundwater monitoring wells were constructed up- and downgradient from the LTU. The 1989 demonstration involved the sequential application and treatment of two, 12-inch thick lifts of contaminated screened soil to the LTU. Each lift contained approximately 800 cubic yards of soil.

The first lift had been biologically pretreated (periodic tilling and irrigation) during storage in the former waste pit area before it was applied to the LTU. The results from treating this lift were anticipated to be representative of LTU treatment performance under typical operating conditions. The second lift was freshly excavated and contained recognizable fragments of essentially pure naphthalene crystals. The results from treating this lift were anticipated to be representative of LTU treatment performance under "worst-case" conditions.

The first lift was applied in July 1989. The lift was periodically tilled and irrigated as necessary. Composite soil samples were collected from four quadrants in the LTU every other week and the samples were analyzed for PAH and PCP. Leachate samples were also collected and analyzed whenever sufficient leachate accumulated. Finally, groundwater samples were periodically collected from the monitoring wells located up- and downgradient from the LTU and analyzed for PAH and PCP. Strict QA/QC procedures were mandated by EPA during the demonstration. Treatment of the first lift continued until analytical results indicated that remedial contaminant goals (Table 22.2) had been largely attained in the lift. The time requirement for treatment of this lift was approximately one month.

The second lift was applied in August 1989 and treatment of the lift continued until early November (approximately 3 months). The higher initial contaminant concentrations as well as the cooler ambient temperatures of the fall season apparently resulted in an increase in the time period needed for treatment of the second lift compared to the first.

Figures 22.9 and 22.10 present graphical summaries of the analytical results for the demonstration.[12] Although the vertical scales on several of the graphs make

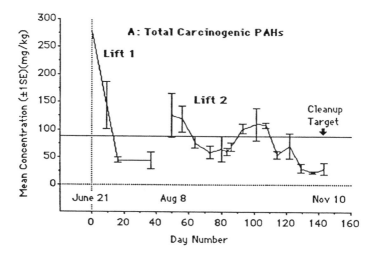

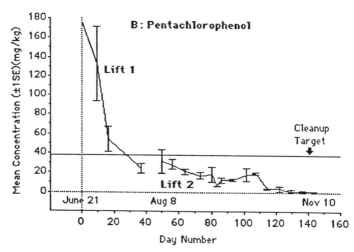

Figure 22.9. Mean concentrations of organic contaminants in soils treated in the Land Treatment Unit, 1989: (A) Total carcinogenic PAHs (polycyclic aromatic hydrocarbons; (B) Pentachlorophenol. Error bars = ± 1 Standard Error (SE). Units; mg/kg. Cleanup Targets: 88 mg/kg for total carcinogenic PAHs, 37 mg/kg for pentachlorophenol.

it difficult to examine the end points of treatment for certain targeted contaminants, all remedial goals were attained by the end of treatment of each lift except for pyrene in the first lift (Figure 22.10). At the end of treatment of this lift, the mean pyrene concentration was 9.2 mg/kg with the goal being 7.3 mg/kg.

The results of leachate analyses indicated that the contaminants were not migrating toward the liner during land treatment. In addition, the groundwater monitoring

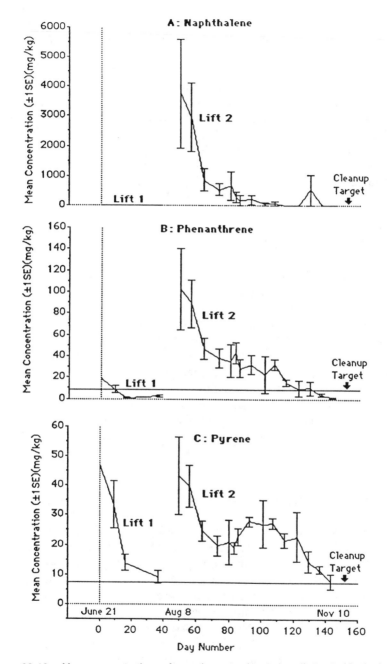

Figure 22.10. Mean concentrations of organic contaminants in soils treated in the Land Treatment Unit, 1989: (A) Naphthalene (Cleanup Target: 8 mg/kg); (B) Phenanthrene (Cleanup Target: 8 mg/kg); (C) Pyrene (Cleanup Target: 7.3 mg/kg). Error bars = ± 1 Standard Error (SE). Units: mg/kg.

results did not indicate contaminant migration outside of the lined unit. Furthermore, air quality investigations during treatment indicated that little of the contamination was volatilizing during soil manipulations. Finally, soil samples were collected periodically and analyzed for indications that microorganisms were present and active in the treated lifts. Total and viable counts were performed on the samples and appreciable numbers of total and viable microorganisms were observed during lift treatments (data not shown).

These results demonstrated that biodegradation was the major mechanism of contaminant loss. The combined results of these analyses were subsequently used to prepare a Petition for a No Migration Variance to the land disposal restrictions under RCRA. The approval of this petition by the EPA was vital for the continuation of the land treatment program of this remedial plan and the petition is discussed below.

No Migration Petition, 1990

Although EPA had formally approved the remedial plan for the site, the land disposal restrictions promulgated under the Resource Conservation and Recovery Act (RCRA) were deemed applicable to the site's soils and would restrict application of the soils to land after August 8, 1990. To obtain permission to continue soil treatment in the LTU after this date, it was necessary to formally petition the EPA by filing a "No Migration Petition" with the agency. In the petition, data must be provided to demonstrate that no migration of the contaminants from the LTU was occurring during treatment.

The No Migration Petition was prepared using the data collected in the 1989 Land Treatment Demonstration and submitted to EPA in February 1990.[12] The data demonstrated that no contaminant migration had occurred, and the EPA formally approved the petition in October 1990. Approval of the petition by the EPA was among the first granted in the Superfund Program.

Groundwater Extraction and Treatment System, 1989–1990

Nonaqueous phase liquids (NAPLs) occur in the upper aquifer in the vicinity of the former waste pit. The presence of the NAPLs will undermine the performance of the in situ bioremediation program, as they undergo biodegradation slowly. Therefore, a program was implemented in late 1989 that focuses on extracting highly contaminated groundwater and NAPLs that occur in the upper aquifer in the vicinity of the former waste pit area.

The program was initiated by the installation of two 6-inch diameter extraction wells in the upper aquifer just downgradient from the former waste pit area (see Figure 22.8 for the approximate locations of the wells).

To treat recovered groundwater, construction of an aboveground biologically based (bioreactor) treatment system to treat the extracted fluids began in the winter of 1989 in the vicinity of the former waste pit (Figure 22.8). Construction of the treatment system was completed in the spring of 1990, the extraction wells were connected to the system by subsurface piping, and operational testing of the treatment system began in March 1990.

System Components and Operation

The treatment system consists of an oil/water separator, a nutrient addition tank, and two aerobic, fixed-film bioreactor vessels (Figure 22.11). The system is designed to treat a flow rate of up to 40 gpm with a minimum, combined bioreactor retention time of approximately 8 hours. Operation of the system is controlled and continuously monitored for system flow rate and water temperatures, dissolved oxygen concentrations, and pH levels in each bioreactor from a central location.

The system separates NAPLs from the water phase and collects them in a storage tank for later recycling. The water phase is then passed by the nutrient addition tank where a dilute solution of inorganic nutrients is added to the water stream. The water is preheated prior to entering the first bioreactor by an in-line heat-exchanger system. Steam is used as the primary heat source, which is then used to heat a glycol loop which feeds the in-line heat exchanger. The water is heated to approximately 30°C. The water then enters an upflow, aerobic bioreactor vessel in which biological treatment begins. The water next passes through the second upflow, aerobic bioreactor vessel. The effluent of the second bioreactor is passed to a distribution system that directs the water to one of several sites of discharge.

In the winter, the treated water is discharged to large, subgrade infiltration galleries located upgradient from the former waste pit area. The galleries contain a portion of the rock debris generated by the screening operations of contaminated soil materials conducted in 1989. Within the galleries, the bioreactor effluent trickles down over the rock surfaces, whereupon it enters the saturated zone just upgradient from the former waste pit.

Disposal of the bioreactor effluent in this way serves two primary purposes. The surfaces of the rock debris in the infiltration galleries contain sorbed organic contaminants. As the effluent passes over the rock surfaces, biodegradation of the sorbed contaminants is stimulated. In addition, as the effluent passes into the saturated zone beneath the galleries, it is anticipated that biodegradation of contaminants below the water table will be stimulated.

In warmer seasons, the bioreactor effluent may be used to either irrigate the soil lifts in the LTU or the abovegrade piles of residual rock debris stockpiled on the site. Again, the surfaces of the rock debris contain sorbed contamination, and the intent of rock irrigation is to stimulate biodegradation of the sorbed contaminants. The effluent may also be used to irrigate the screened soils stockpiled in the former waste pit area to stimulate biodegradation rates in the stockpiles.

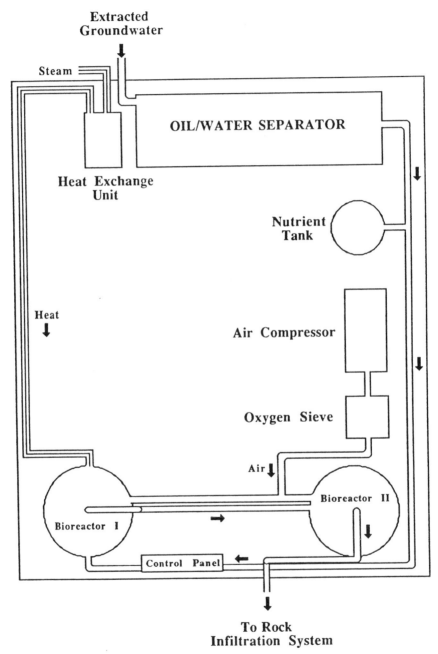

Figure 22.11. Schematic diagram of aboveground system for treatment of heavily contaminated groundwater.

Operational Testing

In the spring of 1990, operational testing of the bioreactor system produced preliminary indications of the performance capabilities of the system. System performance was evaluated at flow rates of 5, 10, and 15 gpm. Water samples were periodically collected from sampling locations on the influent line to the first bioreactor (influent location), the line between the two bioreactors (intermediate location), and the effluent line from the second bioreactor (effluent location). The samples were analyzed in the onsite chemical laboratory for PAH and PCP using the methods described previously. Strict EPA-mandated QA/QC procedures were followed.

Figure 22.12 presents a summary of the PAH data during the spring operational testing phase (data summarized from Reference 13). As can be seen, the influent total PAH concentrations were variable and ranged from several thousand to >25,000 μg/L. The mean ($\pm$ 1 standard error) influent concentration

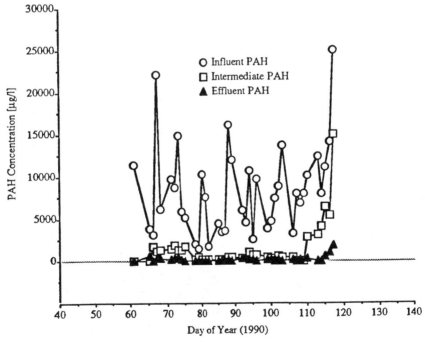

Figure 22.12.   Total PAH concentrations over time in the influent to the first bioreactor (influent, open circles), in the effluent of the first bioreactor (intermediate, open squares), and in the effluent from the second bioreactor (effluent, closed triangles) during the operational testing period of the bioreactor treatment system, spring 1990. Only detected compounds were summed for total PAH calculation. PAH analysis by gas chromatography. EPA-approved sampling techniques and quality control/quality assurance procedures were used during all sample collections.

of total PAH during the testing period was 8,609 ± 862 μg/L. An average reduction in total PAH concentration of 85% took place in the first (heated) bioreactor during the testing period, as the mean PAH concentration in samples collected from the intermediate location was 1,282 ± 377 μg/L. Additional PAH removal took place in the second bioreactor and effluent concentrations of the treatment system averaged 263 ± 70 μg/L throughout the testing period. The overall average percent reduction in the concentrations of PAH compounds induced by the bioreactor system during the testing period was 97%, and an increase in flow rate through the system did not appreciably reduce system performance.

The performance of the bioreactor system in treating PCP during this initial testing period was reduced compared to the performance in treating PAH (Figure 22.13). Influent PCP concentrations were variable and averaged 4,228 ± 552 μg/L during the test period. On average, only 10% of the influent PCP was removed in the first bioreactor, as PCP concentrations in the samples collected from the intermediate location averaged 3,808 ± 439 μg/L.[13] A similar level of PCP removal occurred in the second bioreactor, as the average concentration

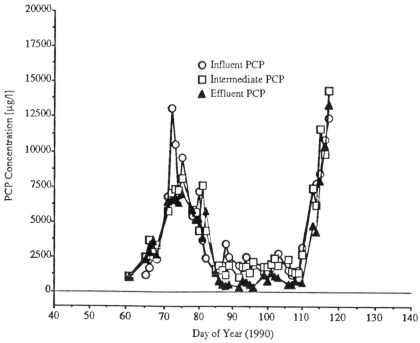

Figure 22.13.   PCP concentrations over time in the influent to the first bioreactor (influent, open circles), in the effluent of the first bioreactor (intermediate, open squares), and in the effluent from the second bioreactor (effluent, closed triangles) during the operational testing period of the bioreactor treatment system, spring 1990. PCP analysis by gas chromatography. EPA-approved sampling techniques and quality control/quality assurance procedures were used during all sample collections.

of PCP in the effluent samples of the bioreactor system was 3,348 ± 481 μg/L. As such, the overall average percent reduction in PCP concentration produced by the bioreactor system during the testing period was 21%.

The limited extent of PCP degradation relative to PAH degradation in this test period suggested that the microbial communities that had developed in the fixed films within the bioreactors were efficient at metabolizing PAH compounds but not PCP. Subsequent bioreactor testing conducted in early 1991 has produced effluent samples containing very low to nondetectable levels of PAH and PCP compounds and indicate high effectiveness in removal of both PAH and PCP compounds (Table 22.3). These observations indicate that the microbial communities within the bioreactors have recently developed the abilities to efficiently metabolize both PAH and PCP compounds, and that the performance of the bioreactor system in reducing contaminant concentrations has improved with maturity.

Table 22.3. Summary of Recent Performance of Bioreactor System[a]

Date	Contaminant	Influent Concentration[b]	Concentration After Bioreactor 1	Concentration After Bioreactor 2	Combined % Reduction
12/12/90	Total PAH[c]	14,440	570	N.D.[e]	~100
	PCP[d]	5,120	1,000	N.D.	~100
12/19/90	Total PAH	3,180	N.D.	N.D.	~100
	PCP	1,624	713	N.D.	~100
01/08/91	Total PAH	10,200	N.D.	N.D.	~100
	PCP	12,960	3,270	1,880	85.4
01/16/91	Total PAH	36,434	1,093	N.D.	~100
	PCP	4,965	652	N.D.	~100
02/02/91	Total PAH	68,984	1,096	84	99.8
	PCP	11,514	633	139	98.8

[a] Analytical data provided by the onsite laboratory.
[b] All concentrations in units of μg/L.
[c] Total PAH = Combined concentrations of detected priority pollutant polycyclic aromatic hydrocarbon (PAH) compounds; analysis by gas chromatography.
[d] PCP = Pentachlorophenol; analysis by gas chromatography.
[e] N.D. = Not Detected; approximate detection limit for individual constituents is 1 μg/L.

In Situ Demonstration Program, 1990

The complete full-scale, in situ bioremediation system for the upper aquifer will consist of three separate injection systems. One will be sited at the head of the contaminant plume (the source area injection system), the second will be sited in the vicinity of the existing pilot-scale system (the intermediate injection system), and the third will be sited along the downgradient boundary of the lumber mill (the boundary injection system). Each injection system will consist of a series of injection sites connected by subsurface piping to a water-amendment system where oxygen and nutrients will be added to the injected water.

The source area injection system will consist of several injection sites placed around the head of the contaminant plume and it will be used to perfuse oxygen and nutrients through the aquifer region that underlies the former waste pit area. In this way, biodegradation rates in the heavily contaminated aquifer sediments in this aquifer region will be stimulated. Eventually, this region of the upper aquifer will be remediated, and once this occurs, the major source of groundwater contamination will have been removed.

The injection sites of the intermediate and boundary injection systems will be sited along transects that run roughly perpendicular to the direction of regional groundwater flow (i.e., across the contaminant plume). Injection of the oxygen- and nutrient-amended water along the intermediate and boundary injection transects is anticipated to serve two purposes. First, the elevated oxygen and nutrient concentrations will stimulate biodegradation of the PCP and PAH contaminants both in the groundwater and sorbed to aquifer sediments in the immediate vicinity of each injection site. As the contaminants are metabolized, oxygen consumption rates in the immediate vicinity of the injection well screens will decline, and zones of elevated dissolved oxygen (oxic zones) will form in these regions of the upper aquifer. As injection proceeds, the oxic zones will expand and stimulate contaminant biodegradation at increasing distances downgradient from the injection sites. The expansion of the oxic zone is expected to remediate the aquifer regions through which the front of the zone passes.

Second, formation and expansion of the oxic zones produced by each injection site in the transect will eventually create an elongate oxic zone in the upper aquifer that runs perpendicular to the direction of regional groundwater flow (i.e., across the contaminant plume). Based on the results of the pilot-scale study conducted in 1987–1988, we anticipate that the extended oxic zones created by the intermediate and boundary injection systems will be biologically active zones that will metabolize dissolved PCP and PAH contaminants passing into the zones from upgradient aquifer regions.

Proper siting of adjacent injection sites for the intermediate and boundary injection systems is crucial for the formation of the elongated oxic zone along each transect. Installation and operation of too many injection sites is costly and wasteful, whereas installation of too few may be insufficient to create the elongate zone. Because the stratigraphy of the upper aquifer is complex and because the hydraulic characteristics of the upper aquifer vary considerably over short distances, we have adopted a phased approach for the implementation of the full-scale, in situ bioremediation system for the upper aquifer. In this approach, a full-scale injection site of one of the three injection systems will be installed and operated during a demonstration program. Monitoring data gathered during the demonstration program is then used to select appropriate locations for the other injection sites of the injection system. In this way, we gather information on the response of the upper aquifer to full-scale injection, gain insights to the operation and performance of each injection system, and avoid unnecessary installations of injection sites.

In late 1989, installation of the first phase of the full-scale, in situ bioremediation system for treatment of the upper aquifer was begun. Because the pilot-scale system was in place and operational, the first demonstration program conducted involved the intermediate injection system. The pilot-scale injection system was upgraded by increasing its flow capacity, a full-scale injection site (which consists of two injection wells) was constructed, and four monitoring wells were installed in the vicinity of the full-scale injection site.

A six-month test was then performed from January through June 1990 to evaluate the hydrogeological and biogeochemical influences of the operation of the full-scale injection site on this region of the upper aquifer. These data were used to define the number and appropriate spacings of the individual injection sites needed to produce adequate in situ biological treatment of this region of the upper aquifer.

Upgrade of the Pilot-Scale System

A second water filter was added to the pilot-scale injection system to allow the system to handle a higher design flow rate. The hydrogen peroxide injection system was expanded to accommodate a higher injection rate, and a continuous nutrient injection system was installed to provide low concentrations of inorganic nitrogen and phosphorus to the injected water. During the upgrade process and the demonstration program, operation of the two pilot injection sites was discontinued so the downgradient influence of the single full-scale injection site could be evaluated.

The full-scale injection site was installed at a location roughly midway between the two injection sites used in the pilot-scale study. It consisted of paired, nested wells, each 4-inches in diameter with a 20-ft wire-wrap screen section. One well was installed with its screen interval sited at depths ranging from 18 to 38 feet bgs; the second, with its screen interval sited at depths ranging from 45 to 65 ft bgs. These depths were selected to distribute the injected water across the thickness of the upper aquifer. Injector pipes were installed down each well and connected to the upgraded injection system by subsurface piping. As in the pilot-scale study, injection was performed under atmospheric pressure and gravity.

Four monitoring wells were installed in locations to augment the existing downgradient network of monitoring wells in the vicinity of the newly installed, full-scale injection site. These wells had screened intervals that allowed an evaluation of the operation of the full-scale injection site on hydrogeological aspects of the upper aquifer both downgradient and at different depths. The wells also allowed an evaluation of the extent of groundwater mounding induced by operation of the full-scale injection well.

Bromide Tracer Study

Prior to the start of the demonstration, a bromide tracer test was conducted to further characterize the hydrogeological condition, degree of hydraulic communication, and injection effects in the vicinity and downgradient from the injection site. The results of the tracer test indicated that the downgradient wells identified for monitoring during the demonstration program were hydraulically connected to the full-scale injection well.[13] It was also observed that some wells were more connected than others in the near-field region. As a result, ground-water flow rates estimated by the tracer test ranged from 50 to 300 ft/day.

The Demonstration Program

In January 1990, flow through the full-scale injection site was begun. Valves on the two injection wells were adjusted to allow the maximum rates of injection for the wells. In a pattern similar to that observed in the pilot-scale study, the deeper injection well (screened lower in the upper aquifer) could not receive as much injected water as the shallower well. The deeper well has a maximum injection flow rate of approximately 20 gpm, whereas the shallower well could inject in excess of 50 gpm. Therefore, a full-scale injection site in this region of the upper aquifer was able to deliver a total injected flow of at least 70 gpm. This flow rate was maintained throughout the course of the demonstration program.

The injected water contained a hydrogen peroxide concentration of approximately 100 mg/L and low concentrations of inorganic nutrients (ammonium chloride and potassium tripolyphosphate). The nutrient concentrations used were stoichiometrically balanced to the concentration of injected oxygen (as peroxide).

Hydrogeological Influence of the Full-Scale Injection Site

Operation of the full-scale injection site during the six-month demonstration period produced a mounding of the water table of the upper aquifer in the vicinity of the injection site that averaged approximately three feet in height.[13] Monitoring wells located downgradient and lateral to the injection site were monitored during operation of the system and during the time period immediately after operation of the injection system was interrupted. Analyses of the recoveries of select monitoring wells to static groundwater levels indicated that some mounding effects may have occurred at distances over 100 feet on either side (i.e., perpendicular to groundwater flow) of the injection site.

Biogeochemical Influence of the Full-Scale Injection Site

The biogeochemical influence of the full-scale injection site was assessed by periodically conducting dissolved oxygen monitoring surveys in the monitoring wells downgradient from and lateral to the injection site. Dissolved oxygen concentrations in wells located in regions potentially under the influence of the injection site were compared to oxygen concentrations in wells located upgradient from the site. A well that exhibited a dissolved oxygen concentration greater than normal concentrations within the contaminant plume (oxygen concentrations typically < 1 mg/L) and in pristine groundwater (typically approximately 5 mg/L) was considered under the biogeochemical influence of the injection site.

The results of the dissolved oxygen surveys indicated that the biogeochemical influence of the full-scale injection site extended over 400 feet laterally and over

1000 feet downgradient from the site.[13] Therefore, the biogeochemical influence of the injection site extended much further in the upper aquifer than the hydrogeological influence.

Conclusions of the In Situ Demonstration Program

Operation of a single, full-scale injection site ($\sim$ 70 gpm) in the intermediate region of the contaminant plume in the upper aquifer for approximately six months produced groundwater mounding effects that extended over 100 feet laterally from the injection site. In addition, evidence of biogeochemical influence of the injection site was observed over 400 feet laterally and over 1000 feet downgradient from the injection site. Furthermore, it is likely that extended operation of a full-scale injection site would increase the hydrogeological and biogeochemical influences significantly further distances in the upper aquifer.

These data indicate that the spacings between adjacent injection sites of the intermediate injection system may be at least 200 feet. The data are being used to develop the design for the intermediate injection system, which is scheduled for complete installation in 1992.

REMEDIAL ACTIONS PLANNED FOR 1991–1992

Soil Treatment

Two or more soil lifts will be applied to the LTU and treated in 1991. A second LTU will also be constructed adjacent to the original in the late spring of 1991. Operation of the two LTUs will double the site's soil treatment capacity. We will also evaluate various means to increase the rate of contaminant degradation in the soil stockpiled in the staging area by performing a bench-scale study this spring. If degradation rates can be enhanced in the staging area, treatment time requirements in the LTUs can be reduced, and more lifts can be processed in the LTUs in one season.

Groundwater Extraction and Treatment System

An additional extraction well will be sited downgradient from the former waste pit area to increase the rate of removal of LNAPL and DNAPL from this region of the upper aquifer. Refinement of the operation of the groundwater treatment system will also take place.

Intermediate Injection System

Installation of the first phase of the intermediate injection system will be completed in 1991. This will involve the installation of at least one additional full-scale injection site to complement the existing system. The existing components of the intermediate injection system (i.e., the full-scale injection site and the two pilot-scale injection sites) have been operating since June 1990, and this system has produced effects in this region of the upper aquifer that combines the effects observed during the pilot-scale study and the full-scale injection site demonstration. We have also developed a monitoring program that is intended to focus on the performance of the intermediate injection system and we will evaluate and refine the program during 1991.

Boundary Injection System

A full-scale injection site will be installed along the downgradient boundary of the lumber mill. This site may receive water amended with oxygen provided by an oxygen generator and inorganic nutrients. Use of an oxygen generator for this region of the aquifer plume is being considered because contaminant concentrations are relatively low compared to upgradient regions of the plume. If implemented, we will evaluate the effectiveness of the oxygen generator in supplying oxygen to the upper aquifer during 1992. A number of offsite wells have been selected for monitoring during this demonstration, and the primary focus of monitoring will be to assess hydrogeological and biogeochemical influences of the operation of a full-scale injection sited in the boundary region of the upper aquifer. Based on the results of the 1990 demonstration program for the intermediate injection system, operation of the boundary injection system is anticipated to have a far-reaching, beneficial impact on the contaminant plume located in offsite, downgradient areas.

Source Area Injection System

In the first phase, a full-scale injection site will be installed in an area just upgradient from the former waste pit area. A hydrogen peroxide-based amendment system will be constructed and serve as the source for the injected water in this region. The hydrogeological and biogeochemical influences of operating the well in this region of the aquifer will be assessed and the results of the assessment will be used to develop the full-scale remedial design for the source area injection system.

PLANNED REMEDIAL ACTIONS

Soil treatment will continue and performance monitoring of the full-scale remediation system will be initiated. As required, components of the remedial system will be expanded. Although it is not possible to predict when remediation of the upper aquifer will occur, we believe that we have designed and implemented a remedial system with a high likelihood of producing aquifer restoration at a faster rate than if traditional remedial approaches had been implemented. Furthermore, by adopting a biological treatment approach for the contaminated soil, we have been able to significantly reduce the costs associated with remediation of the soil contamination. Finally, we have designed and installed an integrated, biologically based, remedial system that serves to treat contaminated soil, rock debris, groundwater, and aquifer sediments in concert. Progress in the remediation of this site will be reported annually.

ACKNOWLEDGMENTS

There are a number of organizations and people who have contributed in the past and are taking active roles in the implementation of the remedial actions at the Libby site. Foremost among these is the current owner of the Libby site, Champion International (Champion). This organization inherited the contamination problem when they merged with the previous operator of the site. Champion has assumed responsibility for site remediation and has exhibited foresight and willingness to apply innovative technologies to correct the environmental problem. Specific Champion personnel who have been instrumental in the implementation of this project include James Carraway, James Davidson, Ralph Heinert, Gerald Cosgriff, and David Cosgriff.

The people of the town of Libby, Montana also deserve mention. They have participated in the refinements of the records of decision and have taken active roles in the implementation of the remedial actions, either as subcontractors or employees of Champion. Their willingness to allow the applications of innovative remedial technologies to clean up this site has contributed greatly to the progress of this project. Among these people is Michael Funk, who serves double duty as the chemistry teacher at Libby High School and the chemical analyst at the onsite laboratory.

Our regulatory agencies also deserve credit. Personnel from the Montana Department of Health and Environmental Sciences, the U.S. Environmental Protection Agency's Region VIII Montana Operations Office, and the U.S. Environmental Protection Agency's Robert S. Kerr Environmental Research Laboratory (RSKERL) have worked closely with Champion and Woodward-Clyde Consultants to produce a remedial action plan that is satisfactory to all. Included among these are Kenneth Wallace (formerly with the EPA), Julie Dalsoglio (the current site manager for the EPA), Scott Huling, Bert Bledsoe, and John Matthews (all with RSKERL).

Finally, numerous Woodward-Clyde employees have made major contributions to the progress of this project. Foremost among these is J. Robert Doyle, who is the project manager. Mr. Doyle has been the leader of this project from its inception and has demonstrated the ability to work effectively with Champion and our regulatory agencies to produce major advancements throughout the course of the project. Along with Champion, Mr. Doyle recognized the importance of conducting pilot and demonstration studies and the results of these studies served as the bases for many of the regulatory decisions handed down for the project. This approach has been a key to the progress of the project. Mr. Doyle also provided editorial comments on the manuscript for this chapter. Other past and present Woodward-Clyde employees who have contributed to the continuing success of this project include William Turner, Vanavan Ekambarum, JoAnn Tischler, Daniel Hawk, Michael Berndtson, Maxence Vermersch, Robert Junkrowski, Martin Smith, Kirk Miller, Scott Andrews, Richard Beyak, Daniel Carroll, David Nicholson, Bryan Gallagher, and Sandra Jones.

This project has been truly multidisciplinary in scope and its continuing success is a tribute to effective, constructive interaction between Champion, Woodward-Clyde Consultants, the regulatory agencies, and the people of Libby, Montana.

REFERENCES

1. "Impact of Wood Treating Operations at Libby, Montana; Phase III Field Investigation," Prepared by Alsid/Carr, Alsid and Associates, and J. R. Carr/Associates, 1985.
2. Phase IV Remedial Investigation Report, Libby Montana Ground Water Contamination Site. Prepared by Woodward-Clyde Consultants, Denver, CO. July 1986.
3. Phase IV, Step 3 Remedial Investigation Report, Libby Montana Ground Water Contamination Site. Prepared by Woodward-Clyde Consultants, Denver, CO. December 1986.
4. Feasibility Study for Site Remediation, Libby Montana. Prepared by Woodward-Clyde Consultants, Denver, CO. November 1988.
5. Work Plan for Remedial Design/Remedial Action, Ground Water Site, Libby, Montana. Prepared by Woodward-Clyde Consultants, Denver, CO. June 1989.
6. Record of Decision: Libby Ground Water Superfund Site, Lincoln County, Montana. U.S. Environmental Protection Agency Region VIII, Montana Operations Office. September 1986.
7. Piotrowski, M. R. "In Situ Biogeochemical Reduction of Hydrocarbon Contamination of Groundwater by Injecting Hydrogen Peroxide: A Case Study in a Montana Aquifer Contaminated by Wood Preservatives," Ph.D. dissertation, Boston University, Boston, MA, 1989. UMI No. 8913768.
8. In-Situ Biofeasibility Summary. Attachment to the Work Plan for Ground Water Tracer and Pilot-Scale In-Situ Biodegradation Stimulation Study at Libby, Montana. Prepared by Woodward-Clyde Consultants, Denver, CO. April 1987.
9. Piotrowski, M. R. "Improving Feasibility Studies," HazMat World 2(6)42–45 (1989).
10. Piotrowski, M. R. "U.S. EPA-Approved, Full-Scale Biological Treatment for Remediation of a Superfund Site in Montana," in E. J. Calabrese and P. T. Kostecki, Eds., Hydrocarbon Contaminated Soils, Volume 1 (Chelsea, MI: Lewis Publishers, Inc., 1991), pp. 433–457.
11. Record of Decision: Libby Ground Water Superfund Site, Lincoln County, Montana. U.S. Environmental Protection Agency Region VIII, Montana Operations Office. December 1988.
12. No Migration Petition Report, Land Treatment Units, Libby, Montana. Prepared by Woodward-Clyde Consultants, Denver, CO. February 1990.
13. Pre-Final Remedial Design Report: Upper Aquifer Operable Unit, Libby Ground Water Site, Libby, Montana. Prepared by Woodward-Clyde Consultants, Denver, CO. September 1990.

CHAPTER 23

Steam Injection to Enhance Removal of Diesel Fuel from Soil and Groundwater

John F. Dablow III, Hughes Environmental Systems, Inc., Manhattan Beach, California

BACKGROUND

Steam injection was developed by the petroleum production industry as a means of enhanced oil recovery (EOR) in the late 1950s. As reservoir pressure dissipated and high viscosity crude oils become more difficult to recover, EOR methods were developed to increase production. The reduction of viscosity and increase of the pressure gradient due to high pressure steam injection resulted in significant improvements in production rates. The application of these EOR techniques to the problem of hydrocarbon contamination of shallow soils and groundwater was first investigated in the early 1980s as concern for environmental restoration dramatically increased.

The initial research by Dr. Kent Udell and others at The University of California–Berkeley investigated the movement of nonaqueous phase liquids (NAPL) through soils. As NAPL migrates from a spill source through the soils, gravity acts as the dominant driving force. As the downward migration proceeds, a significant portion of the NAPL is either adsorbed to soil particles or occupies soil pore spaces. The amount of NAPL present in the soil in this form is termed the residual saturation. The residual saturation is dependent on several soil and NAPL factors, including grain and pore size, NAPL viscosity, interfacial tension and pore water content and can range as high as 60%. Although the vapor phase of volatile NAPL is easily removed from soil by vapor extraction (soil venting) techniques, the residual components are usually very difficult to remove. Water flushing is limited due to the immiscibility of NAPL and particularly the very low solubility of most these materials in water. In typical soils, mass balance evaluations predict that decades of water flushing is required to significantly reduce the residual saturation. Research conducted by Udell indicates that in order to achieve a factor of 10 reduction in the time period of removal, 100 times more water must be removed and treated.[1,2] As an alternative, steam injection was investigated as a means of reducing viscosity and increasing vaporization of NAPL, and consequently greatly enhancing removal efficiencies.

LABORATORY EXPERIMENTAL STUDIES

One-dimensional laboratory experiments using sand as the soil medium, were conducted at UC–Berkeley under the direction of Dr. Udell. The test apparatus consisted of a stainless steel sample chamber, steam generator and pump system, and a vapor condenser and effluent collection system, as shown in Figure 23.1. The sample chamber was heated by ceramic tape heaters to maintain adiabatic conditions. Temperatures at various points were continuously monitored by several thermocouples. Steam was injected into the sample chamber at a constant rate during the steaming period of the test. After steam injection, a vacuum was applied to the sample until completion of the test period. Vapor and liquid effluent were condensed and collected throughout the test for laboratory analysis. Tests were run on a variety of soil types contaminated with trichloroethene, benzene, toluene, gasoline, and diesel fuel and under variable steaming and vacuum extraction conditions. The results of the experiments determined that removal efficiencies due to steam injection varied from 80% to 90% for diesel fuel to

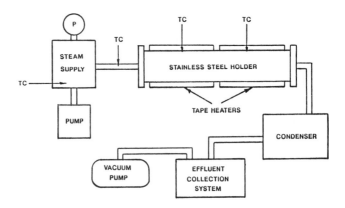

TC - THERMOCOUPLE
AFTER UDELL AND STEWART, 1989

Figure 23.1. Experimental apparatus for steam injection into contaminated soils. (After Udell and Stewart, Reference 3.)

near complete removal for the more volatile components. Several mechanisms were identified that account for the removal efficiencies observed during the experiments, including:

- vaporization of low boiling point contaminants
- enhancement of contaminant evaporation rates
- displacement of NAPL at the steam condensation front
- dilution of initial pore water concentrations
- desorption of contaminants
- boiling of interstitial water and contaminants.

PILOT PLANT

A solvent recycling and acid treatment facility in northern California was chosen as the site for the first pilot plant study in the United States. The pilot study was conducted by UC–Berkeley under grants from the Environmental Protection Agency and the National Institute of Environmental Health Sciences (NIEHS).[3] The site was contaminated principally by xylene, ethylbenzene, acetone, trichloroethene (TCE), and tetrachloroethene (PCE) which leaked from underground storage tanks or was spilled at the surface. The zone of contamination occurred in sandy silt and is constrained vertically by a clay layer located at a depth of 20 feet. Concentrations of the contaminants ranged from 500 to greater than 10,000 ppm.

The pilot system consisted of a steam generator, six injection wells spaced evenly around a 10-foot diameter circle centered on a single extraction well, a jack pump for liquid extraction, and a condenser. Steam was injected at a depth of 16 to 19 feet for a duration of 140 hours. Breakthrough to the extraction well occurred after 32 hours, as evidenced by rapidly rising temperatures in the extraction well. Temperature probes installed throughout the site indicated that the injected steam influenced a zone from 9 to 17 feet below the surface.

Continuous monitoring of the liquid and vapor effluent streams documented several significant relationships. Initial concentrations of high vapor pressure/low boiling point compounds were high and decreased sharply as steaming continued. Conversely, concentrations of low vapor pressure/high boiling point compounds increased continually after steam breakthrough. The concentrations of individual compounds in the soil decreased by roughly 2 orders of magnitude.

DUTCH "STEAM STRIPPING"

The steam injection process ("steam stripping") was developed along a parallel path and has been used successfully in The Netherlands since 1985. The Dutch process is similar to that developed by Udell in that it utilizes wet or dry steam to strip and simultaneously vaporize contaminants from the soil while a vacuum is applied. The vapors are extracted to the surface, condensed to form a liquid, and collected for disposal.

The process has been completed at sites with a variety of soil conditions that have been contaminated by gasoline, and diesel fuel. At sites impacted by gasoline, initial hydrocarbon concentrations of 10,000 to 15,000 ppm have been reduced to less than 5 ppm in periods of 4 to 9 months. Diesel fuel concentrations have been reduced from 20,000 ppm to less than 1,000 ppm in 2 months.

Due to the local site conditions, the area impacted by hydrocarbons tends to be small and shallow (groundwater depths of less than 15 feet). Therefore, portable steam systems utilizing special boilers and steam lances that are jetted into place are used rather than large-scale fixed plants. The volume of soil treated typically ranges from 1,000 to 5,000 cubic yards.

FULL-SCALE APPLICATION

Steam injection was selected to remediate the soil and groundwater contaminated by 135,0000 gallons of diesel fuel spilled from a ruptured product delivery pipe at a site in Huntington Beach, CA. The shallow alluvial soils at the site consist of interbedded, highly permeable sand, moderately permeable silt and nearly impermeable clay. An impermeable clay layer at a depth of 40 feet acts as a barrier that separates the contaminated perched groundwater and floating diesel fuel from the underlying aquifer. Groundwater monitoring of the lower aquifer confirms that diesel fuel has not penetrated the clay layer.

The selection of the injection and extraction well locations and construction specifications was a key result of the design engineering phase of the project. In addition to the subsurface conditions, the site logistics had a major impact on the design criteria. Daily operations of the facility consist of high truck traffic volume from 6:00 a.m. until 4:00 p.m., Monday through Saturday. Since operations could not be suspended or relocated economically, all site work was accomplished during off-hours. A detailed understanding of the soil lithology was developed from the initial site characterization and subsequent verification soil borings. This understanding was used to determine the steam contact zones and corresponding injection well slot configuration. A soil venting evaluation of the soil air permeability and laboratory steam tests of representative samples was used to select the extraction and injection well configuration and spacing. As a result of the engineering design, 41 injection wells and 35 extraction wells were installed to remediate the 2.3 acre plume, while minimizing the impact on the daily operations.

A schematic flow diagram of the steam injection system is shown in Figure 23.2. Low pressure, saturated steam is produced by two 25-million-BTU-per-

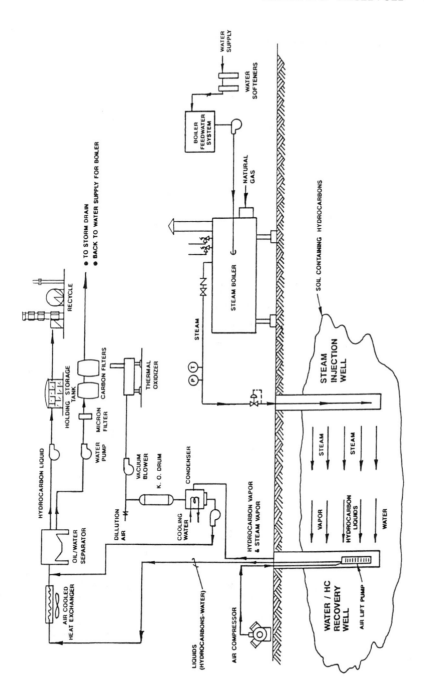

Figure 23.2.   Steam-enhanced soil remediation system.

hour boilers and injected into the soil at pressures ranging from 10 to 70 psi. A boiler feedwater conditioning system and shell and tube heat exchange system were designed to minimize boiler maintenance and utilize heat from the extracted groundwater to preheat the boiler feedwater, thereby preventing shock to the boiler system. Boiler feedwater is provided by an onsite production well.

As the steam front moves through the subsurface, hydrocarbon vapors and a hydrocarbon liquids/water mixture are generated by a variety of mechanisms, as previously discussed. A specially designed dual purpose extraction well is capable of extracting both liquids and vapors from the subsurface as shown in Figure 23.3.

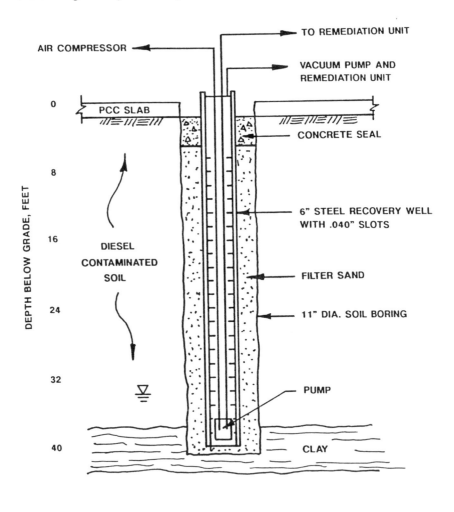

Figure 23.3. Water and vapor recovery well (typical).

Liquids are extracted by positive air displacement pumps, while vapors are evacuated by a 1,200 scfm blower. Vapors are treated in a packed bed thermal oxidizer to meet local air quality standards. The extracted liquids are pumped into an oil/water separator, with the diesel fuel being collected for offsite recycling. Contaminated groundwater and condensate are treated by a series of filters and carbon adsorption before discharge into the local storm drain in accordance with National Pollutant Discharge Elimination System (NPDES) permit conditions. It is estimated that 18 months will be required to attain cleanup levels of less than 1,000 ppm in the soil and 100 ppm in the perched water.

REFERENCES

1. Hunt, J. R., N. Sitar, and K. S. Udell. "Nonaqueous Phase Liquid Transport and Cleanup 1. Analysis of Mechanisms," *Water Resour. Res.* 24:1247–1258 (1988).
2. Hunt, J. R., N. Sitar, N., and K. S. Udell. "Nonaqueous Phase Liquid Transport and Cleanup 2. Experimental Studies," *Water Resour. Res.* 24:1259–1269 (1988).
3. Udell, K. S., and L. D. Stewart. "Mechanisms of In Situ Remediation of Soil and Ground Water Contamination by Combined Steam Injection and Vacuum Extraction," Paper presented at the Symposium on Thermal Treatment of Radioactive and Hazardous Waste at the American Institute of Chemical Engineers Annual Meeting, San Francisco, California, November 6, 1989.

CHAPTER 24

Biological Treatment:
Soil Impacted with Crude Oil

Robert S. Skiba, Texaco U.S.A., Ventura, California
Nancy Gilbertson, The Earth Technology Corporation, Long Beach, California
James J. Severns, The Earth Technology Corporation, Long Beach, California

With fluctuating crude oil mineral values and real estate values, crude oil and natural gas properties are involved in property transfers with increasing frequency. Concurrently, there has been an increased awareness of potential environmental liabilities on these properties by sellers, purchasers, and lenders. Suddenly, the historical use and existing environmental conditions have become one of the most important issues in a property transfer. When existing environmental conditions are uncertain, the timing and success of the property transfer are also uncertain. This chapter presents a case history of a soil cleanup project conducted to allow for transfer of a crude oil and natural gas production property. A summary of technical and regulatory aspects is included.

BACKGROUND ON PROPERTY

The property to be considered is an approximately 30-acre site located in the County of Los Angeles, California. This property has been used for oil and gas production operations since around 1930. Currently, the property use is limited to oil and gas production. The property has a relatively high real estate value because it is undeveloped but is located in a highly urbanized area.

As with many oil and gas production properties that were operational prior to 1960, past uses have involved multiple field operators using land depressions and excavations for collection of crude oil, waste water, drilling fluids, drippings, overflows, and other miscellaneous fluids. Several land depressions which appeared to be used for this purpose were identified in past aerial photographs of the property. The aerial photographs also identified two areas where former oil production tank batteries were located. One existing operating tank battery also was present on the property at the initiation of property sale.

INITIATION OF ENVIRONMENTAL CLEANUP

An offer was made to purchase the surface and mineral rights on the property in 1986. The sales agreement included an indemnification from the buyer to the seller for environmental liabilities on the property. The agreement also provided that prior to close of escrow, the property would be restored with the seller and buyer sharing the costs of restoration. In effect, the buyer offered to release the seller from liability but both the seller and buyer agreed that the property be cleaned up prior to escrow close.

At the time that a sales agreement was negotiated, very limited information about the environmental conditions on the property was known. However, based on an environmental site assessment conducted after agreement on the sale, a large quantity of hydrocarbon-contaminated soil was identified. Initial estimates indicated that around 13,000 cubic yards of mainly crude oil-contaminated soil could require cleanup. Later assessments would prove this to be a small portion of hydrocarbon-impacted soils at the site.

SELECTION OF CLEANUP METHOD

Selection of the cleanup method for this property involved the evaluation of several criteria. The cleanup method selected was to:

- minimize cost and future liability
- have had demonstrated past success on similar projects
- be a technically feasible approach for the contamination present
- provide confidence that the cleanup would be satisfactory
- achieve expected cleanup goal in a reasonable time
- provide flexibility to handle additional contaminated soil if encountered
- be acceptable to appropriate regulatory agencies
- be logistically possible on the property.

Biological land treatment was selected as the cleanup method for soils contaminated with crude oil on the property. The onsite management of the contaminated soils and degradation of hydrocarbons offered a way of minimizing costs and future liability. Success of past similar projects and treatability studies on the contamination from the site indicated that biological treatment would be effective in degrading the contamination present.[1] The treatability studies also demonstrated that actual treatment could be achieved in a reasonable amount of time and could be consistent with regulatory requirements.

Excavation of contaminated soil in a biological land treatment project offered a way of exposing contamination vertically and horizontally to ensure that significant contamination did not remain in place. The property had enough open space to allow for construction of treatment plots with flexibility to expand if needed. Regulatory agencies would allow biological land treatment as long as the appropriate permits and approvals were obtained.[2]

BIOLOGICAL TREATMENT PROJECT REGULATORY ISSUES

The regulatory issues in a biological treatment project are highly dependent on the location of the property where the treatment will occur. Based on the property location, the regulatory agencies that were involved in this project included the state of California Regional Water Quality Control Board—Los Angeles Region (RWQCB), the South Coast Air Quality Management District (SCAQMD), and the local city agency. The state of California Department of Health Services reviewed the project work plan but did not have a direct involvement because the contamination was a nonhazardous waste. Table 24.1 contains a list of the main agency involvement and the general requirements of each.

Table 24.1.   Agencies Involved in Project and Areas of Concern

Agency	Permit Granted	General Requirements
RWQCB	Waste discharge requirements	Established cleanup goals Soil monitoring Groundwater monitoring
SCAQMD	Air pollution	Ambient air monitoring Soil vapor screening
City	Grading permit	Compaction reports Plan drawings
DHS	None/advisory role	Regulate hazardous waste only

REQUIRED MONITORING

Soil, groundwater, and air monitoring was conducted during the excavation, treatment, and backfilling activities. Following is a summary of the monitoring requirements for each component.

Soil Monitoring

Soil sampling and analysis was conducted to determine when contaminated soil was excavated and to demonstrate when the treatment process was complete. Soil samples were analyzed for total petroleum hydrocarbons in the field using a portable infrared analyzer. Confirmatory testing was conducted on samples from the site at a certified hazardous waste laboratory. The confirmation testing requirements included analysis of most samples for total petroleum hydrocarbons (EPA Method 418.1). Many samples were analyzed for additional parameters, as summarized in Table 24.2.

Soil samples from the treatment process were collected weekly to monitor treatment progress. Soil sample locations were selected at random using a 100-foot grid system in the approximately 12 acres of treatment area.

Table 24.2.   Types of Samples, Methods Used, and Conditions Triggering Testing

Parameter	Method	Conditions
Total petroleum hydrocarbons	8015 (modified)	Random and when volatile components suspected
Priority pollutants	8240 and 8270	Random
Heavy metals	Various	Random
Aromatic volatile organics	8020	Random and when volatile components suspected
Bacteria plate count		Soils in treatment plots

Groundwater Monitoring

Several groundwater monitoring wells were required by the RWQCB to determine groundwater conditions. Groundwater samples were required to be collected and analyzed quarterly. The groundwater monitoring program included the following parameters:

- water elevation
- total dissolved solids
- pH
- total petroleum hydrocarbons
- lead
- hexavalent chromium
- total fuel hydrocarbons
- volatile organics
- semivolatile organics

Air Monitoring

A meteorological station was maintained throughout the project to record wind speed and direction, temperature, and precipitation. Air monitoring was conducted during the excavation, treatment, and backfilling. Ambient air was monitored at two permanent air stations and at various locations using a portable flame ionization detector (FID). The permanent monitoring stations were located at the expected upwind and downwind locations relative to the excavation and treatment areas. The portable FID was used to measure volatile hydrocarbon levels at perimeter locations throughout the site and at the excavations during soil removal. Samples collected at the permanent monitoring stations were required to be analyzed for benzene, toluene, xylene, and halogenated hydrocarbons by a certified air laboratory.

TECHNICAL ASPECTS OF BIOLOGICAL TREATMENT

Biological Process

Biological treatment involves the use of microorganisms to degrade the hydrocarbons present. Process elements which are necessary for a successful bioremediation include oxygen, water, nutrients, moderate temperature, and controlled pH. If one or more of these elements are not within acceptable levels for an appreciable amount of time, the process can be slowed or halted.

Air is required to provide the bacteria with the oxygen needed to convert the hydrocarbons to carbon dioxide and water. Water is required for the bacteria to live and move about. Nutrients and minerals support the diet of the bacteria. Temperature is important for the movement and activity of the bacteria. Extreme pH conditions can be toxic to the microorganisms.

Hydrocarbons are degraded directly by the bacteria or indirectly through cometabolism.[3] Cometabolism occurs from the enzymes produced by the bacteria. The ultimate end product is carbon dioxide and water.

Treatability Study Results

Two treatability studies and one control were conducted on composite contaminated soil samples from the site. The first treatability study was conducted with the addition of microorganisms. The microorganisms selected for the test are known to flourish in a slightly alkaline environment and degrade petroleum hydrocarbons. The second treatability study was conducted with the indigenous bacteria and varying levels of nutrients added. The indigenous bacteria were killed in the control so that nonbiological losses of petroleum hydrocarbons could be quantified.

After 10 days, the study which included microorganism addition showed a 41% reduction in petroleum hydrocarbons. The second study using indigenous bacteria showed a 90% reduction in 35 days. Based on these results, treatment with the indigenous bacteria was considered adequate and was selected for use on this project.

Bacteria Plate Counts

Bacteria plate counts were conducted weekly to monitor the growth and activity of the microorganisms in the treatment plot. Bacteria plate counts within the range of 10^6 to 10^9 colony-forming units are considered to reflect an acceptable bacteria growth rate for successful biodegradation. Bacteria counts below 10^6 could indicate that bacteria were not receiving enough nutrients or food source. Counts above 10^9 could indicate that the bacteria were too populated and could toxify themselves.

Nutrients/Oxygen/Temperature Conditions

The need to add nitrogen, phosphorous, and potassium to stimulate bacterial growth was minimal. A maximum of one nutrient addition was required for each lift of soil treated.

Oxygen deficiencies were not encountered. Oxygen was readily supplied during the discing, tilling, and turning of the soil. The soil was rarely if ever saturated to the point that oxygen was displaced by water.

The temperature during the operations was usually very moderate, even during the winter season. The temperature variations that did occur had little if any long-term effect on the success of the microorganisms in degrading the hydrocarbons.

Treatment Plot Design and Operation

The treatment plots were designed to comply with regulatory requirements and optimize operations. Each treatment plot was bermed to prevent run-on and run-off. The berms were constructed to support any onsite equipment to give access to the entire treatment area. Each treatment plot had an approximate 1% slope to collect rainwater. A 1-foot thick clay liner was placed below one treatment plot area for treatment of soil that may be encountered with a elevated benzene concentrations.

Once the treatment plots were constructed, contaminated soil approximately 18 inches thick was spread in the treatment plot with scrapers. The soil was disced, tilled, and turned over on a daily basis (excluding rain days and weekends). A water truck was used to maintain a soil moisture content of approximately 10%.

RESULTS SUMMARY

Biological land treatment proved to be a successful way to manage contamination at this oil and gas production property. During the project, approximately 120,000 yards of contaminated soil was treated in the land treatment plots to below the cleanup goal of 1,000 ppm total petroleum hydrocarbons. In general, remaining hydrocarbon levels in treated soil were in the 200 ppm total petroleum hydrocarbons range or lower. Cleanup goals were achieved in less than 2 months for each lift of soil treated. The treated soil was used as fill material onsite. No significant odor problems occurred during the project. Groundwater monitoring confirmed that no impact to groundwater occurred due to the biological land treatment process.

As a result of the success of the biological land treatment, the soil conditions were restored to allow the seller and buyer to proceed with the transfer of the property.

REFERENCES

1. Dragun, J. *The Soil Chemistry of Hazardous Materials* (Silver Spring, MD: The Hazardous Materials Control Research Institute, 1988), pp.415–420.
2. Kostecki, P. T., and E. J. Calabrese. *Petroleum Contaminated Soil,* Volume 1 (Chelsea, MI: Lewis Publishers, Inc., 1989), pp. 121–123.
3. Grubbs, R. B., and A. Molnna. In Situ Treatment of Troublesome Organics, Fresno, CA: CWPCA Industrial and Hazardous Waste Information Exchange, 1987.

CHAPTER 25

Passive Recovery Trench Design Under the Influence of Tidal Fluctuations

C. Y. Chiang, P. D. Petkovsky, R. A. Ettinger, I. J. Dortch, Shell Development Company, Houston, Texas

C. C. Stanley, Shell Oil Company, Houston, Texas

R. W. Hastings, Brown and Caldwell Consultants, Los Angeles, California

M. W. Kemblowski, Department of Civil and Environmental Engineering, Utah State University, Logan, Utah

INTRODUCTION

The hydrocarbon distribution facility is located in the port of Los Angeles and occupies the southwestern end of the man-made peninsula (Figure 25.1). The northern portion of the peninsula was filled in the early 1900s and the southern portion was completed in the late 1960s.

Approximately half of the northern portion of the facility is contaminated by either free product or soluble plumes. The major objective of this study was to design an intercepting (passive) trench to prevent both soluble and free product from migrating toward the ship channel. The design considers the dynamic properties of the vertically averaged hydraulic conductivity ($\overline{K}$) and the specific yield (S_y) of the aquifer under the influence of tidal fluctuations. The specific yield in an unconfined aquifer is defined as the amount of water released (retained) from storage per unit fall (rise) in the groundwater surface per horizontal area.

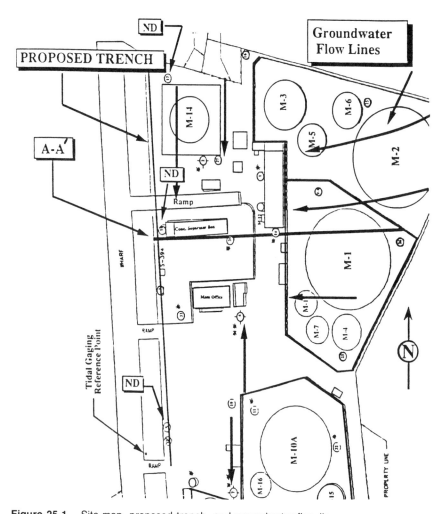

Figure 25.1.   Site map, proposed trench, and groundwater flow lines.

GROUNDWATER FLOW EQUATIONS
AND PARAMETERS ESTIMATION

Boussinesq Equation

The one-dimensional unsteady flow in a phreatic (i.e., free surface) aquifer is governed by the Boussinesq equation:[1]

$$\frac{\partial}{\partial x}\left(\overline{K}L_1\frac{\partial h}{\partial x}\right) + N = S_y\frac{\partial h}{\partial t} \tag{1}$$

where: h = the apparent groundwater level in monitoring wells for an un-
             confined aquifer (L)
       L_1 = ground surface
       N = net groundwater recharge rate per unit area (LT^{-1})
       S_y = specific yield for an unconfined aquifer (dimensionless)
       K = vertically averaged hydraulic conductivity (LT^{-1}),

$$\overline{K} = \frac{k\gamma}{\mu_w L_1} \int_0^{L_1} k_{rw}\,(\theta_w(z))\, dz \qquad (2)$$

       k   = intrinsic permeability (L^2)
       k_{rw} = relative permeability of water (dimensionless)
       m_w = dynamic water viscosity (M/LT)
       γ = specific weight (F/L^3)
       θ_w = water saturation (dimensionless)

The dependence of saturation on depth in Equation 2 is defined by the capillary-pressure saturation relationship. Recall that the relationship between saturation and height above the water table can be derived through translating the capillary pressure into capillary rise through the relation

$$h = P_c/\Delta\rho\,g \qquad (3)$$

Specific Yield

Under equilibrium conditions, the volume of water yielded by drainage of pores can be evaluated by plotting the volumetric water content θ above the water table at the beginning and at the end of a certain water-table drop Δh (Figure 25.2). The amount of water yielded by the pores at each depth increment dz is equal to $(\theta_a - \theta_b)dz$, where θ_a and θ_b are the volumetric water contents (expressed as volume fraction) at that depth before and after the water-table drop. The total volume of water released by the pores is

$$\int_0^{L_1} (\theta_a - \theta_b)\, dz \qquad (4)$$

which is the hatched area in Figure 25.2. The specific yield is then calculated by the following equation

$$S_y = \frac{\partial \int_0^{L_1} (\theta_a - \theta_b)\,dz}{\partial h} \qquad (5)$$

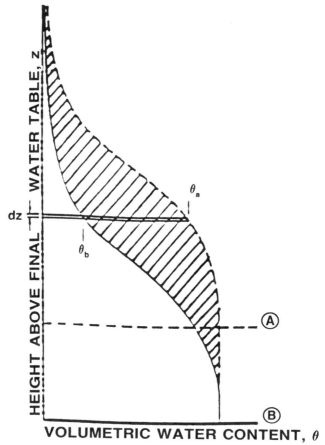

Figure 25.2. Determining specific yield from equilibrium water-content distributions above two water-table positions (A and B) (after H. Bouwer[6]).

Equation 5 is analogous to the specific moisture capacity ($d\theta/d\varphi$, where φ is the pressure head) in the Richards equation that describes the unsaturated flow.

It is of interest to note that under dynamic conditions, when the water table is lowered relatively quickly, the changes in the moisture distribution are not instantaneous and reach a new equilibrium only after a certain time interval that depends on the soil properties. A time lag will also take place when the rising tide causes the water table to rise.[1] Thus, when considering a small time interval, the apparent specific yield is dependent on the rates of change of the water-table heights. Boulton[2] developed an equation to analyze the delayed yield from transient pump tests. The rate of delayed yield (S'_y) at any time t, due to an increase in drawdown Δs during the time interval $\Delta\tau$, ($\tau < t$) is

$$\frac{dS'_y}{dt} = \Delta sa S'_y \, e^{-a(t - \tau)} \tag{6}$$

where a is an empirical constant.

Since the dynamic relationship between the changes in pressure head (or capillary pressure) and moisture content greatly influences the specific yield, it is fruitful to review the hysteretic nature of this relationship, particularly in light of the dynamic tidal fluctuations.

Capillary Pressure Hysteresis

The capillary pressure of two fluids (water and air) in a single pore is given by the Young-Laplace equation:

$$P_c = \frac{2\sigma \cos \alpha}{R_T} \tag{7}$$

where P_c is the capillary pressure, σ is the interfacial tension between the wetting and nonwetting phases, α is the contact angle between the two fluids and the solid surface, and R_T is the mean radius of curvature of the interface. Increasing the pressure difference between the wetting phase and nonwetting phase in a porous medium causes a higher capillary pressure and forces the wetting phase to reside in smaller pores. This results in a dependence of capillary pressure on saturation of the porous medium as shown in Figure 25.3.

The solid line in this figure shows the capillary pressure measurement during a drainage or reduction in wetting phase saturation process. Upon reversing the direction of flow (initiating imbibition), a new capillary pressure curve is observed as shown by the dashed line in Figure 25.3. This curve does not follow the original drainage capillary pressure-saturation relationship, indicating a hysteresis (i.e., flow direction dependence) of capillary pressure. An additional change in direction (called the secondary drainage) results in a capillary pressure functionality shown in Figure 25.3 by the dotted line that eventually overlaps the primary drainage capillary pressure curves. All other changes in flow direction will result in capillary pressure curves between the primary imbibition and secondary drainage curves.

Two mechanisms have been proposed to explain the hysteresis observed in the capillary-pressure saturation relations. First, a fluid redistribution[3] will occur when changing from the drainage to the imbibition cycles, trapping some of the nonwetting fluid in pores and consequently altering the capillary pressure curve. Another phenomena that is believed to induce the flow-direction dependence is contact-angle hysteresis.[3] From Equation 7 it is clear that changes in the contact angle can have a significant influence on the capillary pressure.

When comparing the primary drainage and imbibition capillary pressure curves, it is clear that the fluid distribution effect for hysteresis is playing a significant

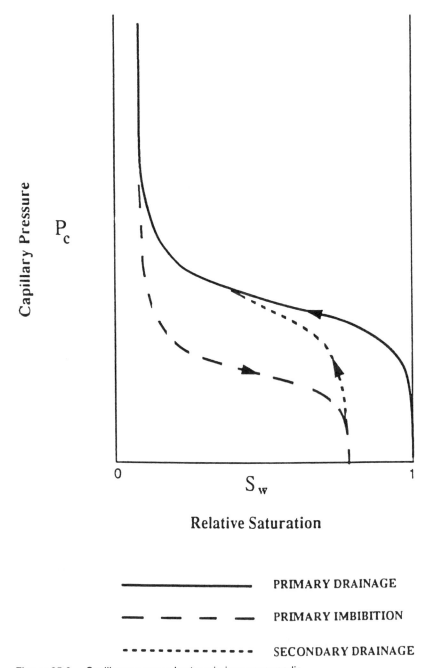

Figure 25.3.   Capillary pressure hysteresis in porous media.

role. In fact, a redistribution occurs that traps a fraction of the nonwetting phase, resulting in a new maximum wettingphase saturation as shown in Figure 25.3. It is unclear, however, which factor is the dominant hysteresis mechanism following the primary imbibition cycle. For our analysis it was not necessary to know which factor is dominant.

The Relationship Between Relative Permeability and Water Table Fluctuations

Since the relative permeability of a medium is related to the fluid distribution at a particular saturation (it is more precise to say that relative permeability is a function of the fluid distribution in the pore structure of the medium; the saturation is simply a measured property that is a result of the fluid distribution), it is no surprise that a flow dependence has also been observed in relative permeability measurements. However, Killough[4] states that relative permeability hysteresis will not be evident when the saturation varies between fully saturated and the historical minimum wetting-phase saturation. For example, consider a porous medium that is initially saturated at $S_{w,max}$ and then drained to some lower saturation, $S_{w,min}$. As long as the saturation remains between $S_{w,max}$ and $S_{w,min}$ any further imbibition or drainage relative permeabilities will be functions only of saturation, not of the flow direction. In the case that we are considering, the saturation should always remain in this regime, and one could argue that hysteresis of the relative permeability will not be evident.

The above states that the relative permeability does not depend on the direction of groundwater flow after the primary imbibition cycle has occurred. However, when the water table is moving up in the low permeability aquifer, the water is also initially flowing through the well, but at a higher velocity, due to the potential gradient and the well's high "permeability." In the meantime, the formation saturation changes lag behind. When the pressure in groundwater decreases, the situation is reversed and the monitoring well may act as a conduit, allowing the water to flow faster down the well than in the formation.[5] Thus there exists a time lag between the pressure head (as measured in a monitoring well) and that in the formation.

This time lag would cause the vertically averaged hydraulic conductivity to be tidal/groundwater level-dependent. Since the well bore functions as a conduit, it is clear that the K value would be lower during the high tide period because the formation moisture distribution lags behind. To verify the concept, slug tests were performed at two wells (wells 18 and 37) adjacent to the ship channel during both high and low tide periods. The data were analyzed by the method developed by Bouwer and Rice.[6] Table 25.1 shows the results. The results indicate that K values in the soil adjacent to the wells are higher during low tide periods.

Table 25.1.   Results of Slug Tests Performed at Wells 18 and 37

Well ID	Tidal Period	K(10^{-4} cm/sec)
18	high	1.1
18	low	2.9
37	high	4.0
37	low	4.4

LABORATORY STUDY OF THE SPECIFIC YIELD

Experimental Procedure and Results

To ascertain the influence of pressure gradient and flow direction on the specific yield, a simple laboratory-scale model shown in Figure 25.4 was used. This experimental setup consists of a 5.1-cm (2.0-in.) diameter, 91-cm (3-ft) long, glass column packed with 35 to 45 mesh, cleaned sand. The capillary height as determined from Leverett scaling[3] is approximately 15 cm (5.8 in.) and the absolute permeability calculated from the Kozeny model[3] is approximately 120 μm^2. The dry sand was packed to a height of 85 cm; then water was introduced from the bottom of the column to saturate the sand to a height of 42.5 cm. The potentiometric head of the water in the sand column was controlled by an external water source connected to the column bottom. The volume of water contained in the water level controller was maintained at a constant value by continuously pumping water from a reservoir and allowing excess water to overflow at a constant

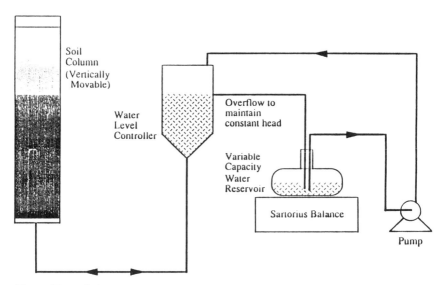

Figure 25.4.   Schematic representation of the laboratory-scale model to measure the specific yield.

specified height. The amount of water released from (or added to) the sand column was determined by measuring the weight of the water reservoir.

The specific yield was measured by adjusting the vertical position of the sand column (keeping the water level in the overflow vessel at a constant height) and then measuring the amount of water drained into (or removed from) the water reservoir. Although measurements were taken as a function of time, only the steady-state results are presented here. The specific yield is calculated as the ratio of the volume of water released from (or taken up by) the sand column per unit cross-sectional area and the change in height of the potentiometric head relative to the bottom of the sand column. This is written:

$$S_y = \frac{\text{weight change}}{(\text{height change}) \ (\text{column x-sect area})}$$

Specific yield measurements were taken over two regimes. The first, which we have termed the transient directional flow regime, involved experiments where the direction of flow was changed from the previous condition (for example, changing from imbibition to drainage). The second range, termed here the steady directional flow regime, included experiments involving a continuation of flow direction (i.e., the previous specific yield measurement was taken under the same flow cycle conditions).

Figure 25.5 shows the results of our steady-state specific yield experiments. First, note that the specific yield measurements taken in the steady directional flow regime (closed symbols) reflect an essentially constant value of about 0.2. This value is typical to that reported for a medium sand. Note that no hysteretic effects are evident; the imbibition (squares) and drainage (circles) all show the same approximate value for the specific yield. The open symbols show the specific yield values obtained upon change of flow direction (i.e., in the transient directional flow regime). Clearly, the values shown for the transient directional flow regime are substantially below the steady directional flow values. We assert that this is due to the hysteretic nature of the capillary pressure curve.

Two points should be made concerning the accuracy of these measurements. First, there is a bit of scatter in the steady flow regime values shown in Figure 25.5. This scatter is likely due to the crude measurement methods used both in determining the weight of water transferred to and from the sand column. Also, there was one systematic source of error: all experiments were performed in a hood with a face velocity of approximately 80 ft/min. This air flow can cause a measurable evaporation from the reservoir and, consequently, the drainage experiments likely underestimate the actual specific yield. Similarly, the imbibition experiments will tend to overestimate the specific yield of the sand. To address this systematic error, the evaporation rate of the static system was measured and found to be approximately 0.05 g/hr. As a typical specific yield measurement takes two hours, and at least 15 g of water were transferred to or from the reservoir, we feel that any error induced by evaporation may be ignored. Consequently,

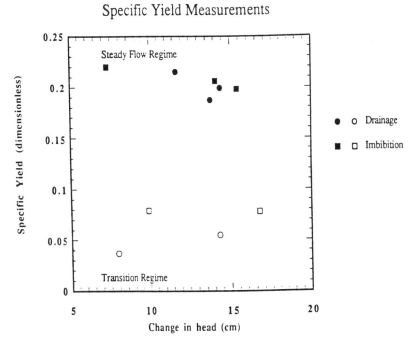

Figure 25.5.   Results of steady-state specific yield experiments.

even with the scatter evident in the steady flow regime, the difference between the transition regime and steady flow regime is quite real and not a result of experimental error.

One final note should be made concerning the significance of the transition regime and how it applies to the modeling of subsurface flow influence by tides. The semi-diurnal nature of the tide will result in periodic changes in flow direction. Consequently, values for the specific yield when modeling this situation will not be in the standard steady-flow range, but much lower values may be expected as the specific yield will be in the transition range. An idea that we have not verified yet is that the transition range is important as long as the head change is below the capillary height. For the modeling at Mormon Island with a tight soil matrix, the capillary height of the formation is likely always greater than any changes in the water level, indicating the transition regime might be valid throughout the modeled conditions.

FIELD DATA ANALYSIS

Semi-Diurnal Tides

Semi-diurnal tides at the Los Angeles ship channel were recorded hourly for a 24-hour period (Figure 25.6). Typically, tidal heights are taken to be the

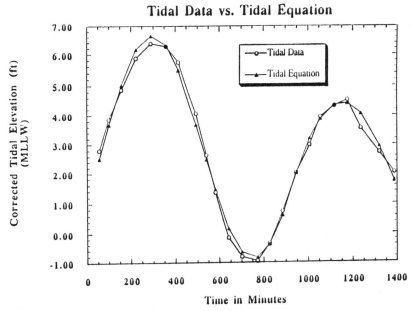

Figure 25.6. Comparison between observed tidal fluctuation data and the harmonic equation.

solution of the differential equation describing simple harmonic oscillations

$$\frac{d^2h_d(t)}{dt^2} + \omega^2 h_d(t) = 0 \qquad (8)$$

where h_d is the harmonic tidal level and $\omega = 2\pi/t$ is the natural period. The general solution to Equation 8 is

$$h_d(t) = c_1\cos\omega t + c_2\sin\omega t + c_3 \qquad (9)$$

where c_1, c_2, and c_3 are constant coefficients.

An alternate form of the solution is

$$h_d(t) = B_i + \Sigma_i A_i \sin(\omega_i t + \phi_i) \qquad (10)$$

where: A = amplitude above mean tidal elevation $= \sqrt{(c_1)^2 + (c_2)^2}$
    B = mean tidal elevation $= c_3$
    ϕ = phase angle given by $\tan\phi = \dfrac{c_1}{c_2}$

Figure 25.6 shows the comparison between the observed data and the simulated result using the general form of Equation 10. Excellent match is achieved by the following equations

$$h_d = 2.89 + 3.76 \sin\left(\frac{2\pi t}{911} - 0.477\right) \qquad \text{for } t \leq 950 \text{ minute}$$

$$h_d = 1.58 + 2.86 \sin\left(\frac{2\pi t}{962} + 0.307\right) \qquad \text{for } t \geq 950 \text{ minute}$$

These equations were used for the downgradient boundary condition.

Tidal Influences on Groundwater

Figures 25.7 and 25.8 show the pressure responses in monitoring wells along the line A-A' (Figure 25.1) caused by tidal fluctuations. These figures clearly show the tidal wave directly influences the groundwater level, and there exists a lag time for the tidal wave to propagate through the formation.

Assuming the tide fluctuates according to the uniform wave equation (Equation 8) and assuming the aquifer is under a confined condition, the analytical solution to Equation 1 (without the recharge term) is[7]

$$h = \Sigma_i A_i \exp(-x(\pi S_y/t_o \overline{K} L_1)^{0.5}) \sin$$
$$(2\pi t/t_o - x(\pi S_y/t_o \overline{K} L_1)^{0.5}) + B + Gx \qquad (11)$$

where t_o is the tidal period and G is the bulk groundwater gradient. The reduction of the amplitude inland is given by the first part of Equation 11,

$$h_r = A \exp(-x(\pi S_y/t_o \overline{K} L_1)^{0.5}) \qquad (12)$$

The time lag (t_r) between the high tide and the peak of the groundwater level (or low tide and the low point in the groundwater level) is found by equating the sine argument of Equation 11 to zero and solving for t. The solution is given by

$$t_r = x(t_o S_y/4\pi \overline{K} L_1)^{0.5} \qquad (13)$$

The above equation can be applied to unconfined flow as an approximation, if the range of tidal fluctuations is small compared with the saturated aquifer thickness

The time lag (t_r) between the high tide and the peak of the groundwater level at well 18 is observed to be 55 minutes; Equation 13, however, shows the lag time to be 100 minutes. The difference may be due to the applicability of this equation to an unconfined condition, as discussed in the preceding section.

Inland Groundwater Motion Established by Tidal-Groundwater Interactions

Figure 25.9 shows the groundwater pressure responses with time at wells 18 and 34; well 34 is located at 113 feet further inland from well 18. This figure shows only a short interval of 168 minutes (from 520 to 698 minutes, the shaded

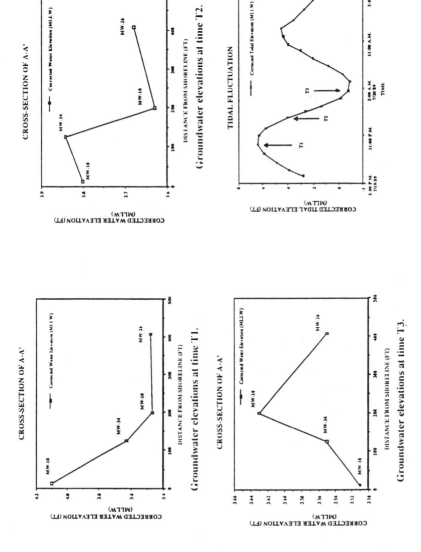

Figure 25.7. Groundwater elevations along A–A′ at the indicated times.

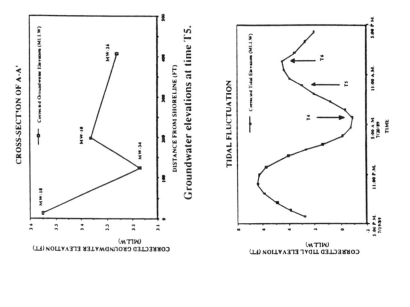

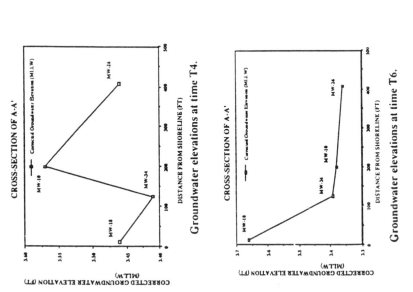

Figure 25.8. Groundwater elevations along A–A′ at the indicated times.

area) during which the groundwater flow direction is seaward, i.e., the groundwater pressure at well 34 is higher than that at well 18. If we assume the apparent wave velocity ($v = (4\pi \overline{K}L_1/t_oS_y)^{0.5}$) is 0.218 ft/minute (calculated by dividing the distance between the shoreline and well 18 by the observed lag time of 55 minutes), then in that 168 minute duration the apparent wave would be pushed back in the direction of the ocean by a distance of approximately 37 feet. Through these oscillatory motions, a Maginot line approximately 80 feet from the ocean is developed. That implies no solute or free product located inland of this Maginot line that is able to be transported beyond this Maginot line by the advective motion, provided that residual product is not present in the soil matrix. Groundwater quality analyses at monitoring wells adjacent to the shoreline (wells 18 and 37) have not recorded any detectable concentrations of soluble hydrocarbon since these wells were being installed in 1988. Although it is possible through groundwater level fluctuations dissolved oxygen can be trapped in groundwater and causes the soluble organics to aerobically biodegrade, these analyses provide support that the Maginot line concept is well founded.

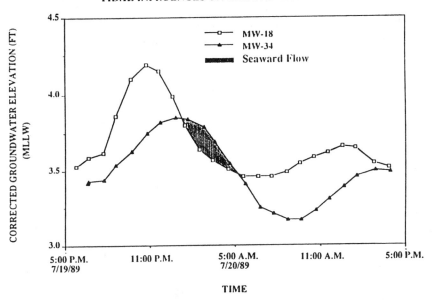

Figure 25.9. Groundwater pressure response to tidal fluctuations at wells MW-18 and MW-34.

If free product was present at well 34, a soluble plume may migrate a short distance beyond this Maginot line due to the dissolution from the residual product trapped in the soils to the ocean side of the Maginot line.

The question yet to be addressed is the total flow balance. From a preliminary data analysis, the bulk of the groundwater flow in the northern portion of the site is southerly. The vertical component of the groundwater flow has not been characterized due to insufficient data, particularly under recharge conditions from heavy rainfalls.

PASSSIVE INTERCEPTING TRENCH DESIGN

Groundwater Flow Model Calibration

Based on the tidal-dependent hydraulic conductivity values, lag time in pressure propagation, and changes in the specific yield affected by tidal fluctuations, sensitivity analyses were performed on Equation 1 to match the groundwater level fluctuations with time at well 18. Figure 25.10 shows a reasonably good match between the observed data and a simulated result achieved after many trial-and-error analyses. To produce this match, parameters were varied according to the following rudiments: (1) $\overline{K}$ values (on the average of 2.5×10^{-4} cm/sec) are

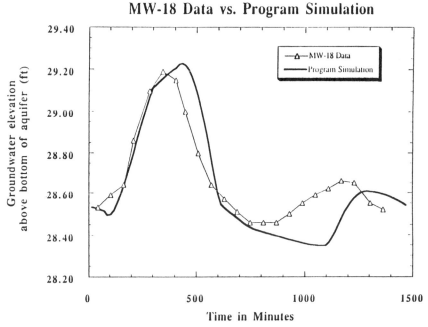

Figure 25.10. Comparison between the observed groundwater elevation at well MW-18 and a simulated result.

higher during the low tide period than that (on the average of 1.4×10^{-4} cm/sec) during the high tide period, (2) S_y values (on the average of 0.05) during the time period with steep pressure gradient, as measured in the well, are in general much less than that (between 0.1 and 0.2) reported in the literature, (3) S_y values (on the average of 0.15) during the time period with mild pressure gradient are typically higher than that during other time periods. The mild pressure gradient in a monitoring well allows the formation capillary height to catch up, and therefore these higher S_y values are caused by the delayed yield. All of these three rudiments are consistent with the theoretical considerations, field slug tests, and laboratory results described in previous sections. These calibrated parameters were used to design the intercepting trench as described next.

Interception Trench-Aquifer Interface

To design the intercepting trench depth near the central part of the property, we assume that the aquifer is hydraulically connected with the trench. The partial penetration of the trench causes vertical groundwater flow in the vicinity of the trench and additional head loss between the aquifer and the trench. The additional head loss (Δs_o) is defined as the difference in the heads between fully penetrating and partially penetrating trenches. This difference is caused by higher groundwater velocities in the immediate vicinity of the partially penetrating trenches and may be determined by a flow-net analysis depicted in Figure 25.11:[8]

$$\Delta s_o \approx \frac{1}{2} \frac{q_o}{Kh} \frac{2}{3} h \approx \frac{q_o}{K} \frac{1}{3} \tag{14}$$

where q_o is the outflow to the trench. The additional head loss may be simulated by introducing a flow connection between the aquifer and the trench.[9] The

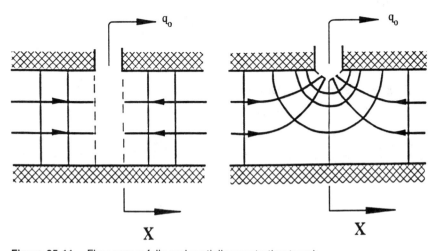

Figure 25.11. Flow near a fully and partially penetrating trench.

parameter that describes this connection is called conductance and is defined by the ratio of the flow rate q_o to the additional head loss. Shestakow[10] provided graphs for estimating the conductance for a rectangular trench (Figure 25.12). From these curves the conductance may be estimated with respect to the channel and aquifer geometry and the aquifer hydraulic conductivity. The conductance is calculated as $R = K/r$ where r is the penetration factor and is given on Figure 25.12.

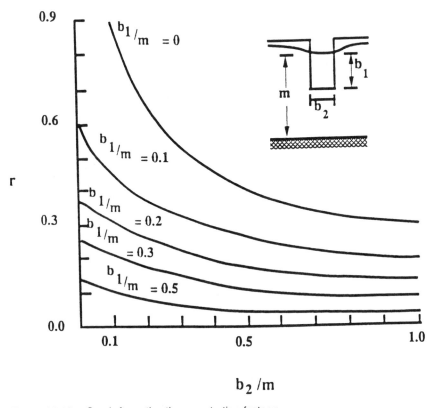

Figure 25.12.   Graph for estimating penetration factor r.

To design the passive interception trench, the vertical component of the groundwater flow in the near-field of the trench is considered through the flow balance in the vicinity of the trench

$$q\ (t) = q_o\ (t) + q'(t) \tag{15}$$

where q is the total groundwater flow (per unit length of the trench) from the upstream side of the interception trench, q_o is the flow intercepted by the trench, and q' is the flow bypassing the trench.

q (t) = i (t) T

i (t) = groundwater gradient upstream from the trench before trench installation

T = aquifer transmissivity, $T = K L_1$

q_o (t) = (h_{aq} (t) − h_{tr} (t)) R,

h_{aq} (t) = potentiometric head in the aquifer below the trench

h_{tr} (t) = potentiometric head in the trench

q' (t) = (h_{aq} (t) − h_d (t))T/L

L = distance between the trench and the downgradient boundary

h_d (t) = pressure head at the boundary downgradient from the trench and is the tidal fluctuations at the ship channel that is discussed in a previous section

The design criteria for the depth of the trench to capture the plume is that the ratio of the groundwater seepage into the trench, q_o, to the total groundwater flow from the upgradient side, q, must be greater than one to ensure a complete capture. Based on this criteria, the trench depth was solved by combining Equations 1 and 15. A block-centered finite difference numerical scheme was used to solve the combined equation, and the resulting system of algebraic equations was solved by the Gauss-Jordan elimination method. The upgradient boundary condition was specified either as a constant head or as a constant flux. The internal trench boundary condition was of the Dirichlet type.

The numerical solution showed that a trench depth of eight feet below grade, which is one feet below the mean groundwater elevation, would provide a complete capture of the plume.

CONCLUSIONS

The following conclusions are made concerning modeling the groundwater response to tidal fluctuations and the subsequent trench design:

1. Modeling groundwater flow subject to tidal fluctuation influences is strongly dependent on the flow direction and time dependent nature of both hydraulic conductivity and specific yield.
2. Slug tests indicate that the vertically averaged hydraulic conductivity is dependent on the tidal elevations. Hydraulic conductivities measured at low tides may be as much twice the values determined at high tides.

3. Laboratory experiments have shown that the steady-state specific yield is dependent on flow direction. Measured steady-state specific yield values in a sand pack show that the specific yield immediately following a change in flow direction is about 20% of the effective porosity. We assert that this flow-direction dependence is due to capillary pressure hysteresis.

4. Previous studies have also shown a time dependence for the specific yield. While we incorporated the lag time of the specific yield in an ad hoc manner, future modeling will require more insight into this time dependence nature.

5. Monitoring well pressure responses measured perpendicular to the shoreline indicate inland motion of the groundwater is more significant than seaward flow. Consequently, a boundary line is established 80 ft from the shoreline beyond which no contaminant will likely pass.

6. A passive groundwater intercepting trench 8-ft deep should be sufficient to prevent migration of contamination toward the ship channel.

REFERENCES

1. Bear, J., *Dynamics of Fluids in Porous Media,* (New York, NY: Elsevier Publishing Co., 1972).

2. Boulton, N. S., "Analysis of Data From Nonequilibrium Pumping Tests Allowing for Delayed Yield from Storage," *Proceedings of the Institution of Civil Engineers,* 26(6693):1963.

3. Collins, R. E. *Flow of Fluids Through Porous Materials,* (New York, NY: Van Nostrand Reinhold Co., 1961).

4. Killough, J. E., "Reservoir Simulation with History-Dependent Saturation Functions," *SPE Journal* 26, February 1976, pp. 37–48.

5. Kemblowski, M. W., and C. Y. Chiang, "Hydrocarbon Thickness Fluctuations in Monitoring Wells," *Ground Water* 28(2):244–252 (1990).

6. Bouwer, H., and R. C. Rice. "A Slug Test for Determining Hydraulic Conductivity of Unconfined Aquifers with Completely or Partially Penetrating Wells," *Water Resour. Res.,* 12:423–428 (1976).

7. Ingersall, L. R., O. J. Zobel, and A. C. Ingersall, *Heat Conditions: with Engineering and Geological Applications,* (New York, NY: McGraw Hill Book Company, 1948).

8. Huisman, L. *Groundwater Recovery,* (New York, NY: Winchester Press, 1972).

9. Kemblowski, M. W., and P. A. Macfarlane, "Numerical Analysis of Stream-Aquifer Systems—Model Development and Application," accepted for publication, *Ground Water,* 1991.

10. Shestakow, V. M., *Theoretical Basis of Assessing Bank Storage, Pumping and Drainage* (Moscow: Izdat, MGU).

CHAPTER 26

NAPL Removal from Groundwater by Alcohol Flooding: Laboratory Studies and Applications

Glen R. Boyd and **Kevin J. Farley,** Environmental Systems Engineering, Clemson University, Clemson, South Carolina

INTRODUCTION

Leaks of gasoline and industrial solvents from underground storage tanks pose serious threats to subsurface water supplies. The hydrocarbon mixtures which constitute these spills often exhibit low aqueous solubilities and persist in groundwater as separate, immiscible phases. These liquids are known as nonaqueous phase liquids (NAPLs). The acronym DNAPL is used to represent NAPL contaminants denser than the aqueous phase.

At a spill site, the extent of NAPL contamination depends on the fluid and mineral properties and the magnitude of the driving force.[1,2] Remediation strategies for removing NAPL contamination from groundwater typically involve the application of pump-and-treat technology. At most sites, pump-and-treat operations alone do not effectively remove the total contamination. Large residual concentrations of the contaminant are usually left behind in the formation, due to trapping of NAPL globules by capillary forces. These trapped globules can occupy 14% to 30% of the total pore volume[2] and can serve as long-term sources of groundwater contamination. Continued waterflooding (pump-and-treat) reduces the NAPL saturation by the slow process of dissolution, with complete contaminant removal often requiring years or decades.[3]

Other remedial approaches are presently being considered for removing NAPL residuals in a more timely and cost-effective manner. For example, treatment by in situ volatilization has been successful for remediation of volatile organic contaminants, particularly from the vadose zone. This process, however, may not be feasible for low volatility compounds or for DNAPLs which penetrate deep into the saturated zone. Alternatively, bioremediation programs can be implemented to remove water soluble and highly biodegradable contaminants, but the process tends to be less efficient for recalcitrant NAPL contaminants.

Chemical floods (e.g., surfactants, alcohols) can also be used to remove NAPL globules by reducing the interfacial tension (IFT) between the fluid phases. A number of petroleum industry investigators[4-6] have shown that residual oil globules can be mobilized and recovered after waterflooding by injecting alcohol solutions. Although tests have been successful in the laboratory, the alcohol flooding process is generally not profitable, and therefore not widely used in the petroleum industry. However, in the environmental field, the objective of a remediation effort is to restore a contaminated site to its original state; for this application, alcohol flooding may offer an environmentally safe and economically attractive groundwater restoration alternative.

In addition to NAPL mobilization, alcohol floods can simultaneously remove adsorbed contaminants by functioning as a cosolvent for hydrophobic material in the aqueous phase. Several studies on cosolvency effects have been conducted in the environmental field confirming that alcohol floods are effective in enhancing solubilization (and decreasing partitioning) of dissolved organic contaminants (e.g., Rao et al.,[7,8] El-Zoobi et al.,[9] Nkedi-Kizza et al.,[10,11] and Fu and Luthy[12]). Another advantage of alcohol flooding is that low molecular weight alcohols are less likely to adsorb to mineral surfaces due to their low carbon number and high molecular polarity; thus, alcohol solutions are easily displaced from the aquifer by subsequent water flushes. Finally, any alcohol that remains in the subsurface should be easily biodegraded by native microorganisms.

In our preliminary studies[13,14] we demonstrated that alcohol flooding can be used to mobilize and remove residual light and dense NAPL globules from porous media. These studies also showed that the injected alcohol concentrations were reduced to biodegradable levels by subsequent flushes with water through the treated zone. A number of issues, however, still need to be addressed to determine if alcohol flooding can provide a cost-effective alternative for groundwater remediation without posing undue risks to the environment. In this chapter, we address the factors influencing residual NAPL removal from saturated groundwaters by alcohol flooding and present preliminary design criteria for the application of the process. As part of this analysis, we examine the displacement efficiency of the injected alcohol by a subsequent waterflood and the potential migration of mobilized DNAPLs from contamination zones.

REMOVAL PROCESSES

The removal efficiency of NAPLs from contaminated zones depends on microscopic and macroscopic processes. On a microscale, fluid and soil characteristics influence the balance of viscous, gravitational, and capillary forces. The relative strength and direction of each force plays an important role during free product displacement and residual globule mobilization. Figure 26.1 schematically depicts

a) NAPL Displacement by Waterflooding

b) NAPL Displacement by the Bypass Mechanism Prior to Snap-off

c) Trapped NAPL Globules after Snap-off

d) Residual NAPL Globule Displacement by Alcohol Flooding

Figure 26.1.   Idealized behavior of NAPL contaminants in porous media during waterflooding and alcohol flooding.

the behavior of the NAPL contaminant in a water-wet formation during water and alcohol flooding operations. Figure 26.1a shows the displacement of NAPL by waterflooding from a heavily contaminated saturated zone. As water is pumped into the zone, the injected water moves faster through the smaller pores due to the effects of capillary forces.

As the waterflood continues, the NAPLs in the larger pores take the shape of small threads due to the influence of viscous forces (Figure 26.1b). As the threads get smaller, the magnitude of the capillary forces at pore constrictions (points A and B in Figure 26.1c) exceed the viscous forces and the NAPL snaps off, forming separate NAPL globules. The resultant NAPL globules remain trapped in the pore spaces because the capillary forces resisting NAPL flow are greater than the viscous forces acting to push the globules through the pore constrictions.

After waterflooding, trapped residual NAPL globules can be mobilized and removed from the subsurface by exceeding a critical viscous-to-capillary force ratio. This can be accomplished by either increasing injection velocities or by reducing the IFT between the fluid phases. The critical injection velocity can be achieved in the laboratory, but the equivalent velocity usually cannot be accomplished in the field.[15] However, in both the laboratory and the field, the IFT between the immiscible fluids can be reduced by the addition of a cosolvent (e.g., alcohol) to the flooding phase. By reducing the IFT between the immiscible phases, viscous (and gravitational) forces cause the trapped NAPL globules to become elongated (Figure 26.1d) and they are then easily displaced through the constricting pores. In this process, a portion of the NAPL contaminant is dissolved in the advancing front. As a result, the alcohol acts as both a chemical and solvent flood.

The sequence and direction of fluid injections will affect the ability of the injected fluid to contact the resident fluid. Although the efficiency of this contact is complicated by the presence of macroscopic heterogeneities, the bypassing of a resident fluid by an injected fluid can also occur in homogeneous formations due to viscous and/or gravitational instabilities. This bypassing of the resident fluid is typically referred to as fingering.[16]

The concept of fingering is shown schematically for a dipping aquifer in Figure 26.2. As the injected fluid displaces the resident fluid, a perturbation of length ϵ develops along the displacement front. This perturbation may be caused, for example, by an isolated nonuniformity in the matrix structure. The conditions under which ϵ will grow can be evaluated in terms of density difference and the mobility ratio. If the perturbation grows in time, fingering will develop, resulting in reduced sweep efficiencies of the flooding fluid. Mathematical descriptions for NAPL removal are presented in the following section.

THEORETICAL DEVELOPMENT

The flow of two immiscible fluids in porous media can be described mathematically by combining the capillary pressure relationship with the mass

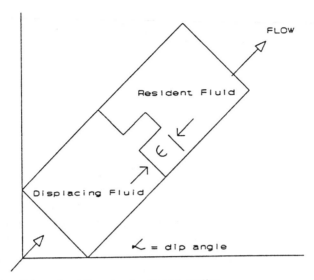

Figure 26.2.   Schematic of fingering for dipping aquifer.

conservation and the Darcy equations for each phase. For a NAPL contaminant in a linear flow system of unit cross-sectional area, the equations are given as follows:

$$\frac{\partial S_o}{\partial t} - \frac{\partial v_w}{\partial x} = 0 \tag{1}$$

$$\frac{\partial S_o}{\partial t} + \frac{\partial v_o}{\partial x} = 0 \tag{2}$$

$$v_w = \frac{-k}{\phi} \frac{k_{rw}}{\mu_w} \left[\frac{\partial P_w}{\partial x} + \rho_w g \sin \alpha \right] \tag{3}$$

$$v_o = \frac{-k}{\phi} \frac{k_{ro}}{\mu_o} \left[\frac{\partial P_o}{\partial x} + \rho_o g \sin \alpha \right] \tag{4}$$

$$P_o = P_w + P_c \tag{5}$$

Equations 1 through 4 represent the mass conservation and the Darcy equations for the aqueous and NAPL (oleic) phases, respectively. Equations 3 and 4 are based on pressure gradients and gravitational effects (where the angle α is measured from the horizontal plane). Equation 5 describes the pressure difference between the two phases in a water wet medium due to capillarity.

Equations 1 through 5 can be solved simultaneously to provide the following expression for NAPL (or water) saturation in porous media:

$$\frac{\partial S_o}{\partial t} + v_w \frac{\partial(F)}{\partial x} + \frac{k\Delta\rho g \sin \alpha}{\phi\mu_w} \frac{\partial(Fk_{rw})}{\partial x} - \frac{k}{\phi\mu_w} \frac{\partial}{\partial x}\left[Fk_{rw}\frac{\partial P_c}{\partial x}\right] = 0 \quad (6)$$

where:

$$F = \left[\frac{\mu_o k_{rw}}{k_{ro}\mu_w} + 1\right]^{-1}$$

The first term in Equation 6 represents the change in NAPL saturation with respect to time; the second term represents the viscous displacement by the flooding phase; the third term represents the differential transport of fluids due to density differences (i.e., gravitational effects); and the fourth term represents capillary suction of the wetting phase and is the primary cause of capillary end effects.

Introducing the following dimensionless parameters for distance, time, and capillary pressure (as presented by Leverett[17])

$$X = \frac{x}{L}$$

$$T = \frac{\sigma t}{L\mu_w}$$

$$P_c^* = \frac{P_c}{\sigma}\left(\frac{k}{\phi}\right)^{1/2}$$

Equation 6 can be expressed as

$$\frac{\partial S_o}{\partial T} + N_c \frac{\partial(F)}{\partial X} + \frac{N_B}{\phi} \frac{\partial(k_{rw}F)}{\partial X} - \frac{1}{L}\left(\frac{k}{\phi}\right)^{1/2} \frac{\partial}{\partial X}\left(k_{rw}F\frac{\partial P_c^*}{\partial X}\right) = 0 \quad (7)$$

In this equation, N_B and N_C represent dimensionless ratios known as the Bond number (i.e., the ratio of gravity to capillary forces) and the Capillary number (i.e., the ratio of viscous to capillary forces), respectively. The ratios are given as follows:

$$N_B = \frac{k\Delta\rho g \sin \alpha}{\sigma} \quad (8)$$

$$N_C = \frac{v_w\mu_w}{\sigma} \quad (9)$$

Equation 7 shows the conditions under which the effects of viscous, gravitational, and capillary forces contribute to immiscible fluid flow. Capillary forces are analogous to dispersion forces in that the capillary forces cause a smearing of the advancing front. However, unlike dispersion forces in single-phase solute flow, the capillary zone of influence does not continue to grow with time. The length of the zone of influence (or transition zone) depends on the properties of the porous medium and the fluids present.

In the field, the distance between injection and production wells is typically greater than the capillary zone of influence and the effects of capillary suction on fluid flow are considered negligible (Buckley-Leverett assumption). However, for a bench-scale model, the zone of influence may be significant. Accordingly, scaling of laboratory models should be addressed before conducting multiphase fluid studies to ensure that laboratory results adequately represent field conditions.

For scaling analysis, Equation 7 can be simplified by making the following assumptions: (1) as the minimum waterflood saturation, S_{or}^*, is approached, k_{rw} approaches a constant value and $k_{rw}(dF/dX) >> F(dk_{rw}/dX)$; and (2) for cases when the capillary suction term is significant, dP_c^*/dX can be approximated by a linear gradient. Under these conditions, the equation is reduced to:

$$\frac{\partial S_o}{\partial T} + N_c \frac{\partial(F)}{\partial X} + N_B \frac{k_{rw}}{\phi} \frac{\partial(F)}{\partial X} - \frac{k_{rw}}{L}\left(\frac{k}{\phi}\right)^{\frac{1}{2}} \frac{\partial P_c^*}{\partial X} \frac{\partial(F)}{\partial X} = 0 \quad (10)$$

A column scaling criterion can then be determined by comparing the viscous and gravitational terms to the capillary suction term.[18] This results in a scaling criterion, N_s, given as:

$$N_s = \frac{L}{k_{rw}}\left(\frac{\phi}{k}\right)^{\frac{1}{2}} \left[N_C + \frac{k_{rw}N_B}{\phi}\right] > 0.9 \quad (11)$$

A similar expression has previously been presented by Kyte and Rapoport[19] for horizontally-oriented columns. Here, the investigators showed that for waterwet cores the effects of capillary forces become negligible when

$$LV\mu_w > 3 \text{ cm}^2 \cdot cp \cdot min^{-1} \quad (12)$$

By incorporating the data reported by the authors for their experiments (L = 9.27 cm; ϕ = 0.236; k = 5.32 × 10^{-9} cm²; σ = 23 dyne/cm) and by assuming k_{rw} = 0.7, the criterion established by Kyte and Rapoport[19] can be expressed in terms of the mobilization number as $N_s > 0.9$. This equation is equivalent to our scaling criterion Equation 11 for the condition of a zero Bond number corresponding to a horizontally-oriented column (i.e, α = 0).

Equations 11 and 12 indicate that as the value of the left-hand side increases, the effects of capillary suction become negligible. Under these conditions, laboratory results provide a good representation of field conditions. Thus, the laboratory operating conditions expressed by Equation 11 may be met by increasing the total length of the column, increasing the applied viscous force, or orienting the column to take advantage of the gravitational force. For example, TCE displaced in the downflow direction by water through a soil packing with a porosity of 40% and a permeability of 2.5 × 10^{-6} cm² (k_{rw} = 0.7) at a velocity of

5.5 m/day can be reduced to the residual waterflood saturation in a column with L = 27 cm. However, if the column is operated in the horizontal position, a total length L = 865 cm will be required to establish capillary stabilized flow.

Residual NAPL Mobilization and Removal

Microscopic models for NAPL mobilization and removal can be developed by balancing the viscous, gravitational, and capillary forces acting on a NAPL globule.[18] The criteria can be expressed as the mobilization number, N_{CB}, written as:

$$N_{CB} = \left[N_C + \frac{k_{rw}N_B}{\phi} \right]$$
(13)

When N_{CB} exceeds some critical value, NAPL mobilization is expected to occur. Equation 13 is similar to an empirical relationship reported by Morrow et al.[4] for light NAPL removal from glass beads. Using a Bond number based on permeability, this relationship is given as:

$$N_C + 0.45N_B > 4 \times 10^{-5}$$
(14)

for the mobilization of trapped NAPL. The coefficient for N_B empirically determined by Morrow et al.[4] falls within the range 0.25 to 1.75 expected for glass beads.[18]

Sweep Efficiency

An efficient sweep of the contamination zone is characterized by a plug flow-type displacement of the resident fluid. Gravitational and viscous instabilities between the fluids, however, may develop during the displacement process resulting in fingering of the injectant and reduced sweep efficiency. For a one-dimensional column, gravitational instability can be expressed in terms of the density difference between the injected and resident fluids ($\Delta\rho$) and the column orientation as $\Delta\rho g \sin \alpha$. When the term $\Delta\rho g \sin \alpha > 0$, the positioning of the fluids relative to each other is stable. When $\Delta\rho g \sin \alpha < 0$, the positioning of the fluids is unstable and gravitational fingering is likely to occur. This effect, however, can be mitigated by increased capillary or viscous forces.

The mobility ratio is another measure of potential fingering and reduced sweep efficiency. The mobility of each fluid, k_{ri}/μ_i, is based on the proportionality in Darcy's law relating fluid velocity to pressure gradient. Mobility ratio is therefore defined as

$$M = \frac{k_{rd}/\mu_d}{k_{rr}/\mu_r}$$
(15)

For cases where the fluids are completely miscible, this ratio reduces to the simplified form $M = \mu_r/\mu_d$.

Mobility ratio is widely used in the petroleum industry as a measure of viscous fingering. Craig[20] summarized the relationships between sweep efficiencies and mobility ratios for a variety of injection and recovery well patterns. In general, sweep efficiency improves as the mobility ratio decreases below a value of 1.

Mobility ratios and gravity terms describing our laboratory column scenarios were used in a perturbation analysis to evaluate the influence of viscous and gravitational fingering on sweep efficiencies. The analysis included water, isopropyl alcohol (IPA), and trichloroethylene (TCE) injected in three directions (upflow, horizontal, downflow). As shown in Table 26.1, results indicate that the displacement of TCE from the column by either water or alcohol flooding was most stable in the downflow direction and that the displacement of IPA by water was most stable in the upflow direction.

Table 26.1. Stability Criteria

Displacement Process	Upflow	Horizontal Flood	Downflow
H$_2$O Displacing TCE	V > 12 m/d	stable[a]	stable
IPA Displacing H$_2$O	V > 17 m/d	stable[a]	stable
H$_2$O Displacing IPA	V < 24 m/d	unstable	unstable

[a]Gravity segregation may occur under field conditions.

EXPERIMENTAL PROCEDURE

Laboratory experiments were conducted with glass beads packed in a 2.5-cm diameter glass chromatography column. Each end of the column was fitted with a Teflon-tipped plunger which provided column length flexibility. Glass frits (0.6 cm thick) were embedded in each of the plunger tips to prevent clogging of tubing and fittings. Trichloroethylene and deaerated distilled water were used to represent DNAPL and water, respectively. Isopropyl alcohol was used to enhance the mobilization and removal of residual DNAPL globules. Selected properties for these fluids are summarized in Table 26.2.

Table 26.2. Summary of Fluid Properties

Name	Viscosity (cp)	Density (g/cm^3)	Surface Tension (dyn/cm)	Interfacial Tension (with water) (dyn/cm)	Water Solubility (mg/L)	K$_{ow}$
Trichloroethylene[a]	0.59	1.46	29	35	1100	200
Isopropyl Alcohol (2-propanol)	2.86	0.785	22[c]	—	∞	1[b]
Water	1.00	1.00	72[c]	—	—	—

[a] Hunt et al.[21]
[b] Pinal et al.[22]
[c] Weast.[23]

In order to ensure homogeneous packing and initially water-saturated conditions, the glass beads were introduced into the column as a slurry and allowed to settle through a 5-cm column of water. After introducing approximately 2 cm of packing, the matrix material was mixed under the column of water to remove any entrapped air bubbles. The column was then tapped to promote particle settling and establish packing homogeneity. The measured porosity and permeability of the glass bead packing were 40% and 2.2×10^{-6} cm², respectively.

The packed columns were attached to a series of liquid separators and water reservoirs capable of withstanding pressure differences up to 50 psi. The laboratory apparatus is shown in Figure 26.3. DNAPL contamination was achieved by applying a constant pressure gradient of approximately 21 (25 psi for a column length L = 83 cm) for a minimum of 1.5 pore volumes (PVs) of flow. Free product displacement by waterflooding was performed at a constant velocity of 5.5 m/d (V = 20 cm/hr) by using a peristaltic pump. After establishing a stabilized residual NAPL saturation (based on scaling parameters discussed above), alcohol injections and follow-up water flushes were pumped through the column at the same flowrate.

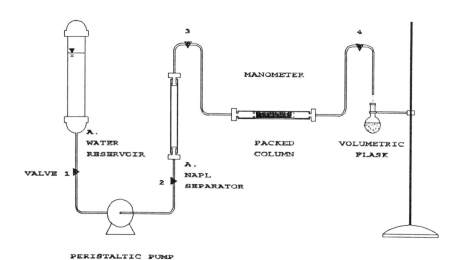

Figure 26.3. Schematic of laboratory apparatus used for columns packed with glass beads and a representative South Carolina soil.

After contamination with TCE, all of the effluent from the column was collected in volumetric flasks. Each sample was diluted in methanol and analyzed by flame ionization gas chromatography (GC-FID) to a minimum detection level of 100 mg/L. If the TCE concentration was greater than 100 mg/L at the end of a simulation, a second 2-PV slug of concentrated IPA was injected into the column and allowed to remain in contact with the entire apparatus for a minimum time period of 12 hours. The TCE recovered from the system was used to complete the mass balance for the experiment.

Visual observations of NAPL and solvent behavior in the glass beads were conducted as preliminary experiments using organic soluble dyes, Automate Red B (2-naphthalenol [(phenylazo)phenyl] azo alkyl derivatives) and Automate Blue 8 (1,4-dialkylamino anthraquinone), courtesy of Morton Thiokol, Inc. Since the effects of the dyes on the physical and chemical behavior of the system were not investigated, the dyes were not used during experiments conducted for quantitative analyses.

In addition to the column studies performed with glass beads, similar experiments were conducted[24] using a low organic carbon content (0.06%), fine-grain silicate soil from Charleston, South Carolina (SC). A detailed characterization of the SC soil is given in Brickell.[25] The aggregate material was prepared by initially sieving to 1.0 mm grain size (18 US mesh) to remove large grains and debris. The sieved material was then introduced into the column by the 'inverted method' described by Chu et al.[26] to minimize the effects of particle segregation. Once in place, the soil was saturated by attaching a water source to the bottom of the column and a vacuum to the top.

The porosity and permeability of the representative SC soil were 44% and 3.3 $\times$ 10^{-7} cm^2, respectively. The soil samples were contaminated with TCE by applying a pressure gradient equivalent of 0.2. All fluid injections were performed at constant velocity (v = 12 cm/hr) and all effluent samples were collected for analysis by GC-FID.

RESULTS AND DISCUSSION

Glass Bead Experiments

Figures 26.4 through 26.7 show the laboratory results obtained from simulations of TCE contamination and recovery scenarios by water displacement and IPA flooding. Data presented for each simulation begin at the time of initial water-flooding and continue until the alcohol is reduced to nondetectable concentrations. For each simulation, the first plot shows the DNAPL saturation remaining in the column as a function of cumulative fluid injection. The second plot shows TCE and IPA effluent concentrations beginning at the time of initial alcohol injection. Results are presented for glass bead packings and are later compared to similar experiments conducted in our laboratory by Patwardhan[24] using a representative SC soil.

Table 26.3 summarizes the operating conditions and scaling parameters characterizing the displacement processes for experiments I through IV.[18] Included in Table 26.3 are the scaling criterion for the free product recovery (Equation 11) and the NAPL mobilization number for IPA flooding (Equation 13). Also tabulated for each of the displacement processes are mobility ratios and density differences

Table 26.3. Summary of Operating Conditions, Scaling and Mobility Criterion for Glass Bead Experiments

Experiment No.	Free Product Recovery			IPA Flood			Followup Waterflood	
	Mobility Ratio, M	$\Delta\rho g \sin \alpha$ $(g \cdot cm^{-2} \cdot s^{-2})$	Scaling No. N_s	Mobility Ratio, M	$\Delta\rho g \sin \alpha$ $(g \cdot cm^{-2} \cdot s^{-2})$	Mobilization No. N_{CB}	Mobility Ratio, M	$\Delta\rho g \sin \alpha$ $(g \cdot cm^{-2} \cdot s^{-2})$
I	0.3	450	2.5	0.2	660	2.7×10^{-2}	2.6	−210
II	0.3	450	2.5	0.2	560	1.6×10^{-3}	3.3	105
III	0.3	450	2.5	0.5	470	9.5×10^{-5}	1.1	21
IV	0.3	0	0.1	0.2	0	1.8×10^{-3}	2.6	0

Experiment I: 100% IPA Injection. Free product recovery, IPA flood, and follow-up waterflood were conducted in the downflow direction.
Experiment II: 50% IPA Injection. Free product recovery and IPA flood were conducted in the downflow direction; follow-up waterflood was conducted in the upflow direction.
Experiment III: 10% IPA Injection. Free product recovery and IPA flood were conducted in the downflow direction; follow-up waterflood was conducted in the upflow direction.
Experiment IV: 100% IPA Injection. Free product recovery, IPA flood, and follow-up waterflood were conducted in the horizontal direction.

which serve as indicators of sweep efficiency. Note for each experiment that viscous displacement was stable (M < 1) during free product recovery and IPA floods, but viscous displacement was unstable (M > 1) during final waterfloods.

Results are shown in Figure 26.4 for experiment I which was conducted under ideal operating conditions for free product displacement and IPA flooding (i.e., flooding in the downward direction). Figure 26.4a shows TCE saturation in the column as a function of cumulative fluid injection. The column was initially saturated to 74% and the free product was displaced by waterflooding, leaving a residual TCE saturation, S_{or}^*, of 12%. After 6 PVs of cumulative water injection, a 1-PV slug of concentrated IPA was injected into the column. Within 2 PVs of fluid injection, the TCE saturation was reduced to a nondetectable concentration.

Figure 26.4b shows the measured concentrations of TCE and IPA in the column effluent commencing at the time of IPA injection. After approximately 0.4 PV of injection, TCE concentration increased suddenly from aqueous solubility to a maximum concentration. This occurred prior to the appearance of IPA in the effluent, indicating the mobilization of residual TCE as a separate phase. As the IPA concentration increased, the effluent samples evolved from a milky emulsion into a single-phase solution whereby the TCE was completely dissolved in the alcohol phase. Within 2 PVs of IPA injection, TCE was reduced to nondetectable concentrations.

Figure 26.4b also shows that approximately 7 PVs of a subsequent waterflood was required to reduce the injected IPA to a nondetectable concentration. This inefficient IPA displacement was attributed to the unfavorable fluid injection sequence, i.e., the displacement of IPA by water in the downflow mode is expected to result in low sweep efficiency due to both viscous and gravitational instabilities.

Results for experiment II are shown in Figure 26.5. In this experiment, a 1-PV slug of 50% IPA (rather than a 100% IPA slug) was injected into the column. All fluids were injected in gravitationally stable directions, including the follow-up waterflood. Figure 26.5a indicates that the treatment successfully reduced the TCE to less than 1 PV percent saturation. The data in Figure 26.5b indicate that within 3 PVs of IPA injection, TCE was reduced to nondetectable concentrations.

Figure 26.6 shows results for experiment III which was conducted by injecting a 1-PV slug of a 10% solution of IPA (rather than a 50% IPA solution). The data indicate that the IPA treatment did not significantly reduce the contaminant saturation (Figure 26.6a), but it did double TCE solubility (Figure 26.6b). Based on these results and the NAPL mobilization number developed earlier (Equation 13), residual DNAPL globules trapped in glass beads are expected to be mobilized when $N_{CB} > 10^{-4}$. This conclusion is consistent with data presented by Morrow et al.[4] and is discussed in more detail later in this section.

During experiments II and III (Figures 26.5 and 26.6, respectively) the injected IPA solutions were displaced from the column in the upflow direction. Upflow injection is expected to improve the waterflood sweep efficiency even though viscous fingering remains unstable. A comparison of the data in Figures 26.5b and

26.6b to previous results for experiment I (Figure 26.4b) indicates that the up-flow injection orientation was more efficient for the follow-up waterflood.

When the alcohol flooding process is applied to the field, the choice of flooding direction may be limited by operational and regulatory constraints. For example, the treatment of a residual DNAPL by injecting alcohol in the downflow direction may be considered risky due to the increased potential for offsite migration of the mobilized DNAPL. Therefore, we are also considering the efficiency of the alcohol flooding process when restricted to horizontal and upflow directions. Results for the horizontal direction are reported below.

Laboratory results are shown in Figure 26.7 for a horizontal column treated with a 1-PV slug of concentrated IPA (experiment IV). The column length and injection velocities were the same as those in previous experiments. However, without the aid of gravitational forces during the waterflood phase, viscous forces were not sufficient to meet the scaling criterion for stabilized flow (Table 26.3). If the scaling criterion is not met, the free product displacement process is expected to be less efficient due to capillary suction effects. As shown in Figure 26.7a, the minimum waterflood saturation was not reached within a reasonable period of time and the IPA injection process was begun at a higher DNAPL saturation.

Data obtained from the horizontal column experiments indicated that the IPA mobilization of TCE globules was less efficient than the vertical downflow injection. During downflow experiments, gravitational forces contributed to displacement efficiency in two ways: (1) by assisting viscous forces in overcoming capillary forces, and (2) by maintaining gravitational stability, thereby improving sweep efficiency. During the horizontal experiment, it is quite likely that the density difference between the injected and resident phases resulted in gravitational segregation of the fluids. In fact, we observed over-running of IPA in horizontal columns during preliminary experiments conducted with dyed fluids. The data in Figure 26.7b also indicate that TCE was reduced to solubility concentration after 3 PVs of total fluid injection. This process was less efficient than the downflow displacement process shown earlier in Figure 26.4b.

Soil Experiments

Additional experiments were conducted by Patwardhan[24] using columns packed with a representative SC soil. The procedures employed were similar to the experiments discussed above. In all cases, except for the 50% IPA experiment, results obtained with SC soils were consistent with glass bead experiments. For the 50% IPA experiment, the 1-PV slug reduced TCE to 25% of its residual saturation rather than completely removing it as in the experiment using glass beads. Lower TCE removal efficiency may be attributed to adsorption on clay surfaces (the clay content of the soil was 16%) and variability in pore channel sizes due to a broader grain size distribution (the uniformity coefficient for the soil was 16).

(a)

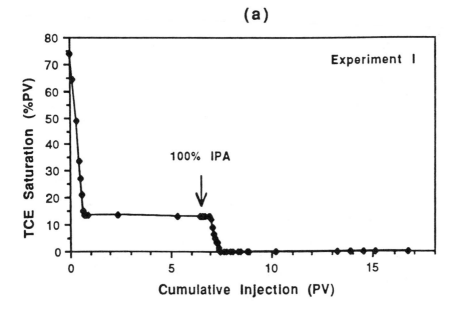

(b)

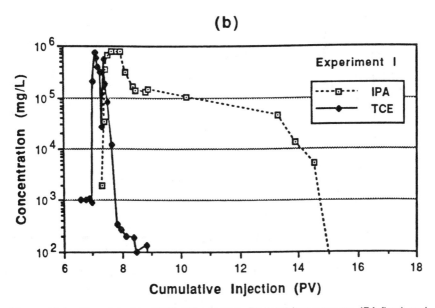

Figure 26.4.   Experiment I: 100% IPA Injection. Free product recovery, IPA flood, and follow-up waterflood were conducted in the downflow direction. (a) TCE saturation versus cumulative injection and (b) TCE and IPA concentration versus cumulative injection.

(a)

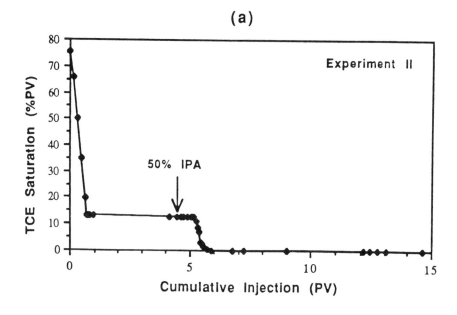

(b)

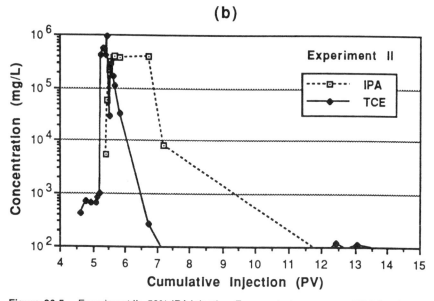

Figure 26.5. Experiment II: 50% IPA Injection. Free product recovery and IPA flood were conducted in the downflow direction; follow-up waterflood was conducted in the upflow direction. (a) TCE saturation versus cumulative injection and (b) TCE and IPA concentration versus cumulative injection.

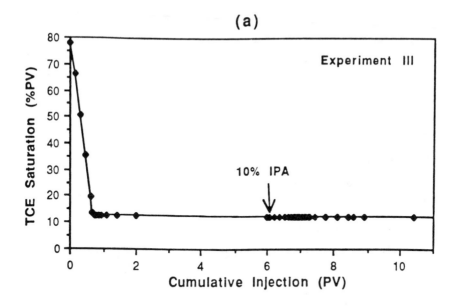

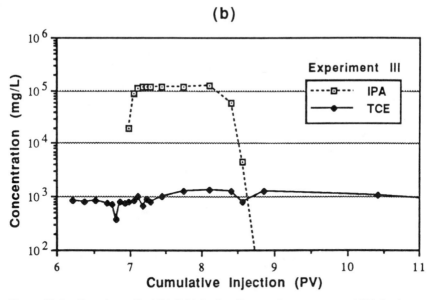

Figure 26.6. Experiment III: 10% IPA Injection. Free product recovery and IPA flood were conducted in the downflow direction; follow-up waterflood was conducted in the upflow direction. (a) TCE saturation versus cumulative injection and (b) TCE and IPA concentration versus cumulative injection.

(a)

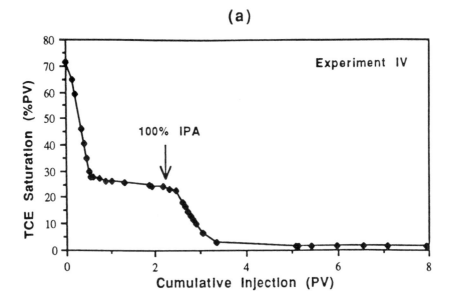

(b)

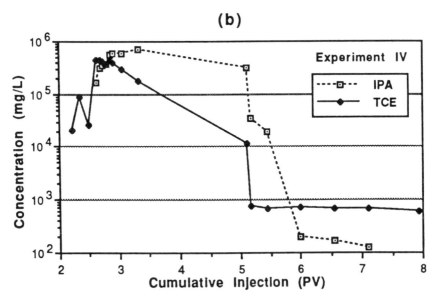

Figure 26.7. Experiment IV: 100% IPA Injection. Free product recovery, IPA flood, and follow-up waterflood were conducted in the horizontal direction. (a) TCE saturation versus cumulative injection and (b) TCE and IPA concentration versus cumulative injection.

Comparison of Results

In our DNAPL mobilization studies using TCE, gravitational forces favor both microscopic displacement and macroscopic sweep efficiency. On a microscale, the combined effect of viscous plus gravitational forces maximize DNAPL mobilization in the downflow direction. On a macroscale, the density of the IPA solution favors stable displacement of the resident fluid in the downflow direction and the process approximates plug flow-type displacement. The downflow displacement of a DNAPL should therefore provide the most efficient removal, even more efficient than the upflow displacement of a light NAPL which is affected by gravitational instability.

A summary of our laboratory results is presented in Figure 26.8 where the data are shown as normalized residual NAPL saturation, S_{or}/S_{or}^*, plotted as a function of the mobilization number given in Equation 13. Also included in Figure 26.8 are trapping and mobilization curves for isooctane recovered from glass beads.[4] These curves were based on experiments conducted with isopropyl alcohol injected in the upflow direction. The difference between NAPL trapping and mobilization is based on the continuity of the immiscible phase at the start of the alcohol flood. Morrow et al.[4] showed that the removal of a continuous phase NAPL (trapping) is more efficient than the removal of a discontinuous phase NAPL (mobilization) by isopropyl alcohol flooding. The difference was shown experimentally by varying the capillary number for laboratory column alcohol floods where $N_C > > N_B$.

For our experiments with glass bead packings, TCE was injected into the column and then reduced to a minimum waterflood saturation before starting the alcohol

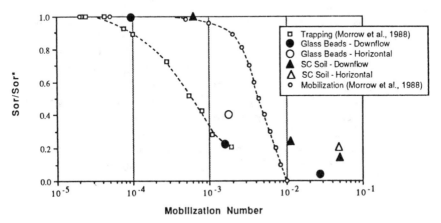

Figure 26.8. Normalized NAPL saturation versus mobilization number (N_{CB}) for glass beads and a representative South Carolina soil.

flood. For this initial condition, the residual NAPL globules were assumed to be discontinuous and the alcohol flood was expected to remove the TCE globules by the mobilization mechanism. The data points in Figure 26.8 represent the average TCE saturations remaining in the column after 1 PV of cumulative alcohol injection (since TCE removal by injection after 1 PV is likely due to dissolution rather than immiscible displacement). Since ternary fluid property values for TCE, IPA, and water were not available to us, mobilization numbers were calculated based on estimated values for viscosity, density, and interfacial tension.[18] More work is currently underway to compile fluid property data for ternary systems.

Our laboratory results for downflow experiments with glass bead packings were expected to correlate with the mobilization curve as reported by Morrow et al.[4] and shown in Figure 26.8. Instead, the results obtained approximated the trapping curve. The reason for this difference is not clear, but may be attributed to the higher aqueous solubility of TCE or the accuracy of the estimated ternary fluid property values used to calculate N_{CB}.[18] In general, however, residual TCE saturation was reduced when N_{CB} was greater than 10^{-4} and completely removed from the glass beads when $N_{CB} > 10^{-2}$. These results are in general agreement with results presented by Morrow et al.[4]

For the glass bead experiment conducted in the horizontal direction, the injected alcohol was not stabilized by gravitational forces and the NAPL displacement process was less efficient than experiments conducted in the downflow direction. The results shown in Figure 26.8 indicate a higher residual TCE saturation remaining in the column at the end of this experiment, indicating a lower removal efficiency compared to alcohol floods in the downflow direction.

Results obtained from laboratory experiments conducted with SC soil[24] were similar to the glass bead experiments. The data indicate that residual TCE remaining in the column was reduced to lower saturations for alcohol floods conducted at higher mobilization numbers. Unlike the glass bead experiments, results obtained from the SC soil experiments closely approximated NAPL mobilization (rather than NAPL trapping) as reported by Morrow et al.[4] For these experiments, residual TCE saturation was reduced when N_{CB} was greater than approximately 10^{-3}. More research is presently being conducted to examine this apparent difference between the SC soil and glass beads. The difference may be attributed to factors such as soil grain-size distribution, mineral composition, and parameter estimates used to calculate the mobilization numbers. In general, however, the results obtained for the SC soil fall within the range of NAPL removal effectiveness as given by the mobilization and trapping curves of Morrow et al.[4]

Field Applications

The NAPL mobilization number given by Equation 13 can be used to predict the conditions required to mobilize residual NAPL contaminants in the subsurface. For example, for uniform-sized glass beads, mobilization of residual NAPL occurs when $N_{CB} > 10^{-4}$. Complete removal of NAPL residual requires a mobilization number greater than 10^{-2}. These results can be extended to both positive and negative values of the mobilization number (with a mirror image of the displacement curve on the negative axis of Figure 26.8). If the value for N_{CB} is positive, mobilized NAPL globules are expected to migrate in the direction of the flood. However, if the value for N_{CB} is negative, the gravitational force is dominant and is acting opposite the direction of viscous forces. In this case, mobilized NAPL globules are expected to migrate in the opposite direction of the alcohol flood. (Such behavior is more likely to occur in high permeability formations, or for heterogeneous systems, in high permeability streaks where N_B is expected to be significant.) This situation may pose undue environmental risk with the potential migration of NAPL globules away from the contaminated site. In these cases, complete dissolution of the NAPL contaminant may be the preferred remediation strategy.

The data presented in Figure 26.8 may also be used as a planning tool to improve NAPL removal efficiency by cosolvent injection. For example, consider the case of a site contaminated with TCE. Horizontal wells have been drilled under the contaminated zone and regulations dictate that injected fluids must move in the upflow direction. Data presented in Figure 26.8 indicate that NAPL removal efficiency improves with favorable gravitational orientation. Thus, instead of selecting a cosolvent (e.g., IPA) with a density much lighter than water, an alternative cosolvent may be selected which has a specific gravity closer to a value of unity. By choosing such an alternative cosolvent, the injectant is less likely to finger into the resident fluid and the overall displacement efficiency is expected to increase. Note that these same principles may be applied to other injected fluids, e.g., surfactants.

CONCLUSION

Results obtained from one-dimensional laboratory column experiments with glass beads and a representative SC soil indicated that residual trichloroethylene globules left in a porous medium after waterflooding (pump-and-treat) may be removed by injecting an isopropyl alcohol (2-propanol) solution through the contaminated zone. In the glass bead experiments, mobilization of TCE occurred when the mobilization number, N_{CB}, given in Equation 13 was greater than 10^{-4} and complete removal occurred when N_{CB} was greater than 10^{-2}. The efficiency of the removal process was improved when the sequence and orientation of the fluids were injected under favorable conditions (based on viscous and gravitational stability). Subsequent waterfloods through the treated zone can reduce the alcohol to biodegradable concentrations.

The mobilization number (Equation 13) may also be used to predict the direction of the migration of mobilized NAPL globules. A positive value for N_{CB} indicates that the net force acting on residual NAPL globules will be in the same direction as injectant flow. A negative value for N_{CB} indicates that gravitational forces dominate and NAPL movement will not follow the axis of injectant flow. This may result in offsite migration of mobilized DNAPL, posing undue environmental risk. In these cases, complete dissolution of the DNAPL contaminant may be the preferred remediation strategy. More research is needed to verify these fundamental fluid relationships for other fluid/matrix systems.

ACKNOWLEDGMENTS

South Carolina soil column experiments were conducted by Sanjay Patwardhan. Funding was provided in part by the South Carolina Water Resources Research Institute, the Clemson University Research Office, and Sigma Xi, The Scientific Research Society.

NOMENCLATURE

k = specific permeability, cm^2
k_r = relative permeability, dimensionless
L = total length, cm
P = pressure, Pa
P_c = capillary pressure, Pa
S = saturation, dimensionless
S_{or} = residual NAPL saturation, dimensionless
S_{or}^* = minimum waterflood saturation, dimensionless
t = time, sec
v = linear velocity, cm/sec
V = Darcy velocity, cm/sec
x = distance, cm
α = angle positive upward from the horizontal, deg
μ = viscosity, g/cm·s
ρ = density, g/cm^3
$\Delta\rho$ = $\rho_w - \rho_o$
σ = interfacial tension, dynes/cm
ϕ = porosity, dimensionless

$$F = \left[\frac{\mu_o k_{rw}}{k_{ro}\mu_w} + 1 \right]^{-1}$$

$$N_B = \frac{k\Delta\rho g \sin \alpha}{\sigma}$$

$$N_C = \frac{v\mu_w}{\sigma}$$

$$N_{CB} = \text{mobilization number, } N_C + \frac{k_{rw}N_B}{\phi}$$

$$P_c^* = \frac{P_c}{\sigma}\left(\frac{k}{\phi}\right)^{1/2}$$

$$T = \frac{\sigma t}{L\mu_w}$$

$$X = \frac{x}{L}$$

Subscripts

o = organic phase
w = aqueous phase
d = displacing phase
r = resident phase

REFERENCES

1. Schwille, F. *Dense Chlorinated Solvents in Porous and Fractured Media,* trans. by J. F. Pankow, (Chelsea, MI: Lewis Publishers, Inc., 1988).
2. Wilson, J. L., S. H. Conrad, W. R. Mason, W. Peplinski, and E. Hogan. "Laboratory Investigations of Residual Liquid Organics from Spills, Leaks, and the Disposal of Hazardous Wastes in Groundwater," EPA/600/6-90/004, New Mexico Institute of Mining and Technology, Socorro, NM, 1990.
3. Mackay, D. M., and J. A. Cherry. "Groundwater Contamination: Pump-and-Treat Remediation," *Env. Sci. Tech.,* 23(6):630–636 (1989).
4. Morrow, N. R., I. Chatzis, and J. J. Taber. "Entrapment and Mobilization of Residual Oil in Bead Packs," *SPERE,* 3(3):927–934 (1988).
5. Morrow, N. R., and B. Songkran. "Effect of Viscous and Bouyancy Forces on Nonwetting Phase Trapping in Porous Media," *Surface Phenomena in Enhanced Oil Recovery,* D. O. Shah, Ed. (New York, NY: Plenum Press, 1982), pp. 387–411.
6. Gatlin, C., and R. L. Slobod. "The Alcohol Slug Process for Increasing Oil Recovery," *Trans. AIME,* 219:46 (1960).
7. Rao, P. S. C., A. G. Hornsby, D. P. Kilcrease, and P. Nkedi-Kizza. "Sorption and Transport of Toxic Organic Substances in Aqueous and Mixed Solvent Systems," *J. Env. Qual.* 14:376–383 (1985).
8. Rao, P. S. C., L. S. Lee, and R. Pinal. "Cosolvency and Sorption of Hydrophobic Organic Chemicals," *Env. Sci. Tech.,* 24:647–654 (1990).
9. El-Zoobi, M. A., G. E. Ruch, and F. R. Groves. "Effect of Cosolvents on Hydrocarbon Partition Coefficients for Hydrocarbon Mixtures and Water," *Env. Sci. Tech.,* 24(9):1332–1338 (1990).
10. Nkedi-Kizza, P., P. S. C. Rao, and A. G. Hornsby. "Influence of Organic Cosolvents on Sorption by Hydrophobic Organic Chemicals by Soils," *Env. Sci. Tech.,* 19(10):975–979 (1985).
11. Nkedi-Kizza, P., P. S. C. Rao, and A. G. Hornsby. "Influence of Organic Cosolvents on Leaching of Hydrophobic Organic Chemicals Through Soils," *Env. Sci. Tech.,* 21(11):1107–1111 (1987).
12. Fu, J. K. and R. G. Luthy. "Effects of Organic Solvent on Sorption of Aromatic Solutes onto Soils," *J. Env. Eng.,* 112:346–366 (1986).

13. Boyd, G. R., and K. J. Farley. "Residual NAPL Removal from Groundwater by Alcohol Flooding," *Trans. Amer. Geophys. Union,* 71(17):500 (1990).
14. Boyd, G. R., and K. J. Farley. "Factors Influencing Nonaqueous Phase Liquid Removal from Groundwaters by the Alcohol Flooding Technique," in *Concepts in Manipulation of Groundwater Colloids for Environmental Restoration* J. F. McCarthy, Ed. (Chelsea, MI: Lewis Publishers, 1992).
15. Wilson, J. L., and S. H. Conrad. "Is Physical Displacement of Residual Hydrocarbons a Realistic Possibility in Aquifer Restoration?" Proceedings, NWWA/API Conference on Petroleum Hydrocarbons and Organic Chemicals in Ground Water—Prevention, Detection and Restoration. National Water Well Association, Worthington, OH, 1984.
16. Lake, L. W. *Enhanced Oil Recovery.* (Englewood Cliffs, NJ: Prentice-Hall, Inc., 1989).
17. Leverett, M. C. "Capillary Behavior in Porous Solids," *Trans. AIME,* 142:159–169 (1941).
18. Boyd, G. R. "Factors Influencing Nonaqueous Phase Liquid Removal from Groundwater by Alcohol Flooding," PhD dissertation, Clemson University, SC, 1991.
19. Kyte, J. R., and L. A. Rapoport. "Linear Waterflood Behavior and End Effects in Water-Wet Porous Media," *Trans. AIME,* 213:423–426 (1958).
20. Craig, F. F., Jr. *The Reservoir Engineering Aspects of Waterflooding.* Third Printing, Society of Petroleum Engineers of American Institute of Mining, Metallurgical and Petroleum Engineers, New York, 1971.
21. Hunt, J. R., N. Sitar, and K. S. Udell. "Nonaqueous Phase Liquid Transport and Cleanup. 2. Experimental Studies," *Water Res. Res.,* 24(8):1259–1269 (1988).
22. Pinal, R., P. S. C. Rao, L. S. Lee, P.V. Cline, and S. H. Yalkowsky. "Cosolvency of Partially Miscible Organic Solvents on the Solubility of Hydrophobic Organic Chemicals," *ES&T,* 24:639–647 (1990).
23. Weast, R. C., Ed. *Handbook of Chemistry and Physics.* (Boca Raton, FL: CRC Press, Inc., 1986).
24. Patwardhan, S. "The Removal of Dense Immiscible Contaminants from a Representative South Carolina Soil by Alcohol Flooding," Master's thesis in preparation, Clemson University, SC.
25. Brickell, J. L. "The Effects of Surfactants on the Adsorption of Organic Contaminants from Aquifer Materials," PhD dissertation, Clemson University, SC, 1989.
26. Chu, T. Y., D. T. Davidson, and A. E. Wickstrom. "Permeability Test for Sands," Symposium on Permeability of Soils, ASTM Special Technical Publication No. 163:43–55, 1954.

CHAPTER 27

Cost-Effective Alternative Treatment Technologies for Reducing the Concentrations of Methyl Tertiary Butyl Ether and Methanol in Groundwater

Kim N. Truong and **Charles S. Parmele,** IT Corporation, Knoxville, Tennessee

INTRODUCTION

This chapter describes an engineering assessment of eight treatment technologies conducted by IT Corporation (IT) for the American Petroleum Institute (API) for removing methyl tertiary butyl ether (MTBE) or methanol and benzene, toluene, xylene (BTX) in the three following groundwater treatment scenarios:

- removal of MTBE to 10 parts per billion (ppb) by eight technologies
- removal of methanol to 10 ppb by four selected technologies
- removal of methanol to 1,000 ppb by two selected technologies at several combinations of BTX inlet concentrations and flow rates.

The following eight technologies were evaluated for removing 20 parts per million (ppm) of MTBE from groundwater that also contained 20 ppm total BTX:

- air stripping with aqueous-phase carbon adsorption
- air stripping with off-gas incineration and aqueous-phase carbon adsorption
- air stripping alone
- heated air stripping

- steam stripping
- diffused aeration
- aboveground biological treatment
- ultraviolet (UV)-catalyzed oxidation using hydrogen peroxide (H_2O_2) and ozone (O_3).

Four of these technologies were evaluated for removing low concentrations of methanol (20 ppm) from groundwater that also contained 20 ppm total BTX:

- heated air stripping
- steam stripping
- aboveground biological treatment
- UV-catalyzed oxidation using H_2O_2 and O_3.

Aboveground biological treatment and UV-catalyzed oxidation using H_2O_2 and O_3 were evaluated for removing higher concentrations of methanol from groundwater. A combination of parameters was evaluated to test the effect of different flow rates and BTX inlet concentrations on the costs of removing high concentrations of methanol from groundwater (see Table 27.1).

Table 27.1.   Parameters Evaluated for Costs of Removing Methanol from Groundwater

Methanol Concentration (ppm)	BTX Concentration (ppm)	Water Flow (gpm)
1,000	20	25
1,000	20	250
1,000	100	250

DESIGN APPROACH

The full report has the same title as this chapter. It will be available from the API. The results of a literature review, as well as a consideration of the technical and regulatory requirements which the treatment systems must meet, were used to establish bases for preliminary design and cost estimates for each technology. Values for the design parameters were assumed in cases where values were not found in the literature.

Using the bases established by the considerations discussed above, preliminary designs consisting of a process flow diagram and an equipment list were developed for each technology.[1] Order-of-magnitude capital costs were then developed for each technology using a composite installation factor applied to purchased equipment costs obtained from vendor budgetary quotations or from IT's file of recently purchased equipment. The composite factor included engineering and other indirect costs adjusted as appropriate for the amount of vendor engineering (which is sometimes included in purchased equipment cost), the type of construction contracts, etc. The factors used in this method were based on historical data. Operating costs and treatment costs were the sum of operating costs, fixed charges (based on 30% of capital cost per year for depreciation, maintenance, and insurance), and labor. Capital and operating costs generated using the described methods were considered to be accurate to $\pm 35\%$.

The design bases and treatment costs for each design case are discussed in the following sections of this evaluation. These bases were established by API and IT to closely represent a real-world application. It should be noted at the outset that the cost estimates presented in this evaluation were based on similar assumptions and thus should be used only as a basis for comparison and screening of various technologies. A more accurate estimate of the actual cost for implementation of alternatives would require a detailed engineering design.

The cost of systems to pretreat the feed to any of the treatment technologies was outside the scope of this evaluation because if pretreatment were required, it would be required for all options. In addition, the costs of pretreatment will vary by option. For example, filtration of solids may be required in advance of carbon adsorption, air stripping, or steam stripping to prevent fouling of the carbon or packing. Precipitation for iron removal may be required in advance of packed air stripper columns to minimize scale formation, but it probably will not be needed with diffused aeration and biological oxidation. Biological treatment and $UV/H_2O_2/O_3$ oxidation may or may not require pretreatment for removal of metals and/or suspended solids. Effluent collection was not included in the design. Costs for disposing of residues from the various treatment processes such as activated carbon, distillate from steam stripping, and biosludge from biological treatment were developed by assuming that the residues were hazardous wastes.

The unit costs listed in Table 27.2 were used for all treatment system cost estimates and were based on 1990 data. The total of $1.20/pound activated carbon cost included $0.85/pound activated carbon purchase, $0.10/pound activated carbon shipping, and $0.25/pound activated carbon disposal ($500/ton).

Table 27.2. Treatment System Unit Costs (Based on 1990 Data)

Activated carbon	$1.20/pound
Cooling water	$0.10/1,000 gallons
Electricity	$0.075/kilowatt hour (kWh)
Natural gas	$3.5/million Btu
Fuel oil	$0.70/gallon
H_2O_2	$0.65/pound
Power for UV lights	$0.075/kWh
UV lights	$60.00/light
Nutrients (20% solution)	$0.5/gallon
Methanol	$0.45/gallon
Disposal of organics	$0.30/pound
Disposal of biosludge	$0.14/pound

UV lights should be replaced once a year. The initial installation of UV lights was included in the capital cost of the unit. Groundwater was assumed to be at 55°F; treatment period was two years. Electrical power and labor were assumed to be available on site.

DISCUSSION OF TECHNOLOGIES

Air Stripping with Aqueous-Phase Carbon Adsorption

Air stripping and aqueous-phase carbon adsorption are used in tandem in this design to provide effective treatment of the contaminated water stream. Air stripping is used to remove the bulk of the contaminants to approximately 500 ppb, while carbon adsorption is used as a polishing step to achieve the required effluent criteria. The design and costs of the air stripper and carbon adsorber are determined by the removal of MTBE because MTBE is not as easily stripped or adsorbed as BTX.

The air stripping system consists of a 2-foot-diameter by 30-foot-high column containing high-efficiency, randomly dumped packing. Water is pumped to the top of the column where distributors ensure that channeling does not occur. A demister is installed at the top of the packed section to prevent fog formation. The exhaust air containing stripped organics (MTBE and BTX) is emitted from the top of the air stripper to the atmosphere. The stripped water is pumped to the carbon adsorption system.

The function of the carbon adsorber is to remove dissolved organics from the contaminated water stream that are not removed by the air stripper. The adsorption system consists of two columns four feet in diameter by eight feet high and loaded with granular activated carbon. Virgin carbon is used so that low (ppb) effluent level can be achieved. The fixed-bed adsorbers are operated in series and downflow in the normal lead/polish sequence. When the carbon in the lead adsorber has been exhausted, the carbon is replaced and it is placed back in service as the polish adsorber. This lead/polish alternating sequence is continued as the carbon in each adsorber becomes spent. The flow diagram for this process is shown in Figure 27.1.

The bases used for the air stripping and carbon adsorption treatment system design are presented in Table 27.3.

The required packaging height (Z) of the air stripper column was calculated by using the following equation:

$$Z = HTU \times NTU$$

The number of theoretical units (NTU) was calculated by using the following equation from Reference 3.

$$NTU = S/(S - 1) \, Ln \, [(S - 1)/(f \times S) + (1/S)]$$

where: f   = effluent concentration/feed concentration

HTU   = height of transfer unit is provided by the packing vendor

stripping factor S   = (mole air/mole water) × Henry's law constant

Henry's law constant = mole fraction/mole fraction

Total pressure   = 1 atmosphere

(To change units of Henry's Law constant from atm − m³ · mole to mole fraction/mole fraction, multiply by 55,556/total pressure in atmosphere.)[4]

In design cases where a column taller than 30 feet would be required, multiple columns in series were used. Thirty feet was chosen as a maximum economical height for a single column to avoid problems with poor liquid distribution. Poor distribution is frequently encountered in tall columns and can significantly reduce removal efficiency. Taller columns can also require more stringent construction and support standards.

Air Stripping with Off-Gas Incineration and Aqueous-Phase Carbon Adsorption

Thermal fume incineration was chosen as the air pollution control technique to be used to treat the off-gas from an air stripper in this case. The exhaust air contaminated with BTX and MTBE flows from the top of the air stripper column to the incinerator. Natural gas is used as auxiliary fuel to fire a flame to heat the exhaust air to 1600°F. The gas stream has a residence time of 0.3 to 0.75 seconds in the combustion chamber. The heat duty of the incinerator is 1.2 million Btu/hour. The air stripper and the aqueous-phase carbon adsorber are the same as described in previous section.

Air Stripping Alone

The air stripping system is used to remove MTBE and BTX from 20 ppm each to 10 ppb each. Because aqueous-phase carbon adsorption is not used as a polishing step, the system requires two 2-foot-diameter by 30-foot-high air stripper columns operating in series. Water is pumped to the top of the first column and the partially stripped water from the sump of the first column is pumped to the top of the second column. The flow diagram would be similar to Figure 27.1 except that the carbon system is replaced by another air stripper.

Heated Air Stripping

Heated air stripping is used to remove MTBE or methanol from 20 ppm to the 10 ppb level. A boiler is included in the design to heat the feed to 205°F. The boiler pressure is less than 15 pounds per square inch gauge (psig) to avoid

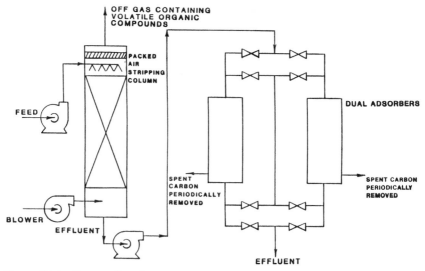

Figure 27.1.   Process flow diagram: air stripping with aqueous-phase carbon adsorption treatment system.

the need for a licensed boiler operator. The system for removing MTBE consists of two 2-foot-diameter by 23-foot-high columns containing high-efficiency, glass-filled-polypropylene, randomly dumped packing. The system for removing methanol consists of two 2.5-foot-diameter by 30-foot-high columns. Heated water is pumped to the top of the column and a blower introduces air at the bottom of the column providing countercurrent contact between the air and water. The exhaust air containing stripped organics (BTX and MTBE or methanol) is emitted from the top of the air stripper to the atmosphere. The bases used for the heated air stripping treatment system design are presented in Table 27.3.

One technical difficulty with the heated air stripping is the significant heat loss associated with the water that evaporates into the exhaust from the air stripper. This causes a temperature drop of 83°F in the water between the inlet and outlet of the stripper thus reducing the effect of heating the water in the first place. This significant cooling complicates the design of the stripper because the mass transfer calculations usually assume constant temperature in the stripper. The design was based on an average temperature of 122°F in the stripper.

Steam Stripping

A boiler is required to heat the feed to 212°F in this design case. Live steam is injected at the bottom of the column. The overhead is a volatile solute and steam. The steam stripping system for MTBE removal consists of one 1-foot-diameter by 40-foot-high column containing 30 feet of high-efficiency, randomly dumped ceramic packing. The steam stripping system for methanol removal consists of three 1-foot-diameter by 40-foot-high columns. In these two cases the

Table 27.3. Design Bases for the Air Stripping, Carbon Adsorption, and Heated Air Stripping Treatment Systems

	Air Stripping	Carbon Adsorption	Heated Air Stripping
Influent concentration	20 mg/L MTBE 20 mg/L BTX	500 ppb MTBE	20 mg/L MTBE or 20 mg/L methanol and 20 mg/L BTX
Effluent criteria	500 ppb MTBE (because the air stripper is followed by the carbon adsorption system)	10 ppb MTBE, 10 ppb BTX	10 ppb MTBE or 10 ppb methanol and 10 ppb BTX
Air-to-water ratio: MTBE removal[a]	20 cubic feet per minute (cfm)/gallons per minute (gpm)		20 cfm/gpm
Methanol removal			60 cfm/gpm (estimated)
Liquid loading[b]	Less than 10 gpm/ft²	2 to 7 gpm/ft²	
Materials of construction: Column	Fiberglass-reinforced plastic (FRP)	Plasite™-lined or rubber-lined carbon steel	FRP
Packing	Plastic, high-efficiency, low-pressure drop, randomly dumped packing		Glass-filled polypropylene, high-efficiency, low-pressure drop, randomly dumped packing
Auxiliary equipment	Carbon steel or ductile iron	Carbon steel or ductile iron	Carbon steel or ductile iron
Height of transfer unit (HTU) corrected for liquid loading, temperature, and treating ppm levels of: MTBE	53.6 inches (from vendor data)		38.0 inches (from vendor data)
Methanol			18.2 inches (from vendor data)
Adsorption capacity[a]: BTX		0.03 gram (g) BTX/g-carbon	
MTBE		0.004 g MTBE/g-carbon	
Ratio of column diameter to packing size[b]	8:1 to 24:1		8:1 to 24:1

[a] From Reference 2.
[b] From vendor and Reference 2.

height of the stripper columns exceeds IT's recommended height because the vendor recommended these heights to minimize the capital costs. To reduce the steam consumption, high temperature effluent in the sump at the bottom of the steam stripper is pumped through a heat exchanger to preheat the feed. Heated water is pumped to the top of the column where distributors ensure that channeling does not occur. Live steam flows up from the bottom of the column, providing countercurrent contact between the steam and water. The steam containing BTX and MTBE or methanol exits from the top of the steam stripper to one condenser, which uses ambient temperature cooling water to condense steam and BTX. Insoluble BTX is separated from the condensate in the decanter. The majority of the condensate (98%) is refluxed back to the top of the column.

Vapor containing MTBE or methanol is condensed in a second condenser that uses refrigerated water. The treated water is collected in a sump at the bottom of the column. A typical flow diagram for this process is shown in Figure 27.2.

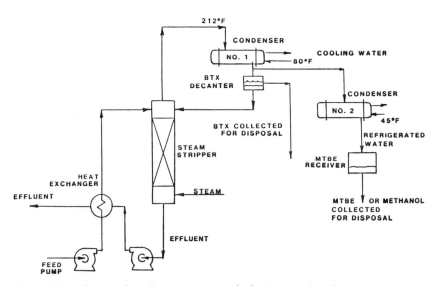

Figure 27.2.   Process flow diagram: steam stripping treatment system.

The bases presented below were used for the steam stripping treatment system design.

- The steam stripper is a skid-mounted unit supplied by one vendor.
- The steam source is a boiler with a pressure less than 15 psig to avoid the need for a licensed boiler operator.
- The boil-up for MTBE removal system is 0.02 pound overhead/pound feed (from vendor).
- The boil-up for methanol removal system is 0.17 pound overhead/pound feed (from vendor).

Diffused Aeration

In this system, ambient air from the blower is introduced through the diffusers in the bottom of a series of tanks containing the contaminated groundwater. Air that contains MTBE and BTX removed from the water is discharged to the atmosphere. The design approach follows a removal mechanism that is first order in the concentration of MTBE. The first order rate constant is a function of the Henry's law constant and ratio of the air flow rate to the volume of the tank. This approach also gives a desirable air-to-water ratio[5] of greater than 300 cubic feet per minute (cfm)/cfm. The design equations are:

$$C_{ao}/C_{ae} = [1 + K_{st} (HRT)]^n$$

and

$$K_{st} = 3.71 \times 10^{-3} (H_c)^{1.045} \times \frac{Q_{air}}{V} \quad \text{(Reference 6)}$$

where:
C_{ao}	= inlet concentration (mg/L)
C_{ae}	= effluent concentration (mg/L)
HRT	= hydraulic retention time of one stage (hour)
n	= number of stages in series
K_{st}	= stripping rate constant (1/hour)
H_c	= Henry's law constant (Torr-L/gram-mole)
V	= volume of the liquid in the reactor (L)
Q_{air}	= air flow rate (L/minute)
Q_{air}/V	= (L/minute)/L
Unit of 3.71×10^{-3}	= (L × minute/L) (gram-mole/Torr-L) (1/hour)

To change units of H_c from atm · m³/gram · mole to Torr-L/gram-mole, multiply by 760,014.

The system is designed in multiple stages to reduce costs by decreasing the total quantity of air and the total volume of tankage required to achieve the desired effluent concentrations. Figure 27.3 presents the effect of the number of aeration stages on the aeration tank size and air flow rate required to give the MTBE desired effluent concentration. Six stages operating in series were chosen for this design case because the point of diminishing returns was reached and both capital and operating costs were minimized. Capital costs were minimized because there was no further significant decrease in total volume. Similarly, operating costs were minimized because the total air flow needed flattened out at about six stages. Feed water is pumped to the first tank. The water then gravity overflows to each successive tank in series. The inlets are located substantially below the surface water level to ensure that the contents of the tanks are completely mixed. In the

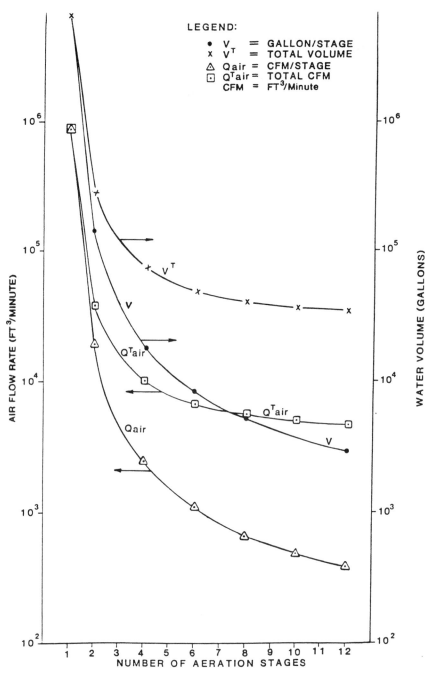

Figure 27.3. The effect of number of aeration stages on tank size and air flow rate for removing MTBE.

case where space is available, the series of tanks can be replaced by one large tank that is divided into a series of compartments. FRP tanks are used in the design. A typical flow diagram for this process is shown in Figure 27.4.

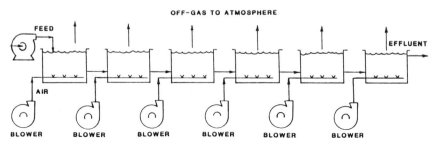

Figure 27.4.   Process flow diagram: diffused aeration treatment system.

Aboveground Biological Treatment

Aboveground biological treatment involves the treatment of contaminated water by microorganisms that metabolize the contaminants. The biological treatment process for MTBE removal was designed by using a suspended growth system in a one-stage activated sludge process. The activated sludge system consists of a tank that contains the contaminated water and a diffused aeration system to introduce air to the system. Nutrients are also introduced to the aeration tank.

A preliminary process design and cost estimate were developed for comparison purposes in this engineering assessment even though it is not likely that this technology can achieve 10 ppb MTBE or 10 ppb methanol in the effluent. The estimate was prepared to quantify the economic incentive that might exist to overcome the technical limitations.

The main technical limitation is the inability to develop and maintain a viable biomass in treated water that has practically no substrate to support metabolism. This problem can be put into perspective by considering the fact that the *inlet* conditions for this treatment system are equivalent to the *outlet* conditions for conventional biological wastewater treatment plants.

Several approaches to maintaining a viable biomass were considered, including adding a surrogate substrate, using an extended aeration lagoon or a two-stage activated sludge system. None of these approaches are practical solutions to this problem. Methanol is frequently used as a surrogate substrate by initially adding 2.0% to 2.5% methanol intermittently to develop the biomass. The methanol addition is then decreased to approximately 0.5% to maintain the biomass. Even if the low effluent criteria for MTBE and BTX could be met by this technique, it is not likely that an effluent containing methanol at a minimum of several hundred ppm could be discharged.

Modifying the reactor configuration for removing MTBE was also considered. However, an extended aeration lagoon was ruled out because it would occupy more space than is likely to be available. A two-stage bioreactor was also considered but this configuration still does not resolve the dilemma between generating effluent containing very low concentrations of organic chemicals in the second stage and providing enough substrate to support the metabolism of the biomass.

In spite of the technical limitations discussed above, a benchmark was established for the costs for a biological treatment system for removing MTBE by using a conventional one-stage biological treatment system. This was done to determine if there was economic incentive to solve the technical problems discussed above. Conventional design approaches were used and methanol was added to maintain the biomass. The effluent concentrations from this system were estimated to be 365 ppb MTBE and 72 ppb BTX. It was not practical to modify the design to achieve 10 ppb MTBE in the effluent.

The same design was used to establish the benchmark for a biological treatment system to remove low and high concentrations of methanol except that a two-stage biological reactor was used. This was done to ensure that the effluent criterion for BTX of 10 ppb was achieved even though it was not likely that 10 ppb methanol could be achieved in the low methanol concentration case.

The design conditions for the biological treatment system that removes MTBE and BTX are established by the physical properties of MTBE because MTBE is less biodegradable than BTX. The design conditions for the biological treatment system that removes methanol and BTX are established by the physical properties of BTX because benzene, toluene, and xylene are less biodegradable than methanol.

A typical flow diagram for this process is shown in Figure 27.5. Water and activated sludge flow by gravity from the aeration tank to two clarifiers operating in series. Some of the concentrated activated sludge at the bottom of the first clarifier is recycled back to the aeration tank to maintain a viable biomass level.

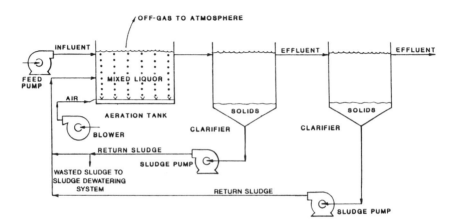

Figure 27.5. Process flow diagram: aboveground biological treatment system.

The second clarifier is used as a polishing step to achieve the required effluent clarity. All of the sludge that settles in the second clarifier is recycled to the aeration tanks.

The bases presented in Table 27.4 were used for the biological treatment system design.

Table 27.4.   Bases for the Biological Treatment System Design

• Concentration of biological solids in the reactor, mixed liquor suspended solids (MLSS):	3,000 mg/L[7,8,9]
• Hydraulic retention time (HRT) for MTBE removal case:	6.7 hours (one bio-reactor)
• HRT for methanol removal cases (based on BTX):	10.7 hours (two bio-reactors)
• Recycle ratio (for all cases):	0.43 gal/gal
• Biological solids retention time (BSRT):	Greater than 2 days
• Growth yield coefficient (Y):	0.65 pound MLSS/pound biological oxygen demand (BOD)[5]
• MTBE utilization rate constant (K):	0.064 L/mg-day (assumed)
• BTX utilization rate constant (K):	0.317 L/mg-day (estimated from Reference 7)
• Methanol utilization rate constant:	1.44 to 2.5 L/mg-day (assumed)

UV-Catalyzed Oxidation Using Hydrogen Peroxide and Ozone

Chemical oxidation modifies the chemicals in the waste stream either by completely oxidizing the organics to carbon dioxide and water or by partial oxidation of the organics to detoxify them. The main oxidizing agents in this design are ozone (O_3) and peroxide (H_2O_2). Vendors recommended that both these agents be used in combination with UV lights to increase the effectiveness of oxidizing MTBE, methanol, and BTX in groundwater. The skid-mounted package for MTBE removal has a stainless steel oxidation reactor approximately three feet wide, six feet long, and six feet high. The reactor can accommodate up to seventy-two 75-watt low-pressure UV lamps.

Ozone is produced in an ozone generator from air and diffused through porous ceramic spargers in the base of the reactor. Contaminated water passes through the reactor in a plug-flow pattern and is contacted by UV light, O_3, and H_2O_2. The same reactor and oxidant dosages were recommended by the vendor to be used for the removal of low concentrations of methanol. The flow diagram for this process is shown in Figure 27.6.

To verify values of design parameters for oxidizing MTBE, UV-catalyzed oxidation using H_2O_2 and O_3 was evaluated during a laboratory bench-scale test.

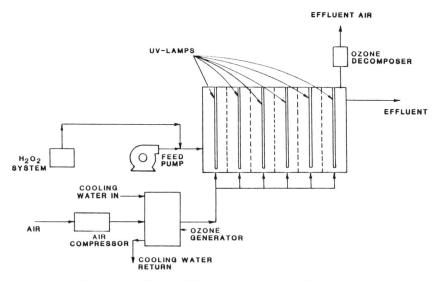

Figure 27.6. Process flow diagram: UV-catalyzed oxidation using hydrogen peroxide and ozone treatment system.

The results from the bench-scale $UV/H_2O_2/O_3$ oxidation test were used as the basis for the cost estimates of the full-scale UV-oxidation treatment system designs. The results of the test are presented in Figure 27.7. As shown in Figure 27.7, the removal of MTBE during the oxidation is well below the estimated removal due to air stripping alone. This estimate was verified by analyzing MTBE in the exhaust from the reactor. Less than 2% of the MTBE charged to the reactor was present in the exhaust.

The bases presented in Table 27.5 were used for the UV-catalyzed oxidation treatment system.

Table 27.5. Bases used for the UV-Catalyzed Oxidation Treatment System

• A packaged, skid-mounted unit supplied by one vendor was used for this cost estimate	
• Ozone and H_2O_2 source:	O_3 generator and the H_2O_2 feed system to be included in the treatment system
• Ozone dosage[1]:	2.9 grams O_3/gram (MTBE and BTX) (from bench-scale test)
• Ozone concentration in air[1]:	2 percent by weight
• H_2O_2 dosage[1]:	3.72 grams H_2O_2/gram (MTBE + BTX) (from bench-scale test)
• Retention time:	Less than 1 hour (from vendor)

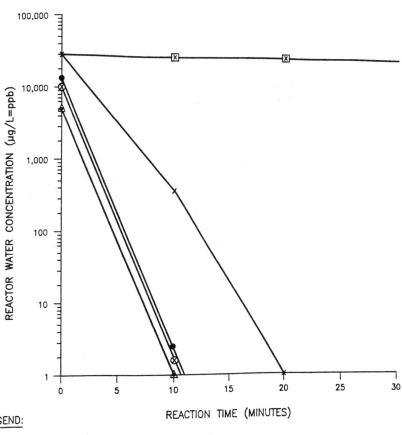

Figure 27.7. UV/H$_2$O$_2$/ozone oxidation of methyl tertiary butyl ether (MTBE) and benzene, toluene, xylene (BTX).

TREATMENT COSTS FOR REMOVING MTBE

Tables 27.6 and 27.7 summarize operating and treatment costs for the technologies studied for removing MTBE from groundwater to allow a side-by-side cost comparison.

The eight treatment technologies should be grouped according to whether or not they emit chemicals to the air. Therefore, the treatment costs for air stripping with aqueous-phase carbon adsorption, air stripping alone, heated air stripping, and diffused aeration should be compared for the group that emits chemicals to the atmosphere. Similarly, air stripping with off-gas incineration and aqueous-phase carbon adsorption, steam stripping, biological oxidation, and UV-catalyzed oxidation using H_2O_2 and O_3 make-up the group that does not emit chemicals to the atmosphere.

Air stripping alone and air stripping followed by aqueous-phase carbon adsorption are the most cost-effective treatment technologies with virtually the same treatment costs of \$9/1,000 gallons and \$10/1,000 gallons, respectively. This is because the costs savings associated with deleting the carbon adsorption system are offset by the costs for the second air stripper needed when air stripping alone is used to generate an effluent concentration of 10 ppb MTBE. Air stripping followed by aqueous-phase carbon adsorption will probably be the best choice in situations where other contaminants with low Henry's law constants are present.

The total treatment costs for removing MTBE by heated air stripping is \$20/1,000 gallons. The same design approach was used for heated air stripping and air stripping except that in the heated air stripping case, the feed was heated to 205°F. The treatment costs for heated air stripping are higher than for the air stripping alone or air stripping with aqueous-phase carbon adsorption because the cost savings associated with deleting the carbon adsorption system or the second air stripper do not offset the increased energy and capital costs associated with the heated air stripper. This preliminary design for the heated air stripping also identified the temperature drop in the water of 83°F as it fell through the packed column as a significant technical issue that was not solved during this engineering assessment. If the total treatment costs had been more competitive, the design would have been refined more to better account for the changing temperature.

The total treatment cost for removing MTBE by diffused aeration is \$19/1,000 gallons with untreated stripping vapors emitted to the atmosphere. The treatment costs are higher than for air stripping alone or air stripping with aqueous-phase carbon adsorption because the capital costs are higher and more energy is required for the air blowers. Capital costs are higher because a series of six tanks and blowers is a less efficient way to strip contaminated water with air than a packed column is. Energy costs are higher because a larger volume of air at a higher pressure drop is required (7,200 cfm versus 1,000 cfm and 100 inches versus 9 inches water column, respectively).

For the group of treatment technologies that does not emit chemicals to the atmosphere, the UV/H_2O_2/O_3 catalyzed oxidation and the air stripping with off-gas incineration and aqueous-phase carbon adsorption have the lowest treatment

Table 27.6. Comparison of Alternative Treatment Technologies for Removing Methyl Tertiary Butyl Ether from Groundwater that Emits Chemicals[a]

Costs	Unit Cost	Air Stripping[b] with Aqueous-Phase Carbon Adsorption		Air Stripping Alone		Heated Air Stripping		Diffused Aeration	
		Daily Consumption	$/day	Daily Consumption	$/day	Daily Consumption	$/day	Daily Consumption	$/day
Operating costs									
Carbon[c]	$1.20/pound	38 pounds	46						
Cooling water	$0.10/1,000 gallons								
Electricity	$0.075/kilowatt hour (kWh)	6 horsepower (HP)	8	7 HP	9	7.5 HP	10	176 HP	236
Natural gas	$3.5/million Btu								
Fuel oil	$0.70/gallon					514	360		
Hydrogen peroxide	$0.65/pound								
Power for UV lights	$0.075/kWh								
UV lights[d]	$60.0/light								
Nutrients (20% solution)	$0.5/gallon								
Methanol	$0.45/gallon								
Disposal of organic	$0.30/pound								
Disposal of biosludge	$0.14/pound								
Total Operating Costs:	$/day		54		9		370		236
Fixed Charges[e]	30% DFC/year	($137 K)	113	($121 K)	100	($194 K)	160	($411 K)	338
Labor	$25/hour	8 hours	200	8 hours	200	8 hours	200	4 hours	100
Total Treatment Costs	$/day		367		309		730		674
Treatment Costs	$/1,000 gallons		10		9		20		19

[a] Bases: 25 gallons per minute of groundwater at 55°F containing 20 parts per million (ppm) methyl tertiary butyl ether (MTBE), 20 ppm total benzene, toluene, and xylene (BTX); effluent criteria: 10 parts per billion (ppb) MTBE, 10 ppb BTX. Costs were based on 1990 data.

[b] Stripper effluent or adsorber feed concentration: 500 ppb MTBE.

[c] Carbon: $0.85/pound carbon purchase, $0.10/pound carbon shipping, $0.25/pound carbon disposal ($500/ton), $1.20/pound carbon total.

[d] Replacement once/year, including installation (from vendor information).

[e] Fixed charges—30 percent/year of direct fixed capital (DFC) shown in parentheses under daily consumption for depreciation, taxes, and maintenance.

Table 27.7. Comparison of Alternative Treatment Technologies for Removing Methyl Tertiary Butyl Ether from Groundwater that Does Not Emit Chemicals[a]

Costs	Unit Cost	Air Stripping[b] with Off-Gas Incineration and Aqueous-Phase Carbon Adsorption Daily Consumption	$/day	Steam Stripping Daily Consumption	$/day	Aboveground[c] Biological Treatment Daily Consumption	$/day	UV-Catalyzed Oxidation Using Hydrogen Peroxide and Ozone Daily Consumption	$/day
Operating costs									
Carbon[d]	$1.20/pound	38 pounds	46						
Cooling water	$0.10/1,000 gallons			26 × 10³ gallons	3			15 × 10³ gallons	2
Electricity	$0.075/kilowatt hour (kWh)	6 horsepower (HP)	8	4 HP	5	20 HP	27	470 kWh	35
Natural gas	$3.5/million Btu	22 × 10⁶ Btu	77						
Fuel oil	$0.70/gallon			283 gallons	198				
Hydrogen peroxide	$0.65/pound							45 pounds	29
Power for UV lights	$0.075/kWh							130 kWh	10
UV lights[e]	$60.0/light							72 lights	12
Nutrients (20% solution)	$0.5/gallon					49 gallons	25		
Methanol	$0.45/gallon					135 gallons	61		
Disposal of organic	$0.30/pound			120 pounds	36				
Disposal of biosludge	$0.14/pound					847 pounds	119		
Total Operating Costs:	$/day		131		242		232		88
Fixed Charges[f]	30% DFC/year	($237 K)	195	($480 K)	394	($432 K)	355	($290 K)	238
Labor	$25/hour	10 hours	250	8 hours	200	10 hours	250	8 hours	200
Total Treatment Costs	$/day		576		836		837		526
Treatment Costs	$/1,000 gallons		16		23		23		15

[a]Bases: 25 gallons per minute of groundwater at 55°F containing 20 parts per million (ppm) methyl tertiary butyl ether (MTBE), 20 ppm total benzene, toluene, and xylene (BTX); effluent criteria: 10 parts per billion (ppb) MTBE, 10 ppb BTX. Costs were based on 1990 data.

[b]Stripper effluent or adsorber feed concentration: 500 ppb MTBE.

[c]Estimated effluent concentration: 365 ppb MTBE, 72 ppb BTX.

[d]Carbon: $0.85/pound carbon purchase, $0.10/pound carbon shipping, $0.25/pound carbon disposal ($500/ton), $1.20/pound carbon total.

[e]Replacement once/year, including installation (from vendor information).

[f]Fixed charges—30 percent/year of direct fixed capital (DFC) shown in parentheses under daily consumption for depreciation, taxes, and maintenance.

costs. The total treatment cost of the UV/H_2O_2/O_3 oxidation is \$15/1,000 gallons. This cost was developed by factoring a budgetary quotation from a vendor for a skid-mounted system. The operating cost was based on the data from the UV/ H_2O_2/O_3 bench-scale test.[1] Even though the UV-catalyzed oxidation system has the advantage of no emissions to the atmosphere, the by-products from the oxidation of MTBE and BTX in the effluent are still unknown. The treatment cost of the air stripping with off-gas incineration and aqueous-phase carbon adsorption is virtually the same as the cost of the UV/H_2O_2/O_3 oxidation. Adding off-gas incineration to the air stripper/carbon adsorber increases the treatment costs to \$16/1,000 gallons because of the capital cost and energy costs for the incinerator.

The treatment costs for the aboveground biological treatment system were \$23/1,000 gallons even though the technical deficiencies discussed previously were temporarily overlooked during the estimation of the costs. The high costs are caused by the extraordinary measures that are needed to maintain a viable biomass and the sludge dewatering system in the capital cost. It should be recognized that these high costs are for a system that cannot meet the effluent criteria for MTBE and BTX. The effluent will also have a significantly higher biological oxygen demand (BOD) because of the methanol that is added to maintain a viable biomass. The high treatment costs show that biological oxidation is not a preferred application for the removal of MTBE from groundwater. Because the total treatment costs were too high, the preliminary design was not refined to address this issue.

The treatment cost for steam stripping was \$23/1,000 gallons even though an optimistically low boil-up of 2% was used. These high treatment costs are due to the cost for the steam and the capital cost for the system. The capital costs were developed by factoring a budgetary quotation from a supplier of a package system. A royalty of 15% was included in the purchased cost because this is a patented stripping technology designed for moderately volatile compounds.

TREATMENT COSTS FOR REMOVING
LOW CONCENTRATIONS OF METHANOL

Table 27.8 summarizes operating and treatment costs for the technologies studied for removing low concentrations of methanol from groundwater.

The design conditions for the biological treatment system for removing methanol and BTX are established by the physical properties of BTX because benzene, toluene, and xylene are less biodegradable than methanol. The design conditions for the other treatment systems are established by the physical properties of methanol because methanol is harder to remove from the water than benzene, toluene, and xylene.

The total treatment costs for UV-catalyzed oxidation using hydrogen peroxide and ozone are \$15/1,000 gallons. These costs are the lowest total treatment costs for any of the four treatment technologies that were evaluated for removing methanol. The costs for the UV-catalyzed oxidation were lowest even though the

Table 27.8. Comparison of Alternative Treatment Technologies for Removing Low Concentrations of Methanol from Groundwater[a]

Technology		Heated Air Stripping		Steam Stripping		Aboveground[b] Biological Treatment		UV Catalyzed Oxidation Using Peroxide and Ozone	
Costs	Unit Cost	Daily Consumption	$/day	Daily Consumption	$/day	Daily Consumption	$/day	Daily Consumption	$/day
Operating costs									
Cooling water	$0.10/1,000 gallons			212 × 10³ gallons	21			15 × 10³ gallons	2
Electricity	$0.075/kilowatt hour (kWh)	12.5 horsepower (HP)	17	8 HP	11	33 HP	44	470 kWh	35
Natural gas	$3.5/million Btu								
Fuel oil	$0.70/gallon	514 gallons	360	771 gallons	540				
Hydrogen peroxide	$0.65/pound							45 pounds	29
Power for UV lights	$0.075/kWh							130 kWh	10
UV lights[c]	$60.0/light							72 lights	12
Nutrients (20% solution)	$0.5/gallon					47 gallons	24		
Methanol	$0.45/gallon					135 gallons	61		
Disposal of organic	$0.30/pound			1,020 pounds	306				
Disposal of biosludge	$0.14/pound					813 pounds	114		
Total Operating Costs:	$/day		377		878		243		88
Other Costs:									
Fixed Charges[d]	30% DFC/year	($220 K)	181	($683 K)	562	($471 K)	387	($290 K)	238
Labor	$25/hour	8 hours	200	8 hours	200	10 hours	250	8 hours	200
Total Treatment Costs	$/day		758		1,640		880		526
Treatment Costs	$/1,000 gallons		21		46		24		15

[a] Bases: 25 gallons per minute of groundwater at 55°F containing 20 parts per million (ppm) methanol, 20 ppm total benzene, toluene, and xylene (BTX); effluent criteria: 10 parts per billion (ppb) methanol, 10 ppb BTX. Costs were based on 1990 data.

[b] Two-stage aeration system; estimated effluent concentration: 100 to 1,000 ppb methanol, 10 ppb BTX.

[c] Replacement once/year, including installation (from vendor information).

[d] Fixed charges—30 percent/year of direct fixed capital (DFC) shown in parentheses under daily consumption for depreciation, taxes, and maintenance.

total treatment costs for heated air stripping, steam stripping, and aboveground biological treatment were developed by temporarily overlooking severe technical limitations for each technology.

Both stripping technologies have high total treatment costs ($21/1,000 gallons for heated air stripping and $46/1,000 gallons for steam stripping) because of excessive energy requirements and capital costs. This is because methanol has a *very* low Henry's law constant. It is about 4% of the Henry's law constant for MTBE, which is already a marginally low value. A lower Henry's law constant for methanol translates to a bigger heated air stripper that must use higher air-to-water ratios for air stripping. The air-to-water ratio used for this application was 60 cfm/gpm or three times more air than in the heated air stripping system for removing MTBE.

This higher air flow increased the amount of heat lost to the air exhaust, which cooled off the water in the air stripper even more than it did in the process for removing MTBE. This means that water in the stripper would not be hot enough to facilitate stripping methanol with air even if the water enters the air stripper at the boiling point. It might be possible to equip an air stripper with heat exchangers to keep the system hot enough, but this would only increase the already uneconomical capital costs and energy costs.

Steam stripping is the only sensible way to keep a stripper hot enough to remove methanol. Unfortunately, the Henry's law constant for methanol is so low that a very high boil-up ratio of 0.17 pound steam/pound feed is required. This makes energy and cooling water costs higher and increases capital cost so that the total treatment costs are excessive even if the technology was feasible.

Although methanol is readily biodegradable, it is not technically or economically feasible to achieve the effluent criterion of 10 ppb methanol with a biological treatment process. The technical limitations of this process are highlighted by the fact that more methanol must be added to the system in order to provide enough substrate to maintain the biomass. The design was prepared by assuming that the practical details can be worked out to achieve the effluent quality even though methanol must be added to the system. Even with this assumption, the total treatment costs were $24/1,000 gallons for biological oxidation. This is not cost-effective compared to UV-catalyzed oxidation.

TREATMENT COSTS FOR REMOVING
HIGH CONCENTRATIONS OF METHANOL

Tables 27.9 and 27.10 summarize the total operating and treatment costs of the biological treatment and $UV/H_2O_2/O_3$ oxidation, respectively, for removing high concentrations of methanol from groundwater to allow a side-by-side cost comparison.

As mentioned in the cases of removing low concentrations of methanol, the design conditions for the UV-catalyzed oxidation treatment systems for removing methanol and BTX were established by the physical properties of methanol

Table 27.9.   Treatment Costs of the Biological Treatment System for Removing High Concentrations of Methanol[a]

Design Case	Unit Cost	20 ppm Methanol[b,c] 25 gpm, 20 ppm BTX		1,000 ppm Methanol[d] 25 gpm, 20 ppm BTX		1,000 ppm Methanol[d] 250 gpm, 20 ppm BTX		1,000 ppm Methanol[d] 250 gpm, 100 ppm BTX	
		Consumption Units/Day	Costs $/Day	Consumption Units/Day	Costs $/Day	Consumption Units/Day	Costs $/Day	Consumption Units/Day	Costs $/Day
Operating Costs:									
Electricity	$0.075/kilowatt hour	33 horsepower (HP)	44	33 HP	44	260 HP	349	300 HP	403
Nutrients (20% solution)	$0.5/gallon	47 gallons	24	47 gallons	24	474 gallons	237	551 gallons	276
Methanol	$0.45/gallon	135 gallons	61	0 gallons	0	0 gallons	0	0 gallons	0
Disposal of biosludge	$0.14/pound	813 pounds	114	813 pounds	114	8,143 pounds	1,140	9,477 pounds	1,327
Total operating cost			243		182		1,726		2,006
Other Costs:									
Fixed charges[e]	30% DFC/year	($471,000)	387	($452,000)	372	($1,314,000)	1,080	($1,377,000)	1,132
Labor	$25/hour	10 hours	250	10 hours	250	10 hours	250	10 hours	250
Total treatment cost	$/day		880		804		3,056		3,388
Treatment cost	$/1,000 gallons		24		22		8.50		9.40

[a] Bases: Groundwater at 55°F. Costs were based on 1990 data.
[b] Two-stage aeration system; estimated effluent concentration: 10 ppb BTX, 100 to 1,000 ppb methanol. Effluent criteria: 10 ppb BTX, 10 ppb methanol.
[c] From the removal of low concentrations of methanol, Table 27.8.
[d] Two-stage aeration system; estimated effluent concentration: 10 ppb BTX, 100 to 1,000 ppb methanol. Effluent criteria: 10 ppb BTX, 1,000 ppb methanol.
[e] Fixed charges—30%/year of direct fixed capital (DFC) shown in parentheses under daily consumption for depreciation, taxes, and maintenance.

Table 27.10.   Treatment Costs of the UV-Catalyzed Oxidation Treatment System for Removing High Concentrations of Methanol[a]

Design Case	Unit Cost	20 ppm Methanol[b,c] 25 gpm, 20 ppm BTX		1,000 ppm Methanol[d] 25 gpm, 20 ppm BTX		1,000 ppm Methanol[d] 250 gpm, 20 ppm BTX		1,000 ppm Methanol[d] 250 gpm, 100 ppm BTX	
		Consumption Units/Day	Costs $/Day	Consumption Units/Day	Costs $/Day	Consumption Units/Day	Costs $/Day	Consumption Units/Day	Costs $/Day
Operating Costs:									
Electricity	$0.075/kilowatt hour (kWh)	470 kWh	35	12,000 kWh	900	120,000 kWh	9,000	129,400 kWh	9,706
Cooling water	$0.10/1,000 gallons	15×10^3 gallons	2	370×10^3 gallons	37	$3,700 \times 10^3$ gallons	370	$3,990 \times 10^3$ gallons	399
Hydrogen peroxide	$0.65/pound	45 pounds	29	1,140 pounds	741	11,400 pounds	7,410	12,290 pounds	7,990
Power for UV lights	$0.075/kWh	130 kWh	10	3,300 kWh	248	33,000 kWh	2,480	35,590 kWh	2,675
UV lights[e]	$60.0/light	72 lights	12	1,836 lights	302	18,360 lights	3,020	19,800 lights	3,257
Total operating cost			88		2,228		22,280		24,027
Other Costs:									
Fixed charges[f]	30% DFC/year	($290,000)	238	($1,800,000)	1,479	($13,900,000)	11,400	($15,000,000)	12,329
Labor	$25/hour	8 hours	200	8 hours	200	8 hours	200	10 hours	250
Total treatment cost	$/day		526		3,907		33,880		36,556
Treatment cost	$/1,000 gallons		15		109		94		101

[a] Bases: Groundwater at 55°F. Costs were based on 1990 data.
[b] Effluent criteria: 10 parts per billion (ppb) methanol, 10 ppb total benzene, toluene, and xylene.
[c] From the removal of low concentrations of methanol, Table 27.8.
[d] Effluent criteria: 1,000 ppb methanol, 10 ppb BTX.
[e] Replacement once/year, including installation (from vendor information).
[f] Fixed charges—30%/year of direct fixed capital (DFC) shown in parentheses under daily consumption for depreciation, taxes, and maintenance.

because methanol was harder to remove from the water than BTX. The design conditions for the biological treatment system were established by the physical properties of BTX because benzene, toluene, and xylene were less biodegradable than methanol.

At low water flow rate and low concentration of methanol and BTX, the total treatment costs for UV-catalyzed oxidation using hydrogen peroxide and ozone are the lowest. The cost is $15/1,000 gallons compared to $24/1,000 gallons when using biological treatment as mentioned in the previous section.

At a high water flow rate and a high concentration of methanol, biological treatment is more cost-effective compared to UV-catalyzed oxidation as shown in Tables 27.9 and 27.10. The total treatment costs are $22, $8.50, and $9.40 per 1,000 gallons for the biological treatment compared to $109, $94, and $101 per 1,000 gallons, respectively, for the UV/H_2O_2/O_3 oxidation. The capital costs of the UV/H_2O_2/O_3 oxidation systems were developed by factoring the budgetary quotations. The increased capital costs are due to significantly larger ozone generators and bigger reactors. The operating cost was based on the data of the UV/H_2O_2/O_3 bench-scale test for MTBE described in Reference 1. The oxidant dosage was based on dosage per pound of chemical oxidized. The extraordinarily high capital and oxidant costs in the UV/H_2O_2/O_3 oxidation treatment show that this technology is not economically feasible for removing high concentrations of methanol from groundwater.

SUMMARY AND CONCLUSIONS

Eight treatment technologies were evaluated for removing MTBE from groundwater. Four of these technologies were evaluated for removing low concentrations of methanol from groundwater. Aboveground biological treatment and UV/H_2O_2/O_3 oxidation were evaluated for removing high concentrations of methanol at low and high water flow rates. In the biological treatment system for removing methanol and BTX, the design requirements were established by the physical properties of BTX, because benzene, toluene, and xylene were less biodegradable than methanol. In the other groups of treatment technologies, the design requirements for removing the oxygenated organics (MTBE or methanol) were more stringent than the design requirements for removing the BTX that was also present. This is because separating BTX from water is much easier than removing MTBE or methanol in these technologies.

The total treatment costs for each technology for removing MTBE and methanol are summarized in Tables 27.6, 27.7, 27.8, 27.9, and 27.10. The reader is cautioned against using these data without understanding the technical details that support the data. This is particularly true for those applications of technologies in which the total treatment costs were prepared even though significant technical issues were not resolved. This caveat applies to the application of heated air stripping and aboveground biological oxidation for removing MTBE from groundwater. The caveat also applies to the use of heated air stripping, steam stripping, and aboveground biological oxidation for removing low concentrations of methanol from groundwater.

The eight technologies considered for removing MTBE should be divided into two groups: one that emits vaporized organics to the atmosphere and one that does not. The most cost-effective technologies of the four technologies in the former group are air stripping alone and air stripping with aqueous-phase carbon adsorption. The other technologies in this group are variations of air stripping, but both heated air stripping and diffused aeration have higher treatment costs. When the situation calls for removing MTBE from groundwater without emitting organic chemicals to the atmosphere, the most cost-effective technologies are UV-catalyzed oxidation using H_2O_2 and O_3 and air stripping with off-gas incineration and aqueous-phase carbon adsorption.

Only one of the technologies evaluated for removing low concentrations of methanol from groundwater is technically feasible. That technology is UV-catalyzed oxidation using H_2O_2 and O_3. The two stripping technologies are not technically feasible because the low Henry's law constant for methanol translates to excessive requirements for air and energy. Biological oxidation is not technically feasible because of the unresolved dilemma between the goal of generating an effluent with essentially no organic contamination and the need to provide enough substrate to maintain a viable biomass.

Of the two technologies evaluated for removing high concentrations of methanol from groundwater, biological treatment is the only alternative that is feasible. Bench- or pilot-scale study of MTBE, BTX, and methanol biological and kinetic parameters would provide valuable information to the full-scale process design. The extraordinarily high capital and oxidant costs of the $UV/H_2O_2/O_3$ oxidation treatment show that this technology is not economically feasible for removing high concentrations of methanol from groundwater.

ACKNOWLEDGMENTS

IT Corporation would like to thank the members of the American Petroleum Institute (API) Groundwater Technical Task Force who provided assistance throughout the performance of this study. Dr. D. Chen (API), Mr. R. Claff (API), Mr. R. Hinds (Chevron Research Company), Dr. D. Keech (Chevron Oil Field Research), Mr. A. Liguori (Exxon Research and Engineering Company), and Dr. G. Mancini (ARCO) provided guidance during all phases of work. Mr. P. Gates (Mobil Oil Corporation) and Mr. J. Rocco (BP America) also devoted time to the project.

REFERENCES

1. "Cost-Effective, Alternative Treatment Technologies for Reducing the Concentrations of Methyl Tertiary Butyl Ether and Methanol in Groundwater," Final Report, Prepared for American Petroleum Institute, IT Corporation, January 1991.
2. "Treatment System for the Reduction of Aromatic Hydrocarbons and Ether Concentrations in Groundwater," Final Report, Prepared for American Petroleum Institute, IT Corporation, June 1988.
3. Speece, R. E., et al., "Nomograph for Air Stripping of VOC from Water," *J. Environ. Eng.,* 113(2):434–443 (April 1987).
4. Fleming, J. L., "Field Application of Treatment Technologies for Removal of Volatile Organics from Water," Final Report, prepared for the Environmental Protection Agency, Cincinnati, OH, IT Corporation Contract No. 68-03-3069, 1984.
5. Lowry, J. and S. Lowry, "Restoration of Gasoline-Contaminated Household Water Supplies to Drinking Water Quality," Proceedings of the Association of Ground Water Scientists and Engineers Second Annual Eastern Regional Groundwater Conference, National Water Well Association, July 1985, pp. 506–521.
6. Truong, K. and J. W. Blackburn, "The Stripping of Organic Chemicals in Biological Treatment Processes," *Environ. Progress,* 3(3):143–151 (August 1984).
7. Petrasek, A. C., Jr., B. M. Austern, and T. W. Neiheisel, "Removal and Partitioning of Volatile Organic Priority Pollutants in Wastewater Treatment," presented at the ninth U.S.-Japan Conference on Sewage Treatment Technology, Tokyo, Japan, September 13–29, 1983.
8. Blackburn, J. W., et al., "Organic Chemical Fate Prediction in Activated Sludge Treatment Processes," Final Report, prepared for EPA Cincinnati, IT Corporation Contract No. 68-03-3027 and 68-03-3074, September 1984.
9. Benefield, L. D., and C. W. Randall, *Biological Process Design for Wastewater Treatment,* (Englewood Cliffs, NJ: Prentice-Hall, Inc., 1980).

CHAPTER 28

The Use of Modern Onsite Bioremediation Systems to Reduce Crude Oil Contamination on Oilfield Properties*

W. W. Hildebrandt and **S. B. Wilson,** Groundwater Technology, Inc., Ventura, California

INTRODUCTION

During the normal operation of oilfield properties, crude oil or other materials frequently leak or spill onto land surfaces. These degraded soils are commonly associated with oil wells, sumps and pits, tank batteries, gathering lines, and pump stations. Crude oil and drilling muds may also be found in these areas and, like the contaminated soil, they may be considered either "hazardous wastes" or "designated wastes" under current regulatory guidelines. The regulations that pertain to crude oil vary among states.

Using California as an example, crude oil is considered a designated waste[1] and its offsite disposal is subject to regulations. If certain constituents (e.g., benzene) are present in sufficiently high concentrations, the soil degraded by crude oil may be considered a hazardous waste[2] and subject to additional regulations. Typically, crude oil sludge in an old sump will contain oilfield treatment chemicals and, therefore, the sludge will be considered a hazardous waste.

Typically, regulations specify the levels to which contaminated soils and groundwater must be cleaned up. Occasionally, however, a quantitative human health

*Copyright 1991, Society of Petroleum Engineers, Hildebrandt, W. W. and Wilson, S. B.: "On-Site Bioremediation Systems Reduce Crude Oil Contamination," *JPT* (January 1991) 18-22.

risk assessment may be used to determine the significance of the presence of crude oil on oilfield properties.[3] An environmental risk assessment can evaluate the potential health effects from the measured onsite concentrations and can propose levels which would be acceptable to regulators.

When oilfield properties are transferred to new operators, abandoned, or converted to other uses, the presence of degraded soil can significantly reduce the value of the property—or even lead to a cancellation of the transaction. Lending institutions are concerned, too, because of their potential liabilities if the borrower defaults. Hence, current property owners frequently consider remediation or removal of the contaminated soil.

If soils are to be reworked to reduce the concentrations of hydrocarbons or other contaminants, regulatory agencies often must be notified. Subsequently, these agencies may require the installation of groundwater monitoring wells and, further, they may establish long-term reporting and monitoring requirements. Excavation and offsite disposal of the degraded soil is often considered, but this method can be quite expensive. In addition, potential liabilities generally remain with the oilfield operator if the degraded soil is transported to a landfill or other site. Soil bioremediation, on the other hand, eliminates an owner's subsequent liabilities and can be very cost-effective—particularly when performed on site.

BACKGROUND

Bioremediation exploits the natural ability of microorganisms to degrade organic chemical contamination in soil and groundwater. Bioremediation of petroleum hydrocarbons is not a new technology. Landfarming (i.e., the spreading of oily waste over the ground surface for natural decomposition) was the first form of soil and sludge bioremediation, and it has been practiced for close to a century. In more recent years, the art of landfarming has been aided by an understanding of the science of hydrocarbon microbiology, and this has led to today's sophisticated remediation technology.

C. E. Zobell[4] is credited with the first publication in this field which documented bacterial growth on petroleum and the relationships that exist between bacterial consortia and petroleum composition. Since this initial publication, an enormous amount of literature has been developed on the subject of hydrocarbon microbiology and on topics related to its application. A review of the literature shows how bacteria interface with hydrocarbons in aqueous media,[5,6] the relationship of bacteria to mineral-bound hydrocarbons,[7] oxygen and nutrient transport,[8] general petroleum degradation technology,[9-14] and innovative applications of biodegradation technology on crude oil.[15,16] The growing understanding of hydrocarbon microbiology has allowed bioremediation to advance from the early days of backlot landfarms to today's sophisticated remediation technology.

Modern soil bioremediation systems, unlike conventional landfarming techniques, do not rely on large surface areas on which to spread contaminated soil and sludge. Instead, solids (i.e., soil or sludge) are placed in windrows or treatment cells, and atmospheric oxygen is supplied via forced aeration or negative-pressure systems. Simultaneously, inorganic nutrients (fertilizers) are applied either manually or through automated systems which are installed directly within the treatment cells. Numerous details must be addressed in a well-designed treatment cell to ensure that it will operate effectively. The simplest design for a treatment cell is shown in Figure 28.1. A modern bioremediation treatment cell is presented in Figure 28.2. The details presented in these figures are discussed in subsequent paragraphs.

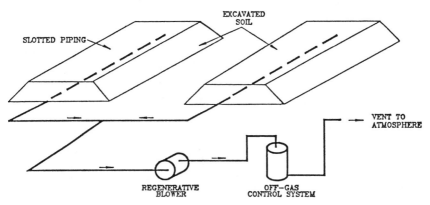

Figure 28.1.   Features of a basic bioremediation system.

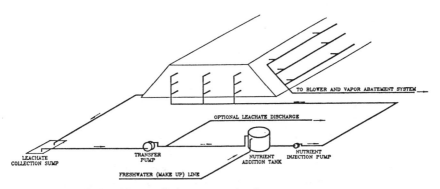

Figure 28.2.   Modern bioremediation treatment cell.

The advantages of modern biotreatment systems are many-fold. These advantages include (1) a significant reduction in the surface area required for treatment, (2) reduced remediation time as a result of improved design and increased system controls, and (3) the ease of applying off-gas treatment controls, if required by local regulations.

The installation and successful operation of a modern soil bioremediation system is based on a design and implementation program which incorporates the following phases:

1. *The preparation of a feasibility/optimization study to determine whether the inorganic, organic, microbiological, and geological conditions are conducive to remediation of the hydrocarbon-contaminated sediments with a biological treatment system.* This phase includes a comprehensive field sampling program, laboratory analysis, and a bench-scale optimization study to determine the conditions that are necessary to complete the remedial program in the most timely and cost-effective manner. Bacterial and nutrient requirements are determined, and soil characteristics are evaluated.

Confirming of the presence of indigenous microorganisms capable of degrading the contaminants of concern is a vital step. This is accomplished by performing bacterial enumeration analyses upon aqueous extractions of the solids, grown with the contaminant of concern as the sole carbon source. The appropriate nutrient requirements are determined by soil microcosm optimization testing, during which various types and ratios of nitrogen and phosphorus sources (and other trace elements) are studied relative to stimulating contaminant degradation.

To identify soil characteristics, tests are performed to study the flow of air and nutrients through the soil/sludge matrix. If inadequate mass transfer exists to stimulate the desired biodegradation effects, flow is tested after the solids have been amended with a series of bulking agents such as rice hulls, wood chips, or silicates. Bulking agents, when properly used, can significantly improve the mass transport throughout the contaminated matrix and, thereby, greatly increase reaction times. The data gathered during this phase are used in the design process and in subsequent phases.

2. *The design and construction of the aboveground biodegradation facilities, including earthwork and the installation of piping, liners, drains, etc.* During this phase, the aeration and nutrient distribution systems are designed, equipment and piping is sized and procured, and contaminated soils and/or sludges are excavated into either windrows or treatment cell configurations. Either a negative-pressure or forced-air system, with slotted piping, is installed in accordance with the design requirements. Nutrients are either mixed with the soil during construction, sprayed during the operation of the system, or supplied through the piping system that is defined by the optimization study and installed during the construction phase.

The dimensions of the treatment system are quite flexible and are generally a function of the site upon which the treatment will occur. Because oxygen and nutrients are generally supplied via forced aeration and pressure-feed systems, respectively, the outside dimensions are governed mostly by site obstacles such as property boundaries and onsite facilities. The mounding of solids at the treatment site and actual construction of the soil cells should not include compaction of the solids. Compaction will only result in decreased mass transfer through the system.

The simplest design for a treatment system is the windrow. The windrow, a proven design utilized extensively in the composting industry, is portrayed in Figure 28.1. Soils to be treated are laid in long row-like piles. Slotted piping is connected to a negative-pressure or forced-air system and is installed so that there is a cylinder-of-influence encompassing each windrow mass. In this manner, oxygen from the atmosphere is either forced from the center of the windrow outward with a forced-air system or drawn from the outside atmosphere inward (toward the piping) in a negative-pressure system. In either case, pressure is applied by a regenerative-type blower which is sized to provide an adequate flow of air through the system. The advantage to the negative-pressure system is that off-gas control can be easily accomplished by covering the system with a perforated cover of appropriate material and placing an off-gas control mechanism such as vapor-phase carbon on the blower effluent stream. Nutrients are either mixed with the soil during windrow construction or are supplied by periodic spraying during the operation of the system.

More elaborate treatment cells can be constructed where areal constraints prohibit multiple windrow systems. These are essentially large soil or sludge piles with aeration and nutrient feed systems as shown in Figure 28.2. Like the windrow system, forced air or negative-pressure aeration systems are utilized; however, unlike the windrow installation, the larger piles require the design and placement of these systems in three dimensions throughout the cell. A nutrient supply system is also employed to percolate aqueous nutrient solutions throughout the treatment cell as defined by the optimization study. Often the site of the treatment installation is graded to a leachate collection system on one end, to facilitate the recovery and subsequent recirculation of water with or without the addition of nutrient amendments.

3. *The operation and maintenance of the biotreatment cell, which includes the aeration system, nutrient supply system, moisture control, and monitoring of the facility.* The condition of the site (e.g., pile covers, liners, berms) is monitored, and field analysis and laboratory assays are performed. Nutrient parameters are monitored in an effort to supply optimum concentrations, and soil moisture is tracked to ensure optimum bioremediation results.

Soil pore moisture is also tracked during operations, since inadequate moisture in the system disturbs the osmotic balance between the bacterial consortium and the medium. Too high a soil moisture content results in soil pore blockage and reduced air transport. Soil pore moisture for optimum bioremediation results varies with soil types; in many cases, 50% field saturation has produced excellent bioremediation results.

Contaminant concentrations are analyzed at regular intervals and compared to starting concentrations to assess progress of the treatment—and to determine the time remaining before closure procedures can be initiated.

4. *Closure of the facility.* This phase consists of three steps, namely (a) documenting that the hydrocarbon content of the soil and/or sludges has been reduced to a concentration that is acceptable to governing bodies, (b) disassembling the treatment cell facility, and (c) demobilizing equipment.

Closure requirements regarding the acceptable level of remaining contaminants vary with state and local agencies. However, requirements are usually in the form of maximum concentrations allowable in the soil/sludge after treatment, and these concentrations are dependent upon the ultimate fate of the material—its fate being disposal in a municipal landfill, use as backfill on the site, etc. Therefore, the procedure for closure consists of documenting any remaining contaminant concentrations in a form that is acceptable to the relevant agencies. Usually, random samples are taken for analyses from throughout the treatment cell. The number of samples can be negotiated, but it is generally based on the volume of solids treated; for example, one sample might be taken for each 75 cubic yards.

CASE STUDIES

Soils contaminated with crude oil have been successfully treated with onsite bioremediation systems at numerous sites. Three case studies, with treated volumes and contaminant concentrations that are representative of those found on oilfield properties, are presented below. Although bioremediation often requires five months or more, the installation and operation of well-designed treatment systems in areas of moderate regulatory action levels has resulted in these sites reaching acceptable hydrocarbon levels in less than ten weeks.

Case Study No. 1:

At a storage facility in south Texas, crude oil contamination of the soil had occurred during years of operation. Approximately 1500 cubic yards of soil were impacted at concentrations near 1000 mg/kg oil and grease, as measured by EPA Method 413.2,[17] modified for soils. The soils at the site were gumbo-type clays with very low permeability.

Studies concluded that conventional onsite treatment techniques such as soil washing were very expensive and cumbersome, and these techniques could still require offsite disposal of the contaminants. Excavation of the soil with offsite disposal was also determined to be very expensive, and it left the operator with the liability associated with the transferring and storing of wastes. It was determined that onsite bioremediation would be successful as a treatment option, and the operator chose to further consider this approach. A feasibility/optimization study was undertaken.

Results of the study showed very positive results, indicating a high potential for success with onsite bioremediation. The nutrients necessary for optimal bioremediation rates were determined to be ammonium chloride and dipotassium phosphate. Although it was initially thought that the clay matrix had low permeability, tests determined that oxygen and nutrient transport would not be significantly restricted; therefore, the addition of bulking agents to the soil to promote maximum bioremediation was found to be unnecessary.

A biotreatment cell, similar to that shown in Figure 28.2, was constructed with approximate dimensions of 100 by 120 feet on a liner of high-density polyethylene sheeting material. Berms were constructed along the perimeter of the cell, ranging in heights from three to five feet. Sediments to be treated were placed on the liner. To ensure that the sediments were adequately porous, a mechanical mixer was used to break up consolidated sediments; this permitted proper aeration of the mound. Two-inch, slotted PVC piping was placed throughout the mound in accordance with the engineering design for optimum aeration, and the piping was gravel-packed to prevent plugging of the slots with sediment.

A one-inch, porous piping system was installed within the treatment cell to permit the addition of both inorganic nutrients and moisture to the biodegradation cell. A mixing tank was installed to feed aqueous solutions of nutrient salts to the piping system. A French drain was constructed along the length of the bioremediation cell to collect leachate, and a recirculation transfer pump equipped with level controls was installed to pump leachate back to the nutrient-addition system. To ensure that the runoff of contaminated rainwater was not a concern during the operation, the cell was covered with polyethylene sheeting. This cover was vented to allow the circulation of air, and it was installed so that rainwater could be diverted to a point outside of the bermed area. The diversion of rainwater in this manner reduced the quantity of contaminated water which had to be disposed of at the site.

As indicated in Figure 28.3, the bioremediation cell constructed at the site performed very well, reducing petroleum contaminant levels within the soil to less than detectable limits within eight weeks. There were a total of five sampling sessions. In each session, samples were analyzed for petroleum concentrations as well as soil moisture. Nutrient concentrations and contaminant-utilizing bacteria counts were also monitored. The average initial contaminant concentrations before system start-up was 920 mg/kg. After approximately three weeks, the contaminant levels had dropped to the 300 mg/kg level. By the eighth week of operation, sampling indicated that contaminants in the soils had been degraded to a level below the detection limit, 50 mg/kg for the analytical method utilized.

Closure was obtained after 20 final samples were collected and analyzed. This sampling program represented one sample for every 75 cubic yards of treated soil. The results of this analysis showed that the average oil and grease value in the treatment cell was 59 mg/kg, which was less than the 75 mg/kg background concentrations. A Request for Closure was submitted to the appropriate agencies, and closure was promptly granted.

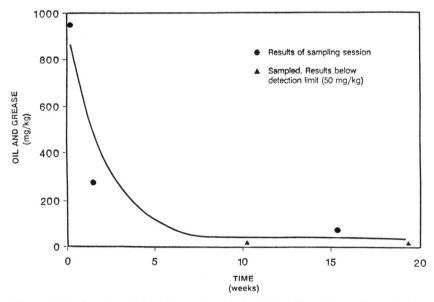

Figure 28.3. Petroleum (oil and grease) reduction during 20-week sampling period.

Case Study No. 2:

The assessment of a property in California, which included shallow borings and laboratory analysis, confirmed the hydrocarbon contamination of approximately 5,000 cubic yards of soil. The analysis of samples, utilizing EPA Method No. 418.1,[17] indicated Total Petroleum Hydrocarbon (TPH) levels up to 4500 mg/kg. The magnitude of the contamination and the pending sale and development of the property required the use of a remediation technique that was expeditious yet cost-effective. An onsite soil bioremediation treatment program was implemented.

A feasibility/optimization study was conducted and it showed good biodegradation potential within the contaminated soil. It was concluded that, to achieve success, it would only be necessary to stimulate the indigenous bacterial population. An analysis of nutrients indicated that ammonium chloride and disodium phosphate would optimize the biodegradation process. The soils were bay muds of low hydraulic conductivity, yet laboratory analyses indicated that the soils exhibited adequate flow capacity to provide for the overall biodegradation reaction, both in terms of air flow and water transport.

A biotreatment system was constructed of a series of negative-pressure windrows. Perforated high-density polyethylene covers were placed on the windrows to minimize the escape of fugitive nuisance odors. Nutrients were added to the system by spraying aqueous solutions upon the soil during the construction phase.

The aeration system, which consisted of slotted pipe, was manifolded to a 3/4-hp regenerative blower. Off-gas controls, in the form of vapor-phase carbon, were applied to the blower effluent.

The biotreatment system was started and 12 sample locations were used to monitor treatment progress. Soil samples were collected on the first day of operation, and the results were used as baseline data to verify subsequent hydrocarbon (TPH) measurements. The baseline concentrations of TPH ranged from 800 mg/kg to 4500 mg/kg. After three weeks, hydrocarbon concentrations were reduced slightly, ranging from 700 mg/kg to 4300 mg/kg. Subsequently, a rapid decrease in hydrocarbon concentrations was detected. In less than six weeks, the results were below 5 mg/kg, which was the method detection limit. The reduction of hydrocarbon contamination at the site is presented in Figure 28.4.

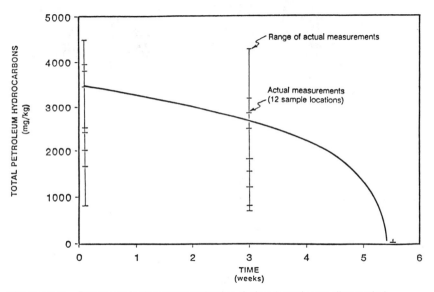

Figure 28.4. Petroleum hydrocarbon reduction during 6-week sampling period.

This case study has been included in this chapter because of supplemental information which demonstrates the effectiveness of well-designed bioremediation techniques. A review of microbial enumeration data from eight locations on the property showed that a general population increase occurred and reached maximum levels five weeks after system start-up. This was synchronous with the time during which the major reduction in TPH occurred. Following the reduction of TPH, there was a reduction in the population of hydrocarbon-utilizing bacteria. This appears to be the result of the diminished organic source. These results are presented in Figure 28.5.

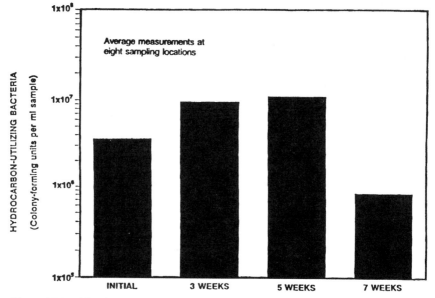

Figure 28.5. Microbial response.

Case Study No. 3:

Approximately 15,000 cubic yards of soil contaminated with crude oil and other hydrocarbons were encountered during excavation work on a property being developed for new land use in central California.[18] Laboratory analyses indicated contamination levels of 42 to 770 mg/kg TPH, utilizing modified EPA Method 8015.[17] Because the local regulatory agency considered soil with concentrations from 100 to 1,000 mg/kg TPH to be a designated waste, it was determined that the contaminated soil could be discharged to a Class I or Class II waste management facility—but this alternative was considered quite expensive, especially since the owner had planned to keep this soil on site and use it as fill material. The soil was a mixture of clay and bay mud.

A bioremediation optimization study was performed on 10 representative soil samples. It was determined that a relatively high background level of hydrocarbon-utilizing bacteria existed in the soil under pH and moisture conditions which were within the optimal range for maximum bioremediation rates. Additional tests were performed to measure available phosphorous and nitrogen levels in the soil, to aid in determining supplemental nutrient requirements. Based on empirical biomass formulas and the hydrocarbon concentrations in the soil, it was concluded that nitrogen was the primary limiting inorganic nutrient. Hence, an aqueous nutrient solution of dipotassium phosphate and ammonium chloride at a pH of 7 was designed to supply sufficient nutrients for complete degradation.

The soil was conditioned and 10 treatment cells were constructed on highly permeable sand beds. Each cell was approximately 6 feet by 110 feet by 9 feet high. Slotted PVC venting pipe was placed in each cell during construction, in accordance with engineering specifications. The inorganic nutrient solution was also applied to the soil during construction. Finally, the cells were covered with polyethylene plastic and the vent piping was manifolded to a high-vacuum blower. The blower effluent was manifolded to a vapor-phase activated-carbon canister.

The treatment cells were sampled after 45 days of operation, and the analytical results indicated that average hydrocarbon contamination had been reduced to 22 ppm, with nearly 60% of the results below the 10 mg/kg detection limit. Soil with these reduced concentrations is not considered a designated waste by regulatory definitions; thus, the soil was used on site as fill material.

This case study has been included in this chapter because additional information, obtained during the project, demonstrates the economic benefit of monitoring off-gases. To monitor microbial activity at this site, carbon dioxide (CO_2) concentrations were measured in the vacuum system effluent. Atmospheric CO_2 levels averaged approximately 350 ppm. Initial CO_2 levels from the treatment cells, however, exceeded 12,000 ppm. After about two weeks these levels approached 4000 ppm and continued to decline slowly. By arbitrarily assuming that 80% of the CO_2 was the result of microbial degradation of the hydrocarbons in the soil, the average degradation rate was calculated by mass balance analysis. This resulted in a prediction of cleanup 45 days after start-up, and it dictated the less expensive sampling schedule which provided the results reported in the preceding paragraph. The CO_2 concentrations measured in the system effluent are presented in Figure 28.6.

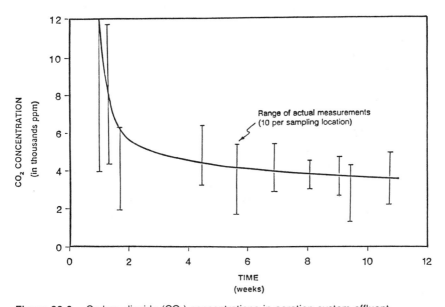

Figure 28.6.   Carbon dioxide (CO_2) concentrations in aeration system effluent.

COSTS

The costs associated with modern soil bioremediation systems are primarily those related to construction activities. The majority of the costs are for earthmoving and construction management. Generally, operation and maintenance costs represent 25% to 30% of the total remediation cost at a site, and this percentage varies inversely with the volume of soil that is treated.

Bioremediation projects have been successfully completed on soils contaminated with 17,000 cubic yards low-gravity crude oil at a total cost of $40 per cubic yard, including excavation costs, in an open, oilfield-type environment. Similarly, the treatment of 10,000 cubic yards of soil containing high-gravity crude has been accomplished at a cost of $35 per cubic yard. It must be recognized, however, that these costs can be significantly higher in some soil types, whenever excavation costs are high due to the presence of pipelines or buildings, or if the crude oil concentrations have become intermixed with groundwater. For example, projects on low-gravity oil in developed areas generally require a high level of control and more stringent operation and maintenance procedures. Such operations have cost as much as $120 per cubic yard for comparable volumes of soil.

CONCLUSIONS

Modern bioremediation systems have many advantages over earlier landfarming techniques. Onsite bioremediation systems are successful at reducing the hydrocarbon contamination in soils to levels acceptable to regulatory agencies. Onsite biotreatment is cost-effective, and it eliminates potential future liabilities of property owners. The successful onsite bioremediation of soils contaminated with crude oil has been accomplished at a total cost of $35 to $40 per cubic yard, including moderate excavation costs, but costs have also approached $120 per cubic yard in areas of more stringent operating conditions.

ACKNOWLEDGMENT

Appreciation is extended to employees of Groundwater Technology, Inc. for their assistance in compiling the data presented in this chapter and, as project managers, for their professional contributions to the success of the bioremediation case studies described herein.

REFERENCES

1. California Code of Regulations, Title 23, Waters, Chapter 3, Water Resources Control Board, Subchapter 15, Discharges of Waste to Land (1989).

2. California Code of Regulations, Title 22, Division 4, Environmental Health, Chapter 30, Minimum Standards for Management of Hazardous and Extremely Hazardous Waste, California Department of Health Services (1989).

3. Sullivan, M. J. "Environmental and Human Health Risk Assessment Methodology for the Evaluation of Environmental Contamination of Crude Oil," paper presented at the 1990 Society of Petroleum Engineers Annual Technical Conference and Exhibition held in New Orleans, LA, September 23–26, 1990.

4. Zobell, C. E. "Action of Microorganisms on Hydrocarbons," *Bacterial Rev.*, 10:1–49 (1946).

5. Gerson, D. F. "The Biophysics of Microbial Growth on Hydrocarbons: Insoluble Sustrates," *International Bioresources J.*, 1:39–53 (1985).

6. Raymond, R. L., Jamison, V. W. and Hudson, J. O. "Oil Degradation in Soil," *Appl. Environ. Microbiol.* 31:522–535 (1976).

7. Wyndham, R. C., and Costeron, J. W. "In Vitro Microbial Degradation of Bituminous Hydrocarbons and In Situ Colonization of Bituminous Surfaces within the Athabasca Oil Sands Deposit," *Appl. Environ. Microbiol.* 41:791–800 (1981).

8. Brown, R. A. "Oxygen Sources for Biotechnological Applications," proceedings of the 1989 Biotechnology Work Group, U.S. Navy Civil Engineering Laboratory/U.S. Army Construction Engineering Laboratory, Monterey, CA, pp. 1–18.

9. Atlas, R. R. "Microbial Degradation of Petroleum Hydrocarbons; An Environmental Perspective," *Microbial Rev.*, 45(1):180–209 (1981).

10. Huddleston, R. L., and Cresswell, L. W. "Environmental and Nutritional Constraints of Microbial Hydrocarbon Utilization in the Soil," in proceedings of the 1975 Engineering Foundation Conference: The Role of Microorganisms in the Recovery of Oil, NSF/RANN, Washington, DC, 1976, pp. 71–72.

11. Kobayashi, H., and Rittman, B. E. "Microbial Removal of Hazardous Organic Compounds," *Environmental Science and Technology*, 16:170A–183A (1982).

12. Loehr, R. C., Heuhauser, E. F., and Martin, J. H., Jr. "Land Treatment of Industrial Waste: Degradation and Impact on Soil Biota," in proceedings of the Sixteenth Mid-Atlantic Industrial Waste Conference, Lancaster, PA, Technomic Publishing Co., 1984.

13. Raymond, R. L., Hudson, J. O., and Jamison, V. W. "Oil Degradation in Soil," *Applied Environ. Microbiol.* 31:522–535 (1976).

14. Whitfill, D. L., and Boyd, P. A. "Soil Farming of Oil Mud Drill Cuttings," paper SPE/IADC 16099 presented at the 1987 SPE/IADC Drilling Conference held in New Orleans, LA, March 15–18.

15. "Exxon Valdez Spill Cleanup: New Lessons Learned," *Ocean Industry* December 1989, pp. 21–24.

16. Hildebrandt, W. W., and Wilson, S. B. "On-Site Bioremediation Systems Reduce Crude Oil Contamination," *Soc. Pet. Eng. J. Pet. Technol.* January 1991, pp. 18–22.

17. "EPA Methods for Chemical Analysis of Water and Waste," EPA 600/4-79-020, Environmental Monitoring and Support Laboratory, Cincinnati, OH, March 1979.

18. Jacobson, J., and Hoehn, G. "Working with Developing Air Quality Regulations: A Case Study Incorporating On-Site Aeration Standards for Vadose Zone Remediation Alternatives," proceedings of the 1987 NWWA/API Conference in Houston regarding Petroleum Hydrocarbons and Organic Chemicals in Groundwater—Prevention, Detection and Restoration.

CHAPTER 29

Use of Government Environmental Records to Identify and Analyze Hydrocarbon Contaminated Sites

Ronald D. Miller, Vista Environmental Information, Inc., San Diego, California

Government environmental agencies offer a wealth of information to the environmental professional. However, lack of coordination between federal, state and local agencies, as well as lack of coordination between various program offices can make use of these records difficult. Notwithstanding this dilemma, government environmental records represent an opportunity to help the environmental professional "leverage" their work product to better assist their clients.

I. WHY GOVERNMENT RECORDS ARE IMPORTANT

PRP Searches

One of the primary uses of government environmental records is the identification of potentially responsible parties (PRPs) who may have contributed to a hydrocarbon contaminated site. For example, your client may be the current target of an agency enforcement action. Hydrocarbon based contaminants may have leached through the soil, thereby contaminating the underlying aquifer. Government environmental records can be used to identify additional sources of potential hydrocarbon contamination to determine the possibility that other parties contributed to a shared, contaminated aquifer.

Previous Investigation/Remedial Activity

Government environmental records can also be used to obtain valuable information regarding previous remedial actions on the subject site. Information, such as the details of the initial site assessment, remedial investigation, well locations, sample results, and other valuable background data regarding a particular site allows an opportunity to gain the insight and information of other individuals who have worked on the site.

Groundwater and Soil Samples

Government environmental records can be used to identify the location and results of groundwater and/or soil samples taken from neighboring properties. This information is helpful in characterizing the extent of potential contamination caused by off-site sources. These records can also be helpful in characterizing the geology and hydrogeology of a particular site.

Agency Perspective

Government environmental records also contain valuable information regarding an enforcement agency's perspective about a particular site or company. This information will be helpful in characterizing the agency's disposition toward the current owner or operator, as well as any previous owners and operators of the property. This type of information can also be used to determine the agency's track record with the acceptance of various alternative remedial technologies.

Groundwater Depth/Direction

Agency files can also contain valuable information on previous borings to determine the geologic, as well as hydrogeologic character of the property. This information can help in determining the general character of a particular site, as well as any potential patterns for migration.

II. WHERE TO FIND RECORDS RELATING TO HYDROCARBON CONTAMINATED SITES

A. Federal Sources

At the federal level, the primary source of information will be the regional office of the U.S. Environmental Protection Agency (U.S. EPA). Within U.S. EPA, information sources are generally available from the various program offices charged with regulating particular activities such as Air, Water, Superfund, and others. However, as a general rule, the following information sources may contain valuable information relating to hydrocarbon active and/or hydrocarbon contaminated sites:

- CERCLIS: Potential sites of environmental contaminations that may require cleanup in the future.
- National Priority List (NPL): Sites requiring cleanup under the Comprehensive Environmental Response, Compensation, and Liability Act (CERCLA).
- Emergency Response Notification System (ERNS): Locates spills of hazardous materials regulated under CERCLA and the Superfund Amendments and Reauthorization Act (SARA).
- Facility Index Data System (FINDS): Contains information concerning particular facilities and the programs they are regulated under, including the Resource Conservation and Recovery Act (RCRA), National Pollutant Discharge Elimination System, and others.
- National Pollutant Discharge Elimination System (NPDES): Water Discharge Permits under the Clean Water Act.
- AIRS: Air Emission Permits under the Clean Air Act.
- Resource Conservation Recovery Act (RCRA): Hazardous waste generators, transporters and treatment, storage, disposal facilities.
- FATES: Pesticide producers regulated under the Federal Insecticide, Fungicide, and Rodenticide Act (FIFRA).
- Toxics Release Inventory (TRI): Indicating sites where toxic releases occur under SARA.

Other valuable federal sources of information include the Department of Transportation, where information can be obtained regarding pipeline accidents and leaks, as well as information on hazardous materials transporters. Finally, the Occupational Safety and Health Administration (OSHA) also has valuable information on facilities that have more than three serious violations or one willful OSHA violation.

B. State Records

Because environmental laws are often implemented at the state level, many state agencies maintain information systems that contain records relating to hydrocarbon contaminated sites:

- Underground Storage Tank (UST) registrations. (It has been estimated that less than 50% of all remaining underground storage tanks have been registered.)

- Leaking Underground Storage Tanks (LUSTs).
- Spills of environmental contaminants and/or hydrocarbon based contaminants maintained by state environmental agencies.
- Landfills: This information normally includes listings of active and inactive landfills, transfer stations, incinerators and other solid waste disposal facilities.
- Hazardous Waste Sites: Many states maintain information systems relative to their state Superfund Site Programs.
- Hydrocarbon Production/Contaminated Sites: Usually, each state Department of Natural Resources (DNR) or Division of Oil and Gas maintains lists of sites used as production facilities. Also, because petroleum is exempted under Superfund, some states have lists of petroleum/hydrocarbon contaminated sites.

C. Local Records

In addition to federal and state records, as identified above, local government agencies may also have valuable information relating to hydrocarbon contaminated sites. First, the county environmental health department may have information relating to contaminated sites, environmentally active sites and other permit holding facilities. In addition, the local fire department or fire marshal's office typically will have records indicating the locations of hazardous and/or explosive materials that generally are composed of hydrocarbon components.

III. HOW TO ACCESS GOVERNMENT ENVIRONMENTAL RECORDS

At the federal level, access to government information is regulated by the federal Freedom of Information Act (FOIA). This law, passed in the late 1960s, allows public access to government agency information. To comply with requirements under FOIA, the information request must contain the following elements:

A. Identify, specifically, the record or information to be disclosed. (Government agencies cannot be forced to "create" records, nor can they be forced to search records in a way in which they are not filed.)
B. Identify clearly the person requesting the records and indicate a maximum fee that the requestor is willing to pay.
C. Indicate that the requestor is expecting a response within 10 days, as required by law. Requestor should cite that the request is being made pursuant to the Freedom of Information Act.
D. If there is any question as to who, within a particular agency, has possession of the information, it is advisable to send the request to the highest person in the relevant group within the entire organization.

Information maintained by state environmental agencies is often subject to state public records acts. Although these acts contain specific issues relative to that particular state's disclosure law, the state can be no more restrictive than allowable under the federal Freedom of Information Act. In most cases, state information will be at least as readily available, if not more available, than federal information.

However, simply because the requestor properly submits a request for information, several compliance issues are likely to arise. First, government agencies typically take far beyond the 10-day period to respond. For example, the U.S. EPA routinely responds with a postcard indicating that they have received the request, assigned a Request Identification Number (RIN), and that they will be responding to you shortly with the requested information. Usually requests to EPA are answered in three to five weeks.

Some agencies also require the submission of fees in advance of processing the request. This requirement effectively then adds to the total turnaround time necessary to access the records. Furthermore, some agencies require use of a specific request form, and/or require that the requestor appear in person in their offices before they will process an information request.

There are also several important exceptions to release requirements. Exempted information is not required to be disclosed and includes:

- Information maintained by a public agency outside the "ordinary course of business," provided that the public interest in withholding records clearly outweighs the public interest in disclosure;
- Records pertaining to pending litigation or enforcement actions;
- Personnel, medical, or similar files;
- Information obtained from a particular party which was conditioned upon the agency holding that information in confidence (unless otherwise required by law);
- Records of complaints to, or investigations conducted by, the state or local police agencies;
- Files relating to employees of the agency itself.

Therefore, unless specifically exempted by particular federal or state laws, information is legally required to be disclosed to the public. Further, in the environmental area, public policy clearly weighs in favor of disclosure of such information.

IV. YOUR RESULTS AND CORRESPONDENCE AS A PUBLIC RECORD

As an environmental professional, you should be aware of your obligations to your clients to keep certain information confidential. For example, if you identify an existing or a potential problem on your client's property, you will need to either keep the information in confidence, as required under your client contract, or disclose the information to a public agency as required by various state or local laws.

Regardless of disclosure obligations, it is critical to remember that in your correspondence, reports, and other communication with environmental agencies, the information that you convey is likely to end up as a public record. Accordingly, it is important to remember your client's interest in protecting their trade secrets and other sensitive information. You may be able to obtain an agreement from the agency to keep submitted information confidential.

Therefore, because information that you generate is likely to become part of the public record in the future, care must be taken to disclose only that which is necessary, and nothing more. In too many cases environmental professionals overreport, or include too much information in reports submitted to an environmental agency.

V. EXAMPLES

As addenda to this chapter you can refer to various examples of valuable information contained in environmental agency files. As you can see, sample site assessments, boring logs, groundwater flow maps, underground storage tank registrations, and fire permits indicate the value of some of this information.

VI. CONCLUSION

Government records have great potential to assist environmental professionals in the identification and analysis of hydrocarbon contaminated sites. While the process of obtaining the information can become burdensome, the value of the information clearly outweighs them. Certainly, with the development of commercial government record information services, such as VISTA, environmental professionals will be able to leverage their effort to provide maximum benefit to their clients. Using government records efficiently and appropriately will enhance the site assessment work, as well as other analytical processes required to perform services on your client's behalf. Therefore, by "leveraging" existing information with your current initiative at end, your clients and our industry will be better served.

(see attachments beginning on next page)

PRELIMINARY SITE ASSESSMENT OF

SOIL AND GROUNDWATER CONTAMINATION

ARCO AM/PM Facility No. 5064
2889 East Valley Boulevard
West Covina, California

for

ARCO PETROLEUM PRODUCTS CO.
P.O. Box 6411
Artesia, California 90702-6411

J.N. 8R4232R

March 4, 1988

Prepared by

Schaefer Dixon Associates
251 Tennessee Street
Redlands, California 92373

Geotechnical Engineers   251 Tennessee Street   (714) 793-2691
                                          Redlands, California 92373

Schaefer Dixon Associates

Schaefer J. Dixon
Ellis J. Jones
Robert J. Lynn
Paul Davis
William J. Monahan
James J. Weaver
John J. Butelo
Joseph F. Montagna
James M. Bell
John H. Foster
Dennis V. Long
Kyle D. Emerson

J.N. 8R4232R

March 4, 1988

ARCO Petroleum Products Company
P.O. Box 6411
Artesia, California 90702-6411

Attention: Mr. Joe Tully

Re: ARCO MP&G Tune-up Facility No. 5064
      2889 East Valley Boulevard
      West Covina, California

Subject: Preliminary Site Assessment of Soil and
           Groundwater Contamination

Gentlemen:

At the request of Mr. Joe Tully, we have performed a preliminary site assessment of the above referenced facility. The intent of our work was to establish the presence, concentration and relative extent of soil and groundwater contamination associated with the existing underground petroleum storage tanks. The report prepared was to address the impact of any identified contamination on the proposed tank removal and reinstallation program. Recommendations were to be provided for mitigation and/or additional evaluation of any contamination warranting such action. The basic findings of our study are presented below for review.

EXECUTIVE SUMMARY

To date, six exploratory borings have been drilled at the referenced facility. The exact locations of these borings are specified on the plot plan included as Exhibit 2. The borings were drilled with an eight or ten-inch rotary flight hollow stem auger to depths ranging from 19 to 38 feet. Within boring no's BH-1, BH-2 and BH-3, a four-inch diameter PVC casing was set within the annular space of the auger and completed to the well specifications presented on Exhibit 11 attached.

Soil samples were collected within these borings at five foot intervals to the total depth of the borings. Boring logs were developed based upon these samples collected and the drilling cuttings and are presented within the body of

J.N. 8R4232R
ARCO Petroleum Products Company
March 4, 1988
Page 2

the report. Water samples were collected from the three monitoring wells installed after appropriate production. All samples collected were obtained in accordance with Los Angeles County specified procedures.

All samples collected from the field evaluation were delivered to a state certified laboratory under our representatives direct custody. Selected samples were analyzed within this laboratory to determine the concentrations of nonhalogenated and aromatic volatile organics following EPA Test Method 8015 and 8020 respectively. Samples collected from around the waste-oil tank were analyzed to determine the concentrations of total petroleum hydrocarbon concentrations following EPA Test Method 418.1. Certain additional selected samples were analyzed to determine the concentrations of organic lead. The results are attached as Exhibit 10.

The results of the field evaluation indicated the presence of slight to definite petroleum odors associated with volatile organic measurements within the sediment above the groundwater table within boring no's BH-1, BH-2 and BH-4. Within BH-3, definite odors and volatile organic measurements were obtained at essentially the groundwater table. Within BH-5 and BH-6, contamination identified as petroleum odors or organic vapors were not detected in any sample obtained.

The chemical analysis performed indicated the absence of nonhalogenated volatile organics and organic lead in all soil samples analyzed. Total petroleum hydrocarbons were detected within BH-4 at a depth of 13½ to 14-feet at 162 parts per million (ppm). All water samples analyzed detected the presence of gasoline contamination and aromatic volatile organic compounds in varying concentrations.

Schaefer Dixon Associates

J.N. 8R4232R
ARCO Petroleum Products Company
March 4, 1988
Page 3

The basic conclusions of our work have indicated that soil contamination associated with the underground storage tanks at this facility have generated groundwater contamination. This groundwater contamination contains concentrations of contaminants above state and local action levels. The contamination is apparently isolated in the soils in the area directly surrounding and beneath the underground storage tanks with minimal lateral detected dispersion.

The basic hydrologic conditions of this site indicate that the groundwater present is perched on the yorba member of the Puente formation. Based upon our borings, it appears that the thickness of alluvium overlying the Puente formation varies from 13 to 22 feet below the ground surface. Varying water depths were obtained from within the three wells installed. Tentative groundwater contours were developed based upon the information obtained. Preliminary information indicates the aquifer maintains a low transmisivity and storage capacity generating the deviations in the groundwater surface.

It is apparent based upon the information developed to date, that soil contamination will be present within the tank cluster to the depth of the groundwater table. Groundwater contamination has apparently migrated away from the storage tanks and beyond the limits of the three wells installed to date to concentrations above current action levels. No contamination was identified in the area of the proposed new tank installation which would require attention during construction. Formalized recommendations for additional work and/or mitigation are proposed within the body of the report following.

Schaefer Dixon Associates

J.N. 8R4232R
ARCO Petroleum Products Company
March 4, 1988
Page 4

It is our hope the information we have presented within this
brief summary and in the attached report, will be sufficient
for your review at this time. Should you have any
questions, please feel free to contact this office at your
convenience.

Respectfully,

SCHAEFER DIXON ASSOCIATES

Kyle D. Emerson, CEG #1271
Associate Geologist

KDE:jks
Addressee (2)
cc: Mr. Joe Tully
     Mr. Andy Paszterko
     California Regional Water Quality Control Board
         Los Angeles Region   Attn: Mr. Josh Workman
     County of Los Angeles, Department of Public Works
         Waste Management Section

Schaefer Dixon Associates

BORING SUMMARY NO. BH-1

ELEVATION: N/1   **DATE DRILLED:** 01-19-88

DEPTH IN FEET	SAMPLES	BLOW COUNT PER FOOT	FIELD MOISTURE % DRY WEIGHT	DRY DENSITY LB./CU. FT.	RELATIVE COMPACTION %	UNIFIED SOIL CLASSIFICATION	MATERIAL DESCRIPTION			
1							4″ asphaltic concrete Fine to medium clayey sand (no odor)	Firm	Slightly Moist	Red/ Brown
2										
3										
4	G&B	52					Fine to medium slightly clayey sand		Slightly Moist	Red/
5										
6										
7										
8							Fine to medium very clean sand with trace of clay & pea grains Reading: 20–50 ppm (slight odor)	Dense	Slightly Moist	Brown
9	G&B	30								
10										
11										
12										
13										
14	G&B	37					Fine to medium sandy clay (slight odor)	Very	Moist	Brown
15										
16										
17										
18										
19	G&B	43					Fine sandy clay with red/brown oxide along soil partings	Very Dry	Slightly Moist to Dry	Gray
20										
21										
22										
23							TOTAL DEPTH OF BORING = 19.5 feet			
24										
25										

G&B Nominal 2-inch California modified
Readings taken with h.n.u photoionization analyzer

ARCO MP&G Tune-up Facility No. 5064 2889 East Valley Boulevard West Covina, California	EXHIBIT NUMBER 3

CENTRUM ANALYTICAL LABORATORIES

CERTIFIED HAZARDOUS WASTE TESTING LABORATORY • CHEMICAL AND BIOLOGICAL ANALYSES

CLIENT : Pioneer Consultants   DATE RECEIVED : 01/25/88
SITE   : Arco-Sentous   DATE ANALYZED : 02/06/88
SAMPLE : 1A   SAMPLE AMOUNT : 500 uls
MATRIX : Water   STANDARD ID   : VOA41

EPA METHOD 8020 (602)

CAS #	COMPOUND:	CONC: UG/L (ppb)	DETECTION LIMIT:
71-43-2	BENZENE	1400	0.5
108-88-3	TOLUENE	190	0.5
95-47-6	TOTAL XYLENES	1400	1.0

EXHIBIT 10A

290 TENNESSEE STREET • REDLANDS, CA 92373 • (714) 798-9336

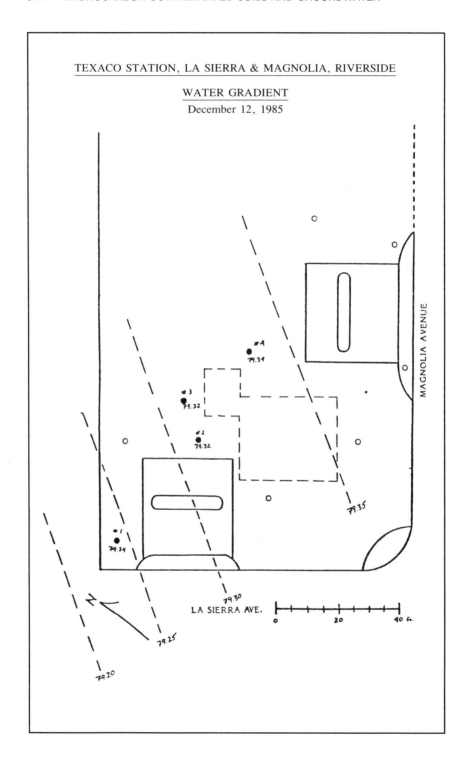

TEXACO STATION, LA SIERRA & MAGNOLIA, RIVERSIDE

WATER GRADIENT
December 12, 1985

Name **LABREA & SECOND SERVICE**   Address **LA BREA SOUTH 201**   No. __9809__

HONORABLE BOARD OF FIRE COMMISSIONERS OF THE CITY OF LOS ANGELES   Code __3 - A__

Date __Nov. 13, 1934__

Gentlemen:

I __herewith request permission to operate and/or maintain (in conformance with the Ordinances of the City of Los Angeles and under the supervision of the Chief Engineer of the Fire Department or his duly authorized representative).

Air Vehicle Factory	Auto Parking Station
Air Vehicle Hangar	Public Filling Station
x Auto Filling Station	Public Oil Depot
	Tenant Garage

On Premises designated as __201 South La Brea__   On __West__   side of Street

Between _____ and _____

and to install and/or maintain in connection therewith TANKS & DISPENSING APPARATUS as follows:

TANKS

No.	Capacity	Contents	Make & Symbol	Location
2	3000	Gasoline	Comwell	OP 15' So of pump island — 4' UG
1	1850	"	"	do
1	1000	"	"	do
2	550	"	"	do

__6__ Tanks __9950__ GALS. UNDERGROUND STORAGE _____ Portable Buggies of _____ Gallons Total Capacity

Applicant is a CORPORATION—ASSOCIATION—PARTNERSHIP—INDIVIDUAL (Indicate by placing an X above type of organization)

                                                    x

Known as __GILBERT KLEIN__   DBA   **LA BREA & SECOND SERVICE**

Contractor _____   Phone _____   Address _____

Signature _____   Title __Prop.__

Mall Address __Same__   Insp. Date _____   Inspector _____

Recommend APPROVAL—DISAPPROVAL   Violation of Ordinance No. _____ Section _____

REMARKS **Formerly Bloome-Cooperman   Appl. 3481**

rmb

_____

_____

_____

_____

_____

Chief Engineer

CHAPTER 30

Development of Remediation Endpoints for Gasoline in Soils and Groundwater Using Risk Assessment, Transport and Dispersion Models

Kenneth Thomas, Groundwater Technology, Inc., Torrance, California
Ruth Custance and **Michael J. Sullivan,** Envirologic Data, Inc., Ventura, California

INTRODUCTION AND BACKGROUND

Risk assessments have been gaining greater acceptance by regulatory agencies in determining the potential environmental health risks associated with petroleum hydrocarbon releases. Earlier evaluations have focused primarily on determining action—no action with respect to remedial activity. However, for most sites, some remedial work will have to be performed. Therefore, the focus of risk assessments today is to develop health-based cleanup goals which balance the strict requirements of protecting the environment while controlling the increasing cost of remediation.

The need to understand the rates of contaminant mobility along with possible exposure pathways is essential in determining potential human health risk limits so that remedial alternatives can be chosen correctly. This is the approach that was taken at a manufacturing facility in Southern California which had an underground fuel tank that leaked premium unleaded gasoline into a shallow perched aquifer. The risk assessment was part of a comprehensive remedial action plan which included a detailed data evaluation, remedial alternative screening, and a risk assessment which used SESOIL (Seasonal Soil Compartment model) along with AT123D (Analytical Transient Dispersion model).

SITE CHARACTERISTICS

The subsurface contamination at this site was a result of leaking in a 10,000-gallon underground premium unleaded gasoline tank system. The tank and piping system were removed in early 1985, and a limited volume of contaminated soil around the tank area was removed. Three groundwater observation wells were then installed around the backfilled tank excavation. Free-phase liquid hydrocarbons were observed in two of the three wells. A remediation plan was developed to address both the free-phase and dissolved hydrocarbons in the groundwater. A groundwater recovery treatment system ("dual pump-and-treat" system) was put into operation in early March 1987, and has been in operation since (Figure 30.1).

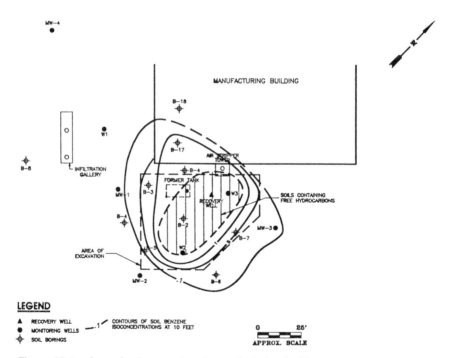

Figure 30.1.   Area of soil excavation at manufacturing facility.

The subsurface beneath the site is characterized by dark stiff clays containing variable amounts of secondary gypsum cement to about 19 feet below grade. Groundwater at the site occurs in two vertically connected hydrostratigraphic zones: an upper low permeability saturated zone (aquiclude) beginning at 10 to 12 feet below grade, and a lower aquifer that extends beyond the maximum depth of investigation. The low permeability stiff clays comprising the aquiclude result in severely retarded plume movement, both as it attempts to disperse and as the treatment system attempts to capture it.

The treatment system has significantly reduced the levels of both phase-separated and dissolved petroleum hydrocarbons in water. However, due to the physical characteristics of the site (i.e., dense, clayey soils) the operation of the pump-and-treat system reached a point of minimum efficiency. Impacted soils and groundwater still contained petroleum hydrocarbon concentrations in excess of established allowable regulatory levels. To compound cleanup efforts, contaminated soils and groundwater were found under the southwest corner of the facility, essentially eliminating more aggressive, remediation techniques.

RISK ASSESSMENT APPROACH

Based on the concentration of benzene in soils and groundwater, and its toxicity and mobility relative to other gasoline components, the risk assessment focused on potential exposures from benzene. The purpose of the risk assessment was to recommend remediation endpoints for benzene in soil and groundwater which are protective of the public health.

Potential exposure scenarios associated with the site were reviewed. It was identified that potential health risks may result from the use of impacted groundwater and from exposure to benzene volatilizing from subsurface soil. No direct contact with the soils was likely to occur due to conditions at the site; i.e., subsurface contamination and the presence of an asphalt cap on the soil.

Impacted groundwater was not currently being utilized for any purposes. Basic water quality parameters (e.g., total dissolved solids, conductivity, sulfates, and chlorides) for the groundwater onsite were well above secondary drinking water standards, indicating that the water was not potable. Although the groundwater exceeded secondary drinking water standards, consumption of groundwater from a hypothetical well located downgradient of the site was conservatively assumed to occur. Three municipal water wells used for domestic or agricultural purposes, located within 3 miles of the site, were identified. When agricultural and domestic (ingestion) water-related scenarios are compared, domestic uses are associated with greater potential exposure. Therefore, the focus of the evaluation was on domestic use of groundwater, assuming that agricultural use of that same water would have been protected for in the evaluation.

Benzene concentration levels proposed for the site were based on Initial Safe Concentrations (ISCs). ISCs take into account attenuation of benzene concentrations due to degradation, migration, dilution, and other factors. The ISCs for groundwater were developed to protect a person ingesting groundwater from a hypothetical well downgradient of the site. The ISCs for soil were derived from potential exposure to vapors in indoor air resulting from volatilization of benzene from subsurface soils.

Two models were used to evaluate the impact of the concentration of benzene in groundwater and soils on a hypothetical downgradient groundwater well. The models considered both soil and groundwater as potential benzene sources. The goals of the environmental fate modeling are presented in Table 30.1. The first model, SESOIL,[1] predicted the vertical migration of benzene from soil to groundwater. The second model, AT123D,[2] predicted the downgradient migration of dissolved benzene in the aquifer. Both of these models used site-specific parameters when available. When site-specific parameters were not available, conservative estimates of those parameters were used to ensure protection of the public health.

Table 30.1.   Goals of Environmental Fate Modeling

Groundwater:	To establish allowable concentration in site groundwater that would not adversely impact a hypothetical downgradient well
Soil:	To establish allowable concentration in site soil that would not result in health risk resulting from vapor migration from subsurface into indoor air
	To determine if benzene left in remediated soil would adversely impact groundwater

Exposures Associated with Impacted Groundwater

An exposure scenario was developed assuming all free-phase hydrocarbons were removed.

The evaluation of groundwater use assumed that a hypothetical well existed approximately 435 feet downgradient from the former tank pit located on the site. Although there were no current uses of the aquifer in this area, this hypothetical scenario was constructed for groundwater protection.

The benzene concentration in the hypothetical downgradient well was based on a mass loading of benzene at an upgradient location. The upgradient location was the former tank pit on the property. The calculated upgradient loading was the mass of benzene which may enter the groundwater and migrate toward the hypothetical downgradient well and not exceed the allowable average concentration of benzene at the tap (California Applied Action Level). This upgradient loading mass becomes the basis for the site-specific remediation endpoint.

The allowable loading mass is used to estimate the health-based allowable on-site groundwater benzene concentration. The site assessment data generated by Groundwater Technology[3] provided information on the horizontal and vertical extent of contamination at the site. Knowledge of the area of contamination and porosity of the saturated zone was used to calculate the total volume of water beneath the site. The allowable loading mass of benzene divided by the volume of groundwater results in the average concentration which may be left onsite in groundwater and not exceed the allowable benzene concentration at the tap. This average groundwater concentration is the health-based benzene concentration criteria for the site. It was recommended that all measured groundwater must be below this criteria for adequate protection of public health.

Estimation of Groundwater Concentration at Hypothetical Well

Groundwater modeling was performed using AT123D to determine an allowable loading mass that may be left in groundwater underneath the site without adversely impacting a downgradient well. The assumptions and parameters used in the model are presented in Table 30.2.

Table 30.2. Assumptions and Parameters Used in Model

Parameter	Assumption	Reference
No continuing source	All free phase hydro-carbons are removed	Remedial Action Plan (Reference 3)
Hypothetical residential well location	At property line, 435 ft from former tank pit area	Site-specific data
Loading rate	650 kg	From monitoring data
Decay constant	4.2 years	Two times anaerobic bio-degradation half-life (References 4 and 5)
Hydraulic conductivity	0.00051 m/hr	Site-specific data
Hydraulic gradient	0.01	Site-specific data
Bulk density	1.45	Site-specific data
Effective porosity	0.27	Silty-clay soils (Reference 1)
Aquifer depth	50 ft	(Reference 4)
Model run duration	50 years	ONRL, 1981 (Reference 4)
Longitudinal dispersivity	50 m	For silty-sand (Reference 4)
Lateral dispersivity	0.5 m	(Reference 4)
Vertical dispersivity	0.5 m	(Reference 4)

A source area was chosen that assumed that the farthest downgradient monitoring well was the zero line, and therefore, the leading edge of contamination at the site. At the present time, neither benzene nor TPH has been detected in this well. By assuming the zero-line is at this well, the actual area of contamination is most likely overestimated. This overestimates the current mass of benzene on the site, which adds additional conservatism to the results.

Concentrations in the aquifer at a distance from the loading area were a function of advection, mechanical dispersion, retardation due to adsorption, and anaerobic biodegradation, as described by the analytical three-dimensional groundwater transport model, AT123D.[2] AT123D simulations were conducted for benzene for a duration of 50 years or until the peak concentration was reached. Output results describe the maximum concentrations of benzene along the assumed direction of groundwater flow to the west-northwest.

The highest average 30-year concentration was calculated from the modeled concentrations for benzene in groundwater at the well location at the property boundary. This average was conservatively assumed to be the concentration throughout the exposure duration. This protects individuals who may ingest groundwater during the peak groundwater concentration period. Thirty-year averages were used as residential exposure point concentrations associated with benzene concentrations in the aquifer because this represents the 90th percentile duration of time spent at a single residence.[6] Maximum concentrations were predicted to occur at the well location at Year 38 of model simulation. Predicted concentrations were calculated for every year of the model run.

A Risk-Specific Concentration (RSC) of 1.0 ppb benzene, which is currently the California DHS action level, was used as the maximum allowable average concentration that could be present at the hypothetical well over the exposure duration. When the initial loading rate of 650 kg (derived from monitoring information) was used, the resulting maximum 30-year concentration was 0.05 ppb. A loading rate of 13,000 kg resulted in the highest 30-year average to be 1.0 ppb. The loading rate of 13,000 kg was then used to calculate an allowable concentration in the aquifer beneath the site. This allowable concentration was established so that the average 30-year concentration at the well location would not exceed the RSC of 1 ppb. From the modeling results, the retarded Darcy velocity for benzene was estimated to be 5.8×10^{-2} m/hr. The predicted migration rate was consistent with site conditions and past monitoring data.

Exposures Associated With Impacted Soils

The remedial goal for soil must be protective of all evaluated routes of exposure associated with that matrix. In the case of soil at the site, there were two possible exposure scenarios: (1) vaporization into buildings and (2) leachate entering groundwater and migrating to a downgradient hypothetical well.

The vapor migration model calculated soil pore-gas concentrations (predicted using fugacity equations) and estimates of diffusion through soil and into buildings. Diffusion rates were then coupled with indoor air-exchange rates to predict indoor air concentrations and exposure.

The SESOIL model was used to predict the mass of benzene entering the groundwater which would not result in an excess of the loading mass corresponding to the RSC. In this case, the mass loading from both SESOIL (soil) and AT123D (groundwater) were added to estimate the total mass loading to the aquifer. The total mass loading from soil and groundwater should be the mass that, if present in the aquifer at the site, will not adversely impact the hypothetical downgradient well. In other words, the risk assessment set an allowable soil level which protects a person consuming drinking water from a hypothetical downgradient well.

Evaluation of Potential Exposure to Inhalation of Benzene in Indoor Air from Subsurface Contamination

Benzene had been detected in soils near the facility during the site assessment. Additionally, although the high clay content, etc., suggest that volatilization may not be significant, volatilization of benzene into indoor air from subsurface contamination was conservatively evaluated. This exposure scenario evaluated potential inhalation exposure to an individual who was assumed to work in a building located directly over the area of contamination. Contamination was assumed to be present at 5 feet below the surface, which was approximately at the depth of the former tank.

A factor contributing to conservative estimates of indoor benzene concentrations resulting from migration from the groundwater and soils is the assumption that a constant source of benzene is present. Mechanisms such as chemical oxidation, and horizontal migration are likely to account for decreases in both the source strength with time, and therefore, the amount of benzene reaching the foundation. In addition, biodegradation was not incorporated in the model adding additional conservatism. Another factor increasing the conservatism of the model is the assumption that the wind always moves toward the building and the site is uncapped. Furthermore, it was assumed that steady-state conditions apply between inside air and outside air.

Estimation of Migration of Benzene From Soil to Groundwater

The potential for benzene to migrate to groundwater from site soils exists. Chemical and soil characteristics as well as climate affect the leaching potential of soils. Figure 30.2 presents the important parameters in evaluating the environmental fate of benzene in soil using environmental modeling.

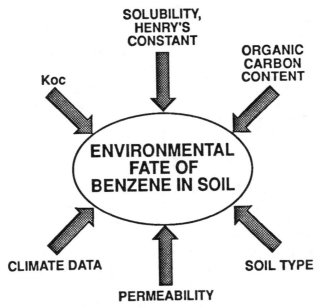

Figure 30.2. Parameters in evaluation of environmental fate of benzene in soil using environmental modeling.

An estimate of the total amount of benzene that could reach the groundwater was obtained from SESOIL. Files used to model migration of benzene in site soils included monthly climate data, chemical loading data, soil-specific data, and chemical-specific data. During calibration of the SESOIL program, many parameters were conservatively estimated, which resulted in an increase in the amount and rate of benzene penetration through the soil column.

These parameters included the soil intrinsic permeability, soil disconnectedness index, organic carbon partitioning coefficient (Koc), and the degradation rate coefficient. A site-specific intrinsic permeability value of 2.4×10^{-12} cm^2 was measured by Groundwater Technology[3] and was initially entered into the model. A value three orders of magnitude greater was used in the final simulation to facilitate rainwater percolation and increase the soil moisture. By using a higher permeability, additional conservatism was introduced into the analysis. Additional parameters used in SESOIL are presented in Table 30.3.

Table 30.3.   Additional Parameters Used in SESOIL

Parameter	Assumption	Reference
Chemical Characteristics:		
K_{oc}	23	(Reference 7)
Solubility	1780 μg/mL	(Reference 8)
Air diffusion coefficient	0.088 cm^2/sec	(Reference 9)
Degradation rate	365 days	One-half anaerobic half-life (Reference 10)
Loading rate	1.0 ppm in first five feet of soil	From site-data
Soil Characteristics:		
Bulk density	1.45	Site-specific data
Effective porosity	0.35	Chosen during calibration
Intrinsic permeability	1×10^{-9} cm^2	For silty or sandy-clay; chosen during calibration
Disconnectedness	11	(Reference 1)
Fraction organic carbon	0.0022	Area study
Climate Data:	From L.A. Civic Center	(Reference 1)

Two scenarios were evaluated using SESOIL: (1) contamination in site soils was assumed to exist throughout the soil column (pre-remediation), and (2) contamination in site soils was assumed to exist in five feet of the soil column (post-remediation). The results indicated that for both scenarios benzene will reach the groundwater (at a depth of 13 feet) by Year 1 and 4, respectively. However, the total mass loading over time for both scenarios would be 1.4 kg and 0.06 kg benzene, respectively.

Benzene concentrations in the leachate were predicted to be greater than 1 ppb benzene for a period of 10 years during the model simulation time. It should be emphasized that the actual concentration in the groundwater will be lower, due to dilution. The contribution to mass loading of benzene into groundwater from soil was then compared to the allowable mass loadings estimated by AT123D. The allowable mass loading estimated by AT123D is 13,000 kg. Presently, it was conservatively estimated that 650 kg of benzene exists in the aquifer below the site. According to model predictions, the maximum mass that may enter the groundwater from soil with an assumed concentration of 1.5 ppm is 1.4 kg. This amount would not significantly add to the amount of benzene that may be left in the groundwater underneath the site.

CONCLUSIONS

An evaluation was performed to determine remediation endpoints for benzene in soil and groundwater. It was determined, using the groundwater model AT123D, that current benzene concentrations in onsite groundwater would not adversely impact a downgradient well. In addition, the amount of benzene predicted by SESOIL to leach from soil into groundwater, assuming onsite soil is at the highest measured concentration, was insignificant when compared to what may remain in the groundwater. It was also determined that predicted indoor air concentrations, using the highest measured soil concentration, would not pose a significant health risk. Therefore, it was concluded that the presence of benzene in soils and groundwater at current site concentrations would not significantly affect the health of potentially exposed individuals.

Based on the results of the risk assessment, a remedial action plan was developed which proposed that the groundwater extraction system be turned off and the groundwater monitored over time. In addition, soil cleanup levels were recommended at the highest onsite concentration; therefore, contaminated soils existing under the facility could remain in place. Currently, the local water control board has approved the remedial action plan recommendations for groundwater. The recommendations for soil are currently being reviewed by the local public health agency.

REFERENCES

1. *"SESOIL," A Seasonal Soil Compartment Model (Documentation)*. Environmental Protection Agency, EPA Contract No. 68-01-6271, prepared in association with Arthur D. Little, 1984.
2. *AT123D: Analytical Transient One-, Two-, and Three-Dimensional Simulation of Waste Transport in the Aquifer System*. Environmental Sciences Division, Oak Ridge National Laboratory, Publication Number 1439. Oak Ridge, TN, 1981.
3. Groundwater Technology, Inc. Subsurface Investigation—For Confidential Client—Maintenance Building Site. Torrance, California. May 8, 1990.
4. *Toxicological Profile for Benzene*. Draft. Agency for Toxic Substances and Disease Registry. Oak Ridge National Laboratory, Oak Ridge, TN, 1987.
5. *Chemical Fate Rate Constants for SARA Section 313 Chemicals and Superfund Health Evaluation Manual Chemicals*. Office of Toxic Substances, U.S. Environmental

Protection Agency, Washington, DC; Environmental Criteria and Assessment Office. EPA. Cincinnati, OH 45268. Prepared for EPA by Syracuse Research Corporation, Syracuse, NY, 1989a.

6. *Exposure Factors Handbook.* Exposure Assessment Group. Office of Health and Environmental Assessment, U.S. Environmental Protection Agency. Washington, DC. EPA/600/8-89/043, 1989b.

7. Rogers R. D., J. C. McFarlane, and A. J. Cross. "Absorption and Description of Benzene in Two Soils and Montmorillonite Clay," *Environ. Sci. Technol.* 14:457–460 (1980).

8. *The Merck Index.* M. Windholz, S. Budavari, R. Blumetti, and E. Otterbein, Eds. Tenth Edition. Merck and Co., Inc. Rahway, NJ, 1983.

9. Silka, L. R. "Simulation of Vapor Transport Through the Unsaturated Zone, Interpretation of Soil-Gas Surveys," *Groundwater Monit. Rev.* 8(2):115–123 (1988).

10. Wilson, B. H., G. B. Smith, and J. F. Rees. "Biotransformation of Selected Alkylbenzenes and Halogenated Hydrocarbons in Methanogenic Aquifer Material: A Microcosm Study," *Environ. Sci. Technol.* 20(10):997–1002 (1986).

CHAPTER 31

The Effects of Toxicological Uncertainties in Setting Cleanup Levels for Hazardous Waste Sites

Mike Alberts and **Fred Weyman,** Radian Corporation, Sacramento, California

INTRODUCTION

How clean is clean? How safe is safe? These are questions that must be asked and answered when setting soil or groundwater cleanup levels at hazardous waste sites. Quantitative health risk assessment is often the tool used to answer the questions by defining appropriate health-based cleanup levels when other state or federal guidelines do not exist. One of the potential problems risk assessors and risk managers face during this process is uncertainty. Is the uncertainty adequately characterized or quantified? Are the risk assessment results too conservative, resulting in cleanup criteria that are overly protective and very costly to attain? Uncertainty analysis, and at the least a sensitivity analysis, should be a critical part of any risk assessment. This chapter focuses on uncertainty in the risk assessment process and one method for quantifying that uncertainty.

THE ROLE OF UNCERTAINTY IN HEALTH RISK ASSESSMENT

Many variables used in health risk assessments have a great deal of uncertainty. In the past, health risk assessments relied primarily on worst-case assumptions and parameters that, when multiplied together in a series of calculations, compounded the error and lead to a very conservative bottom line. This process is still being used today in many risk assessment applications in California (air toxics "hot spots" and permitting of new air emission sources) and other states. There is an advantage to this approach; one can say with confidence that the true risk is not any greater than the calculated risk. The obvious disadvantage is error; the worst-case risk estimate may contain a tremendous amount of error. Two options have commonly been used to address uncertainty in a worst-case analysis: (1) a qualitative discussion of the many conservative assumptions and data is included in the risk assessment text; or (2) a plausible case is presented using more realistic assumptions and data. Option 1 provides no insight into the magnitude of uncertainty or the compounding effects of worst-case assumptions. Option 2 provides some indication of uncertainty by presenting a range of risks, but does not provide any indication of the distribution of risk between the two extremes or the magnitude of uncertainty associated with either result.

The Environmental Protection Agency (EPA), in their latest risk assessment guidelines (RAGs) for Superfund[1] addressed some of the problems associated with a worst-case analysis by advocating a Reasonable Maximum Exposure (RME) scenario for developing baseline and remediation risk assessments. In an RME analysis, a mix of average and upperbound values are used in calculating exposure to a contaminant. For example, in calculating dermal exposure to soil, average values are used for soil concentration and exposed surface area and upperbound values are recommended for other variables. When statistical data are available, the upperbound should be the 90th or 95th percentile. When statistical data are not available, variable values should be selected such that the end result will approximate a 95th percentile. The EPA acknowledges that selecting a combination of variables to achieve a reasonable maximum exposure estimate is not a simple task when numerous variables are used in the calculations and risks must be summed over several pathways and contaminants. To ease some of this burden, the EPA has provided specific values and methods to be used in the RAGs document.

The RME approach advocated by the EPA is a positive step in alleviating the problems associated with the traditional worst-case analysis. In essence, it's a precursor to the next logical step to quantifying uncertainty, a full probabilistic analysis.

Probabilistic simulation, such as a Monte Carlo analysis, represents a third method of conducting a risk assessment and examining uncertainty in the process of inferring risk. It is seldom used for this purpose, and seldom recognized as an acceptable methodology. However, its advantages over the worst-case approach or the two-scenario approach are considerable and worthy of full consideration.

MONTE CARLO RISK ASSESSMENT

A risk assessment conducted using a Monte Carlo, Latin Hypercube, or similar probabilistic analysis offers numerous advantages over the conventional methods described above. The compounding effects on risk from using a series of worst-case assumptions and data are avoided and a quantitative estimate of uncertainty can be obtained. Additionally, probabilistic analysis recognizes that there is no single right answer for most of the risk assessment variables. Rather, many of the variables have a range of values that may be appropriate. Its chief disadvantages are complexity, availability of statistical data for variables, and the difficulty of agency review and public understanding.

In a Monte Carlo or similar analysis, variables in the risk assessment are represented as frequency distributions rather than single point estimates. Hundreds or thousands of probability-weighted risk estimates are determined by independently sampling each variable's distribution and performing risk calculations for each independent data set. Caution is needed to avoid introducing error by independently sampling dependent variables. It is not practical or necessary to specify distributions for all variables. Only those variables for which data exist have a wide degree of uncertainty associated with common point estimates, or are important contributors to risk, need be included. This should include cancer slope factors since they are likely responsible for a majority of the uncertainty in cancer risk. The output of the Monte Carlo simulation is a distribution of risk estimates. A cumulative frequency distribution can be plotted from the output to define any desired risk percentile (50th, 95th, 99th, etc.).

Under the RME approach advocated by the EPA, the risk assessor must make judgments regarding the appropriate variable values to achieve a 90th or 95th upper confidence limit of risk. This can be an arduous task when several contaminants and exposure routes are involved in the calculation of risk. With the Monte Carlo analysis, the upperbound estimate can be determined readily with much more precision.

CASE STUDY: TRADITIONAL VERSUS MONTE CARLO RISK ASSESSMENT RESULTS

To examine the benefits of using a Monte Carlo simulation for risk assessment, and more specifically the effects of toxicological uncertainty on risk estimates, two simple case studies were conducted. The first case study involves benzene contamination in drinking water and exposure through ingestion (for simplicity, dermal contact and other indirect pathways such as inhalation in a shower are neglected). The second case study is based on an air release of benzo(a)pyrene and exposure via inhalation, plant and soil ingestion, and dermal contact. To provide a clear indication of the effects of cancer slope factor uncertainty on risk estimates, the Monte Carlo simulations for the second case study were performed in several steps; the first step varied only the exposure assessment variables and the succeeding steps of the simulation included both exposure assessment variables and the slope factors. Results from each step were then contrasted with the risks calculated using EPA recommendations in the RAGs and California Department of Health Services (DHS) recommendations in their draft waste site risk assessment guidelines. With the exception of cancer slope factors, the draft DHS guidelines are equivalent to EPA. To date, DHS has developed cancer slope factors for approximately 13 substances. Typically, the DHS estimates are higher than EPA.

Case Study #1

Benzene contamination in drinking water and exposure via water ingestion is the basis for the first case study. A contamination level of 5 parts per billion (ppb) was assumed for the calculations of lifetime cancer risk. Statistical distributions for the key exposure variables (ingestion rate and exposure period) are shown in Table 31.1 along with EPA and DHS recommended values. In their Superfund RAGs, the EPA recommends a 90th percentile ingestion rate of 2 liters per day for the RME calculations and an adult average of 1.4 liters per day for an average case. California DHS recommends the same figures. For simplicity, the EPA average and 90th percentile values were used to generate the drinking water ingestion distribution for the Monte Carlo analysis. Exposure period, defined as the number of years actually exposed, was the other key ingestion variable included in the Monte Carlo simulation. The EPA has recommended a 90th percentile estimate of 30 years based on 1980 census data. DHS appears to have adopted the EPA recommendation. By combining 1980 census data with other Census Bureau studies,[2,3] an alternative distribution of approximately 8.4 years with a 95th percentile of approximately 25 years was calculated. These values agree reasonably well with the mean and 90th percentile cited by the EPA.

Table 31.1.   Drinking Water Ingestion Variables

Variable	EPA-RME	DHS	Monte Carlo[a]
Water ingestion	2	2	1.4 (1.05, 1.75)
Exposure period	30	30	8.3 (4.8, 14.4)

[a] Value shown is the mean value. Numbers in ( ) indicate plus one and minus one standard deviation.

Uncertainty in the slope factor for benzene was examined in four scenarios. The distribution for each scenario are indicated in Table 31.2. In the first scenario, the DHS recommended slope factor was used as the basis for calculating a distribution. DHS relied on an animal study and the Linear Multistage (LMS) model to generate an upper 95th percentile slope factor estimate of 0.17 kg-day/mg. The most likely estimate, not recommended by DHS, is 0.082 kg-day/mg. A lognormal distribution was assumed for the DHS results.

The second scenario combined DHS and EPA estimates of the benzene slope factor. The EPA has based its benzene slope factor of 0.029 kg-day/mg on the geometric mean of four values derived from an occupational human epidemiology study. In developing a distribution for this scenario, it was assumed that the DHS and EPA values had an equal probability of being correct and therefore had equal

Table 31.2. Benzene Slope Factor Distributions

Scenario	Potency Distribution (kg-day/mg)[a]
DHS	0.082 (0.055, 0.12)
EPA	0.029 (0.02, 0.04)
DHS-EPA	0.0488 (0.025, 0.094)
EPA-ABELSON	0.00995 (0.0029, 0.034)

[a] Values shown are mean values. Values in ( ) indicate minus one and plus one standard deviation.

weighting in the composite. A log-normal distribution was also assumed for the DHS-EPA composite. A third benzene slope factor scenario was derived solely from the four EPA estimates assuming a log-normal distribution.

In an epidemiological study, one key source of uncertainty is quantifying the dose rate to the cohort. Abelson et al.[4] examined the benzene exposure data from the epidemiology study cited by the EPA and has concluded that exposure was overestimated for the cohort by a factor of 3- to 24-fold. Assuming cancer potency is linear with dose (recognizing that this may be an oversimplification, but conservative), scenario 4 combines the EPA estimate of potency (four individual estimates) with the assumption that exposure was overstated by 3- to 24-fold. The composite distribution for this scenario is shown in Table 31.2.

Case Study #2

This case study examines the risk from a continuous air release of benzo(a) pyrene from a large boiler. An emission rate of 70 pounds per year was assumed. Exposure pathways for this case study included inhalation, soil and plant ingestion, and dermal contact. Statistical distributions for key exposure variables for each pathway are listed in Table 31.3 along with the values recommended by the EPA and DHS. Inhalation parameters were held constant for this analysis due to a correlation with body weight and body surface area.

Table 31.3. B(a)P Exposure Variables

Variable	EPA-RME/DHS	Monte Carlo[a]
Soil ingestion (mg/day)	100	50 (1.65, 154.4)
Plant ingestion (g/day)	NS (116)	38 (1, 85)
Interception fraction (%)	NS (40)	0.4 (0.02, 6.6)
Dust loading[b] (mg/cm^2)	1.45	0.25 ($<<1$, 0.93)
Exposure period (yrs)	30	8.3 (4.8, 14.4)

[a] Values shown are mean values. Values shown in ( ) are minus one and plus one standard deviation.
[b] The mean dust loading rate minus one standard deviation approaches zero in the log-normal distribution.

Exposure via ingestion of fruits and vegetables was based on a typical back-yard garden in an urban area. Neither the EPA RAGs nor the DHS guidelines specify ingestion rates. Therefore, values commonly applied in risk assessments were used in calculating exposure. Other plant variables, such as interception fraction, cleaning efficiency, weathering constants, and root uptake factors, are also not specified in the regulatory guidelines and have been set to the commonly used values shown in Table 31.3. Algorithms for calculating exposure from each pathway are equivalent to those in the EPA RAGs.

A distribution for soil ingestion was derived from tracer study data reported by Sedman.[5] The DHS estimate of soil ingestion (150 mg/day) was also derived from Sedman by applying a safety factor equivalent to one standard deviation. The EPA value (100 mg/day for an adult) was similarly derived.

Soil loading on skin and exposure period were the key dermal contact variables for which distributions were defined. Body surface area represents a source of significant uncertainty, but due to its dependence on body weight, it was held constant at an average value of 3100 cm^2 (this is consistent with the surface area of an adult's arms and hands and represents a reasonable case). The distribution for the dust loading on skin was calculated from a review of dermal contact data by Clement.[6] Exposure period was based on the same distribution specified in the benzene case study.

Two scenarios were examined for B(a)P slope factor uncertainty. The slope factors for inhalation and ingestion appearing in the EPA Health Effects Assessment Summary Tables (third quarter 1990) are 6.1 kg-day/mg and 11.5 kg-day/mg, and represent the 95th percentile upper confidence limit derived from an animal study. The mean value or most likely estimate (MLE) of the slope factor for inhalation is extremely small (on the order of 10^{-7}) in comparison to the upperbound value, due to a zero response in the lowest dose group. The MLE for ingestion was determined to be 3.11 kg-day/mg. In both cases, EPA potency extrapolations were calculated using the LMS model.

In the first slope factor uncertainty scenario, the mean and upper 95th percentile values derived by the EPA were used assuming a log-normal distribution. When the MLE value for inhalation is combined with the upper 95th percentile to develop the distribution, the standard deviation becomes very large and causes difficulties in the Monte Carlo Simulation. Therefore, the most likely value was conservatively assumed to be 0.00611 kg-day/mg. The distributions are indicated in Table 31.4.

Slope factor distributions obtained from the GEN-T software package coupled with the Multistage results developed by the EPA formed the basis for the second scenario. Four models within GEN-T (Multihit, Probit, Logit, and Weibull) were applied to the same animal data used by the EPA. These results, shown in Table 31.4, were combined with the LMS results to obtain a composite distribution.

When combining output from several extrapolation models, there are various schemes that can be applied. For this simple case study, it was assumed that the four GEN-T models and the LMS model had an equal probability of being correct, and therefore, were all weighted equally.

Table 31.4. B(a)P Slope Factor Distributions

Scenario	Potency Distributions (kg-day/mg)[a]
DHS/EPA:	
Oral	3.11 (1.4, 6.9)
Inhalation	6.11×10^{-3} (9.2×10^{-5}, 0.407)
GEN-T results[b] (Oral only):	
Multihit	2.8×10^{-8}
Probit	2.9×10^{-11}
Logit	2.0×10^{-7}
Weibull	4.3×10^{-5}
Combined[c]:	1.9×10^{-8} (1.4×10^{-10}, 0.025)

[a] Values shown are mean values. Values shown in ( ) are minus one and plus one standard deviation.
[b] GEN-T results are most likely estimates of the slope factor.
[c] Combined distribution of GEN-T and LMS values for oral slope factor.

CASE STUDY RESULTS

A Monte Carlo simulation was performed for each case study. To illustrate the effects of toxicological uncertainty, the second case study was conducted in two steps. The first step varied only the ingestion parameters. The second step varied both ingestion and slope factors. In both cases, the results were compared with the traditional deterministic approach to calculating risk. One thousand simulations were used for each Monte Carlo run and should be sufficient to provide a stable output.

Case Study #1

Results from Case Study #1 (benzene in drinking water are illustrated in Figure 31.1. As a baseline for comparison, the cancer risks calculated using the EPA RME approach and the California DHS requirements are 2×10^{-6} and 10×10^{-6}, with the DHS risk being higher due to the higher DHS slope factor for benzene. Monte Carlo results range from approximately 0.5×10^{-6} to just under 3×10^{-6} for the 95th percentile and from 0.1×10^{-6} to 0.8×10^{-6} for the 50th percentile. A 100-fold difference exists between the worst-case DHS approach and the 50th percentile of risk determined using the EPA-Abelson slope factor scenario. In comparing the EPA RME approach to the 95th percentile risk from the Monte Carlo, the RME risk, which is to represent a 90th or 95th percentile, is 3-fold higher. Although this difference is not unreasonable, it could have a significant impact on cleanup alternatives and costs.

It is interesting to note the slight increase in 95th percentile risk between the EPA and EPA-Abelson scenarios. This is likely the result of assuming log-normal distributions coupled with a wider distribution for EPA-Abelson relative to EPA.

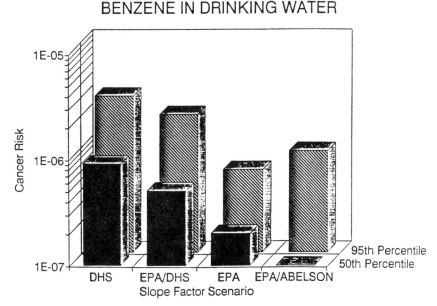

Figure 31.1.   Cancer risk results for case study #1.

Had a different distribution been selected, the EPA-Abelson scenario would likely have shown a lower 95th percentile risk, as would be expected.

Case Study #2

Results for the second case study are illustrated in Figure 31.2. Included in the figure is the EPA risk estimate following the RME approach. For variables that EPA has not designated a value, the upper 95th percentile from the Monte Carlo distribution was applied in the calculations. Under the RME approach, the cancer risk for the benzo(a)pyrene release would be approximately 90×10^{-6}. Including the ingestion parameters in a Monte Carlo, holding the slope factors constant, the risk decreases to 30×10^{-6} at the 95th percentile and just under 2×10^{-6} at the 50th percentile. In this instance, the ingestion parameters have a large influence on uncertainty and risk. Plant ingestion rates, interception fraction, and exposure period have the most influence. As for benzene, the RME approach results in a risk estimate 3 times greater than the Monte Carlo 95th percentile.

When the slope factors are included in the Monte Carlo simulation, using the EPA distributions obtained from the LMS model (labeled EPA on Figure 31.2), risks drop approximately 10-fold at the 50th percentile and about 4-fold at the 95th percentile. Although the 4-fold decrease in risk at the 95th percentile is not substantial, the resulting risk estimate is below a significance criteria used by numerous regulatory agencies. As expected, the combination of GEN-T and LMS

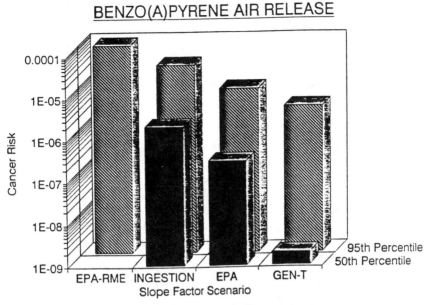

Figure 31.2.   Cancer risk results for case study #2.

model results used in the second B(a)P slope factor scenario resulted in substantially lower risks at the 50th percentile (2×10^{-9}), but only marginally lower at the 95th percentile (3×10^{-6}). The combined EPA/GEN-T distribution specified for the oral slope factor had a very large standard deviation, leading to a 95th percentile estimate that was not much different than predicted by the LMS model. However, had the LMS model not been included in the GEN-T composite, the 95th percentile would have been much closer to the 50th percentile estimate.

In total, the risk estimates for the second case study span over 4 orders of magnitude. This degree of uncertainty in the risk results is substantial, yet is seldom discussed or displayed in the risk assessment. Although policy requires the RME approach to risk assessment and the use of the LMS model upperbound slope factor estimate, Monte Carlo results should be considered as an additional data point in the risk management decisionmaking process.

CONCLUSIONS

Health risk assessment requires numerous assumptions and judgments describing how someone might be exposed, at what level, and over what period. In addition, we must make judgments and assumptions regarding the toxicity of a substance and how it may affect an individual. All of these assumptions introduce error into the risk assessment process. Oftentimes, worst-case assumptions are used to be certain that the risk is not underestimated. To the extent that a worst-case risk is desired, and fully understood, a worst-case approach is reasonable. However, if the results are used as a basis for decisionmaking, the worst-case approach by itself, without any indication of uncertainty, may be inadequate. To address uncertainty in a simplistic manner, a plausible case is commonly presented to provide some indication of the range of possible risk estimates and the magnitude of the uncertainty. The EPA has taken steps under the 1989 Superfund Guidelines to eliminate the compounding effects or worst-case assumptions by advocating a Reasonable Maximum Exposure (RME) scenario, composed of a mix of upperbound (90th or 95th percentile) and average values. This chapter has examined a third option, a probabilistic analysis of major variables. Although controversial, the probabilistic simulation included slope factors, recognizing that these variables contain the greatest degree of uncertainty in the risk assessment process.

Results from the two case studies indicated the degree of uncertainty that may be inherent in the risk assessment process. For benzene in drinking water, the risk spanned two orders of magnitude. Over 4 orders of magnitude (a factor of 10,000) difference was noted in the benzo(a)pyrene air release case study. In most risk assessments, this degree of uncertainty is seldom presented or discussed. For B(a)P, and to a lessor extent benzene, the uncertainty in the slope factor estimates had a significant impact on cancer risk estimates. It must be stressed that slope factor uncertainty, as evaluated in the B(a)P case study, addresses only the uncertainty in the mathematical models used to extrapolate animal data to humans. It does not in any way address the uncertainty associated with using a high dose animal study to extrapolate to low doses in humans. Similarly for benzene, the uncertainty analysis examined only the uncertainty in the epidemiological results and does not address the uncertainty of applying high dose occupational data to low environmental doses.

REFERENCES

1. Risk Assessment Guidance for Superfund. Volume 1, Human Health Evaluation Manual (Part A). Interim Final. U.S. Environmental Protection Agency, EPA/540/1-89/002, 1989.
2. U.S. Bureau of the Census, Current Population Reports, Series P-20, No. 430, Geographical Mobility: March 1986 to March 1987, U.S. Government Printing Office, Washington, DC, 1989.
3. Long, L., C. J. Tucker, and W. L. Urton. "Measuring Migrating Distances: Self-Reporting and Indirect Methods," *Am. Stat. J.* 83:674–678 (1988).
4. Abelson, P. H. "Testing for Carcinogens with Rodents," *Science* 249: 1357 (1990).
5. Sedman, R. M. "The Development of Applied Action Levels for Soil Contact: A Scenario for the Exposure of Humans to Soil in a Residential Setting," *Environ. Health Persp.* 79:291–313 (1989).
6. Multipathway Health Risk Assessment Input Parameters Guidance Document, Clement Associates Inc., February, 1988.

Glossary of Acronyms

ACGIH	American Conference of Governmental Industrial Hygienists
AEL	Angus Environmental Limited
API	American Petroleum Institute
ARAR	applicable or relevant and appropriate requirement
AST	aboveground storage tank
BCF	bioconcentration factor
BGS	below ground surface
BLM	Bureau of Land Management
BTX	benzene, toluene, and xylene
CAA	Clean Air Act
CAP	corrective action plan
CBC	circulating bed combustor (or incinerator)
CCME	Canadian Council of Ministers of the Environment
CDC	Centers for Disease Control
CEC	cation exchange capacity
CERCLA	Comprehensive Environmental Response, Compensation, and Liability Act (Superfund)
CDHS	California Department of Health Services
C-H	carbon-hydrogen
CHESS	Council for Health and Environmental Safety of Soils
CROW	Contained recovery of Oily Wastes
CV	critical value
CWA	Clean Water Act
DAPC	Department of Air Pollution Control
DEQ	Department of Environmental Quality
DHS	Department of Health Services
DNAPL	dense, nonaqueous phase liquid
DNR	Department of Natural Resources
DRE	destruction and removal efficiency
ECD	Environmental Cleanup Division
EOR	enhanced oil recovery
EPA	Environmental Protection Agency
EPRI	Electric Power Research Institute
EPTC	Extraction Procedure Toxicity Characteristic
ERNS	Emergency Response Notification System
FDA	Food and Drug Administration
FIFRA	Federal Insecticide, Fungicide, and Rodenticide Act
FINDS	Facility Index Data System
FOIA	Freedom of Information Act

FRP	fiberglass reinforced plastic
FS	feasibility study
GC	gas chromatography
HRS	Hazard Ranking System
HRT	hydraulic retention time
HSW	Hazardous and Solid Waste
HSWA	Hazardous and Solid Waste Amendments
IFOV	instantaneous field of view
IFT	interfacial tension
IPA	isopropyl alcohol
IR	infrared
ISC	initial safe concentration
IT	IT Corporation
LADD	lifetime average daily dose
LMS	Linear Multistage
LNAPL	light nonaqueous phase liquid
LTDU	land treatment demonstration unit
LTU	land treatment unit
LUST	leaking underground storage tanks
MCL	Maximum Contaminant Level
MDT	minimum detectable temperature
MEI	maximally exposed individual
MIS	mobile incineration system
MLE	most likely estimate
MPN	most probable number
MRTD	minimum resolvable temperature difference
MTBE	Methyl tert-butyl ether
NAPL	nonaqueous phase liquid
NCSRP	National Contaminated Sites Remediation Program
NETAC	National Environmental Technology Application Center
NFOV	narrow field of view
NIEHS	National Institute of Environmental Health Sciences
NOAEL	no-observed-adverse-effect-level
NPDES	National Pollutant Discharge Elimination System
NPL	National Priority List
OSHA	Occupational Safety and Health Administration
PAH	polynuclear (or polycyclic) aromatic hydrocarbon
PC	personal computer
PCB	polychlorinated biphenyl
PCE	tetrachloroethene
PCP	pentachlorophenol
PRP	potentially responsible party
PV	pore volume
RAG	risk assessment guideline

RAP	remedial action plan
RC	rapid curing
RCRA	Resource Conservation and Recovery Act
RfD	reference dose
RFP	Request for Proposal
RI/FS	Remedial Investigation/Feasibility Study
RME	Reasonable Maximum Exposure
ROD	Record of Decision
RQ	respiratory quotient
RsD	risk-specific dose
RSC	risk-specific concentration
RV	recreational vehicle
SARA	Superfund Amendments and Reauthorization Act
SC	slow curing
SCAQMD	South Coast Air Quality Management District
SCE	Southern California Edison
SDR	sediment delivery ratios
SEM	scanning electron microscopy
SITE	Superfund Innovative Technology Evaluation
SOP	standard operating procedure
SRU	soil remediation unit
SVC	semivolatile organic compound
TC	toxicity characteristics
TCDD	2,3,7,8-tetrachlorodibenzo-p-dioxin
TCE	trichloroethylene
TCLP	Toxicity Characteristic Leachate Procedure
TEM	transmission electron microscopy
TIS	transportable incineration system
TLV	threshold limit value
TPH	total petroleum hydrocarbon
TRI	Toxics Release Inventory
TSCA	Toxic Substances Control Act
TSP	total suspended particulates
USDA	United States Department of Agriculture
USLE	universal soil loss equation
UST	underground storage tanks
USTCS	underground storage tank cleanup section
UV	ultraviolet
VCS	Voluntary Cleanup Section
VOC	volatile organic compound
WFOV	wide field of view
WRI	Western Research Institute
XPS	X-ray photoelectron spectroscopy
XRD	X-ray diffraction

List of Contributors

Michael T. Alberts, Radian Corporation, 10395 Old Placerville Road, Sacramento, CA 95827

Michael R. Anderson, Oregon Department of Environmental Quality, 811 S.W. Sixth Avenue, Portland, OR 97204

Janick F. Artiola, Soil and Water Chemistry Laboratory, University of Arizona, Tucson, AZ 85721

James F. Barker, Waterloo Centre for Groundwater Research, University of Waterloo, Ontario, Canada N2L3G1

Renáñ Bass, ChemRisk Division, McLaren/Hart Environmental Engineering, 901 St. Louis Street, Springfield, MO 65806

Allen Biaggi, Nevada Division of Environmental Protection, Bureau of Waste Management, Underground Storage Tank Branch, 123 West Nye Lane, Capitol Complex, Carson City, NV 89710

Robert W. Blanton, Bureau of Equipment Management, Department of General Services, City of Richmond, 2901 N. Boulevard, Richmond, VA 23220

Mark Bonnell, Environment Canada, Place Vincent Massey, 351 St. Joseph Blvd., Hull Quebec K1AOH3

Glen R. Boyd, ENVIRON Corporation, 1 Park Plaza, Suite 700, Irvine CA 92714

Amanda Brady, Environment Canada, Place Vincent Massey, 351 St. Joseph Blvd., Hull Quebec K1AOH3

Edward J. Calabrese, Environmental Health Sciences Program, School of Public Health, University of Massachusetts, Amherst, MA 01003

David J. Carty, K.W. Brown & Associates, Inc., 500 Graham Road, College Station, TX 77845

Chen Y. Chiang, Shell Development Company, Westhollow Research Center, P.O. Box 1380, Houston, TX 77251

Ruth Custance, Envirologic Data, 4572 Telephone Road, Suite 914, Ventura, CA 93003

John F. "Jay" Dablow III, Hughes Environmental Systems, Inc., Building 420, MS 2N206, Manhattan Beach, CA 90266

Robert A. Dixon, Sweet-Edwards/EMCON, Inc., 15055 S.W. Sequoia Parkway, Suite 140, Portland, OR 97224

Ira J. Dortch, Shell Development Company, Westhollow Research Center, P.O. Box 1380, Houston, TX 77251

Robert A. Ettinger, Shell Development Company, Westhollow Research Center, P.O. Box 1380, Houston, TX 77251

Kevin J. Farley, Environmental Systems Engineering, Clemson University, Clemson, SC 29634-0919

Joe M. Fernandez, 3692 Grazing Lane, Loomis, CA 95650

W. T. Frankenberger, Jr., Center for Environmental Microbiology, Inc., 1660 Chicago Avenue, Suite M-2, Riverside, CA 92507

Connie Gaudet, Environment Canada, Place Vincent Massey, 351 St. Joseph Blvd., Hull Quebec K1AOH3

Nancy Gilbertson, The Earth Technology Corporation, Commercial Waste Management Division, 100 West Broadway, Suite 5000, Long Beach, CA 90802

Robert W. Hastings (*NOTE:* At the time Chapter 25 was written, with Brown and Caldwell Consultants, Atlanta, GA.)

Evan C. Henry, Environmental Services, Bank of America, Second Floor, One City Boulevard West, Orange, CA 92668

Warren W. Hildebrandt, Groundwater Technology, Inc., 4820 McGrath Street, Suite 100, Ventura, CA 93003

C. E. Hubbard, Waterloo Centre for Groundwater Research, University of Waterloo, Ontario, Canada, N2L3G1

James D. Jernigan, Environmental Affairs and Safety Department, Amoco Corporation, 200 E. Randolph Drive, Chicago, IL 60601 (associated with Chem-Risk Division, McLaren/Hart Environmental Engineering at the time Chapter 2 was prepared)

Lyle A. Johnson, Jr., Western Research Institute, P.O. Box 3395, University Station, Laramie, WY 82071-3395

Marian W. Kemblowski, Department of Civil and Environmental Engineering, Utah State University, Logan, Utah 84322

Paul T. Kostecki, Northeast Regional Environmental Public Health Center, School of Public Health, University of Massachusetts, Amherst, MA 01003

Jon La Mori, NOVATERRA [formerly Toxic Treatments (USA)], 373 Van Ness Avenue, Suite 210, Torrance, CA 90501

Phil La Mori, NOVATERRA [formerly Toxic Treatments (USA)], 373 Van Ness Avenue, Suite 210, Torrance, CA 90501

L. A. Lemon, Waterloo Centre for Groundwater Research, University of Waterloo, Ontario, Canada N2L3G1

Alfred P. Leuschner, Remediation Technologies, Inc., 9 Pond Lane, Damonmill Square, Concord, MA 01742

Roy A. Litzenberg, The Earth Technology Corporation, Commercial Waste Management Division, 100 West Broadway, Suite 5000, Long Beach, CA 90802

Raymond C. Loehr, Civil Engineering Department, ECJ 8.6, University of Texas, Austin, TX 78712

Donald Mackay, Department of Chemical Engineering and Applied Chemistry, University of Toronto, Canada M5S 1A4

Tom McDowell, Siallon Corporation, 12121 Wilshire Blvd., Suite 1000, Los Angeles, CA 90025

Mary E. McLearn, Electric Power Research Institute, 3412 Hillview Avenue, Palo Alto, CA 94303

Ronald D. Miller, Vista Environmental Information, Inc., 5060 Shoreham Place, Suite 300, San Diego, CA 92122

Jim Najima, Nevada Division of Environmental Protection, Bureau of Waste Management, Underground Storage Tank Branch, 123 West Nye Lane, Capitol Complex, Carson City, NV 89710

Richard H. Oliver, The Earth Technology Corporation, Commercial Waste Management Division, 100 West Broadway, Suite 5000, Long Beach, CA 90802

Erin F. Parker, Environmental Science & Engineering, Inc., P.O. Box 1703, Gainesville, FL 32602

Charles S. Parmele, IT Corporation, 312 Directors Drive, Knoxville, TN 37923

Dennis J. Paustenbach, ChemRisk—A Division of McLaren/Hart Environmental Engineering, 1135 Atlantic Avenue, Alameda, CA 94501

Paul D. Petkovsky, Shell Development Company, Westhollow Research Center, P.O. Box 1380, Houston, TX 77251

Michael R. Piotrowski, Woodward-Clyde Consultants, 1550 Hotel Circle North, San Diego, CA 92108

Jeffrey Powell, TPS Technologies, Inc., 2070 S. Blossomtrail, Apopka, FL 32703

Lynne M. Preslo, ICF Kaiser Engineers, Inc., 10 Universal City Plaza, Universal City, CA 91608

William G. Rixey, Shell Development Company, Westhollow Research Center, P.O. Box 1380, Houston, TX 77001

Melitta Rorty, Roy F. Weston, Inc., 1350 Treat Blvd., Suite 200, Walnut Creek, CA 94596

Raymond A. Scheinfeld, Roy F. Weston, Inc., West Chester, PA 19380

James J. Severns, The Earth Technology Corporation, Commercial Waste Management Division, 100 West Broadway, Suite 5000, Long Beach, CA 90802

Wan Ying Shiu, Department of Chemical Engineering and Applied Chemistry, University of Toronto, Canada M5S 1A4

Robert S. Skiba, Texaco Exploration and Production, Inc., Ventura Area Office, P.O. Box 811, School Canyon Road, Ventura, CA 93002

Curt C. Stanley, Shell Oil Company, P.O. Box 2099, Houston, TX 77252

Michael J. Sullivan, Envirologic Data, 4572 Telephone Road, Suite 914, Ventura CA 93003

J. David Thomas, Radian Corporation, 2455 Horsepen Road, Suite 250, Herndon, VA 22071

Kenneth Thomas, Groundwater Technology, Inc., 20,000 Mariner Avenue, Suite 200, Torrance, CA 90503

Kim N. Truong, IT Corporation, 312 Directors Drive, Knoxville, TN 37923

William A. Tucker, Environmental Science & Engineering, Inc., P.O. Box 1703, Gainesville, FL 32602

K. A. Vooro, Waterloo Centre for Groundwater Research, University of Waterloo, Ontario, Canada N2L3G1

Fred Weyman, Radian Corporation, 10395 Old Placerville Road, Sacramento, CA 95827

Scott B. Wilson, Groundwater Technology, Inc., 20,000 Mariner Avenue, Suite 200, Torrance, CA 90503

Michael Wong, Environment Canada, Place Vincent Massey, 351 St. Joseph Blvd., Hull, Quebec K1AOH3

Index

Aboveground treatment, 295, 471–473, 490–491. *See also* specific types
Acceptable risk, 23–26
Acclimated microorganisms, 221
Accuracy determination, 196–198
Acetone, 403
Action Levels, 91, 98, 100
Activity coefficients, 119, 142
Aeration, 255, 258–263
Aerial infrared surveys. *See* Infrared surveys
Agar, 248
Aggregate stability, 244
Agricultural scenario, 15–16, 31–32
Air monitoring, 413
Air Pollution Control Permit, 69
Air stripping, 464–468
Air/water partition coefficients, 142
Alcohol flooding, 437–458
applications of, 456–457
experimental procedure in, 445–447
field applications of, 456–457
glass beads in, 447–450
removal processes and, 439–440
results of, 447–457
sweep efficiency and, 444–445
theory of, 440–445
Alcohols, 143. *See also* specific types
Algae, 244
Alkanes, 244
n-Alkanes, 244
Ammonia, 184
Aqueous phase carbon adsorption, 464–465
Aquifer studies, 376–380
Aquifer-trench interface, 433–435
Asphalt, 232–233, 327

Bacteria, 245, 279, 413, 414. *See also* specific types
BAM. *See* Behavior Assessment Model
Barite, 187

Barium chloride, 187
BCF. *See* Bioconcentration factor
Bearings, 159
Beer's law, 173
Behavior Assessment Model (BAM), 36
Benzene, 81, 108, 123, 124, 142, 254, 413. *See also* Benzene, toluene and ethylene (BTE); Benzene, toluene and xylene (BTX); Benzene, toluene, ethylbenzene and xylene (BTEX)
concentration of, 148–149
distribution of, 146–147
migration of, 523–525
molecular mass of, 142
potential exposure to, 522–523
rates of loss of, 147
in recovery well, 297
solubility of, 139, 142, 143
Benzene, toluene and ethylbezene (BTE), 123, 124, 126–129. *See also* Benzene; Ethylbenzene; Toluene
Benzene, toluene, ethylbenzene and xylene (BTEX), 103–113, 125, 130, 335, 338. *See also* Benzene; Ethylbenzene; Toluene; Xylene
bioremediation and, 215, 252, 254
differential analytical results, 184
dissolved, 103–113
fate of, 103–113
oxygenated fuels and, 115, 116
prediction of concentrations, 116–125
remediation of, 103–113, 229
retardation and, 134
solubility of, 104–107
in subsurface, 104
thermal treatment and, 365, 367
Benzene, toluene and xylene (BTX). *See also* Benzene; Toluene; Xylene
aboveground biological treatment and, 471, 472
air stripping and, 464
de minimis criteria for, 35–37

diffused aeration and, 469
reduction in, 229
removal of, 461, 462
steam stripping and, 468
Benzoic acid, 244
Bioassays, 232, 245
Bioaugmentation, 279–282
Bioavailability, 21–22
Bioconcentration factor (BCF), 26
Biodegradation, 217. *See also*
    Bioremediation; Degradation
enhancement of, 231–232
of glucose, 255
heavy metals and, 245–246
kinetics of, 257–258
measurement of, 255
of oils, 237, 245–246
of phenol, 255
potential for, 237
Biokinetics, 255. *See also* Kinetics
Biological diversity, 237
Biological treatment, 68–69, 214
aboveground, 471–473
crude oil-impacted soil and, 409–415
monitoring and, 411–413
regulation and, 411
results of, 415
technical aspects of, 413–414
Bioluminescent bacteria, 245
Bioreactor design, 250–251
Bioremediation, 213–222, 237–284. *See*
    also Biodegradation; Remediation
acclimated microorganisms and, 221
advantages of, 213, 239, 489
aeration and, 258–263
bench-scale treatability studies of,
    240–282
biodegradation kinetics and, 257–258
bioreactor design and, 250–251
case studies of, 492–497
costs of, 498
CROW© Process for. *See* CROW©
    Process
of crude oil, 487–498
defined, 213, 237, 238
degradation and, 217, 218
degradation test protocol in, 215–216
description of, 237
environmental parameters and, 258–282

feasibility of, 238–240, 490
heavy metals and, 245–246
history of, 488–492
intermediates from, 244
loading rates and, 221
materials in, 215–217
methods in, 215–217
microbial population counts and,
    246–250
mineralization and, 255–257
moisture and, 219, 264–266
nutrients and, 273–279
pH and, 263–264
physicochemical properties and,
    240–244
reactor design and, 250–251
sample collection and, 240
at Superfund site, 371–400
    action plan for, 383–384
    aquifer study at, 376–380
    completed actions in, 384–397
    demonstration program at, 393–397
    extent of contamination and,
        372–376
    future plans for, 397–399
    goals of, 382–383
    groundwater extraction and,
        388–393, 397
    operational testing and, 391–393
    pilot-scale studies of, 376–381, 395
    site description, 372
    tempertaure and, 266–273
    toxicity studies and, 244–245
    TPH depletion and, 251–254
Biota, 149–150
Boussinesq equation, 418–419
Bridge deck structural surveys, 159
Bromide tracer study, 395–396
BTE. *See* Benzene, toluene, and
    ethylbenzene
BTEX. *See* Benzene, toluene,
    ethylbenzene, and xylene
BTX. *See* Benzene, toluene, and xylene
Buckley-Leverett assumption, 443
Building energy conservation surveys,
    159

CAA. *See* Clean Air Act
Calculation errors, 186–187

Canadian approach to cleanup levels,
    49–64
  National Classification System in,
    53–55
  public consultation in, 50–51
Capacity term, 139
Capillary forces, 437, 442, 444. *See
    also* specific types
Capillary pressure hysteresis,
    421–423
Capillary-pressure saturation relations,
    421
Carbonates, 256
Carbon dioxide, 184, 255, 256, 257
Carbon monoxide, 115, 256
Carcinogens, 24–26. *See also* specific
    types
Catalog engineering, 6
Cation exchange capacity (CEC), 186
CEC. *See* Cation exchange capacity
CERCLA. *See* Comprehensive
    Environmental Response,
    Compensation and Liability Act
Chemical potential, 137, 138–139
Chemical properties. *See*
    Chemicophysical properties
Chemical treatment, 69
CHESS. *See* Council for the Health and
    Environmental Safety of Soils
Chlorobenzene, 81, 312
Chlorophenols, 215
Chromatography, 216, 217, 255,
    345–347, 445, 446, 447. *See also*
    specific types
Circuit board quality control, 159
Circulating bed incinerators, 230–231
Clean Air Act, 92
Clean Air Act Amendments of 1990,
    115
Cleanup levels
  Canadian approach to, 49–64
  National Classification System in,
    53–55
  public consultation in, 50–51
  establishment of, 51–53
  standard selection for, 91–100
Clean Water Act (CWA), 92, 93, 98
Colloids, 150–151
Composting, 272

Comprehensive Environmental
    Response, Compensation and
    Liability Act, 25, 91, 92–93, 94,
    95, 99, 100, 528
  bioremediation and, 238
  government records and, 503
  remediation and, 224
Contained Recovery of Oily Wastes.
    See CROW© Process
Corrective action, 85, 91, 98, 99, 100
Cosolubility, 116
Cosolvents, 147, 150–151
Council for the Health and
    Environmental Safety of Soils
    (CHESS), 12, 39
Creosote, 344
Critical value, 197
CROW© Process, 343–357
  biotreatment and, 349–357
  results of, 345–349
  tested materials in, 344–345
Crude oil, 409–415, 487–498
Crystallography, 334
CWA. *See* Clean Water Act
Cytochromes, 250

Darcy's law, 444
1,2-DCA, 229
1,2-DCP. *See* 1,2-Dichloropropane
Degradation, 244. *See also*
    Biodegradation
  first-order, 254
  kinetics of, 254
  losses from, 148
  microbial, 237, 244
  of pentachlorophenol, 393
  studies of, 217, 218
  test protocol for, 215–216
Dehydrogenase, 245
De minimis criteria, 11–39
  acceptable levels of risk and, 23–26
  in agricultural scenario, 15–16, 31–32
  for BTX, 35–37
  for carcinogens, 24–26
  defined, 13–14
  establishment of, 26–37
  exposure parameters and, 17–23
  in groundwater protection, 16, 34–35
  in industrial scenario, 14–15, 27–31

exposure parameters and, 17–23
in groundwater protection, 16, 34–35
in industrial scenario, 14–15, 27–31
for noncarcinogens, 24
for nonvolatiles, 14
in recreational scenario, 15, 27–31
in residential scenario, 14, 27–31
state guidelines and, 37–38
for TCDD, 26–35
for volatiles, 14
wildlife protection and, 16, 34–35
Dense, nonaqueous phase liquids
    (DNAPL), 374, 375, 437, 446,
    447, 449, 450, 455
Dermal contact rate, 19–20
Destruction and removal efficiencies
    (DRE), 230
Detoxifier Process, 311–312
1,2-Dichloroethane, 81
1,2-Dichloropropane, 81, 83
Diesel, 81, 245, 335
    steam injection for removal of,
    401–407
Differential analytical results, 181–199
    causes of, 182–191
    consequences of, 191–196
    environmental consequences of, 191–192
    error types in, 183–187
    evaluation of, 182–191
    example problems in, 187–191
    industry consequences of, 195–196
    regulatory agency acceptance of,
    191–196
    resolution of, 196–198
Differential equations, 137, 427
Differential equilibrium, 121–122,
    123–125, 129
Diffraction, 329
Diffused aeration, 469–475
Diffusion, 147, 255
Digestion, 184–185
Dilution, 147, 177–178
Dioxins, 16
Dirichlet condition, 435
Dispersion, 254, 517–525
Displacement, 137
Dissolved oxygen, 379, 396
Dissolved phase, 225
Dissolved plumes, 103–113

Distribution coefficients, 138
DNAPL. *See* Dense, nonaqueous phase
    liquids
Documentation errors, 186–187
Drainage, 421
DRE. *See* Destruction and removal
    efficiencies
Dutch steam stripping, 403–404

Effective solubility, 139
Electrical energy transmission system
    audits, 158
Electron microscopy, 329, 331–333.
    See also specific types
Emulsification, 147
Emulsions, 150–151
Energy conservation surveys, 159
Energy evolution, 159
Energy levels, 164
Energy transmission system audits, 158
Environmental fate, 22–23, 26–27, 137
    of methanol, 103–113
Equilibrium, 131, 139, 152, 419
    BTEX concentrations and, 116–125
    differential, 121–122, 123–125, 129
    displacement from, 137
    flash calculations for, 121–122, 124
    fluid-phase, 118–120
    liquid-liquid phase, 119
    local, 254
    phase, 124
    simulation of, 120
    ultimate value for, 150
Equilibrium partitioning behavior, 116
Ethane oxidizers, 257
Ethanol, 116
Ethylbenzene, 123, 124, 403. *See also*
    Benzene, toluene, ethylbenzene and
    xylene (BTEX); Benzene, toluene
    and ethylbenzene (BTE)
Evaporation mass transfer coefficient, 149
Excavation, 208–209
Exponential regression, 299–301
Exposure duration, 21
Exposure parameters, 17–23
Exposure potential, 239
Extraction, 295
    groundwater, 226–229, 388–393, 397
    vacuum, 254

Facility permitting, 69–70
Fate, 22–23, 26–27, 137
 of methanol, 103–113
Fertility of soil, 244
Fick's law, 139
FID. *See* Flame ionization detection
Fingering, 440, 444
First-order kinetics, 254, 258
Flame ionization detection (FID), 413,
 446, 447
Flash calculations, 121–122, 124
Flue efficiency audits, 158
Fluid-phase equilibrium, 118–120
Fluid redistribution, 421
Fluoresceine diacetate, 250
Flushed water treatment, 350
FOIA. *See* Freedom of Information Act
Fraction of soil from contaminated
 source, 21
Freedom of Information Act (FOIA),
 504
Free-phase hydrocarbon, 130–132
Freon, 173, 177, 253
Fugacity, 137, 139
Furans, 16

Gas chromatography, 216, 217, 255,
 345–347, 446, 447
Gasoline, 81, 115, 123, 245
 in groundwater, 517–525
 methanol in, 103–113
 remediation endpoints for, 517–525
 in soils, 335–337, 517–525
Gasoline stations, 5
Gauss-Jordan elimination method, 435
GC. *See* Gas chromatography
Glass beads, 447–450
Glucose biodegradation, 255
Government environmental records,
 501–515
 access to, 504–505
 correspondence in, 505–506
 examples of, 506, 507–515
 federal sources of, 502–503
 importance of, 501–502
 local sources of, 504
 location of, 502–504
 state sources of, 503–504
Granulation, 244

Groundwater
 alcohol flooding for NAPL removal
  from. *See* Alcohol flooding
 depth of, 502
 direction of, 502
 extraction of, 226–229, 388–393, 397
 flow of, 418–423, 432–433
 gasoline in, 517–525
 monitoring of, 412
 oxygenated fuel effects on, 115–135
 protection of, 16, 34–35
 remediation of, 295–309
 remediation endpoints gasoline in,
  517–525
 samples of, 502
 tidal influences on, 428–432
Growth-inhibitory factor, 244
Growth regulatory compounds, 245

Half-life, 148, 217
Harmonic oscillations, 427
Hazardous and Solid Waste
 Amendments (HSWA), 76, 223
Hazardous waste sites
 in situ treatment of, 343–357
  biotreatment and, 349–357
  results of, 345–349
  tested materials for, 344–345
 toxicological uncertainties in setting
  cleanup levels for, 527–536
Hazard Ranking System (HRS), 92
Health risk assessment. *See* Risk
 assessment
Heated air stripping, 465–468
Heavy metals, 16, 245–246. *See also*
 specific types
Heavy oils, 251, 257
Henry's constant, 22, 23, 138, 253
Herbicides, 15. *See also* specific types
Hexachlorocyclohexane, 255
Hexadecane, 250, 255
n-Hexadecane, 244
Homogeneity, 255
Horizontal wells, 457
HRS. *See* Hazard Ranking System
HSWA. *See* Hazardous and Solid Waste
 Amendments
Hydrocarbon-degrading microorganisms,
 248

Hydrocarbon-to-hydrocarbon structure-activity-property relationships, 138
Hydrocarbon-oxidizers, 248, 249
Hydrogen peroxide, 344, 473–475
Hydrogen sulfide, 184
Hydrolysis, 250
Hysteresis, 421–423

Incineration, 230–231, 465
Industrial scenario, 14–15, 27–31
Infill, 6
Infinite dilution, 147
Infrared imagers, 160
Infrared surveys, 157–167
    applications of, 158–164, 169–172
    calibration in, 176
    case studies of, 162–164
    considerations in, 164–165
    defined, 157
    dilutions in, 177–178
    in environmental assessments, 159, 162–164
    field, 169–179
    history of, 157
    limitations of, 161
    procedures in, 175–178
    system architecture for, 160–161
    terms used in, 165–167
    theory of, 172–176
    TPH analysis with, 169–179
Inhibition capillary pressure curves, 421
Injection system, 397–398
Inocula development, 350, 351
Interception trench-aquifer interface, 433–435
Interfacial tension, 438
2-(p-Iodophenyl)-3-p-phenyl)-5-phenyl tetrazolium, 250
IPA. See Isopropyl alcohol
Isopropyl alcohol, 445, 447, 449, 450. See also Alcohol flooding

Kerosene, 245
Kinetics
    biodegradation, 255, 257–258
    degradation, 254
    first-order, 254, 258
    sigmoidal, 258
Kozeny model, 424

Laboratory differential analytical results. See Differential analytical results
Laboratory performance errors, 185–186
Laboratory treatability studies, 214
LADD. See Lifetime average daily dose
Land disposal restrictions, 234
Landfarming, 239, 240, 241–243
Landfilling, 208–209
Landspreading, 240
Land Treatment Unit (LTU), 383, 385, 386, 397
Latin Hypercube, 528
Lauric acid, 244
Lead, 11, 115
LeBas correlation, 142
Lifetime average daily dose (LADD), 29, 30
Light nonaqueous phase liquids (LNAPL), 374
Linear partitioning, 254
Linear regression, 297, 298
Lipid membranes, 244
Lipids, 149. See also specific types
Liquid-liquid phase equilibrium, 119
LNAPL. See Light nonaqueous phase liquids
Loading rates, 221
Local equilibrium, 254
Logarithmic regression, 297
LTU. See Land Treatment Unit

M85 spill simulation, 125–132
Macro-nutrients, 248
Maginot line concept, 431, 432
Margules relationship, 119
Marine Protection, Research, and Sanctuaries Act, 92
Mass balance calculations, 143–147
Mass spectrometry, 216
Mass transfer coefficient, 149
Matrix spikes, 187
Maximum contaminant levels (MCLs), 24, 98
MCLs. See Maximum contaminant levels
Mechanical systems audits, 159
Metals, 16, 245–246. See also specific types

Methanol, 103–113, 116, 117, 461–485
  aboveground biological treatment and,
    471–473
  air stripping and, 464–468
  concentrations of, 125–129
  costs of removal of, 479–484
  diffused aeration and, 469–475
  fate of, 103–113
  in gasoline, 103–113
  high concentrations of, 481–484
  low concentrations of, 134, 479–481
  maximum concentrations of, 123, 128
  mobility of, 109–112
  mole fraction of, 122
  off-gas incineration and, 465
  persistence of, 109–112
  remediation of, 103–113
  steam stripping and, 466–468
  technologies for removal of, 464–465
  UV-catalyzed oxidation and, 473–475
Methylnaphthalene, 244
Methyl tertiary butyl ether (MTBE),
    115, 116, 143, 461–485
  aboveground biological treatment and,
    471–473
  air stripping and, 464–468
  costs of removal of, 476–479
  diffused aeration and, 469–475
  off-gas incineration and, 465
  steam stripping and, 466–468
  technologies for removal of, 464–465
  UV-catalyzed oxidation and, 473–475
Microbial degradation, 237, 244
Microbial diversity, 244
Microbial population counts, 246–250
Microbiological energy evolution, 159
Microencapsulation, 327–340
Micro-nutrients, 248. *See also* specific
    types
Microorganisms, 221, 237, 248, 344.
    See also specific types
Microscopy, 329, 331–333. *See also*
    specific types
Mineralization, 244, 255–257
Minerals, 413. *See also* specific types
Mirex bait, 15
MIS. *See* Mobile incineration system
Mobile incineration system (MIS), 230
Mobilization studies, 455

Moisture, 219, 264–266, 267–268, 420
Molar volume, 138, 142
Monitoring, 411–413
Monte Carlo methods, 17, 528–535
Most probable number (MPN) method,
    248, 249, 250
Motor oil, 245
MPN. *See* Most probable number
M85 spill simulation, 125–132
MTBE. *See* Methyl tertiary butyl ether
Multi-species sensors, 255

Naphthalene, 244, 250
Naphthenic acid, 245
NAPL. *See* Nonaqueous phase liquid
National Classification System, 53–55
National Contaminated Sites
    Remediation Program (NCSRP),
    50–53, 61–63
National Oil and Hazardous Substances
    Pollution Contingency Plan, 234
National Pollutant Discharge
    Elimination System (NPDES)
    permits, 69, 70, 76, 84, 87
National Priority List (NPL), 92, 93,
    94, 99, 100, 371
NCSRP. *See* National Contaminated
    Sites Remediation Program
Nernst Distribution Law, 138
Nevada soil treatment methods, 67–73
Nitrogen, 273, 344
Nitrous oxide, 184
NOAEL. *See* No-observed-adverse-
    effect-level
No Migration Petition, 388
Nonaqueous phase liquid (NAPL), 225,
    388, 401
  alcohol flooding for removal of. *See*
    Alcohol flooding
  dense, 374, 375, 437, 446, 447, 449,
    450, 455
  light, 374
Noncarcinogens, 24. *See also* specific
    types
Nonlinear methods, 301–302
Nonpolar organic phases, 149
Nonvolatiles, 14. *See also* specific types
No-observed-adverse-effect-level
    (NOAEL), 24

NPDES. *See* National Pollutant
  Discharge Elimination System
NPL. *See* National Priority List
NRTL equations, 119
Nutrients, 248, 273–279, 344, 413,
  414. *See also* specific types

Occupational Safety and Health
  Administration (OSHA), 26, 503
Occupational scenario, 14–15
n-Octanol, 142
Octanol-water partition coefficient, 138,
  142
Off-gas incineration, 465
Oil agar, 248
Oil-degrading microorganisms, 237
Oilfields, 487–498
  case studies of, 492–497
Oil and Hazardous Substances Pollution
  Contingency Plan, 234
Oil industry, 3–9
Oils, 142, 146–147, 245. *See also*
  specific types
  biodegradation of, 237, 245–246
  crude, 409–415
    bioremediation of, 487–498
  emulsified, 150
  heavy, 251, 257
  inventory residual concentration of,
    237
One-dimensional reactors, 345–347
One-dimensional studies, 117, 129–130
Operational testing, 391–393
Optical microscopy, 329
Oregon UST program, 77–88, 201–210
  closure in, 208–209
  difficulties in, 83–85
  economic context of, 201–202
  excavation and, 208–209
  future of, 87–88
  improvements in, 85–87
  landfilling and, 208–209
  political context of, 201–202
  remediation in, 208–209
  site assessment in, 202–207
  site ranking in, 204–207
  site-specific cleanup concentrations in,
    207–208
  soil sampling protocol in, 202–204

technical considerations in, 202
Organic carbon coefficient, 22
Organic carbon-to-water partition
  coefficients, 142
Organic solvents, 142. *See also* specific
  types
OSHA. *See* Occupational Safety and
  Health Administration
Out-of-state waste acceptance, 72
OVA analysis, 336
Oxidation, 226–229, 244, 473–475
Oxygen, 255, 257, 344, 379, 396, 414
Oxygenated fuels, 115–135. *See also*
  specific types
Oxygen diffusion, 255
Oxygen enrichment, 231
Ozone, 115, 473–475

Partial differential equations, 137
Partial pressure, 139
Particle size analysis, 237
Partition coefficients, 138, 142
Partitioning, 137–152, 266
  in biota, 149–150
  cosolvents and, 150–151
  depletion and, 147–150
  equilibrium, 116
  linear, 254
  mass balance calculations and,
    143–147
  soil cleanup criteria and, 151
  sorption in, 142
  theory of, 138–143
  transport and, 147–150, 151
Passive recovery trench design,
  417–436
  field data analysis and, 426–432
  groundwater flow and, 418–423,
    432–433
  parameters estimation in, 418–423
  permeability and, 423
  specific yield and, 424–425
  water table fluctuations and, 423
PCBs, 11, 16, 335, 339–340
PCP. *See* Pentachlorophenol
Pentachlorophenol (PCP), 344, 347,
  374, 378, 380, 381, 385, 391, 392
  degradation of, 393
  dissolved, 394

PERC. *See* Tetrachloroethylene
Permeability, 423, 443
Pesticides, 15. *See also* specific types
Petroleum–degrading microorganisms, 248
pH, 217, 237, 255, 256, 263–264
Phase equilibrium, 124
Phase separation, 139
Phenol, 215, 255
Phenolics, 244
Phosphoprus, 344
Physical properties. *See* Physicochemical properties
Physicochemical properties, 22, 35, 137, 138, 140–141, 225, 226, 237
  bioremediation and, 240–244
Pilot-scale studies, 376–381, 395
Pipe line integrity audits, 158
Plant growth stimulation, 245
Pore volumes, 446
Post-treatment analysis, 336–337
Post-treatment cleanup critera, 311–324
  results of, 312–313, 317–321
  risk assessment and, 314–315
  technology in, 311–312
Potentially responsible parties (PRPs), 77, 80, 81, 83, 95, 501
Precipitation, 139
Pressure hysteresis, 421–423
Primary drainage, 421
Probabilistic simulation, 528
Protozoa, 244
PRPs. *See* Potentially responsible parties
Pump-and-treat, 295, 437

QA. *See* Quality assurance
Quality assurance/quality control, 159, 186–187, 378, 385, 391
Quantitative assessment, 215

Radioautography, 257
RAGs. *See* Risk assessment guidelines
Raoult's law, 119, 147
RCRA. *See* Resource Conservation and Recovery Act
Reactive silicate technology, 327–340
Reactors. *See also* specific types
  design of, 250–251

one-dimensional, 345–347
  slurry, 352–357
Real estate development, 4–5
Recovery wells, 297
Recreational scenario, 15, 27–31
Redox, 255
Reference dose, 23, 24, 98
Reference standards, 183–187
Relative permeability, 423
Remedial Action Plan, 383–384
Remediation, 223–234, 311–324, 335, 338
  applications of, 302–307
  bio-. *See* Bioremediation
  of BTEX, 103–113
  case studies of, 226–233
  data requirements for, 225–226
  Detoxifier Process in, 311–312
  efficacy information for, 226–233
  endpoints of, 517–525
  of gasoline impacted soil, 335–337
  groundwater, 295–309
  in situ methods of, 224
  of methanol, 103–113
  natural, 103–113
  NCSRP framework for, 51–53, 61–63
  non-in situ methods of, 224–225
  of PCB contaminated soil, 339–340
  phase 1 in, 296–301
  process of, 336
  regulatory acceptance of, 233–234
  results of, 301, 312–313, 317–321
  risk assessment and, 314–315
  selection of technologies for, 225
  technologies for, 224–225, 311–312
  thermal treatment in, 229–233
Removal efficiencies, 229, 230
Reporting errors, 186–187
Residential scenario, 14, 27–31
Resource Conservation and Recovery Act (RCRA), 76, 84, 91, 92, 93–96, 99, 100
  bioremediation and, 240, 388
  remediation and, 223, 224, 234
Respiratory quotient, 256
Responsible parties, 77, 80, 81, 83, 95, 317, 501
Retardation, 132–134
RfD. *See* Reference dose

Richards equation, 420
Risk assessment, 24, 239, 314–315,
    487–488, 517–525
  Monte Carlo, 528–535
  uncertainty in, 527–528
Risk assessment guidelines (RAGs), 528
Risk specific dose, 30
RME analysis, 528
Road surveys, 159
Roof moisture surveys, 158–159
Rotary kilns, 230
RP. See Responsible parties
RQ. See Respiratory quotient
RsD. See Risk specific dose
Runoff, 15, 16, 32–34

Safe Drinking Water Act, 92, 93, 98
Salicylic derivatives, 244
Salinity, 237
Sample extraction errors, 184–185
Sample preparation errors, 183–184
SARA. See Superfund Amendments and
    Reauthorization Act
Saturation vapor pressure, 142
Scaling analysis, 443
Scaling laboratory column results,
    129–130
Scanning electron microscopy (SEM),
    329
Screening-level concentrations, 13
SDR. See Sediment delivery ratios
Secondary utilization, 258
Sediment delivery ratios (SDR), 32
Selected Area Electron Diffraction, 329
SEM. See Scanning electron microscopy
Semi-diurnal tides, 426–428
Semivolatile organic compounds
    (SVCs), 311, 312, 316. See also
    specific types
Severity of contamination, 13
Siallon process, 327, 328, 334
Sigmoidal kinetics, 258
Silica cell characterization, 328–329
Silica gel, 248
SITE. See Superfund Innovative
    Technology Evaluation
Skin contact rate, 19–20
Slow-release fertilizers, 273
Slurry reactor treatment, 352–357

Soil characterization, 217
Soil fertility, 244
Soil ingestion rate, 17–19
Soil monitoring, 412
Soil treatment methods in Nevada,
    67–73
Soil venting, 147, 231–232
Solidification, 327
Solid phase density, 143
Solubility, 137–152
  aqueous, 138
  in biota, 149–150
  of BTEX, 104–107
  cosolvents and, 150–151
  depletion and, 147–150
  effective, 139, 144–145
  mass balance calculations and,
      143–147
  in organic carbon, 143
  in pure solute, 142
  soil cleanup criteria and, 151
  of solvents, 115
  theory of, 138–143
  transport and, 147–150, 151
  water, 138, 139
  water-solvent mixture, 143
Solubilization, 150
Solvents, 132–134, 244. See also
    specific types
  organic, 142
  water mixed with, 143
  water soluble, 115
Spectroscopy, 216, 334. See also
    specific types
Statistical methods, 297
Statistical variability, 301
Steam injection, 401–407
Steam stripping, 403–404, 466–468
Structure-activity-property relationships,
    138
Subsurface contamination, 522–523
Superfund. See Comprehensive
    Environmental Response,
    Compensation and Liability Act
    (CERCLA)
Superfund Amendments and
    Reauthroization Act (SARA), 238
Superfund Innovative Technology
    Evaluation (SITE), 239

Surface features, 226
Surfactants, 150, 151, 438. *See also* specific types
Suspended particulates, 14, 20
SVCs. *See* Semi-volatile organic compounds
Sweep efficiency, 444–445

TC. *See* Toxicity characteristics
1,1,1-TCA, 229
TCDD. *See* 2,3,7,8-Tetrachlorodibenzo-p-dioxin
TCE. *See* Trichloroethylene
TCLP. *See* Toxicity Characteristic Leachate Procedure
TEM. *See* Transmission electron microscopy
Temperature, 137
  aerial infrared surveys and, 164, 165
  biological treatment and, 414
  bioremediation and, 255, 266–273
2,3,7,8-Tetrachlorodibenzo-p-dioxin (TCDD), 11, 12, 37
  de minimis criteria for, 26–35
  distribution of, 26–27
  fate of, 26–27
  runoff of, 32–34
Tetrachloroethylene, 312, 318, 321, 403
Thermal imaging, 166–167
Thermal treatment, 68, 229–233, 359–370
  history of, 359–361
  methods in, 361–363
  options in, 363–365
  problems associated with, 365–366
  statistics on, 368–370
Thermoplastic encapsulation methods, 327. *See also* specific types
Thibodeaux-Hwang approach, 254
Three-dimensional studies, 129, 347–349
Threshold limit value (TLV), 39, 149
Tidal fluctuations, 417–436
  field data analysis and, 426–432
  groundwater and, 428–432
  groundwater flow equations and, 418–423
  parameters estimation in, 418–423
  permeability and, 423

semi-diurnal, 426–428
specific yield and, 424–425
water table fluctuations and, 423
TLVs. *See* Threshold limit values
Toluene, 123, 124, 133, 413. *See also* Benzene, toluene, ethylbenzene and xylene (BTEX); Benzene, toluene and ethylene (BTE); Benzene, toluene and xylene (BTX)
Total petroleum hydrocarbons (TPH), 123, 129, 335, 338
  de minimis criteria for, 11–12
  depletion of, 251–254
  disappearance of, 256
  infrared surveys for analysis of, 169–179
  measurement of, 176–177
  Nevada methods for, 68
  regulation of, 360, 363
  thermal treatment and, 365, 367
Total suspended particulates (TSP), 20
Total toxic organics (TTO), 229
Toxicity Characteristic Leachate Procedure (TCLP), 71–72, 84, 85
Toxicity characteristics, 25
  in cleanup standard selection, 91–100
Toxicity potential, 239
Toxicity screening, 215
Toxicity studies, 232, 244–245
Toxicological uncertainties, 527–536
Toxic Substances Control Act (TSCA), 92
Toxic Treatments (USA) (TTUSA), 317
TPH. *See* Total petroleum hydrocarbons
Transformation, 137
Transmission electron microscopy (TEM), 331–333
Transport, 22–23, 70, 137, 151
  depletion by, 147–150
  equations for, 139
  moisture and, 267–268
  remediation endpoints and, 517–525
  simulation of, 120
Treatability studies, 214, 413–414
Trench design. *See* Passive recovery trench design
Trichloroethylene (TCE), 403, 445, 446, 447, 449, 450, 455, 456, 457
Trichlorotrifluoroethane (Freon), 173, 177, 253

TSCA. *See* Toxic Substances Control Act
TSP. *See* Total suspended particulates
TTO. *See* Total toxic organics
TTUSA. *See* Toxic Treatments (USA)
Two-phase systems, 151
Type B cleanup criteria, 11

Ultraviolet-catalyzed oxidation, 473–475
Ultraviolet radiation, 226–229
Uncertainties, 527–536
Underground storage tanks (USTs), 75, 76, 254
    government environmental records on, 503
    history of regulations of, 75
    leaking, 229, 239
    in Oregon. *See* Oregon UST program
    regulation of, 359, 360–361
Universal Soil Loss Equation (USLE), 32
USLE. *See* Universal Soil Loss Equation
UST. *See* Underground storage tanks

Valves, 159
Vapor density, 142
Vapor phase, 225
Vapor pressure, 22, 138, 142
Ventilation flow, 148
Venting, 147, 231–232
Visual radiation, 167
VOCs (volatile organic compounds). *See* Volatiles

Volatiles, 14, 240, 248, 253, 254, 311, 312. *See also* specific types
Volatilization, 184, 231, 251, 254
Volumetric rate, 147
Volumetric ventilation flow, 148

Waste screening, 71
Water-holding capacity, 244
Water quality standards, 98
Water solubility, 138, 139
Water table fluctuations, 423
Water vapor, 184
Waxes, 149, 244
Wildlife protection, 16, 34–35
Wilson equation, 119

XPS. *See* X-ray photoelectron spectroscopy
X-ray crystallography, 334
X-ray diffraction, 329
X-ray photoelectron spectroscopy (XPS), 334
XRD. *See* X-ray diffraction
X-TRAX system, 231
Xylene, 108, 403, 413. *See also* Benzene, toluene, ethylbenzene and xylene (BTEX); Benzene, toluene and xylene (BTX)

Yeasts, 244
Young-Laplace equation, 421

Z values, 139